DISTANT SPEECH RECOGNITION

T0344857

DISTANT SPEECH RECOGNITION

Matthias Wölfel
Universität Karlsruhe (TH), Germany

and

John McDonough
Universität des Saarlandes, Germany

WILEY

A John Wiley and Sons, Ltd., Publication

This edition first published 2009

Registered office

John Wiley & Sons Ltd, The Atrium, Southern Gate, Chichester, West Sussex, PO19 8SQ, United Kingdom

For details of our global editorial offices, for customer services and for information about how to apply for permission to reuse the copyright material in this book please see our website at www.wiley.com.

Wiley also publishes its books in a variety of electronic formats. Some content that appears in print may not be available in electronic books.

Designations used by companies to distinguish their products are often claimed as trademarks. All brand names and product names used in this book are trade names, service marks, trademarks or registered trademarks of their respective owners. The publisher is not associated with any product or vendor mentioned in this book. This publication is designed to provide accurate and authoritative information in regard to the subject matter covered. It is sold on the understanding that the publisher is not engaged in rendering professional services. If professional advice or other expert assistance is required, the services of a competent professional should be sought.

Wiley also publishes its books in a variety of electronic formats. Some content that appears in print may not be available in electronic books.

Library of Congress Cataloging-in-Publication Data

Wölfel, Matthias.
    Distant speech recognition / Matthias Wölfel, John McDonough.
        p. cm.
    Includes bibliographical references and index.
    ISBN 978-0-470-51704-8 (cloth)
  1.   Automatic speech recognition. I. McDonough, John (John W.) II. Title.
    TK7882.S65W64 2009
    006.4′54 – dc22
                                                    2008052791

A catalogue record for this book is available from the British Library

ISBN 978-0-470-51704-8 (H/B)

Typeset in 10/12 Times by Laserwords Private Limited, Chennai, India
Printed and bound in Great Britain by CPI Antony Rowe, Chippenham, Wiltshire

Contents

Foreword

As the authors of *Distant Speech Recognition* note, automatic speech recognition is the key enabling technology that will permit natural interaction between humans and intelligent machines. Core speech recognition technology has developed over the past decade in domains such as office dictation and interactive voice response systems to the point that it is now commonplace for customers to encounter automated speech-based intelligent agents that handle at least the initial part of a user query for airline flight information, technical support, ticketing services, *etc*. While these limited-domain applications have been reasonably successful in reducing the costs associated with handling telephone inquiries, their fragility with respect to acoustical variability is illustrated by the difficulties that are experienced when users interact with the systems using speakerphone input. As time goes by, we will come to expect the range of natural human-machine dialog to grow to include seamless and productive interactions in contexts such as humanoid robotic butlers in our living rooms, information kiosks in large and reverberant public spaces, as well as intelligent agents in automobiles while traveling at highway speeds in the presence of multiple sources of noise. Nevertheless, this vision cannot be fulfilled until we are able to overcome the shortcomings of present speech recognition technology that are observed when speech is recorded at a distance from the speaker.

While we have made great progress over the past two decades in core speech recognition technologies, the failure to develop techniques that overcome the effects of acoustical variability in homes, classrooms, and public spaces is the major reason why automated speech technologies are not generally available for use in these venues. Consequently, much of the current research in speech processing is directed toward improving robustness to acoustical variability of all types. Two of the major forms of environmental degradation are produced by additive noise of various forms and the effects of linear convolution. Research directed toward compensating for these problems has been in progress for more than three decades, beginning with the pioneering work in the late 1970s of Steven Boll in noise cancellation and Thomas Stockham in homomorphic deconvolution.

Additive noise arises naturally from sound sources that are present in the environment in addition to the desired speech source. As the speech-to-noise ratio (SNR) decreases, it is to be expected that speech recognition will become more difficult. In addition, the impact of noise on speech recognition accuracy depends as much on the type of noise source as on the SNR. While a number of statistical techniques are known to be reasonably effective in dealing with the effects of quasi-stationary broadband additive noise of arbitrary spectral coloration, compensation becomes much more difficult when the noise is highly transient in nature, as is the case with many types of impulsive machine noise on factory floors and gunshots in military environments. Interference by sources such as background music or background speech is especially difficult to handle, as it is both highly transient in nature and easily confused with the desired speech signal.

Reverberation is also a natural part of virtually all acoustical environments indoors, and it is a factor in many outdoor settings with reflective surfaces as well. The presence of even a relatively small amount of reverberation destroys the temporal structure of speech waveforms. This has a very adverse impact on the recognition accuracy that is obtained from speech systems that are deployed in public spaces, homes, and offices for virtually any application in which the user does not use a head-mounted microphone. It is presently more difficult to ameliorate the effects of common room reverberation than it has been to render speech systems robust to the effects of additive noise, even at fairly low SNRs. Researchers have begun to make progress on this problem only recently, and the results of work from groups around the world have not yet congealed into a clear picture of how to cope with the problem of reverberation effectively and efficiently.

Distant Speech Recognition by Matthias Wölfel and John McDonough provides an extraordinarily comprehensive exposition of the most up-to-date techniques that enable robust distant speech recognition, along with very useful and detailed explanations of the underlying science and technology upon which these techniques are based. The book includes substantial discussions of the major sources of difficulties along with approaches that are taken toward their resolution, summarizing scholarly work and practical experience around the world that has accumulated over decades. Considering both single-microphone and multiple-microphone techniques, the authors address a broad array of approaches at all levels of the system, including methods that enhance the waveforms that are input to the system, methods that increase the effectiveness of features that are input to speech recognition systems, as well as methods that render the internal models that are used to characterize speech sounds more robust to environmental variability.

This book will be of great interest to several types of readers. First (and most obviously), readers who are unfamiliar with the field of distant speech recognition can learn in this volume all of the technical background needed to construct and integrate a complete distant speech recognition system. In addition, the discussions in this volume are presented in self-contained chapters that enable technically literate readers in all fields to acquire a deep level of knowledge about relevant disciplines that are complementary to their own primary fields of expertise. Computer scientists can profit from the discussions on signal processing that begin with elementary signal representation and transformation and lead to advanced topics such as optimal Bayesian filtering, multirate digital signal processing, blind source separation, and speaker tracking. Classically-trained engineers will benefit from the detailed discussion of the theory and implementation of computer speech recognition systems including the extraction and enhancement of features representing speech sounds, statistical modeling of speech and language, along with the optimal search for the best available match between the incoming utterance and the internally-stored statistical representations of speech. Both of these groups will benefit from the treatments of physical acoustics, speech production, and auditory perception that are too frequently omitted from books of this type. Finally, the detailed contemporary exposition will serve to bring experienced practitioners who have been in the field for some time up to date on the most current approaches to robust recognition for language spoken from a distance.

Doctors Wölfel and McDonough have provided a resource to scientists and engineers that will serve as a valuable tutorial exposition and practical reference for all aspects associated with robust speech recognition in practical environments as well as for speech recognition in general. I am very pleased that this information is now available so easily and conveniently in one location. I fully expect that the publication of *Distant Speech Recognition* will serve as a significant accelerant to future work in the field, bringing us closer to the day in which transparent speech-based human-machine interfaces will become a practical reality in our daily lives everywhere.

Richard M. Stern
Pittsburgh, PA, USA

Preface

Our primary purpose in writing this book has been to cover a broad body of techniques and diverse disciplines required to enable reliable and natural verbal interaction between humans and computers. In the early nineties, many claimed that automatic speech recognition (ASR) was a "solved problem" as the word error rate (WER) had dropped below the 5% level for professionally trained speakers such as in the Wall Street Journal (WSJ) corpus. This perception changed, however, when the Switchboard Corpus, the first corpus of spontaneous speech recorded over a telephone channel, became available. In 1993, the first reported error rates on Switchboard, obtained largely with ASR systems trained on WSJ data, were over 60%, which represented a twelve-fold degradation in accuracy. Today the ASR field stands at the threshold of another radical change. WERs on telephony speech corpora such as the Switchboard Corpus have dropped below 10%, prompting many to once more claim that ASR is a solved problem. But such a claim is credible only if one ignores the fact that such WERs are obtained with *close-talking microphones*, such as those in telephones, and when only a single person is speaking. One of the primary hindrances to the widespread acceptance of ASR as the man-machine interface of first choice is the necessity of wearing a head-mounted microphone. This necessity is dictated by the fact that, under the current state of the art, WERs with microphones located a meter or more away from the speaker's mouth can catastrophically increase, making most applications impractical. The interest in developing techniques for overcoming such practical limitations is growing rapidly within the research community. This change, like so many others in the past, is being driven by the availability of new corpora, namely, speech corpora recorded with far-field sensors. Examples of such include the meeting corpora which have been recorded at various sites including the International Computer Science Institute in Berkeley, California, Carnegie Mellon University in Pittsburgh, Pennsylvania and the National Institute of Standards and Technologies (NIST) near Washington, D.C., USA. In 2005, conversational speech corpora that had been collected with *microphone arrays* became available for the first time, after being released by the European Union projects *Computers in the Human Interaction Loop* (CHIL) and *Augmented Multiparty Interaction* (AMI). Data collected by both projects was subsequently shared with NIST for use in the semi-annual Rich Transcription evaluations it sponsors. In 2006 Mike Lincoln at Edinburgh University in Scotland collected the first corpus of *overlapping speech* captured with microphone arrays. This data collection effort involved real speakers who read sentences from the 5,000 word WSJ task.

In the view of the current authors, ground breaking progress in the field of distant speech recognition can only be achieved if the mainstream ASR community adopts methodologies and techniques that have heretofore been confined to the fringes. Such technologies include speaker tracking for determining a speaker's position in a room, beamforming for combining the signals from an array of microphones so as to concentrate on a desired speaker's speech and suppress noise and reverberation, and source separation for effective recognition of overlapping speech. Terms like filter bank, generalized sidelobe canceller, and diffuse noise field must become household words within the ASR community. At the same time researchers in the fields of acoustic array processing and source separation must become more knowledgeable about the current state of the art in the ASR field. This community must learn to speak the language of word lattices, semi-tied covariance matrices, and weighted finite-state transducers. For too long, the two research communities have been content to effectively ignore one another. With a few notable exceptions, the ASR community has behaved as if a speech signal does not exist before it has been converted to cepstral coefficients. The array processing community, on the other hand, continues to publish experimental results obtained on artificial data, with ASR systems that are nowhere near the state of the art, and on tasks that have long since ceased to be of any research interest in the mainstream ASR world. It is only if each community adopts the best practices of the other that they can together meet the challenge posed by distant speech recognition. We hope with our book to make a step in this direction.

Acknowledgments

We wish to thank the many colleagues who have reviewed parts of this book and provided very useful feedback for improving its quality and correctness. In particular we would like to thank the following people: Elisa Barney Smith, Friedrich Faubel, Sadaoki Furui, Reinhold Häb-Umbach, Kenichi Kumatani, Armin Sehr, Antske Fokkens, Richard Stern, Piergiorgio Svaizer, Helmut Wölfel, Najib Hadir, Hassan El-soumsoumani, and Barbara Rauch. Furthermore we would like to thank Tiina Ruonamaa, Sarah Hinton, Anna Smart, Sarah Tilley, and Brett Wells at Wiley who have supported us in writing this book and provided useful insights into the process of producing a book, not to mention having demonstrated the patience of saints through many delays and deadline extensions. We would also like to thank the university library at Universität Karlsruhe (TH) for providing us with a great deal of scholarly material, either online or in books.

We would also like to thank the people who have supported us during our careers in speech recognition. First of all thanks is due to our Ph.D. supervisors Alex Waibel, Bill Byrne, and Frederick Jelinek who have fostered our interest in the field of automatic speech recognition. Satoshi Nakamura, Mari Ostendorf, Dietrich Klakow, Mike Savic, Gerasimos (Makis) Potamianos, and Richard Stern always proved more than willing to listen to our ideas and scientific interests, for which we are grateful. We would furthermore like to thank IEEE and ISCA for providing platforms for exchange, publications and for hosting various conferences. We are indebted to Jim Flanagan and Harry Van Trees, who were among the great pioneers in the array processing field. We are also much obliged to the tireless employees at NIST, including Vince Stanford, Jon Fiscus and John Garofolo, for providing us with our first real microphone array, the Mark III, and hosting the annual evaluation campaigns which have provided a tremendous impetus for advancing the entire field. Thanks is due also to Cedrick Rochét for having built the Mark III while at NIST, and having improved it while at Universität Karlsruhe (TH). In the latter effort, Maurizio Omologo and his coworkers at ITC-irst in Trento, Italy were particularly helpful. We would also like to thank Kristian Kroschel at Universität Karlsruhe (TH) for having fostered our initial interest in microphone arrays and agreeing to collaborate in teaching a course on the subject. Thanks is due also to Mike Riley and Mehryar Mohri for inspiring our interest in weighted finite-state transducers. Emilian Stoimenov was an important contributor to many of the finite-state transducer techniques described here. And of course, the list of those to whom we are indebted would not be complete if we failed to mention the undergraduates and graduate students at Universität Karlsruhe (TH) who helped us to build an instrumented seminar room for the CHIL project, and thereafter collect the audio and video data used for many of the experiments described in the final chapter of this work. These include Tobias Gehrig, Uwe Mayer, Fabian Jakobs, Keni Bernardin, Kai Nickel, Hazim Kemal Ekenel, Florian Kraft, and Sebastian Stüker. We are also naturally grateful to the funding agencies who made the research described in this book possible: the European Commission, the American Defense Advanced Research Projects Agency, and the Deutsche Forschungsgemeinschaft.

Most important of all, our thanks goes to our families. In particular, we would like to thank Matthias' wife Irina Wölfel, without whose support during the many evenings, holidays and weekends devoted to writing this book, we would have had to survive only on cold pizza and Diet Coke. Thanks is also due to Helmut and Doris Wölfel, John McDonough, Sr. and Christopher McDonough, without whose support through life's many trials, this book would not have been possible. Finally, we fondly remember Kathleen McDonough.

Matthias Wölfel
Karlsruhe, Germany

John McDonough
Saarbrücken, Germany

1

Introduction

For humans, speech is the quickest and most natural form of communication. Beginning in the late 19th century, verbal communication has been systematically extended through technologies such as radio broadcast, telephony, TV, CD and MP3 players, mobile phones and the Internet by voice over IP. In addition to these examples of one and two way verbal human–human interaction, in the last decades, a great deal of research has been devoted to extending our capacity of verbal communication with computers through *automatic speech recognition* (ASR) and speech synthesis. The goal of this research effort has been and remains to enable simple and natural *human–computer interaction* (HCI). Achieving this goal is of paramount importance, as verbal communication is not only fast and convenient, but also the only feasible means of HCI in a broad variety of circumstances. For example, while driving, it is much safer to simply ask a car navigation system for directions, and to receive them verbally, than to use a keyboard for tactile input and a screen for visual feedback. Moreover, hands-free computing is also accessible for disabled users.

1.1   Research and Applications in Academia and Industry

Hands-free computing, much like hands-free speech processing, refers to computer interface configurations which allow an interaction between the human user and computer without the use of the hands. Specifically, this implies that no close-talking microphone is required. Hands-free computing is important because it is useful in a broad variety of applications where the use of other common interface devices, such as a mouse or keyboard, are impractical or impossible. Examples of some currently available hands-free computing devices are camera-based head location and orientation-tracking systems, as well as gesture-tracking systems. Of the various hands-free input modalities, however, *distant speech recognition* (DSR) systems provide by far the most flexibility. When used in combination with other hands-free modalities, they provide for a broad variety of HCI possibilities. For example, in combination with a pointing gesture system it would become possible to turn on a particular light in the room by pointing at it while saying, "Turn on this light."

The remainder of this section describes a variety of applications where speech recognition technology is currently under development or already available commercially. The

Distant Speech Recognition   Matthias Wölfel and John McDonough
© 2009 John Wiley & Sons, Ltd

application areas include intelligent home and office environments, humanoid robots, automobiles, and speech-to-speech translation.

1.1.1   Intelligent Home and Office Environments

A great deal of research effort is directed towards equipping household and office devices – such as appliances, entertainment centers, personal digital assistants and computers, phones or lights – with more user friendly interfaces. These devices should be unobtrusive and should not require any special attention from the user. Ideally such devices should know the mental state of the user and act accordingly, gradually relieving household inhabitants and office workers from the chore of manual control of the environment. This is possible only through the application of sophisticated algorithms such as speech and speaker recognition applied to data captured with far-field sensors.

In addition to applications centered on HCI, computers are gradually gaining the capacity of acting as mediators for human–human interaction. The goal of the research in this area is to build a computer that will serve human users in their interactions with other human users; instead of requiring that users concentrate on their interactions with the machine itself, the machine will provide ancillary services enabling users to attend exclusively to their interactions with other people. Based on a detailed understanding of human perceptual context, intelligent rooms will be able to provide active assistance without any explicit request from the users, thereby requiring a minimum of attention from and creating no interruptions for their human users. In addition to speech recognition, such services need qualitative human analysis and human factors, natural scene analysis, multimodal structure and content analysis, and HCI. All of these capabilities must also be integrated into a single system.

Such interaction scenarios have been addressed by the recent projects *Computers in the Human Interaction Loop* (CHIL), *Augmented Multi-party Interaction* (AMI), as well as the successor of the latter *Augmented Multi-party Interaction with Distance Access* (AMIDA), all of which were sponsored by the European Union. To provide such services requires technology that models human users, their activities, and intentions. Automatically recognizing and understanding human speech plays a fundamental role in developing such technology. Therefore, all of the projects mentioned above have sought to develop technology for automatic transcription using speech data captured with distant microphones, determining who spoke when and where, and providing other useful services such as the summarizations of verbal dialogues. Similarly, the *Cognitive Assistant that Learns and Organizes* (CALO) project sponsored by the US *Defense Advanced Research Project Agency* (DARPA), takes as its goal the extraction of information from audio data captured during group interactions.

A typical meeting scenario as addressed by the AMIDA project is shown in Figure 1.1. Note the three microphone arrays placed at various locations on the table, which are intended to capture far-field speech for speaker tracking, beamforming, and DSR experiments. Although not shown in the photograph, the meeting participants typically also wear close-talking microphones to provide the best possible sound capture as a reference against which to judge the performance of the DSR system.

Figure 1.1   A typical AMIDA interaction. (© Photo reproduced by permission of the University of Edinburgh)

1.1.2   Humanoid Robots

If humanoid robots are ever to be accepted as full 'partners' by their human users, they must eventually develop perceptual capabilities similar to those possessed by humans, as well as the capacity of performing a diverse collection of tasks, including learning, reasoning, communicating and forming goals through interaction with both users and instructors. To provide for such capabilities, ASR is essential, because, as mentioned previously, spoken communication is the most common and flexible form of communication between people. To provide a natural interaction between a human and a humanoid robot requires not only the development of speech recognition systems capable of functioning reliably on data captured with far-field sensors, but also natural language capabilities including a sense of social interrelations and hierarchies.

In recent years, humanoid robots, albeit with very limited capabilities, have become commonplace. They are, for example, deployed as entertainment or information systems. Figure 1.2 shows an example of such a robot, namely, the humanoid tour guide robot *TPR-Robina*[1] developed by Toyota. The robot is able to escort visitors around the Toyota Kaikan Exhibition Hall and to interact with them through a combination of verbal communication and gestures.

While humanoid robots programmed for a limited range of tasks are already in widespread use, such systems lack the capability of learning and adapting to new environments. The development of such a capacity is essential for humanoid robots to become helpful in everyday life. The *Cognitive Systems for Cognitive Assistants* (COSY) project, financed by the European Union, has the objective to develops two kind of robots providing such advanced capabilities. The first robot will find its way around a

---

[1] ROBINA stands for ROBot as INtelligent Assistant.

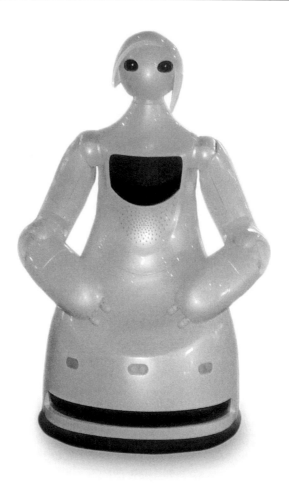

Figure 1.2 Humanoid tour guide robot TPR-Robina by Toyota which escort visitors around Toyota Kaikan Exhibition Hall in Toyota City, Aichi Prefecture, Japan. (© Photo reproduced by permission of Toyota Motor Corporation)

complex building, showing others where to go and answering questions about routes and locations. The second will be able to manipulate structured objects on a table top. A photograph of the second COSY robot during an interaction session is shown in Figure 1.3.

1.1.3 Automobiles

There is a growing trend in the automotive industry towards increasing both the number and the complexity of the features available in high end models. Such features include entertainment, navigation, and telematics systems, all of which compete for the driver's visual and auditory attention, and can increase his cognitive load. ASR in such automobile environments would promote the "Eyes on the road, hands on the wheel" philosophy. This would not only provide more convenience for the driver, but would in addition actually

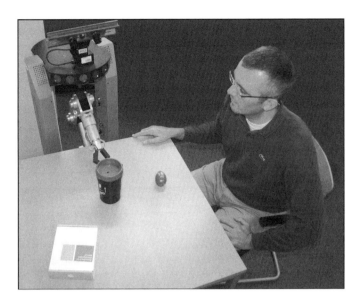

Figure 1.3 Humanoid robot under development for the COSY project. (© Photo reproduced by permission of DFKI GmbH)

enhance automotive safety. The enhanced safety is provided by hands-free operation of everything but the car itself and thus would leave the driver free to concentrate on the road and the traffic. Most luxury cars already have some sort of voice-control system which are, for example, able to provide

- *voice-activated, hands-free calling*
  Allows anyone in the contact list of the driver's mobile phone to be called by voice command.
- *voice-activated music*
  Enables browsing through music using voice commands.
- *audible information and text messages*
  Makes it possible to synthesize information and text messages, and have them read out loud through speech synthesis.

This and other voice-controlled functionality will become available in the mass market in the near future. An example of a voice-controlled car navigation system is shown in Figure 1.4.

While high-end consumer automobiles have ever more features available, all of which represent potential distractions from the task of driving the car, a police automobile has far more devices that place demands on the driver's attention. The goal of Project54 is to measure the cognitive load of New Hampshire state policeman – who are using speech-based interfaces in their cars – during the course of their duties. Shown in Figure 1.5 is the car simulator used by Project54 to measure the response times of police officers when confronted with the task of driving a police cruiser as well as manipulating the several devices contained therein through a speech interface.

Figure 1.4 Voice-controlled car navigation system by Becker. (© Photo reproduced by permission of Herman/Becker Automotive Systems GmbH)

Figure 1.5 Automobile simulator at the University of New Hampshire. (© Photo reproduced by permission of University of New Hampshire)

1.1.4 Speech-to-Speech Translation

Speech-to-speech translation systems provide a platform enabling communication with others without the requirement of speaking or understanding a common language. Given the nearly 6,000 different languages presently spoken somewhere on the Earth, and the ever-increasing rate of globalization and frequency of travel, this is a capacity that will in future be ever more in demand.

Even though speech-to-speech translation remains a very challenging task, commercial products are already available that enable meaningful interactions in several scenarios. One such system from *National Telephone and Telegraph* (NTT) DoCoMo of Japan works on a common cell phone, as shown in Figure 1.6, providing voice-activated Japanese–English and Japanese–Chinese translation. In a typical interaction, the user speaks short Japanese phrases or sentences into the mobile phone. As the mobile phone does not provide enough computational power for complete speech-to-text translation, the speech signal is transformed into enhanced speech features which are transmitted to a server. The server, operated by ATR-Trek, recognizes the speech and provides statistical translations, which are then displayed on the screen of the cell-phone. The current system works for both Japanese–English and Japanese–Chinese language pairs, offering translation in

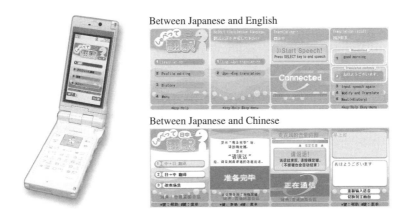

Figure 1.6 Cell phone, 905i Series by NTT DoCoMo, providing speech translation between English and Japanese, and Chinese and Japanese developed by ATR and ATR-Trek. This service is commercially available from NTT DoCoMo. (© Photos reproduced by permission of ATR-Trek)

both directions. For the future, however, preparation is underway to include support for additional languages.

As the translations appear on the screen of the cell phone in the DoCoMo system, there is a natural desire by users to hold the phone so that the screen is visible instead of next to the ear. This would imply that the microphone is no longer only a few centimeters from the mouth; i.e., we would have once more a distant speech recognition scenario. Indeed, there is a similar trend in all hand-held devices supporting speech input.

Accurate translation of unrestricted speech is well beyond the capability of today's state-of-the-art research systems. Therefore, advances are needed to improve the technologies for both speech recognition and speech translation. The development of such technologies are the goals of the *Technology and Corpora for Speech-to-Speech Translation* (TC-Star) project, financially supported by European Union, as well as the *Global Autonomous Language Exploitation* (GALE) project sponsored by the DARPA. These projects respectively aim to develop the capability for unconstrained conversational speech-to-speech translation of English speeches given in the European Parliament, and of broadcast news in Chinese or Arabic.

1.2 Challenges in Distant Speech Recognition

To guarantee high-quality sound capture, the microphones used in an ASR system should be located at a fixed position, very close to the sound source, namely, the mouth of the speaker. Thus body mounted microphones, such as head-sets or lapel microphones, provide the highest sound quality. Such microphones are not practical in a broad variety of situations, however, as they must be connected by a wire or radio link to a computer and attached to the speaker's body before the HCI can begin. As mentioned previously, this makes HCI impractical in many situations where it would be most helpful; e.g., when communicating with humanoid robots, or in intelligent room environments.

Although ASR is already used in several commercially available products, there are still obstacles to be overcome in making DSR commercially viable. The two major sources of degradation in DSR are distortions, such as additive noise and reverberation, and a mismatch between *training* and *test data*, such as those introduced by speaking style or accent. In DSR scenarios, the quality of the speech provided to the recognizer has a decisive impact on system performance. This implies that speech enhancement techniques are typically required to achieve the best possible signal quality.

In the last decades, many methods have been proposed to enable ASR systems to compensate or adapt to mismatch due to interspeaker differences, articulation effects and microphone characteristics. Today, those systems work well for different users on a broad variety of applications, but only as long as the speech captured by the microphones is free of other distortions. This explains the severe performance degradation encountered in current ASR systems as soon as the microphone is moved away from the speaker's mouth. Such situations are known as *distant*, far-field or hands-free[2] speech recognition. This dramatic drop in performance occurs mainly due to three different types of distortion:

- The first is *noise*, also known as *background noise*,[3] which is any sound other than the desired speech, such as that from air conditioners, printers, machines in a factory, or speech from other speakers.
- The second distortion is *echo* and *reverberation*, which are reflections of the sound source arriving some time after the signal on the direct path.
- Other types of distortions are introduced by environmental factors such as *room modes*, the *orientation of the speaker's head*, or the *Lombard effect*.

To limit the degradation in system performance introduced by these distortions, a great deal of current research is devoted to exploiting several aspects of speech captured with far-field sensors. In DSR applications, procedures already known from conventional ASR can be adopted. For instance, *confusion network combination* is typically used with data captured with a close-talking microphone to fuse word hypotheses obtained by using various speech feature extraction schemes or even completely different ASR systems. For DSR with multiple microphone conditions, confusion network combination can be used to fuse word hypotheses from different microphones. Speech recognition with distant sensors also introduces the possibility, however, of making use of techniques that were either developed in other areas of signal processing, or that are entirely novel. It has become common in the recent past, for example, to place a *microphone array* in the speaker's vicinity, enabling the speaker's position to be determined and tracked with time. Through beamforming techniques, a microphone array can also act as a spatial filter to emphasize the speech of the desired speaker while suppressing ambient noise or simultaneous speech from other speakers. Moreover, human speech has temporal, spectral, and statistical characteristics that are very different from those possessed by other signals for which conventional beamforming techniques have been used in the past. Recent research has revealed that these characteristics can be exploited to perform more effective beamforming for speech enhancement and recognition.

---

[2] The latter term is misleading, inasmuch close-talking microphones are usually not held in the hand, but are mounted to the head or body of the speaker.
[3] This term is also misleading, in that the "background" could well be closer to the microphone than the "foreground" signal of interest.

1.3   System Evaluation

Quantitative measures of the quality or performance of a system are essential for making fundamental advances in the state-of-the-art. This fact is embodied in the often repeated statement, "You improve what you *measure*." In order to asses system performance, it is essential to have error metrics or objective functions at hand which are well-suited to the problem under investigation. Unfortunately, good objective functions do not exist for a broad variety of problems, on the one hand, or else cannot be directly or automatically evaluated, on the other.

Since the early 1980s, *word error rate* (WER) has emerged as the measure of first choice for determining the quality of automatically-derived speech transcriptions. As typically defined, an error in a speech transcription is of one of three types, all of which we will now describe. A *deletion* occurs when the recognizer fails to hypothesize a word that *was* spoken. An *insertion* occurs when the recognizer hypothesizes a word that was *not* spoken. A *substitution* occurs when the recognizer *misrecognizes* a word. These three errors are illustrated in the following partial hypothesis, where they are labeled with D, I, and S, respectively:

```
Hyp: BUT ...    WILL SELL THE CHAIN ... FOR EACH STORE SEPARATELY
Utt:     ... IT WILL SELL THE CHAIN ... OR  EACH STORE SEPARATELY
     I       D                          S
```

A more thorough discussion of word error rate is given in Section 14.1.

Even though widely accepted and used, word error rate is not without flaws. It has been argued that the equal weighting of words should be replaced by a context sensitive weighting, whereby, for example, information-bearing keywords should be assigned a higher weight than functional words or articles. Additionally, it has been asserted that word similarities should be considered. Such approaches, however, have never been widely adopted as they are more difficult to evaluate and involve subjective judgment. Moreover, these measures would raise new questions, such as how to measure the distance between words or which words are important.

Naively it could be assumed that WER would be sufficient in ASR as an objective measure. While this may be true for the user of an ASR system, it does not hold for the engineer. In fact a broad variety of additional *objective* or *cost functions* are required. These include:

- The *Mahalanobis distance*, which is used to evaluate the acoustic model.
- *Perplexity*, which is used to evaluate the language model as described in Section 7.3.1.
- *Class separability*, which is used to evaluate the feature extraction component or front-end.
- *Maximum mutual information* or *minimum phone error*, which are used during discriminate estimation of the parameters in a hidden Markov model.
- *Maximum likelihood*, which is the metric of first choice for the estimation of all system parameters.

A DSR system requires additional objective functions to cope with problems not encountered in data captured with close-talking microphones. Among these are:

- *Cross-correlation*, which is used to estimate time delays of arrival between microphone pairs as described in Section 10.1.
- *Signal-to-noise ratio*, which can be used for channel selection in a multiple-microphone data capture scenario.
- *Negentropy*, which can be used for combining the signals captured by all sensors of a microphone array.

Most of the objective functions mentioned above are useful because they show a significant correlation with WER. The performance of a system is optimized by minimizing or maximizing a suitable objective function. The way in which this optimization is conducted depends both on the objective function and the nature of the underlying model. In the best case, a closed-form solution is available, such as in the optimization of the beamforming weights as discussed in Section 13.3. In other cases, an iterative solution can be adopted, such as when optimizing the parameters of a *hidden Markov model* (HMM) as discussed in Chapter 8. In still other cases, numerical optimization algorithms must be used such as when optimization the parameters of an all-pass transform for speaker adaptation as discussed in Section 9.2.2.

To chose the appropriate objective function a number of decisions must be made (Hänsler and Schmidt 2004, sect. 4):

- What kind of information is available?
- How should the available information be used?
- How should the error be weighted by the objective function?
- Should the objective function be deterministic or stochastic?

Throughout the balance of this text, we will strive to answer these questions whenever introducing an objective function for a particular application or in a particular context. When a given objective function is better suited than another for a particular purpose, we will indicate why. As mentioned above, the reasoning typically centers around the fact that the better suited objective function is more closely correlated with word error rate.

1.4   Fields of Speech Recognition

Figure 1.7 presents several subtopics of speech recognition in general which can be associated with three different fields: automatic, robust and distant speech recognition. While some topics such as multilingual speech recognition and language modeling can be clearly assigned to one group (i.e., *automatic*) other topics such as feature extraction or adaptation cannot be uniquely assigned to a single group. A second classification of topics shown in Figure 1.7 depends on the number and type of sensors. Whereas one microphone is traditionally used for recognition, in distant recognition the traditional sensor configuration can be augmented by an entire array of microphones with known or unknown geometry. For specific tasks such as lipreading or speaker localization, additional sensor types such as video cameras can be used.

Undoubtedly, the construction of optimal DSR systems must draw on concepts from several fields, including acoustics, signal processing, pattern recognition, speaker tracking and beamforming. As has been shown in the past, all components can be optimized

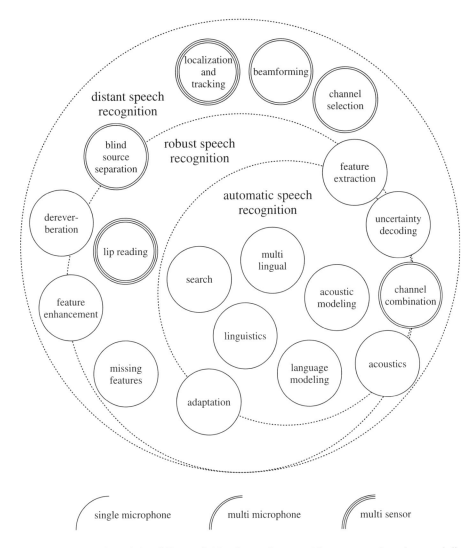

Figure 1.7 Illustration of the different fields of speech recognition: automatic, robust and distant

separately to construct a DSR system. Such an independent treatment, however, does not allow for optimal performance. Moreover, new techniques have recently emerged exploiting the complementary effects of the several components of a DSR system. These include:

- More closely coupling the feature extraction and acoustic models; e.g., by propagating the uncertainty of the feature extraction into the HMM.
- Feeding the word hypotheses produced by the DSR back to the component located earlier in the processing chain; e.g. by feature enhancement with particle filters with models for different phoneme classes.

- Replacing traditional objective functions such as signal-to-noise ratio by objective functions taking into account the acoustic model of the speech recognition system, as in maximum likelihood beamforming, or considering the particular characteristics of human speech, as in maximum negentropy beamforming.

1.5 Robust Perception

In contrast to automatic pattern recognition, human perception is very robust in the presence of distortions such as noise and reverberation. Therefore, knowledge of the mechanisms of human perception, in particular with regard to robustness, may also be useful in the development of automatic systems that must operate in difficult acoustic environments. It is interesting to note that the cognitive load for humans increases while listening in noisy environments, even when the speech remains intelligible (Kjellberg *et al.* 2007). This section presents some illustrative examples of human perceptual phenomena and robustness. We also present several technical solutions based on these phenomena which are known to improve robustness in automatic recognition.

1.5.1 A Priori Knowledge

When confronted with an ambiguous stimulus requiring a single interpretation, the human brain must rely on *a priori* knowledge and expectations. What is likely to be one of the most amazing findings about the robustness and flexibility of human perception and the use of *a priori* information is illustrated by the following sentence, which was circulated in the Internet in September 2003:

> Aoccdrnig to rscheearch at Cmabrigde uinervtisy, it deosn't mttaer waht oredr the ltteers in a wrod are, the olny ipromoetnt tihng is taht the frist and lsat ltteres are at the rghit pclae. The rset can be a tatol mses and you can sitll raed it wouthit a porbelm. Tihs is bcuseae we do not raed ervey lteter by istlef but the wrod as a wlohe.

The text is easy to read for a human inasmuch as, through reordering, the brain maps the erroneously presented characters into correct English words.

A priori knowledge is also widely used in automatic speech processing. Obvious examples are

- the statistics of speech,
- the limited number of possible phoneme combinations constrained by known words which might be further constrained by the domain,
- the word sequences follow a particular structure which can be represented as a *context free grammar* or the knowledge of successive words, represented as an *N-gram*.

1.5.2 Phonemic Restoration and Reliability

Most signals of interest, including human speech, are highly redundant. This redundancy provides for correct recognition or classification even in the event that the signal is partially

Figure 1.8 Adding a mask to the occluded portions of the top image renders the word legible, as is evident in the lower image

occluded or otherwise distorted, which implies that a significant amount of information is missing. The sophisticated capabilities of the human brain underlying robust perception were demonstrated by Fletcher (1953), who found that verbal communication between humans is possible if either the frequencies below or above 1800 Hz are filtered out. An illusory phenomenon, which clearly illustrates the robustness of the human auditory system, is known as the *phonemic restoration* effect, whereby phonetic information that is actually missing in a speech signal can be synthesized by the brain and clearly *heard* (Miller and Licklider 1950; Warren 1970). Furthermore, the knowledge of which information is distorted or missing can significantly improve perception. For example, knowledge about the occluded portion of an image can render a word readable, as is apparent upon considering Figure 1.8. Similarly, the comprehensibility of speech can be improved by adding noise (Warren *et al.* 1997).

Several problems in automatic data processing – such as occlusion – which were first investigated in the context of visual pattern recognition, are now current research topics in robust speech recognition. One can distinguish between two related approaches for coping with this problem:

- *missing feature theory*
  In missing feature theory, unreliable information is either ignored, set to some fixed nominal value, such as the global mean, or interpolated from nearby reliable information. In many cases, however, the restoration of missing features by spectral and/or temporal interpolation is less effective than simply ignoring them. The reason for this is that no processing can re-create information that has been lost as long as no additional information, such as an estimate of the noise or its propagation, is available.
- *uncertainty processing*
  In uncertainty processing, unreliable information is assumed to be unaltered, but the unreliable portion of the data is assigned less weight than the reliable portion.

1.5.3   Binaural Masking Level Difference

Even though the most obvious benefit from binaural hearing lies in source localization, other interesting effects exist: If the same signal and noise is presented to both ears with a noise level so high as to mask the signal, the signal is inaudible. Paradoxically, if either of the two ears is unable to hear the signal, it becomes once more audible. This effect is known as the *binaural masking level difference*. The binaural improvements in observing a signal in noise can be up to 20 dB (Durlach 1972). As discussed in Section 6.9.1, the binaural masking level difference can be related to spectral subtraction, wherein two input signals, one containing both the desired signal along with noise, and the second containing only the noise, are present. A closely related effect is the so-called *cocktail party effect* (Handel 1989), which describes the capacity of humans to suppress undesired sounds, such as the babble during a cocktail party, and concentrate on the desired signal, such as the voice of a conversation partner.

1.5.4   Multi-Microphone Processing

The use of multiple microphones is motivated by nature, in which two ears have been shown to enhance speech understanding as well as acoustic source localization. This effect is even further extended for a group of people, where one person could not understand some words, a person next to the first might have and together they are able to understand more than independent of each other.

Similarly, different tiers in a speech recognition system, which are derived either from different channels (e.g., microphones at different locations or visual observations) or from the variance in the recognition system itself, produce different recognition results. An appropriate combination of the different tiers can improve recognition performance. The degree of success depends on

- the variance of the information provided by the different tiers,
- the quality and reliability of the different tiers and
- the method used to combine the different tiers.

In automatic speech recognition, the different tiers can be combined at various stages of the recognition system providing different advantages and disadvantages:

- *signal combination*
  Signal-based algorithms, such as *beamforming*, exploit the spatial diversity resulting from the fact that the desired and interfering signal sources are in practice located at different points in space. These approaches assume that the time delays of the signals between different microphone pairs are known or can be reliably estimated. The spatial diversity can then be exploited by suppressing signals coming from directions other than that of the desired source.
- *feature combination*
  These algorithms concatenate features derived by different feature extraction methods to form a new feature vector. In such an approach, it is a common practice to reduce the number of features by principal component analysis or linear discriminant analysis.

While such algorithms are simple to implement, they suffer in performance if the different streams are not perfectly synchronized.

- *word and lattice combination*
  Those algorithms, such as *recognizer output voting error reduction* (ROVER) and confusion network combination, combine the information of the recognition output which can be represented as a first best, N-best or lattice word sequence and might be augmented with a confidence score for each word.

In the following we present some examples where different tiers have been successfully combined: Stolcke *et al.* (2005) used two different front-ends, mel-frequency cepstral coefficients and features derived from perceptual linear prediction, for cross-adaptation and system combination via confusion networks. Both of these features are described in Chapter 5. Yu *et al.* (2004) demonstrated, on a Chinese ASR system, that two different kinds of models, one on phonemes, the other on semi-syllables, can be combined to good effect. Lamel and Gauvain (2005) combined systems trained with different phoneme sets using ROVER. Siohan *et al.* (2005) combined randomized decision trees. Stüker *et al.* (2006) showed that a combination of four systems – two different phoneme sets with two feature extraction strategies – leads to additional improvements over the combination of two different phoneme sets or two front-ends. Stüker *et al.* also found that combining two systems, where both the phoneme set and front-ends are altered, leads to improved recognition accuracy compared to changing only the phoneme set or only the front-end. This fact follows from the increased variance between the two different channels to be combined. The previous systems have combined different tiers using only a single channel combination technique. Wölfel *et al.* (2006) demonstrated that a hybrid approach combining the different tiers, derived from different microphones, at different stages in a distant speech recognition system leads to additional improvements over a single combination approach. In particular Wölfel *et al.* achieved fewer recognition errors by using a combination of beamforming and confusion network.

1.5.5 Multiple Sources by Different Modalities

Given that it often happens that no single modality is powerful enough to provide correct classification, one of the key issues in robust human perception is the efficient merging of different input modalities, such as audio and vision, to render a stimulus intelligible (Ernst and Bülthoff 2004; Jacobs 2002). An illustrative example demonstrating the multimodality of speech perception is the *McGurk effect*[4] (McGurk and MacDonald 1976), which is experienced when contrary audiovisual information is presented to human subjects. To wit, a video presenting a visual /ga/ combined with an audio /ba/ will be perceived by 98% of adults as the syllable /da/. This effect exists not only for single syllables, but can alter the perception of entire spoken utterances, as was confirmed by a study about witness testimony (Wright and Wareham 2005). It is interesting to note that awareness of the effect does not change the perception. This stands in stark contrast to certain optical illusions, which are destroyed as soon as the subject is aware of the deception.

---

[4] This is often referred to as the McGurk–MacDonald effect.

Humans follow two different strategies to combine information:

- *maximizing information (sensor combination)*
  If the different modalities are complementary, the various pieces of information about an object are combined to maximize the knowledge about the particular observation.

  For example, consider a three-dimensional object, the correct recognition of which is dependent upon the orientation of the object to the observer. Without rotating the object, vision provides only two-dimensional information about the object, while the haptic[5] input provides the missing three-dimensional information (Newell 2001).
- *reducing variance (sensor integration)*
  If different modalities overlap, the variance of the information is reduced. Under the independence and Gaussian assumption of the noise, the estimate with the lowest variance is identical to the maximum likelihood estimate.

  One example of the integration of audio and video information for localization supporting the reduction in variance theory is given by Alais and Burr (2004).

Two prominent technical implementations of sensor fusion are audio-visual speaker tracking, which will be presented in Section 10.4, and audio-visual speech recognition. A good overview paper of the latter is by Potamianos *et al.* (2004).

1.6   Organizations, Conferences and Journals

Like all other well-established scientific disciplines, the fields of speech processing and recognition have founded and fostered an elaborate network of conferences and publications. Such networks are critical for promoting and disseminating scientific progress in the field. The most important organizations that plan and hold such conferences on speech processing and publish scholarly journals are listed in Table 1.1.

At conferences and in their associated proceedings the most recent advances in the state-of-the-art are reported, discussed, and frequently lead to further advances. Several major conferences take place every year or every other year. These conferences are listed in Table 1.2. The principal advantage of conferences is that they provide a venue for

Table 1.1   Organizations promoting research in speech processing and recognition

Abbreviation	Full Name
IEEE	Institute of Electrical and Electronics Engineers
ISCA	International Speech Communication Association former European Speech Communication Association (ESCA)
EURASIP	European Association for Signal Processing
ASA	Acoustical Society of America
ASJ	Acoustical Society of Japan
EAA	European Acoustics Association

---

[5] Haptic phenomena pertain to the sense of touch.

Table 1.2   Speech processing and recognition conferences

Abbreviation	Full Name
ICASSP	International Conference on Acoustics, Speech, and Signal Processing by IEEE
Interspeech	ISCA conference; previous Eurospeech and International Conference on Spoken Language Processing (ICSLP)
ASRU	Automatic Speech Recognition and Understanding by IEEE
EUSIPCO	European Signal Processing Conference by EURASIP
HSCMA	Hands-free Speech Communication and Microphone Arrays
WASPAA	Workshop on Applications of Signal Processing to Audio and Acoustics
IWAENC	International Workshop on Acoustic Echo and Noise Control
ISCSLP	International Symposium on Chinese Spoken Language Processing
ICMI	International Conference on Multimodal Interfaces
MLMI	Machine Learning for Multimodal Interaction
HLT	Human Language Technology

the most recent advances to be reported. The disadvantage of conferences is that the process of peer review by which the papers to be presented and published are chosen is on an extremely tight time schedule. Each submission is either accepted or rejected, with no time allowed for discussion with or clarification from the authors. In addition to the scientific papers themselves, conferences offer a venue for presentations, expert panel discussions, keynote speeches and exhibits, all of which foster further scientific progress in speech processing and recognition. Information about individual conferences is typically disseminated in the Internet. For example, to learn about the *Workshop on Applications of Signal Processing to Audio and Acoustics*, which is to be held in 2009, it is only necessary to type `waspaa 2009` into an Internet search window.

Journals differ from conferences in two ways. Firstly, a journal offers no chance for the scientific community to gather regularly at a specific place and time to present and discuss recent research. Secondly and more importantly, the process of peer review for an article submitted for publication in a journal is far more stringent than that for any conference. Because there is no fixed time schedule for publication, the reviewers for a journal can place far more demands on authors prior to publication. They can, for example, request more graphs or figures, more experiments, further citations to other scientific work, not to mention improvements in English usage and overall quality of presentation. While all of this means that greater time and effort must be devoted to the preparation and revision of a journal publication, it is also the primary advantage of journals with respect to conferences. The dialogue that ensues between the authors and reviewers of a journal publication is the very core of the scientific process. Through the succession of assertion, rebuttal, and counter assertion, non-novel claims are identified and withdrawn, unjustifiable claims are either eliminated or modified, while the arguments for justifiable claims are strengthened and clarified. Moreover, through the act of publishing a journal article and the associated dialogue, both authors and reviewers typically learn much they had not previously known. Table 1.3 lists several journals which cover topics presented in this book and which are recognized by academia and industry alike.

Table 1.3 Speech processing and recognition journals

Abbreviation	Full name
SP	*IEEE Transactions on Signal Processing*
ASLP	*IEEE Transactions on Audio, Speech and Language Processing* former *IEEE Transactions on Speech and Audio Processing* (SAP)
ASSP	*IEEE Transactions on Acoustics, Speech and Signal Processing*
SPL	*IEEE Signal Processing Letters*
SPM	*IEEE Signal Processing Magazine*
CSL	*Computer Speech and Language* by Elsevier
ASA	*Journal of the Acoustic Society of America*
SP	*EURASIP Journal on Signal Processing*
AdvSP	*EURASIP Journal on Advances in Signal Processing*
SC	*EURASIP and ISCA Journal on Speech Communication* published by Elsevier
AppSP	*EURASIP Journal on Applied Signal Processing*
ASMP	*EURASIP Journal on Audio, Speech and Music Processing*

An updated list of conferences, including a calendar of upcoming events, and journals can be found on the companion website of this book at

http://www.distant-speech-recognition.org

1.7 Useful Tools, Data Resources and Evaluation Campaigns

A broad number of commercial and non-commercial tools are available for the processing, analysis and recognition of speech. An extensive and updated list of such tools can be found on the companion website of this book.

The right data or corpora is essential for training and testing various speech processing, enhancement and recognition algorithms. This follows from the fact that the quality of the acoustic and language models are determined in large part by the amount of available training data, and the similarity between the data used for training and testing. As collecting and transcribing appropriate data is time-consuming and expensive, and as reporting WER reductions on "private" data makes the direct comparison of techniques and systems difficult or impossible, it is highly worth-while to report experimental results on publicly available speech corpora whenever possible. The goal of evaluation campaigns, such as the *Rich Transcription* (RT) evaluation staged periodically by the US *National Institute of Standards and Technologies* (NIST), is to evaluate and to compare different speech recognition systems and the techniques on which they are based. Such evaluations are essential in order to assess not only the progress of individual systems, but also that of the field as a whole. Possible data sources and evaluation campaigns are listed on the website mentioned previously.

1.8 Organization of this Book

Our aim in writing this book was to provide in a single volume an exposition of the theory behind each component of a complete DSR system. We now summarize the remaining

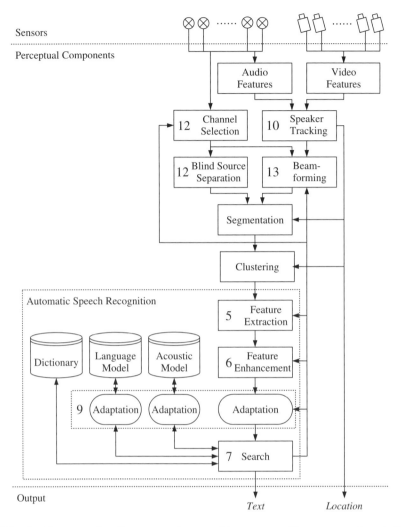

Figure 1.9 Architecture of a distant speech recognition system. The gray numbers indicate the corresponding chapter of this book

contents of this volume in order to briefly illustrate both the narrative thread that underlies this work, as well as the interrelations among the several chapters. In particular, we will emphasize how the development of each chapter is prefigured by and builds upon that of the preceding chapters. Figure 1.9 provides a high-level overview of a DSR system following the signal flow through the several components. The gray number on each individual component indicates the corresponding chapter in this book. The chapters not shown in the figure, in particular Chapters 2, 3, 4, 8 and 11, present material necessary to support the development in the other chapters: The fundamentals of sound propagation and acoustics are presented in Chapter 2, as are the basics of speech production. Chapter 3 presents linear filtering techniques that are used throughout the text. Chapter 4 presents the theory of Bayesian filters, which will later be applied both for speech feature enhancement in Chapter 6 and speaker tracking in Chapter 10. Chapter 8 discusses how the parameters of a HMM can be reliably estimated based on the use of transcribed acoustic data. Such a HMM is an essential component of most current DSR systems, in that it extracts word hypotheses from the final waveform produced by the other components of the system. Chapter 11 provides a discussion of digital filter banks, which, as discussed in Chapter 13, are an important component of a beamformer. Finally, Chapter 14 reports experimental results indicating the effectiveness of the algorithms described throughout this volume.

Speech, like any sound, is the propagation of pressure waves through air or any other liquid. A DSR system extracts from such pressure waves hypotheses of the phonetic units and words uttered by a speaker. Hence, it is worth-while to understand the physics of sound propagation, as well as how the spectral and temporal characteristics of speech are altered when it is captured by far-field sensors in realistic acoustic environments. These topics are considered in Chapter 2. This chapter also presents the characteristics and properties of the human auditory system. Knowledge of the latter is useful, inasmuch as experience has shown that many insights gained from studying the human auditory system have been successfully applied to improve the performance of *automatic* speech recognition systems.

In signal processing, the term *filter* refers to an algorithm which extracts a desired signal from an input signal corrupted by noise or other distortions. A filter can also be used to modify the spectral or temporal characteristics of a signal in some advantageous way. Therefore, filtering techniques are powerful tools for speech signal processing and distant recognition. Chapter 3 provides a review of the basics of digital signal processing, including a short introduction to linear time-invariant systems, the Fourier and z-transforms, as well as the effects of sampling and reconstruction. Next there is a presentation of the discrete Fourier transform and its use for the implementation of linear time-invariant systems, which is followed by a description of the short-time Fourier transform. The contents of this chapter will be referred to extensively in Chapter 5 on speech feature extraction, as well as in Chapter 11 on digital filter banks.

Many problems in science and engineering can be formulated as the estimation of some *state*, which cannot be observed directly, based on a series of features or observations, which can be directly observed. The observations are often corrupted by distortions such as noise or reverberation. Such problems can be solved with one of a number of Bayesian filters, all of which estimate an unobservable state given a series of observations. Chapter 4 first formulates the general problem to be solved by a Bayesian filter, namely, tracking the likelihood of the state as it evolves in time as conditioned on a sequence of observations. Thereafter, it presents several different solutions to this general problem, including the classic Kalman filter and its variants, as well as the class of particle filters, which have much more recently appeared in the literature. The theory of Bayesian filters will be applied in Chapter 6 to the task of enhancing speech features that have been corrupted by noise, reverberation or both. A second application, that of tracking the physical position of a speaker based on the signals captured with the elements of a microphone array, will be discussed in Chapter 10.

Automatic recognition requires that the speech waveform is processed so as to produce feature vectors of a relatively small dimension. This reduction in dimensionality is necessary in order to avoid wasting parameters modeling characteristics of the signal which are irrelevant for classification. The transformation of the input data into a set of dimension-reduced features is called speech feature extraction, acoustic preprocessing or front-end processing. As explained in Chapter 5, feature extraction in the context of DSR systems aims to preserve the information needed to distinguish between phonetic classes, while being invariant to other factors. The latter include speaker differences, such as accent, emotion or speaking rate, as well as environmental distortions such as background noise, channel differences, or reverberation.

The principle underlying speech feature enhancement, the topic of Chapter 6, is the estimation of the original features of the clean speech from a corrupted signal. Usually the enhancement takes place either in the power, logarithmic spectral or cepstral domain. The prerequisite for such techniques is that the noise or the impulse response is known or can be reliably estimated in the cases of noise or channel distortion, respectively. In many applications only a single channel is available and therefore the noise estimate must be inferred directly from the noise-corrupted signal. A simple method for accomplishing this separates the signal into speech and non-speech regions, so that the noise spectrum can be estimated from those regions containing no speech. Such simple techniques, however, are not able to cope well with non-stationary distortions. Hence, more advanced algorithms capable of actively tracking changes in the noise and channel distortions are the main focus of Chapter 6.

As discussed in Chapter 7, search is the process by which a statistical ASR system finds the most likely word sequence conditioned on a sequence of acoustic observations. The search process can be posed as that of finding the shortest path through a search graph. The construction of such a search graph requires several knowledge sources, namely, a language model, a word lexicon, and a HMM, as well as an acoustic model to evaluate the likelihoods of the acoustic features extracted from the speech to be recognized. Moreover, inasmuch as all human speech is affected by coarticulation, a decision tree for representing context dependency is required in order to achieve state-of-the-art performance. The representation of these knowledge sources as weighted finite-state transducers is also presented in Chapter 7, as are weighted composition and a set of equivalence transformations, including determinization, minimization, and epsilon removal. These algorithms enable the knowledge sources to be combined into a single search graph, which can then be optimized to provide maximal search efficiency.

All ASR systems based on the HMM contain an enormous number of free parameters. In order to train these free parameters, dozens if not hundreds or even thousands of hours of transcribed acoustic data are required. Parameter estimation can then be performed according to either a maximum likelihood criterion or one of several discriminative criteria such as maximum mutual information or minimum phone error. Algorithms for efficiently estimating the parameters of a HMM are the subjects of Chapter 8. Included among these are a discussion of the well-known expectation-maximization algorithm, with which maximum likelihood estimation of HMM parameters is almost invariably performed. Several discriminative optimization criteria, namely, maximum mutual information, and minimum word and phone error are also described.

The unique characteristics of the voice of a particular speaker are what allow a person calling on the telephone to be identified as soon as a few syllables have been spoken. These characteristics include fundamental frequency, speaking rate, and accent, among others. While lending each voice its own individuality and charm, such characteristics are a hindrance to automatic recognition, inasmuch as they introduce variability in the speech that is of no use in distinguishing between different words. To enhance the performance of an ASR system that must function well for any speaker as well as different acoustic environments, various transformations are typically applied either to the features, the means and covariances of the acoustic model, or to both. The body of techniques used to estimate and apply such transformations fall under the rubrik *feature and model adaptation* and comprise the subject matter of Chapter 9.

While a recognition engine is needed to convert waveforms into word hypotheses, the speech recognizer by itself is not the only component of a distant recognition system. In Chapter 10, we introduce an important supporting technology required for a complete DSR system, namely, algorithms for determining the physical positions of one or more speakers in a room, and tracking changes in these positions with time. Speaker localization and tracking – whether based on acoustic features, video features, or both – are important technologies, because the beamforming algorithms discussed in Chapter 13 all assume that the position of the desired speaker is *known*. Moreover, the accuracy of a speaker tracking system has a very significant influence on the recognition accuracy of the entire system.

Chapter 11 discusses digital filter banks, which are arrays of bandpass filters that separate an input signal into many narrowband components. As mentioned previously, frequent reference will be made to such filter banks in Chapter 13 during the discussion of beamforming. The optimal design of such filter banks has a critical effect on the final system accuracy.

Blind source separation (BSS) and *independent component analysis* (ICA) are terms used to describe classes of techniques by which signals from multiple sensors may be combined into one signal. As presented in Chapter 12, this class of methods is known as *blind* because neither the relative positions of the sensors, nor the position of the sources are assumed to be known. Rather, BSS algorithms attempt to separate different sources based only on their temporal, spectral, or statistical characteristics. Most information-bearing signals are non-Gaussian, and this fact is extremely useful in separating signals based only on their statistical characteristics. Hence, the primary assumption of ICA is that interesting signals are *not* Gaussian signals. Several optimization criteria that are typically applied in the ICA field include kurtosis, negentropy, and mutual information. While mutual information can be calculated for both Gaussian and non-Gaussian random variables alike, kurtosis and negentropy are only meaningful for non-Gaussian signals. Many algorithms for blind source separation, dispense with the assumption of non-Gaussianity and instead attempt to separate signals on the basis of their non-stationarity or non-whiteness. Insights from the fields of BSS and ICA will also be applied to good effect in Chapter 13 for developing novel beamforming algorithms.

Chapter 13 presents a class of techniques, known collectively as beamforming, by which signals from several sensors can be combined to emphasize a desired source and to suppress all other noise and interference. Beamforming begins with the assumption that the positions of all sensors are known, and that the positions of the desired sources are known or can be estimated. The simplest of beamforming algorithms, the delay-and-sum beamformer, uses only this geometrical knowledge to combine the signals from several sensors. More sophisticated adaptive beamformers attempt to minimize the total output power of an array of sensors under a constraint that the desired source must be unattenuated. Recent research has revealed that such optimization criteria used in conventional array processing are not optimal for acoustic beamforming applications. Hence, Chapter

13 also presents several nonconventional beamforming algorithms based on optimization criteria – such as mutual information, kurtosis, and negentropy – that are typically used in the fields of BSS or ICA.

In the final chapter of this volume we present the results of performance evaluations of the algorithms described here on several DSR tasks. These include an evaluation of the speaker tracking component in isolation from the rest of the DSR system. In Chapter 14, we present results illustrating the effectiveness of single-channel speech feature enhancement based on particle filters. Also included are experimental results for systems based on beamforming for both single distant speakers, as well as two simultaneously active speakers. In addition, we present results illustrating the importance of selecting a filter bank suitable for adaptive filtering and beamforming when designing a complete DSR system.

A note about the brevity of the chapters mentioned above is perhaps now in order. To wit, each of these chapters might easily be expanded into a book much larger than the present volume. Indeed, such books are readily available on sound propagation, digital signal processing, Bayesian filtering, speech feature extraction, HMM parameter estimation, finite-state automata, blind source separation, and beamforming using conventional criteria. Our goal in writing this work, however, was to create an accessible description of all the components of a DSR system required to transform sound waves into word hypotheses, including metrics for gauging the efficacy of such a system. Hence, judicious selection of the topics covered along with concise presentation were the criteria that guided the choice of every word written here. We have, however, been at pains to provide references to lengthier specialized works where applicable – as well as references to the most relevant contributions in the literature – for those desiring a deeper knowledge of the field. Indeed, this volume is intended as a starting point for such wider exploration.

1.9 Principal Symbols used Throughout the Book

This section defines principal symbols which are used throughout the book. Due to the numerous variables each chapter presents an individual list of principal symbols which is specific for the particular chapter.

Symbol	Description
$a, b, c, \ldots$	variables
$A, B, C, \ldots$	constants
a, b, c, A, B, C, $\ldots$	units
$\mathbf{a}, \mathbf{b}, \mathbf{c}, \ldots$	vectors
$\mathbf{A}, \mathbf{B}, \mathbf{C}, \ldots$	matrices
$\mathbf{I}$	unity matrix
j	imaginary number, $\sqrt{-1}$
$.^{*}$	complex conjungate

Symbol	Description
$\cdot^T$	transpose operator
$\cdot^H$	Hermetian operator
$\cdot_{1:K}$	sequence from 1 to K
∇^2	Laplace operator
$\bar{\cdot}$	average
$\tilde{\cdot}$	warped frequency
$\hat{\cdot}$	estimate
$\%$	modulo
λ	Lagrange multiplier
$(\cdot)^+$	pseudoinverse of $(\cdot)$
$\mathcal{E}\{\cdot\}$	expectation value
$/\cdot/$	denote a phoneme
$[\cdot]$	denote a phone
$\|\cdot\|$	absolute (scalar) or determinant (matrix)
$\boldsymbol{\mu}$	mean
$\boldsymbol{\Sigma}$	covariance matrix
$\mathcal{N}(\mathbf{x}; \boldsymbol{\mu}, \boldsymbol{\Sigma})$	Gaussian distribution with mean vector $\boldsymbol{\mu}$ and covariance matrix $\boldsymbol{\Sigma}$
$\forall$	for all
$*$	convolution
δ	Dirac impulse
$\mathcal{O}$	big O notation also called Landau notation
$\mathbb{C}$	complex number
$\mathbb{N}$	set of natural numbers
$\mathbb{N}_0$	set of non-negative natural numbers including zero
$\mathbb{R}$	real number
$\mathbb{R}^+$	non-negative real number
$\mathbb{Z}$	integer number
$\mathbb{Z}^+$	non-negative integer number
$\mathrm{sinc}(z)$	$\triangleq \begin{cases} 1, & \text{for } z = 0, \\ \sin(z)/z, & \text{otherwise} \end{cases}$

1.10   Units used Throughout the Book

This section defines units that are consistently defined throughout the book.

Units	Description
Hz	Herz
J	Joule
K	Kelvin
Pa	Pascal
SPL	sound pressure level
Vs/m^2	Tesla
W	Watt
°C	degree Celsius
dB	decibel
kg	kilogram
m	meter
m^2	square meter
m^3	cubic meter
m/s	velocity
s	second

2

Acoustics

The acoustical environment and the recording sensor configuration define the characteristics of distant speech recordings and thus the usability of the data for certain applications, techniques or investigations. The scope of this chapter is to describe the physical aspect of sound and the characteristics of speech signals. In addition, we will discuss the human perception of sound, as well as the acoustic environment typically encountered in distant speech recognition scenarios. Moreover, there will be a presentation of recording techniques and possible sensor configurations for use in the capture of sound for subsequent distant speech recognition experiments.

The balance of this chapter is organized as follows. In Section 2.1, the physics of sound production are presented. This includes a discussion of the reduction in sound intensity that increases with the distance from the source, as well as the reflections that occur at surfaces. The characteristics of human speech and its production are described in Section 2.2. The subword units or phonemes of which human languages are composed are also presented in Section 2.2. The human perception of sound, along with the frequency-dependent sensitivity of the human auditory system, is described in Section 2.3. The characteristics of sound propagation in realistic acoustic environments is described in Section 2.4. Especially important in this section is the description of the spectral and temporal changes that speech and other sounds undergo when they propagate through enclosed spaces. Techniques and best practices for sound capture and recording are presented in Section 2.5. The final section summarizes the contents of this chapter and presents suggestions for further reading.

2.1 Physical Aspect of Sound

The physical – as opposed to perceptual – properties of sound can be characterized as the superposition of waves of different pressure levels which propagate through compressible media such as air. Consider, for example, one molecule of air which is accelerated and displaced from its original position. As it is surrounded by other molecules, it bounces into those adjacent, imposing a force in the opposite direction which causes the molecule to recoil and to return to its original position. The transmitted force accelerates and displaces the adjacent molecules from their original position which once more causes the molecules

Distant Speech Recognition   Matthias Wölfel and John McDonough
© 2009 John Wiley & Sons, Ltd

to bounce into other adjacent molecules. Therefore, the molecules undergo movements around their mean positions in the direction of propagation of the sound wave. Such behavior is known as a *longitudinal wave*. The propagation of the sound wave cause the molecules which are half a wavelength apart from each other to vibrate with opposite phase and thus produce alternate regions of compression and rarefaction. It follows that the *sound pressure*, defined as the difference between the instantaneous pressure and the static pressure, is a function of position and time.

Our concern here is exclusively with the propagation of sound in air and we assume the media of propagation to be *homogeneous*, which implies it has a uniform structure, *isotropic*, which implies its properties are the same in all directions, and *stationary*, which implies these properties do not change with time. These assumptions are not entirely justified, but the effects due to inhomogeneous and non-stationary media are negligible in comparison with those to be discussed; hence, so they can be effectively ignored.

2.1.1 Propagation of Sound in Air

Media capable of sound transfer have two properties, namely, mass and elasticity. The *elasticity* of an ideal gas is defined by its volume dilatation and volume compression. The governing relation of an *ideal gas*, given a specific gas constant R, is defined by the *state equation*

$$p\frac{V}{M} = R\Theta, \tag{2.1}$$

where p denotes the *pressure*, commonly measured in Pascal (Pa), V the *volume* commonly measured in cubic meters (m^3), M the *mass*, commonly measured in kilograms (kg), and Θ the *temperature* commonly measured in degrees Kelvin (K).[1] For dry air the specific gas constant is $R_{\text{dryair}} = 287.05$ J/(kg · K) where J represents Joule. Air at sea level and room temperature is well-modeled by the state equation (2.1). Thus, we will treat air as an ideal gas for the balance of this book.

The *volume compression*, or *negative dilatation*, of an ideal gas is defined as

$$-\Delta \triangleq -\frac{\delta V}{V},$$

where V represents the volume at the initial state and δV represents the volume variation. The elasticity of an ideal gas is determined by the *bulk modulus*

$$\kappa \triangleq \frac{\delta p}{-\Delta},$$

which is defined as the ratio between the pressure variation δp and the volume compression. An *adiabatic* process is a thermodynamic process in which no heat is transferred to or from the medium. Sound propagation is adiabatic because the expansions and contractions of a longitudinal wave occur very rapidly with respect to any heat transfer. Let C_p

---

[1] Absolute zero is $0\,\text{K} \approx -273.15\,^\circ$ C. No substance can be colder than this.

and C_v denote the specific heat capacities under constant pressure and constant volume, respectively. Given the adiabatic nature of sound propagation, the bulk modulus can be approximated as

$$\kappa \approx \gamma p,$$

where γ is by definition the *adiabatic exponent*

$$\gamma \triangleq \frac{C_p}{C_v}.$$

The adiabatic exponent for air is $\gamma \approx 1.4$.

2.1.2 The Speed of Sound

The wave propagation speed, in the direction away from the source, was determined in 1812 by Laplace under the assumption of an adiabatic process as

$$c = \sqrt{\frac{\kappa}{\rho}} = \sqrt{\kappa \frac{R\Theta}{p}},$$

where the *volume density* $\rho = M/V$ is defined by the ratio of mass to volume. The wave propagation speed in air c_{air} depends mainly on atmospheric conditions, in particular the temperature, while the humidity has some negligible effect. Under the ideal gas approximation, air pressure has no effect because pressure and density contribute to the propagation speed of sound waves equally, and the two effects cancel each other out. As a result, the wave propagation speed is independent of height.

In dry air the wave propagation speed can be approximated by

$$c_{air} = 331.5 \cdot \sqrt{1 + \frac{\vartheta}{273.15}},$$

where ϑ is the temperature in degrees Celsius. At room temperature, which is commonly assumed to be $20\,^{\circ}$C, the speed of sound is approximately 344 m/s.

2.1.3 Wave Equation and Velocity Potential

We begin our discussion of the theory of sound by imposing a small disturbance p on a uniform, stationary, acoustic medium with pressure p_0 and express the total pressure as

$$p_{total} = p_0 + p, |p| \ll p_0.$$

This small disturbance, which by definition is the difference between the instantaneous and atmospheric pressure, is referred to as the *sound pressure*. Similarly, the total density ρ includes both constant ρ_0 and time-varying ρ components, such that,

$$\rho_{total} = \rho_0 + \rho, |\rho| \ll \rho_0.$$

Let **u** denote the fluid velocity, q the volume velocity, and **f** the body force. In a stationary medium of uniform mean pressure p_0 and mean density ρ_0, we can relate various acoustic quantities by two basic laws:

- The law of *conservation of mass* implies,

$$\frac{1}{c^2}\frac{\partial p}{\partial t} + \rho_0 \nabla \mathbf{u} = \rho_0 q.$$

- The law of *conservation of momentum* stipulates,

$$\rho_0 \frac{\partial \mathbf{u}}{\partial t} = -\nabla + \mathbf{f}.$$

To eliminate the velocity, we can write

$$\frac{1}{c^2}\frac{\partial p}{\partial t^2} = \frac{\partial}{\partial t}(\rho_0 q - \rho_0 \nabla \mathbf{u}) = \rho_0 \frac{\partial q}{\partial t} + \nabla^2 p - \nabla \mathbf{f}. \qquad (2.2)$$

Outside the source region where $q = 0$ and in the absence of body force, (2.2) simplifies to

$$\frac{1}{c^2}\frac{\partial^2 p}{\partial t^2} = \nabla^2 p$$

which is the general *wave equation*.

The three-dimensional wave equation in *rectangular coordinates*, where l_x, l_y, l_z define the coordinate axis, can now be expressed as

$$\frac{\partial^2 p}{\partial^2 t^2} = c^2 \nabla^2 p = c^2 \left(\frac{\partial^2 p}{\partial l_x^2} + \frac{\partial^2 p}{\partial l_y^2} + \frac{\partial^2 p}{\partial l_z^2} \right).$$

A *simple* or *point source* radiates a spherical wave. In this case, the wave equation is best represented in *spherical coordinates* as

$$\frac{\partial^2 p}{\partial^2 t^2} = c^2 \nabla^2 p = c^2 \frac{1}{r^2}\frac{\partial}{\partial r}\left(r^2 \frac{\partial p}{\partial r} \right), \qquad (2.3)$$

where r denotes the distance from the source. Assuming the sound pressure oscillates as $e^{j\omega t}$ with angular frequency ω, we can write

$$c^2 \frac{1}{r^2}\frac{\partial}{\partial r}\left(r^2 \frac{\partial p}{\partial r} \right) = -\frac{\omega^2}{c^2} p = -c^2 k^2 p,$$

which can be simplified to

$$\frac{\partial^2 rp}{\partial^2 r^2} + k^2 p = 0. \qquad (2.4)$$

Here the constant k is known as the *wavenumber*,[2] or *stiffness*, which is related to the *wavelength* by

$$\lambda = \frac{2\pi}{k}. \tag{2.5}$$

A solution to (2.4) for the sound pressure can be expressed as the superposition of outgoing and incoming spherical waves, according to

$$p = \underbrace{\frac{A}{r}e^{j\omega t - jkr}}_{\text{outgoing}} + \underbrace{\frac{B}{r}e^{j\omega t + jkr}}_{\text{incoming}}, \tag{2.6}$$

where A and B denote the strengths of the sources. Thus, the sound pressure depends only on the strength of the source, the distance to the source, and the time of observation. In the *free field*, there is no reflection and thus no incoming wave, which implies $B = 0$.

2.1.4 Sound Intensity and Acoustic Power

The *sound intensity* or *acoustic intensity*

$$I_{\text{sound}} \triangleq p\,\phi,$$

is defined as the product of sound pressure p and velocity potential ϕ. Given the relation between the velocity potential and sound pressure,

$$p = \rho_0 \frac{\partial \phi}{\partial t},$$

the sound intensity can be expressed as

$$I_{\text{sound}} = \frac{p^2}{c\,\rho_0}. \tag{2.7}$$

Substituting the spherical wave solution (2.6) into (2.7), we arrive at the *inverse square law* of sound intensity,

$$I_{\text{sound}} \sim \frac{1}{r^2}, \tag{2.8}$$

which can be given a straightforward interpretation. The *acoustic power flow* of a sound wave

$$P \triangleq \int I_{\text{sound}} dS = \text{constant}$$

---

[2] As the wavenumber is used here to indicate only the relation between frequency and wavelength when sound propagates through a given medium, it is define as a scalar. In Chapter 13 it will be redefined as a vector to include the direction of wave propagation.

is determined by the surface S and remains constant. When the intensity I_{sound} is measured at a distance r, this power is distributed over a sphere with area $4\pi r^2$, which obviously increases as r^2. Hence, the inverse square law states that the sound intensity is inversely proportional to the square of the distance.

To consider non-uniform sound radiation, it is necessary to define the *directivity factor* Q as

$$Q \triangleq \frac{I_\theta(r)}{I_{\text{all}}(r)},$$

where I_{all} is the average sound intensity over a spherical surface at the distance r and I_θ is the sound intensity at angle θ at the same distance r. A spherical source has a directivity factor of 1. A source close to a single wall would have a hemispherical radiation and thus Q becomes 2. In a corner of two walls Q is 4, while in a corner of three walls it is 8. The sound intensity (2.8) must thus be rewritten as

$$I_{\text{sound}} \sim \frac{Q}{r^2}.$$

As the distance from the point source grows larger, the radius of curvature of the wave front increases to the point where the wave front resembles an infinite plane normal to the direction of propagation. This is the so-called *plane wave*.

2.1.5 Reflections of Plane Waves

The propagation of a plane wave can be described by a three-dimensional vector. For simplicity, we illustrate this propagation in two dimensions here, corresponding to the left image in Figure 2.1. For homogeneous media, all dimensions can be treated independently. But at the surface of two media of different densities, the components do interact. A portion of the incident wave is reflected, while the other portion is transmitted. The excess pressure p can be expressed at any point in the medium as a function of the coordinates and the distance of the sound wave path ξ as

- for the incident wave: $p_i = A_1 e^{j(\omega t + k_1 \xi_i)}$; $\xi_i = -x \cos \theta_i - y \sin \theta_i$,
- for the reflected wave: $p_r = B_1 e^{j(\omega t - k_1 \xi_r)}$; $\xi_r = x \cos \theta_r - y \sin \theta_r$, and
- for the transmitted wave: $p_t = A_2 e^{j(\omega t - k_2 \xi_t)}$; $\xi_t = -x \cos \theta_t - y \sin \theta_t$.

Enforcing the condition of constant pressure at the boundary $x = 0$ between the two media k_1 and k_2 for all y, we obtain the y-component of the

- pressure of the incident wave $p_{i,y} = A_1 e^{j(\omega t - k_1 y \sin \theta_i)}$,
- pressure of the reflected wave $p_{r,y} = B_1 e^{j(\omega t - y k_1 \sin \theta_r)}$, and
- pressure of the transmitted wave $p_{t,y} = A_2 e^{j(\omega t - y k_2 \sin \theta_t)}$.

These pressures must be such that

$$p_{i,y} = p_{r,y} + p_{t,y}.$$

Similarly, the

- incident sound velocity $v_{i,y} = v_i \cos \theta_i$,
- reflected sound velocity $v_{r,y} = -v_r \cos(180° - \theta_r)$, and
- transmitted sound velocity $v_{t,y} = v_t \cos \theta_t$.

These sound velocities must be such that

$$v_{i,y} = v_{r,y} + v_{t,y}.$$

The well-known law of reflection and refraction of plane waves states that the angle θ_i of incidence is equal to the angle θ_r of reflection. Applying this law, imposing the *boundary conditions*, and eliminating common terms results in

$$k_1 \sin \theta_i = k_1 \sin \theta_r = k_2 \sin \theta_t. \qquad (2.9)$$

From (2.9), it is apparent that the angle of the transmitted wave depends on the angle of the incident wave and the stiffnesses k_1 and k_2 of the two materials.

In the absence of absorption, the incident sound energy must be equal to the sum of the reflected and transmitted sound energy, such that

$$A_1 = B_1 + A_2. \qquad (2.10)$$

Replacing the sound velocities at the boundary with the appropriate value of $p/\rho_0 k$ we can write the condition

$$\frac{A_1}{\rho_1 k_1} \cos \theta_i - \frac{B_1}{\rho_1 k_1} \cos \theta_r = \frac{A_2}{\rho_2 k_2} \cos \theta_t,$$

which, to eliminate A_2 can be combined with (2.10) to give the strength of the reflected source

$$B_1 = A_1 \frac{\rho_2 k_2 \cos \theta_i - \rho_1 k_1 \cos \theta_t}{\rho_2 k_2 \cos \theta_i + \rho_1 k_1 \cos \theta_t}.$$

2.1.6 Reflections of Spherical Waves

Assuming there is radiation from a point source of angular frequency ω located near a boundary, the reflections of the spherical waves can be analyzed by *image theory*. If the point source, however, is far away from the boundary, the spherical wave behaves more like a plane wave, and thus plane wave theory is more appropriate.

The reflected wave can be expressed by a *virtual source* with spherical wave radiation, as in the right portion of Figure 2.1. The virtual source is also referred to as the *image source*. At a particular observation point, we can express the excess pressure as

$$p = \underbrace{\frac{A}{l_1} e^{j(\omega t - k l_1)}}_{\text{directwave}} + \underbrace{\frac{B}{l_2 + l_3} e^{j(\omega t - k(l_2 + l_3))}}_{\text{reflectedwave}}$$

where l_1 denotes the distance between the point source and the observation position, l_2 the distance between the point source and the reflection, and l_3 the distance from the reflection to the observation position.

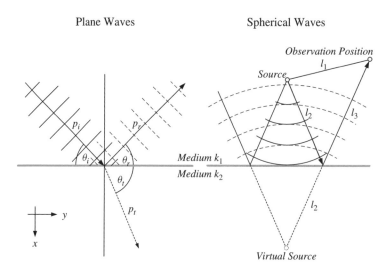

Figure 2.1   Reflection of plane and spherical waves at the boundary of two media

2.2   Speech Signals

In this section, we consider the characteristics of human speech. We first review the process of speech production. Thereafter, we categorize human speech into several phonetic units which will be described and classified. The processing of speech, such as transmission or enhancement, requires knowledge of the statistical properties of speech. Hence, we will discuss these properties as well.

2.2.1   Production of Speech Signals

Knowledge of the vocal system and the properties of the speech waveform it produces is essential for designing a suitable model of speech production. Due to the physiology of the human vocal tract, human speech is highly redundant and possesses several speaker-dependent parameters, including pitch, speaking rate, and accent. The shape and size of the individual vocal tract also effects the locations and prominence of the spectral peaks or *formants* during the utterance of vowels. The formants, which are caused by resonances of the vocal tract, are known as such because they 'form' or shape the spectrum. For the purpose of *automatic speech recognition* (ASR), the locations of the first two formants are sufficient to distinguish between vowels (Matsumura *et al.* 2007). The fine structure of the spectrum, including the overtones that are present during segments of voiced speech, actually provide no information that is relevant for classification. Hence, this fine structure is typically removed during ASR feature extraction. By ignoring this irrelevant information, a simple model of human speech production can be formulated.

The human speech production process reveals that the generation of each *phoneme*, the basic linguistic unit, is characterized by two basic factors:

- the random noise or impulse train excitation, and
- the vocal tract shape.

In order to model speech production, we must model these two factors. To understand the source characteristics, it is assumed that the source and the vocal tract model are independent (Deller Jr *et al.* 1993).

Speech consists of pressure waves created by the airflow through the vocal tract. These pressure waves originate in the lungs as the speaker exhales. The vocal folds in the larynx can open and close quasi-periodically to interrupt this airflow. The result is *voiced speech*, which is characterized by its periodicity. Vowels are the most prominent examples of voiced speech. In addition to periodicity, vowels also exhibit relatively high energy in comparison with all other phoneme classes. This is due to the open configuration of the vocal tract during the utterance of a vowel, which enables air to pass without restriction. Some consonants, for example the "b" sound in "bad" and the "d" sound in "dad", are also voiced. The voiced consonants have less energy, however, in comparison with the vowels, as the free flow of air through the vocal tract is blocked at some point by the articulators.

Several consonants, for example the "p" sound in "pie" and the "t" sound in "tie", are *unvoiced*. For such phonemes the vocal cords do not vibrate. Rather, the excitation is provided by turbulent airflow through a constriction in the vocal tract, imparting to the phonemes falling into this class a noisy characteristic. The positions of the other articulators in the vocal tract serve to filter the noisy excitation, amplifying certain frequencies while attenuating others. A time domain segment of unvoiced and voiced speech is shown in Figure 2.2.

A general linear discrete-time system for modeling the speech production process is shown in Figure 2.3. In this system, a vocal tract filter $V(z)$ and a lip radiation filter $R(z)$ are excited either by a train of impulses or by a noisy signal that is spectrally flat. The local resonances and anti-resonances are present in the vocal tract filter $V(z)$, which overall has a flat spectral trend. The lips behave as a first order high-pass filter $R(z)$, providing a frequency-dependent gain that increases by 6 dB/octave.

To model the excitation signal for unvoiced speech, a random noise generator with a flat spectrum is typically used. In the case of voiced speech, the spectrum is generated by an impulse train with pitch period p and an additional glottal filter $G(z)$. The glottal filter is usually represented by a second order low-pass filter, the frequency-dependent gain of which decreases at 12 dB/octave.

The frequency of the excitation provided by the vocal cords during voiced speech is known as the *fundamental frequency* and is denoted as f_0. The periodicity of voiced speech gives rise to a spectrum containing harmonics nf_0 of the fundamental frequency for integer $n \geq 1$. These harmonics are known as *partials*. A truly periodic sequence,

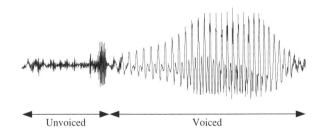

Unvoiced   Voiced

Figure 2.2 A speech segment (time domain) of unvoiced and voiced speech

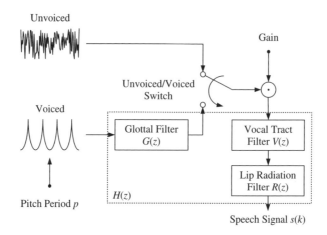

Figure 2.3   Block diagram of the simplified source filter model of speech production

observed over an infinite interval, will have a discrete-line spectrum, but voiced sounds are only locally quasi-periodic. The spectra for unvoiced speech range from a flat shape to spectral patterns lacking low-frequency components. The variability is due to place of constriction in the vocal tract for various unvoiced sounds, which causes the excitation energy to be concentrated in different spectral regions. Due to the continuous evolution of the shape of the vocal tract, speech signals are non-stationary. The gradual movement of the vocal tract articulators, however, results in speech that is quasi-stationary over short segments of 5–25 ms. This enables speech to be segmented into short frames of 16–25 ms for the purpose of performing frequency analysis, as described in Section 5.1.

The classification of speech into voiced and unvoiced segments is in many ways more important than other classifications. The reason for this is that voiced and unvoiced classes have very different characteristics in both the time and frequency domains, which may warrant processing them differently. As will be described in the next section, speech recognition requires classifying the phonemes with a still finer resolution.

2.2.2   Units of Speech Signals

Any human language is composed of elementary linguistic units of speech that determine meaning. Such a unit is known as a *phoneme*, which is by definition the smallest linguistic unit that is sufficient to distinguish between two words. We will use the notation /·/ to denote a phoneme. For example, the phonemes /c/ and /m/ serve to distinguish the word "cat" from the word "mat". The phonemes are in fact not the physical segments themselves, but abstractions of them. Most languages consist of between 40 and 50 phonemes. The acoustic realization of a phoneme is called a *phone*, which will be denoted as [·]. A phoneme can include different but similar phones, which are known as *allophones*. A *morpheme*, on the other hand, is the smallest linguistic unit that has semantic meaning. In spoken language, morphemes are composed of phonemes while in written language morphemes are composed of graphemes. *Graphemes* are the smallest units of written language and might include, depending on the language, alphabetic letters, pictograms, numerals, and punctuation marks.

The phonemes can be classified by their individual and common characteristics with respect to, for example, the place of articulation in the mouth region or the manner of articulation. The International Phonetic Alphabet (IPA 1999) is a standardized and widely-accepted representation and classification of the phonemes of all human languages. This system identifies two main classes: vowels and consonants, both of which are further divided into subclasses. A detailed discussion about different phoneme classes and their properties for the English language can be found in Olive (1993). A brief description follows.

Vowels

As mentioned previously, a vowel is produced by the vibration of the vocal cords and is characterized by the relatively free passage of air through the larynx and oral cavity. For example English and Japanese have five vowels, A, E, I, O and U. Some languages such as German have additional vowels represented by the umlauts Ä, Ö and Ü. As the vocal tract is not constricted during their utterance, vowels have the highest energy of any phoneme class. They are always voiced and usually form the central sound of a *syllable*, which is by definition a sequence of phonemes and a peak in speech energy.

Consonants

A consonant is characterized by a constriction or closure at one or more points along the vocal tract. The excitation for a consonant is provided either by the vibration of the vocal cords, as with vowels, or by turbulent airflow through a constriction in the vocal tract. Some consonant pairs share the same articulator configuration, but differ only in that one of the pair is voiced and the other is unvoiced. Common examples are the pairs [b] and [p], as well as [d] and [t], of which the first member of each pair is voiced and the second is unvoiced.

The consonants can be further split into pulmonic and non-pulmonic. *Pulmonic* consonants are generated by constricting an outward airflow emanating from the lungs along the glottis or in the oral cavity. *Non-pulmonic* consonants are sounds which are produced without the lungs using either *velaric* airflow for phonemes such as clicks, or *glottalic* airflow for phonemes such as implosives and ejectives. The pulmonic consonants make up the majority of consonants in human languages. Indeed, western languages have only pulmonic consonants.

The consonants are classified by the *International Phonetic Alphabet* (IPA) according to the manner of articulation. The IPA defines the consonant classes: nasals, plosives, fricatives, approximants, trills, taps or flaps, lateral fricatives, lateral approximants and lateral flaps. Of these, only the first three classes, which we will now briefly describe, occur frequently in most languages.

Nasals are produced by glottal excitation through the nose where the oral tract is totally constricted at some point; e.g., by a closed mouth. Examples of nasals are /m/ and /n/ such as in "mouth" and "nose".

Plosives, also known as *stop* consonants, are phonemes produced by stopping the airflow in the vocal tract to build up pressure, then suddenly releasing this pressure to create a brief turbulent sound. Examples of unvoiced plosives are /k/, /p/ and /t/ such as in "coal", "bet" or "tie", which correspond to voiced plosives /g/, /b/ and /d/ such as in "goal", "pet" or "die", respectively.

Fricatives are consonants produced by forcing the air through a narrow constriction in the vocal tract. The constriction is due to the close proximity of two articulators. A particular subset of fricatives are the *sibilants*, which are characterized by a hissing sound produced by forcing air over the sharp edges of the teeth. Sibilants have most of their acoustic energy at higher frequencies. An example of a voiced sibilant is /z/ such as in "zeal", an unvoiced sibilant is /s/ such as in "seal". Nonsibilant fricatives are, for example, /v/ such as in "vat", which is voiced and /f/ such as in "fat", which is unvoiced.

Approximants and Semivowels

Approximants are voiced phonemes which can be regarded as lying between vowels and consonants, e.g., [j] as in "yes" [jes] and [ɰ] as in Japanese "watashi" [ɰataɕi], pronounced with lip compression. The approximants which resemble vowels are termed *semivowels*.

Diphthongs

Diphthongs are a combination of some vowels and a gliding transition from one vowel to another one, e.g., /aɪ/ as in "night" [naɪt], /aʊ/ as in "now" [naʊ]. The difference between two vowels, which are two syllables, and a diphthong, which is one syllable, is that the energy dips between two vowels while the energy of a diphthong stays constant.

Coarticulation

The production of a single word, consisting of one or more phonemes, or word sequence involves the simultaneous motion of several articulators. During the utterance of a given phone, the articulators may or may not reach their target positions depending on the rate of speech, as well as the phones uttered before and after the given phone. This assimilation of the articulation of one phone to the adjacent phones is called *coarticulation*. For example, an unvoiced phone may be realized as voiced if it must be uttered between two voiced phones. Due to coarticulation, the assumption that a word can be represented as a single sequence of phonetic states is not fully justified. In continuous speech, coarticulation effects are always present and thus speech cannot really be separated into single phonemes. Coarticulation is one of the important and difficult problems in speech recognition. Because of coarticulation, state-of-the-art ASR systems invariably use context-dependent subword units as explained in Section 7.3.4.

The direction of coarticulation can be forward- or backward-oriented (Deng *et al.* 2004b). If the influence of the following vowel is greater than the preceding one, the direction of influence is called forward or *anticipatory* coarticulation. Comparing the fricative /ʃ/ followed by /i/, as in the word "she" with /ʃ/ followed by /u/ as in the word "shoe" the effect of anticipatory coarticulation becomes evident. The same phoneme /ʃ/ will typically have more energy in higher frequencies in "she" than in "shoe". If a subsequently-occurring phone is modified due to the production of an earlier phone, the coarticulation is referred to as backward or *perseverative*. Comparing the vowel /æ/ as in "map", preceded by a nasal plosive /m/, with /æ/ preceded by a voiceless stop, such as /k/ in "cap", reveals perseverative coarticulation. Nasalization is evident when a nasal plosive is followed by /æ/, however, if a voiceless stop is followed by /æ/ nasalization is not present.

2.2.3 Categories of Speech Signals

Variability in speaking style is a commonplace phenomenon, and is often associated with the speaker's mental state. There is no obvious set of styles into which human speech can be classified; thus, various categories have been proposed in the literature (Eskénazi 1993; Llisterri 1992). A possible classification with examples is given in Table 2.1.

The impact of many different speaking styles on ASR accuracy was studied by Rajasekaran and Doddington (1986) and Paul *et al.* (1986). Their investigations showed that the style of speech has a significant influence on recognition performance. Weintraub *et al.* (1996) investigated how spontaneous speech differs from read speech. Their experiments showed that – in the absence of noise or other distortions – speaking style is a dominant factor in determining the performance of large-vocabulary conversational speech recognition systems. They found, for example, that the *word error rate* (WER) nearly doubled when speech was uttered spontaneously instead of being read.

2.2.4 Statistics of Speech Signals

The statistical properties of speech signals in various domains are of specific interest in speech feature enhancement, source separation, beamforming, and recognition. Although speech is a non-stationary stochastic process, it is sufficient for most applications to estimate the statistical properties on short, quasi-stationary segments. In the present context, quasi-stationary implies that the statistical properties are more or less constant over an analysis window.

Long-term histograms of speech in the time and frequency domains are shown in Figure 2.4. For the frequency domain plot, the uniform DFT filter bank which will subsequently be described in Chapter 11 was used for subband analysis. The plots suggest that super-Gaussian distributions (Brehm and Stammler 1987a), such as the Laplace, K_0 or Gamma density, lead to better approximations of the true *probability density function* (pdf) of speech signals than a Gaussian distribution. This is true for the time as well as the frequency domain. It is interesting to note that the pdf shape is dependent on the length of the time window used to extract the short-time spectrum: The smaller

Table 2.1 Classification of speech signals

Class	Examples
speaking style	read, spontaneous, dictated, hyper articulated
voice quality	breathy, whispery, lax
speaking rate	slow, normal, fast
context	conversational, public, man-machine dialogue
stress	emotion, vocal effort, cognitive load
cultural variation	native, dialect, non-native, American vs. British English

the observation time, the more non-Gaussian is the distribution of the amplitude of the
Fourier coefficients (Lotter and Vary 2005). In the spectral magnitude domain, adjacent
non-overlapping frames tend to be correlated; the correlation increases for overlapping
frames. The correlation is in general larger for lower frequencies. A detailed discussion
of the statistics of speech and different noise types in office and car environments can be
found in Hänsler and Schmidt (2004, Section 3).

Higher Order Statistics

Most techniques used in speech processing are based on second-order properties of speech
signals, such as the power spectrum in the frequency domain, or the autocorrelation
sequence in the time domain, both of which are related to the variance of the signal.
While second-order statistics are undoubtedly useful, we will learn in Chapters 12 and
13 that *higher-order statistics* can provide a better and more precise characterization of
the statistical properties of human speech. The *third-order statistics* can give information
about the *skewness* of the pdf

$$S = \frac{\frac{1}{N} \sum_{n=1}^{N} (x_n - \mu_x)^3}{\left(\frac{1}{N} \sum_{n=1}^{N} (x_n - \mu_x)^2 \right)^{3/2}},$$

which measures its deviation from symmetry. The *fourth-order* is related to the signal
kurtosis, introduced in Section 12.2.3, which describes whether the pdf is peaked or
flat relative to a normal distribution around its mean value. Distributions with positive
kurtosis have a distinct peak around the mean, while distributions with negative kurtosis
have flat tops around their mean values. As we will learn in Chapter 12, subband samples
of speech have high kurtosis, which is evident from the histograms in Figure 2.4. The
kurtosis of each of the non-Gaussian pdfs shown in Figure 2.4 is given in Table 2.2, which

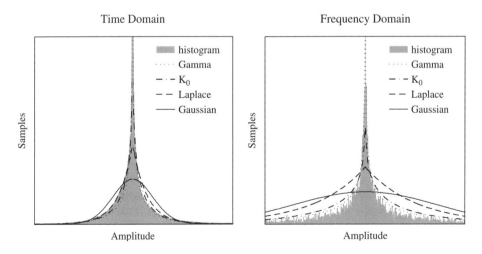

Figure 2.4   Long-term histogram of speech in time and frequency domain and different probability
density function approximations. The frequency shown is 1.6 kHz

Table 2.2 Kurtosis values for several common non-Gaussian pdfs

pdf	equation	Kurtosis
Laplace	$\frac{1}{\sqrt{2}}e^{-\sqrt{2}|x|}$	3
K_0	$\frac{1}{\pi}K_0(|x|)$	6
Γ	$\frac{\sqrt{3}}{4\sqrt{\pi}}\left(\frac{\sqrt{3}|x|}{2}\right)^{-1/2}e^{-\sqrt{3}|x|/2}$	26/3

demonstrates that as the kurtosis of a pdf increases, it comes to have more probability mass concentrated around the mean and in the tail far away from the mean. The use of higher order statistics for independent component analysis is discussed in Section 12.2, and for beamforming in Sections 13.5.2 and 13.5.4.

Higher order statistics are, for example, used in Nemer *et al.* (2002) or Salavedra *et al.* (1994) to enhance speech. Furthermore, it is reported in the literature that *mel frequency cepstral coefficients* (MFCC)s when combined with acoustic features based on higher order statistics of speech signals can produce higher recognition accuracies in some noise conditions than MFCCs alone (Indrebo *et al.* 2005).

In the time domain, the second order is the autocorrelation function

$$\phi[m] = \sum_{n=0}^{N-m} x[n]x[n+m],$$

while the third-order *moment* is

$$M[m_1, m_2] = \sum_{n=0}^{N-\max\{m_1,m_2\}} x[n]x[n+m_1]x[n+m_2].$$

Higher order moments of order M can be formed by adding additional lag terms

$$M[m_1, m_2, \ldots, m_M] = \sum_{n=0}^{N-\max\{m_1,m_2,\ldots,m_M\}} \prod_{k}^{M} x[n-m_k].$$

As mentioned previously, in the frequency domain the second-order moment is the power spectrum, which can be calculated by taking the Fourier transformation of $\phi[m]$. The third-order is referred to as the *bispectrum*, which can be calculated by taking the Fourier transformation of $M[m_1, m_2]$ over both m_1 and m_2.

2.3 Human Perception of Sound

The human perception of speech and music is, of course, a commonplace experience. While listening to speech or music, however, we are very likely unaware of our subjective sensation and the physical reality. Table 2.3 gives a simplified overview between human

Table 2.3 Relation between human perception and
physical representation

Human perception	Physical representation
pitch	fundamental frequency
loudness [sone]	sound pressure level (intensity) [dB]
location	phase difference
timbre	spectral shape

perception and physical representation. The true relationship is more complex as the
different physical properties might affect a single property in human perception. These
relations are described in more detail in this section.

2.3.1 Phase Insensitivity

Under only very weak constraints on the degree and type of allowable phase variations
(Deller Jr *et al.* 2000), the phase of a speech signal plays a negligible role in speech
perception. The human ear is for the most part *insensitive* to phase and perceives speech
primarily on the basis of the magnitude spectrum.

2.3.2 Frequency Range and Spectral Resolution

The sensitivity of the human ear ranges from 20 Hz up to 20 kHz for young people. For
older people, however, it is somewhat lower and ranges up to a maximum of 18 kHz.
Through psychoacoustic experiments, it has been determined that the complex mechanism
of the inner ear and auditory nerve performs some processing on the signal. Thus, the
subjective human perception of pitch cannot be represented by a linear relationship. The
difference in pitch of two pairs of pure tones (f_{a1}, f_{a2}) and (f_{b1}, f_{b2}) are perceived to
be equivalent if the ratio of two frequency pairs is equal, such that,

$$\frac{f_{a1}}{f_{a2}} = \frac{f_{b1}}{f_{b2}}.$$

The difference in pitch is *not* perceived to be equivalent if the *difference* between frequency pairs are equal. For example, the transition from 100 Hz to 125 Hz is perceived
as a much larger change in pitch than the transition from 1000 Hz to 1025 Hz. This is
also evident from the fact that it is easy to tell the difference between 100 Hz and 125 Hz,
while a difference between 1000 Hz and 1025 Hz is barely perceptible. This relative tonal
perception is reflected by the definition of the *octave*, which represents a doubling of the
fundamental frequency.

2.3.3 Hearing Level and Speech Intensity

Sound pressure level (SPL) is defined as

$$L_p \triangleq 20 \log \left(\frac{p}{p_r} \right) \quad \text{[dB SPL]} \tag{2.11}$$

Table 2.4 Sound pressure level with examples and subjective assessment

SPL [dB]	Examples	Subjective assessment
140	artillery	threshold of pain, hearing loss
120	jet takeoff (60 m), rock concert	intolerable
100	siren, pneumatic hammer	very noisy
80	shouts, busy road	noisy
60	conversation (1 m), office	moderate
50	computer (busy)	
40	library, quiet residential	quiet
35	computer (not busy)	
20	forest, recording studio	very quiet
0		threshold of hearing

SPL = sound pressure level

where the reference sound pressure $p_r \triangleq 20\,\mu\text{Pa}$ is defined as the threshold of hearing at 1 kHz. Some time after the introduction of this definition, it was discovered that the threshold is in fact somewhat lower. The definition of the threshold p_r which was set for 1 kHz was retained, however, as it matches nearly perfectly for 2 kHz. Table 2.4 lists a range of SPLs in common situations along with their corresponding subjective assessments, which range from the threshold of hearing to that of hearing loss.

Even though we would expect that a sound with higher intensity to be perceived as louder, this is true only for comparisons at the same frequency. In fact, the perception of loudness of a pure tone depends not only on the sound intensity but also on its frequency. The perception of equivalent loudness for different frequencies (tonal pitch) and different discrete sound pressure levels defined at 1 kHz are represented by equal loudness contours in Figure 2.5. The perceived loudness for pure tones in contrast to the physical measure of SPL is specified by the unit *phon*. By definition one phon is equal to 1 dB SPL at a frequency of 1 kHz. The equal loudness contours were determined through audiometric measurements whereby a 1 kHz tone of a given SPL was compared to a second tone. The volume of the second tone was then adjusted so as to be perceived as equally loud as the first tone. Considering the equal loudness plots in Figure 2.5, we observe that the ear is more sensitive to frequencies between 1 and 5 kHz, than below 1 kHz and above 5 kHz. A SPL change of 6 dB is barely perceptible, while it becomes clearly perceptible if the change is more than 10 dB. The perceived volume of sound is half or twice as loud, respectively, for a decrease or increase of 20 dB.

The average power of speech is only 10 microwatts, with peaks of up to 1 milliwatt. The range of speech spectral content and its approximate level is shown by the dark shape in Figure 2.5. Very little speech power is at frequencies below 100 Hz, while around 80% of the power is in the frequency range between 100 and 1000 Hz. The small remaining power at frequencies above 1000 Hz determines the intelligibility of speech. This is because several consonants are distinguished primarily by spectral differences in the higher frequencies.

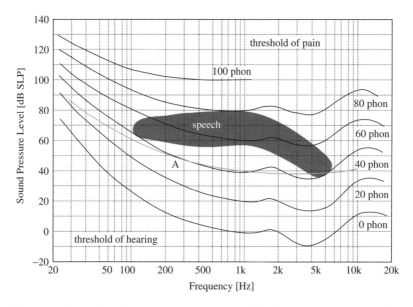

Figure 2.5 Perception of loudness expressed by equal loudness contours according to ISO 226:2003 and the inverse outline of the A-weighting filter

2.3.4 Masking

The term *masking* refers to the fact that the presence of a sound can render another sound inaudible. Masking is used, for example, in MP3 to reduce the size of audio files by retaining only the parts of the signals which are not masked and therefore perceived by the listener (Sellars 2000).

In the case where the masker is present at the same time as the signal it is called *simultaneous masking*. In simultaneous masking one sound cannot be perceived due to the presence of a louder sound nearby in frequency, and thus is also known as *frequency masking*. It is closely related to the movements of the Basilar membrane in the inner ear.

It has been shown that a sound can also mask a weaker sound which is presented before or after the stronger signal. This phenomenon is known as *temporal masking*. If a sound is obscured immediately preceding the masker, and thus masking goes back in time, it is called *backward masking* or *pre-masking*. This effect is restricted to a masker which appears approximately between 10 and 20 ms after the masked sound. If a sound is obscured immediately following the masker it is called *forwards masking* or *post-masking* with an attenuation lasting approximately between 50 and 300 ms.

An extensive investigation into masking effects can be found in Zwicker and Fastl (1999). Brungart (2001) investigated masking effects in the perception of two simultaneous talkers, and concluded that the information context, in particular the similarity of a target and a masking sentence, influences the recognition performance. This effect is known as *informational masking*.

2.3.5   Binaural Hearing

The term *binaural hearing* refers to the auditory process which evaluates the differences of sounds received by the two ears, which vary in time and amplitude according due to the location of the source of the sound (Blauert 1997; Gilkey and Anderson 1997; Yost and Gourevitch 1987).

The difference in the time of arrival at the two ears is referred to as *interaural time difference* and is due to the different distances the sound must propagate before it arrives at each ear. Under optimal conditions, listeners can detect interaural time differences as small as 10 μs. The differences in the amplitude level is called *interaural level difference* or *interaural intensitive difference* and is due to the attenuation produced by the head, which is referred to as the *head shadow*. As mentioned previously, the smallest difference in intensity that can be reliably detected is about 1 dB. Both the interaural time as well as the level differences provide information about the source location (Middlebrooks and Green 1991) and contribute to the intelligibility of speech in distorted environments. This is often referred to as *spatial release of masking*. The gain in speech intelligibility depends on the spatial distribution of the different sources. The largest improvement, which can be as much as 12 dB, is obtained when the interfering source is displaced by 120° on the horizontal plain from the source of interest (Hawley *et al.* 2004).

The two cues of binaural hearing, however, cannot determine the distance of the listener from a source of sound. Thus, other cues must be used to determine this distance, such as the overall level of a sound, the amount of reverberation in a room relative to the original sound, and the timbre of the sound.

2.3.6   Weighting Curves

As we have seen in the previous section, the relation between the physical SPL and the subjective perception is quite complicated and cannot be expressed by a simple equation. For example, the subjective perception of loudness is not only dependent on the frequency but also on the bandwidth of the incident sound. To account for the human ear's sensitivity, frequency-weighted SPLs have been introduced. The so-called *A-weighting*, originally intended only for the measurement of low-level sounds of approximately 40 phon, is now standardized in ANSI S1.42-2001 and widely used for the measurement of environmental and industrial noise. The characteristic of the A-weighting filter is inversely proportional to the hearing level curve corresponding to 40 dB at 1 kHz as originally defined by Fletcher and Munson (1933). For certain noises, such as those made by vehicles or aircraft, alternative functions such as B-, C- and D[3]-weighting may be more suitable. The B-weighting filter is roughly inversely proportional to the 70 dB at 1 kHz hearing level curve. In this work A-, B-, and C-weighted decibels are abbreviated as dB$_A$, dB$_B$, and dB$_C$, respectively. The gain curves depicted in Figure 2.6 are defined by the s-domain transfer functions:

---

[3] This filter was developed particularly for loud aircraft noise and specified as IEC 537. It has been withdrawn, however.

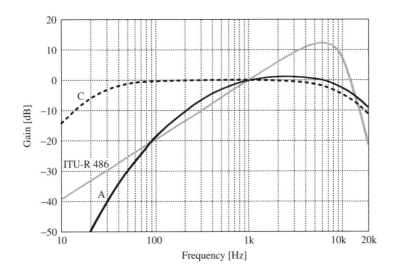

Figure 2.6 Weighting curves for ITU-R 486, A- and C-weighting

- A-weighting

$$H_A(s) = \frac{4\pi^2 12200^2 s^4}{(s + 2\pi 20.6)^2(s + 2\pi 12200)^2(s + 2\pi 107.7)(s + 2\pi 738)}$$

- B-weighting

$$H_B(s) = \frac{4\pi^2 12200^2 s^3}{(s + 2\pi 20.6)^2(s + 2\pi 12200)^2(s + 2\pi 158.5)}$$

- C-weighting

$$H_C(s) = \frac{4\pi^2 12200^2 s^2}{(s + 2\pi 20.6)^2(s + 2\pi 12200)^2}$$

As an alternative to A-weighting, which has been defined for pure tones, the ITU-R 486 noise weighting has been developed to more accurately reflect the subjective impression of loudness of all noise types. ITU-R 486 is widely used in Europe, Australia and South Africa while A-weighting is common in the United States.

2.3.7 Virtual Pitch

The *residue*, a term coined by Schouten (1940), describes a harmonically complex tone that includes higher harmonics, but lacks the fundamental frequency and possibly several other lower harmonics. Figure 2.7, for example, depicts a residue with only the 4th, 5th and 6th harmonics of 167 Hz. The concept of *virtual pitch* (Terhardt 1972, 1974) describes how a residue is perceived by the human auditory system. The pitch that the brain assigns to the residue is *not* dependent on the audible frequencies, but on a range of frequencies that extend above the fundamental. In the previous example, the virtual pitch perceived

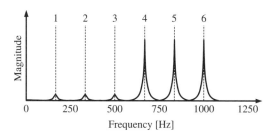

Figure 2.7 Spectrum that produces a virtual pitch at 167 Hz. Partials appear at the 4th, 5th and 6th harmonics of 167 Hz, which correspond to frequencies of 667, 833 and 1000 Hz

would be 167 Hz. This effect ensures that the perceived pitch of speech transmitted over a telephone channel is correct, despite the fact that no spectral information below 300 Hz is transmitted over this channel.

2.4 The Acoustic Environment

For the purposes of DSR, the acoustic environment is a set of unwanted transformations that affects the speech signal from the time it leaves the speaker's mouth until it reaches the microphone. The well-known and often-mentioned distortions are *ambient noise*, *echo* and *reverberation*. Two other distortions have a particular influence on distant speech recordings: The first is *coloration*, which refers to the capacity of enclosed spaces to support standing waves at certain frequencies, thereby causing these frequencies to be amplified. The second is *head orientation and radiation*, which changes the pressure level and determines if a direct wavefront or only indirect wavefronts reach the microphone. Moreover, in contrast to the free field, sound propagating in an enclosed space undergoes absorption and reflection by various objects. Yet another significant source of degradation that must be accounted for when ASR is conducted without a close-talking microphone in a real acoustic environment is speech from other speakers.

2.4.1 Ambient Noise

Ambient noise, also referred to as *background noise*,[4] is any additive sound other than that of interest. A broad variety of ambient noises exist, which can be classified as either:

- *stationary*
  Stationary noises have statistics that do not change over long time spans. Some examples are computer fans, power transformers, and air conditioning.
- *non-stationary*
  Non-stationary noises have statistics that change significantly over relatively short periods. Some examples are interfering speakers, printers, hard drives, door slams, and music.

---

[4] We find the term background noise misleading as the "background" noise might be closer to the microphone as the "foreground" signal.

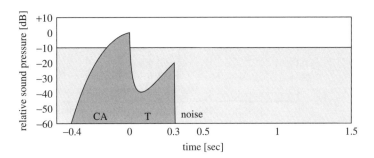

Figure 2.8 Simplified plot of relative sound pressure vs time for an utterance of the word "cat" in additive noise

Most noises are not entirely stationary, nor entirely non-stationary in that they can be treated as having constant statistical characteristics for the duration of the analysis window typically used for ASR.

Influence of Ambient Noise on Speech

Let us consider a simple example illustrating the effect of ambient noise on speech. Figure 2.8 depicts the utterance of the word "cat" with an ambient noise level 10 dB below the highest peak in SPL of the spoken word. Clearly the consonant /t/ is covered by the noise floor and therefore the uttered word is indistinguishable from words such as "cad", "cap", or "cab". The effect of additive noise is to "fill in" regions with low speech energy in the time-frequency plane.

2.4.2 Echo and Reverberation

An *echo* is a single reflection of a sound source, arriving some time after the direct sound. It can be described as a wave that has been reflected by a discontinuity in the propagation medium, and returns with sufficient magnitude and delay to be perceived as distinct from the sound arriving on the direct path. The human ear cannot distinguish an echo from the original sound if the delay is less than 0.1 of a second. This implies that a sound source must be more than 16.2 meters away from a reflecting wall in order for a human to perceive an audible echo. *Reverberation* occurs when, due to numerous reflections, a great many echoes arrive nearly simultaneously so that they are indistinguishable from one another. Large chambers – such as cathedrals, gymnasiums, indoor swimming pools, and large caves – are good examples of spaces having reverberation times of a second or more and wherein the reverberation is clearly audible. The sound waves reaching the ear or microphone by various paths can be separated into three categories:

- *direct wave*
  The direct wave is the wave that reaches the microphone on a direct path. The time delay between the source and its arrival on the direct path can be calculated from the sound velocity c and the distance r from source to microphone. The frequency-dependent attenuation of the direct signal is negligible (Bass *et al.* 1972).

- *early reflections*
  Early reflections arrive at the microphone on an indirect path within approximately 50 to 100 ms after the direct wave and are relatively sparse. There are frequency-dependent attenuations of these signals due to different reflections from surfaces.
- *late reflections*
  Late reflections are so numerous and follow one another so closely that they become indistinguishable from each other and result in a diffuse noise field. The degradation introduced by late reflections is frequency-dependent due to the frequency-dependent variations introduced by surface reflections and air attenuation (Bass *et al.* 1972). The latter becomes more significant due to the greater propagation distances.

A detailed pattern of the different reflections is presented in Figure 2.9. Note that this pattern changes drastically if either the source or the microphone moves, or the room impulse changes when, for example, a door or window is opened.

In contrast to additive noise, the distortions introduced by echo or reverberation are *correlated* with the desired signal by the *impulse response h* of the surroundings through the convolution (discussed in Section 3.1)

$$y[k] = h[k] * x[k] = \sum_{m=0}^{M} h[k]x[k-m].$$

In an enclosed space, the number N of reflections can be approximated (Möser 2004) by the ratio of the sphere volume V_{sphere} with radius $r = ct$, the distance from the source, and the room volume V_{room} by

$$N \approx \frac{V_{\text{sphere}}}{V_{\text{room}}} = \frac{4\pi}{3} \frac{r^3}{V}. \tag{2.12}$$

In a room with a volume of $250\,\text{m}^3$, approximately 85 000 reflections appear within the first half second. The density of the incident impulses can be derived from (2.12) as

$$\frac{dN}{dt} \approx 4\pi c \frac{r^2}{V}.$$

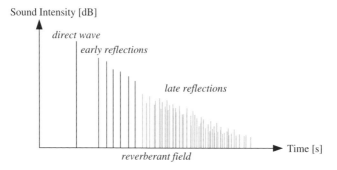

Figure 2.9 Direct wave and its early and late reflections

Thus, the number of reflections grows quadratically with time, while the energy of the reflections is inversely proportional to t^2, due to the greater distance of propagation.

The *critical distance* D_c is defined as the distance where the intensity of the direct sound is identical to the reverberant field. Close to the source, the direct sound predominates. Only at distances larger than the critical distance does the reverberation predominate. The critical distance in comparison to the overall, direct, and reverberant sound fields is depicted in Figure 2.10.

The critical distance depends on a variety of parameters such as the geometry and absorption of the space as well as the dimensions and shape of the sound source. The critical distance can, however, be approximately determined from the *reverberation time* T_{60}, which is defined as the time a signal needs to decay to 60 dB below its highest SPL, as well as the volume of the room. The relation between reverberation time, room volume and critical distance is plotted in Figure 2.11.

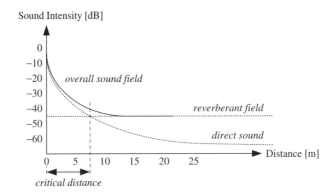

Figure 2.10 Approximation of the overall sound field in a reverberant environment as a function of the distance from the source

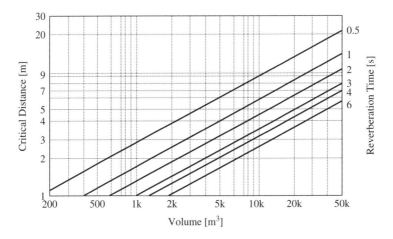

Figure 2.11 Critical distance as a function of reverberation time and volume of a specific room, after Hugonnet and Walder (1998)

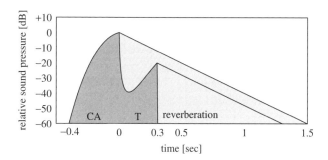

Figure 2.12 Simplified plot of relative sound pressure vs time for an utterance of the word "cat" in a reverberant environment

While T_{60} is a good indicator of how reverberant a room is, it is not the sole determinant of how much reverberation is present in a captured signal. The latter is also a function of the positions of both speaker and microphone, as well as the actual distance between them. Hence, all of these factors affect the quality of sound capture as well as DSR performance (Nishiura *et al.* 2007).

Influence of Reverberation on Speech

Now we consider the same simple example as before, but introduce reverberation with $T_{60} = 1.5$ s instead of ambient noise. In this case, the effect is quite different as can be observed by comparing Figure 2.8 with Figure 2.12. While it is clear that the consonant /t/ is once more occluded, the masking effect is this time due to the reverberation from the vowel /a/. Once more the word "cat" becomes indistinguishable from the words "cad", "cap", or "cab".

2.4.3 Signal-to-Noise and Signal-to-Reverberation Ratio

In order to measure the different distortion energies, namely additive and reverberant distortions, two measures are frequently used:

- *signal-to-noise ratio (SNR)*
  SNR is by definition the ratio of the power of the desired signal to that of noise in a distorted signal. As many signals have a wide dynamic range, the SNR is typically defined on logarithmic *decibel* scale as

$$\text{SNR} \triangleq 10 \log_{10} \frac{P_{\text{signal}}}{P_{\text{noise}}},$$

where P is the average power measured over the system bandwidth. To account for the non-linear sensitivity of the ear, A-weighting, as described in Section 2.3.3, is often applied to the SNR measure.

While SNR is a useful measure for assessing the level of additive noise in a signal as well as reductions thereof, it fails to provide any information of reverberation levels.

SNR is also widely used to measure channel quality. As it takes no account of the type, frequency distribution, or non-stationarity of the noise, however, SNR is poorly correlated with WER.

- *signal-to-reverberation ratio (SRR)*
  Similar to SNR the SRR is defined as the ratio of a signal power to the reverberation power contained in a signal as

$$\text{SRR} \triangleq 10 \log_{10} \frac{P_{\text{signal}}}{P_{\text{reverberation}}} = \mathcal{E} \left\{ 10 \log_{10} \frac{s^2}{(s * h_{\text{r}})^2} \right\}$$

where s is the clean signal and h_{r} the impulse response of the reverberation.

2.4.4 An Illustrative Comparison between Close and Distant Recordings

To demonstrate the strength of the distortions introduced by moving the microphone away from the speaker's mouth, we consider another example. This time we assume there are two sound sources, the speaker, and one noise source with a SPL 5 dB below the SPL of the speaker. Let us further assume that there are two microphones, one near and one distant from the speaker's mouth. The direct and reflected signals take different paths from the sources to the microphones, as illustrated in Figure 2.13. The direct path (solid line) of the desired sound source follows a straight line starting at the mouth of the speaker. The ambient noise paths (dotted lines) follow a straight line starting at the noise source, while the reverberation paths (dashed lines) start at the desired sound source or at the noise source being reflected once before they reach the microphone. Note that in a realistic scenario reflections will occur from all walls, ceiling, floor and other hard objects. For simplicity, only those reflections from a single wall are considered in our examples. Here we assume a sound absorption of 5 dB at the reflecting wall.

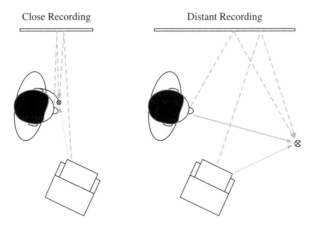

Figure 2.13 Illustration of the paths taken by the direct and reflected signals to the microphones in near- and far-field data capture

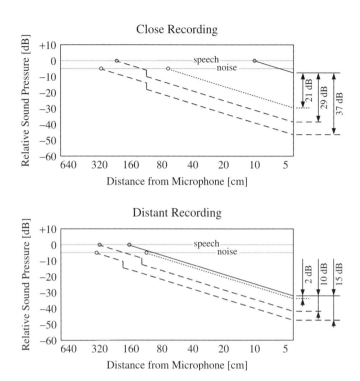

Figure 2.14   Relative sound pressure of close and distant recording of the same sources

If the SPL L_1 at a particular distance l_1 from a point source is known, we can use (2.11) to calculate the SPL L_2 at another distance l_2, in the free-field, by

$$L_2 = L_1 - 20 \log \frac{l_2}{l_1} \quad [\text{dB}].\tag{2.13}$$

With the interpretation of (2.13), namely, *each doubling of the distance reduces the sound pressure level by 6 dB*, we can plot the different SPLs following the four paths of Figure 2.13. The paths start at the different distances from the sound sources and relative SPL. In addition, at the point of the reflection, it is necessary to subtract 5 dB due to absorption. On the right side of the two images in Figure 2.14, we can read the differences of the direct speech signal to the distortion. From the two images it is obvious that the speech is heavily distorted on the distant microphone (2, 10 and 15 dB) while on the close microphone the distortion due to noise and reverberation is quite limited (21, 29 and 37 dB).

2.4.5   The Influence of the Acoustic Environment on Speech Production

The acoustic environment has a non-trivial influence on the production of speech. People tend to raise their voices if the noise level is between 45 and 70 db SPL (Pearsons *et al.* 1977). The speech level increases by about 0.5 dB SPL for every increase of 1 db

SPL in the noise. This phenomenon is known as the *Lombard effect* (Lombard 1911). In very noisy environments people start shouting which entails not only a higher amplitude, but in addition a higher pitch, a shift in formant positions to higher frequencies, in particular the first formant, and a different coloring of the spectrum (Junqua 1993). Experiments have shown that ASR is somewhat sensitive to the Lombard effect. Some ways of dealing with the variability of speech introduced by the Lombard effect in ASR are discussed by Junqua (1993). It is difficult, however, to characterize such alterations analytically.

2.4.6 Coloration

Any closed space will resonate at those frequencies where the excited waves are in phase with the reflected waves, building up a *standing wave*. The waves are in phase if the frequency of excitation between two parallel, reflective walls is such that the distance l corresponds to any integer multiple of half a wavelength. Those frequencies at or near a resonance are amplified and are called *modal frequencies* or *room modes*. Therefore, the spacing of the modal frequencies results in reinforcement and cancellation of acoustic energy, which determines the amount and characteristics of *coloration*. Coloration is strongest for small rooms at low frequencies between 20 and 200 Hz. At higher frequencies the room still has an influence, but the resonances are not as strong due to higher attenuation through absorption. The sharpness and height of the resonant peaks depend not only on the geometry of the room, but also on its sound-absorbing properties. A room filled with, for example, furniture, carpets, and people will have high absorption and might have peaks and valleys that vary between 5 and 10 dB. A room with bare walls and floor, on the other hand, will have peaks and valleys that vary between 10 and 20 dB, sometimes even more. This effect is demonstrated in Figure 2.15. On the left of the figure, the modes are closely-grouped due to the resonances of a symmetrical room. On the right of the figure, the modes are evenly-spaced due to an irregular room shape. Note that additional coloration is introduced by the microphone transfer function.

Given a rectangular room with dimensions (D_x, D_y, D_z) and perfectly reflecting walls, some basic conclusions can be drawn from wave theory. The boundary conditions require pressure maxima at all boundary surfaces, therefore we can express the sound pressure p

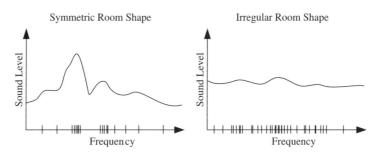

Figure 2.15 Illustration of the effect of geometry on the modes of a room. The modes at different frequencies are indicated by tick marks

as a function of position (l_x, l_y, l_z) according to

$$p(l_x, l_y, l_z) = \sum_{i_x=0}^{\infty} \sum_{i_y=0}^{\infty} \sum_{i_z=0}^{\infty} A \cos\left(\frac{\pi i_x l_x}{D_x}\right) \cos\left(\frac{\pi i_y l_y}{D_y}\right) \cos\left(\frac{\pi i_z l_z}{D_z}\right), i_x, i_y, i_z \in \mathbb{N}_0.$$

As stated by Rayleigh in 1869, solving the wave equation with the resonant frequency $\varrho = 2\pi i$ for $i \in \mathbb{N}_0$, the room modes are found to be

$$f_{\text{mode}}(D_x, D_y, D_z) = \frac{c}{2} \cdot \sqrt{\frac{\varrho_x^2}{D_x^2} + \frac{\varrho_y^2}{D_y^2} + \frac{\varrho_z^2}{D_z^2}}.$$

Room modes with value 1 are called *first mode*, with values 2 are called *second mode* and so forth. Those modes with two zeros are known as *axial modes*, and have pressure variation along a *single* axis. Modes with one zero are known as *tangential modes*, and have pressure variation along two axes. Modes without zero values are known as *oblique modes*, and have pressure variations along all three axes.

The number of resonant frequencies forming in a rectangular room up to a given frequency f can be approximated as (Kuttruff 1997)

$$m \approx \frac{4\pi}{3} \left(\frac{f}{c}\right)^3 V + \frac{\pi}{4} \left(\frac{f}{c}\right)^2 S + \left(\frac{f}{c}\right) \frac{L}{8}, \tag{2.14}$$

where V denotes the volume of the room, $S = 2(L_x L_y + L_x L_z + L_y L_z)$ denotes the combined area of all walls, and $L = 4(L_x + L_y + L_z)$ denotes the sum of the lengths of all walls. Taking, for example, a room with a volume of $250\,\text{m}^3$, and neglecting those terms involving S and L, there would be more than 720 resonances below 300 Hz. The large number of reflections demonstrates very well that only statistics can give a manageable overview of the sound field in an enclosed space. The situation becomes even more complicated if we consider rooms with walls at odd angles or curved walls which cannot be handled by simple calculations. One way to derive room modes in those cases is through simulations based on *finite elements* (Fish and Belytschko 2007).

Figure 2.16 shows plots of the mode patterns for both a rectangular and an irregular room shape. The rectangular room has a very regular mode pattern while the irregular room has a complex mode pattern.

The knowledge of room modes alone does not provide a great deal of information about the actual sound response, as it is additionally necessary to know the phase of each mode.

2.4.7 Head Orientation and Sound Radiation

Common sense indicates that people communicate more easily when facing each other. The reason for this is that any sound source has propagation directivity characteristics which lead to a non-spherical radiation, mainly determined by the size and the shape of the source and the frequency being analyzed. If, however, the size of the object radiating the sound is small compared to the wavelength, the directivity pattern will be nearly spherical.

Figure 2.16 Mode patterns of a rectangular and an irregular room shape. The bold lines indicate the knot of the modes, the thin lines positive amplitudes while the dashed lines indicate negative amplitudes

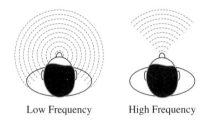

Low Frequency   High Frequency

Figure 2.17   Influence of low and high frequencies on sound radiation

Approximating the head as an oval object with a diameter slightly less than 20 cm and a single sound source (the mouth), we can expect a more directional radiation for frequencies above 500 Hz, as depicted in Figure 2.17. Moreover, it can be derived from theory that different pressure patterns should be observed in the horizontal plane than in the vertical plane (Kuttruff 2000). This is confirmed by measurements by Chu and Warnock (2002a) of the sound field at 1 meter distance around the head of an active speaker in an anechoic chamber, as shown in Figure 2.18. Comparing their laboratory measurements with field measurements (Chu and Warnock 2002b) it was determined that the measurements were in good agreement for spectra of male voices. They observed, however, some differences for female voiced spectra. There are no significant differences in the directivity patterns for male and female speakers, although there are different spectral patterns. Similar directivity patterns were observed for loud and normal voice levels, although the directivity pattern of quiet voices displayed significant differences in radiation behind the head.

As shown by the measurements made by Chu and Warnock as well as measurements by Moreno and Pfretzschner (1978), the head influences the timbre of human speech. Additionally, radiation behind the head is between 5 and 15 dB lower than that measured in front of the head at the same distance to the sound source. Moreover, it has been observed that the direct wavefront propagates only in the frontal hemisphere, and in a way that also depends on the vertical orientation of the head.

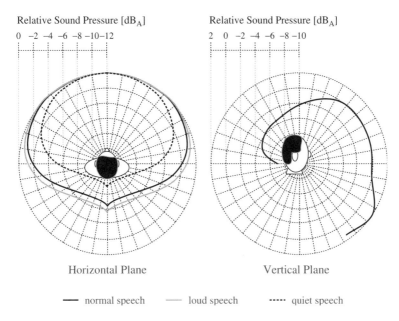

Relative Sound Pressure [dB$_A$]
0 −2 −4 −6 −8 −10−12

Relative Sound Pressure [dB$_A$]
2 0 −2 −4 −6 −8 −10

Horizontal Plane   Vertical Plane

—— normal speech   —— loud speech   ····· quiet speech

Figure 2.18   Relative sound pressure (A-weighted) around the head of an average human talker for three different voice levels. The graphics represent measurements by Chu and Warnock (2002a)

2.4.8   Expected Distances between the Speaker and the Microphone

Some applications such as beamforming, which will be presented in Chapter 13, require knowledge of the distance between the speaker and each microphone in an array. The microphones should be positioned such that they receive the direct path signal from the speaker's mouth. They also should be located as close as possible to the speaker, so that, as explained in Section 2.4.2, the direct path signal dominates the reverberant field. Considering these constraints gives a good estimate about the possible working distance between the speaker and the microphone. In a meeting scenario one or more microphones might be placed on the table and thus a distance between 1 and 2 meters can be expected. A single wall-mounted microphone can be expected to have an average distance of half of the maximum of the length and the width of the room. If all walls in a room are equipped with at least one microphone, the expected distance can be reduced below the minima of the length and the width of the room. The expected distance between a person and a humanoid robot can be approximated by the social interpersonal distance between two people. The theory of *proxemics* by Hall (1963) suggests that the social distance between people is related to the physical interpersonal distance, as depicted in Figure 2.19. Such "social relations" may also play a role in man–machine interactions. From the figure, it can be concluded that a robot acting as a museum guide would maintain an average distance of at least 2 meters from visitors. A robot intended as a child's toy, on the other hand, may have an average distance from its user of less than 1 meter. Hand-held devices are typically used by a single user or two users standing close together. The device is held so that it faces the user with its display approximately 50 cm away from the user's mouth.

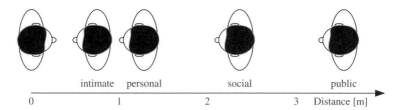

intimate   personal   social   public

0   1   2   3   Distance [m]

Figure 2.19 Hall's classification of the social interpersonal distance in relation to physical inter-personal distance

2.5 Recording Techniques and Sensor Configuration

A microphone is the first component in any speech-recording system. The invention and development of the microphone is due to a number of individuals some of whom remain obscure. One of the oldest documented inventions of a microphone dating back to the year 1860 is by Antonio Meucci, who is now also officially recognized as the inventor of the telephone[5] besides Johann Philipp Reis (first public viewing in October 1886 in Frankfurt, Germany), Alexander Graham Bell, and Elisha Gray. Many early developments in microphone design, such as the carbon microphone by Emil Berliner in 1877, took place at Bell Laboratories.

Technically speaking the microphone is a transducer which converts acoustic sound waves in the form of pressure variation into an equivalent electrical signal in the form of voltage variation. This transformation consists of two steps: The variation in sound pressure set the microphone diaphragm into vibration, so that the acoustical energy is converted to mechanical, which later can be transferred into alternating voltage, so that the mechanical energy can be converted to electrical energy. Therefore, any given microphone can be classified along two dimensions: its mechanical characteristics and its electrical characteristics.

2.5.1 Mechanical Classification of Microphones

The pressure variation can be converted into vibration of the diaphragm in various ways:

- *Pressure-operated microphones* (pressure transducer) are excited by the sound wave only on *one side* of the diaphragm, which is fixed inside a totally enclosed casing. In theory those types of microphones are *omnidirectional* as the sound pressure has no favored direction.

  The force exerted on the diaphragm can be calculated by

$$F = \int_S p \, dS [N],$$

where p is the sound pressure measured in Pascal (Pa) and S the surface area measured in square meters (m^2). For low frequencies, where the membrane cross-section is small

---

[5] *Resolved, that it is the sense of the House of Representatives that the life and achievements of Antonio Meucci should be recognized, and his work in the invention of the telephone should be acknowledged.* – United States House of Representatives, June 11, 2002

compared to the wavelength, the force on the membrane follows approximately the linear relationship $F \approx pS$. For a small wavelength, however, sound pressure with opposite phase might occur and in this case $F \neq pS$.

- *Velocity operated microphones* (pressure gradient transducer) are excited by the sound wave on *both sides* of the diaphragm, which is fixed to a support open at both sides. The resultant force varies as a function of the angle of incidence of the sound source resulting in a *bidirectional* directivity pattern.

    The force exerted on the diaphragm is

$$F \approx (p_{\text{front}} - p_{\text{back}}) \, S \; [N]$$

where $p_{\text{front}} - p_{\text{back}}$ is the pressure difference between the front and the back of the diaphragm.

- *Combined microphones* are a combination of the aforemention microphone types, resulting in a microphone with a *unidirectional* directivity pattern.

2.5.2 Electrical Classification of Microphones

The vibration of the diaphragm can be transferred into voltage by two widely used techniques:

- Electromagnetic and electrodynamic – *Moving Coil or Ribbon Microphones* have a coil or strip of aluminum, a ribbon, attached to the diaphragm which produces a varying current by its movement within a static electromagnetic field. The displacement velocity v (m/s) is converted into voltage by

$$U = Blv$$

where B denotes the electric field measured in Tesla (Vs/m^2) and l denotes the length of the coil wire or ribbon. The coil microphone has a relative low sensitivity but shows great mechanical robustness. On the other hand, the ribbon microphone has high sensitivity but is not robust.

- Electrostatic – *Electret, Capacitor or Condenser Microphones* form a capacitor by a metallic diaphragm fixed to a piece of perforated metal. The alternating movement of the diaphragm leads to a variation in the distance d of the two electrodes changing the capacity as

$$C = \epsilon \frac{S}{d}$$

where S is the surface of the metallic diaphragm and ϵ is a constant. This microphone type requires an additional power supply as the capacitor must be polarized with a voltage V_{cc} and acquires a charge

$$Q = CV_{cc}.$$

Moreover, there are additional ways to transfer the vibration of the diaphragm into voltage:

- Contract resistance – *Carbon Microphones* have been formally used in telephone handsets.
- Crystal or ceramic – *Piezo Microphones* use the tendency of some materials to produce voltage when subjected to pressure. They can be used in unusual environments such as underwater.
- Thermal and ionic effects.

2.5.3 Characteristics of Microphones

To judge the quality of a microphone and to pick the right microphone for recording, it is necessary to be familiar with the following characteristics:

- *Sensitivity* is the ratio between the electrical output level from a microphone and the incident SPL.
- *Inherent (or self) noise* is due to the electronic noise of the preamplifier as well as either the resistance of the coil or ribbon, or the thermal noise of the resistor.
- *Signal to noise ratio* is the ratio between the useful signal and the inherent noise of the microphone.
- *Dynamic range* is the difference in the level of the maximum sound pressure and inherent noise.
- *Frequency response chart* gives the transfer function of the microphone. The ideal curve would be a horizontal line in the frequency range of interest.
- *Microphone directivity*. Microphones always have a non-uniform (non-omnidirectional) response-sensitivity patterns where the directivity is determined by the characteristics of the microphone and specified by the producer. The directivity is determined by two principal effects:

    — the geometrical shape of the microphone.
    — the space dependency of the sound pressure.

    Usually the characteristics vary for different frequencies and therefore the sensitivity is measured for various frequencies. The results are often combined in a single diagram, since in many cases a uniform response over a large frequency range is desirable. Some typical patterns and their corresponding names are shown in Figure 2.20.

2.5.4 Microphone Placement

Selecting the right microphones and placing them optimally both have significant influences on the quality of the recording. Thus, before starting a recording, what kind of data is to be recorded: clean, noisy, reverberant or overlapping speech, just to name a few? From our own experience, we recommend the use of as many sensors as possible, even though at the time of the recording it is not clear for what investigations particular sensors will be needed, as data and in particular hand-labeled data is expensive to produce. It is also very important to use a close-talking microphones for each individual speaker in your sensor configuration to have a reference signal by which the difficulty of the ASR task can be judged.

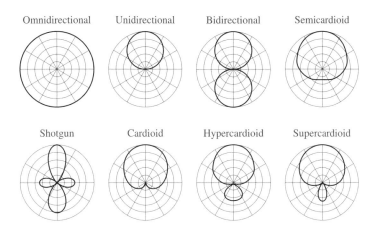

Figure 2.20 Microphone directivity patterns (horizontal plane) including names

Note that the microphone-to-source distance affects not only the amount of noise and reverberation, but also the timbre of the voice. This effect is more pronounced if the microphone has a cardioid pickup instead of an omnidirectional pickup. With increased distance the low frequencies are emphasized more. For clean speech recordings, it is recommended that the microphones should be placed as close as convenient or feasible to the speaker's mouth, which in general is not more than a couple of millimeters. If, however, the microphone is placed very close to the speaker's mouth, the microphone picks up more breath noises and pop noises from plosive consonants, or might rub on the skin of the speaker. In general it is recommended to place the microphone in the direct field. If a microphone is placed farther away from a talker more reflected speech overlaps and blurs the direct speech. At the critical distance D_c or farther, words will become hard to understand and very difficult to be correctly classified. For reasonable speech audio quality, an omnidirectional microphone should be placed no farther from the talker than 30% of D_c while cardioid, supercardioid, or shotgun microphones should be positioned no farther than 50% of D_c. Also be sure to devise a consistent naming convention for all audio channels before beginning your first recording. The sound pressure is always maximized on reflective surfaces and hence a gain of up to 6 dB can be achieved by placing a microphone on a hard surface. However, a microphone placed close to a reflective surface, on the other hand, might cancel out certain frequencies due to the interference between the direct and reflected sound wave and therefore should be avoided.

As discussed in Chapter 13, particular care must be taken for microphone array record-ings as arrays allow spatial selectivity, reinforcing the so-called *look direction*, while attenuating sources propagating from other directions. The spatial selectivity depends on the frequency: for a linear array at low frequency the pattern has a wide beamwidth which narrows for higher frequencies. The microphone array samples the sound field at different points in space and therefore array processing is subject to *spatial aliasing*. At those regions where spatial aliasing occurs the array is unable to distinguish between multiple arrival angles, and large sidelobes might appear. To prevent aliasing for linear arrays, the *spatial sampling theorem* or *half wavelength rule* must be fulfilled:

$$l < \lambda_{\min}/2.$$

As discussed in Chapter 13, the half wavelength rule states that the minimum wavelength of interest λ_{min} must be at least twice the length of the spacing l between the microphones (Johnson and Dudgeon 1993). For randomly distributed arrays the spatial sampling theorem is somewhat less stringent. But, in designing an array, one should always be aware about possible spatial aliasing. Alvarado (1990) has investigated optimal spacing for linear microphone arrays. Rabinkin *et al.* (1996) has demonstrated that the performance of microphone array systems is affected by the microphone placement. In Rabinkin *et al.* (1997) a method to evaluate the microphone array configuration has been derived and an outline for optimum microphone placement under practical considerations is characterized.

A source is considered to be in the *near-field* for a microphone array of total length l, if

$$d < \frac{2l^2}{\lambda},$$

where d is the distance between the microphone array and the source, and λ is the wavelength. An alternative presentation defining the near-field and far-field region for linear arrays considering the angle of incidence is presented in Ryan (1998).

2.5.5 Microphone Amplification

If the amplification of a recording is set incorrectly, unwanted distortions might be introduced. If the level is too high, clipping or overflow occurs. If the signal is too low, too much quantization and microphone noise may be introduced into the captured speech. *Quantization noise* is introduced by the rounding error between the analogue, continuous signal and the digitized, discrete signal. *Microphone noise* is the noise introduced by the microphone itself.

Clipping is a waveform distortion that may occur in the analog or digital processing components of a microphone. Analog clipping happens when the voltage or current exceed their thresholds. Digital clipping happens when the signal is restricted by the range of a chosen representation. For example, using a 16-bit signed integer representation, no number larger than 32767 can be represented. Sample values above 32767 are truncated to the maximum, 32767. As clipping introduces additional distortions into the recorded signal, it is to be avoided at all costs. To avoid clipping, the overall level of a signal can be lowered, or a limiter can be used to dynamically reduce the levels of loud portions of the signal. In general it can be said that it is better to have a quiet recording, which suffers from some quantization noise, than an over-driven recording suffering from clipping. In the case of a digital *overflow*, where the most significant bits of the magnitude, and sometimes even the sign of the sample value are lost, severe signal distortion is to be expected. In this case it is preferable to clip the signal as a clipped signal typically is less distorted than a signal wherein overflows have occurred.

2.6 Summary and Further Reading

This chapter has presented a brief overview of the sound field: the fundamental of sound, the human perception of sound, details about the acoustic environment, statistics of speech signals, speech production, speech units and production of speech signal. A well-written book about sound in enclosures has been published by Kuttruff (2000). Another interesting source is given by Saito and Nakata (1985). Further research into acoustic, speech and noise, psychology and physiology of hearing as well as sound propagation, transducers and measurements are subjects of acoustic societies around the world: the acoustical society of America who publish a monthly journal (JASA), the acoustical society of Japan (ASJ), who also publish in English, and the European acoustics association (EAA).

2.7 Principal Symbols

Symbol	Description
Δ	volume compression, negative dilatation
γ	adiabatic exponent
κ	bulk modulus
λ	wavelength
ϕ	velocity potential
ρ	volume density
ϱ	resonant frequency
Θ	temperature in Kelvin
ϑ	temperature in Celsius
ω	angular frequency, $\omega = 2\pi f$
ξ	distance of the sound wave path
A, B	sound energy
c	speed
C	specific heats capacities
D_c	critical distance
E	energy
f	frequency of body force
f_0	fundamental frequency
F	force
$G(z)$	glottal filter
h	impulse response
l	length, distance, dimensions of a coordinate system
H	transfer function
I	sound intensity
L	sound pressure level
k	wave number, stiffness
m	number of resonant frequencies
M	mass

Symbol	Description
N	number of reflections
p	pressure, pitch period
P	power
q	volume velocity
Q	directivity factor
r	radius
R	specific gas constant
$R(z)$	lip radiation filter
S	surface
T_{60}	reverberation time
t	continuous time
$\mathbf{u}$	fluid velocity
v	velocity
V	voltage or volume
V_s	specific volume
$V(z)$	vocal tract filter

3

Signal Processing and Filtering Techniques

In signal processing the term *filter* is commonly used to refer to an algorithm which extracts a desired signal from an input signal corrupted with noise or other distortions. A filter can also be used to modify the spectral or temporal characteristics of a signal in some advantageous way. Therefore, filtering techniques are powerful tools for speech signal processing and distant speech recognition.

This chapter reviews the basics of *digital signal processing* (DSP). This will include a short introduction of linear time-invariant systems, the Fourier transform, and the z-transform. Next there is a brief discussion of how filters can be designed through pole-zero placement in the complex z-plane in order to provide some desired frequency response. We then discuss the effects of sampling a continuous time signal to obtain a digital representation in Section 3.1.4, as well as the efficient implementation of linear time invariant systems with the discrete Fourier transform in Section 3.2. Next comes a brief presentation of the short-time Fourier transform in Section 3.3, which will have consequences for the subsequent development. The coverage of this material is very brief, in that entire books – and books much larger than the volume now beneath the reader's eyes – have been written about exactly this subject matter.

Anyone with a background in DSP can simply skip this chapter, inasmuch as the information contained herein is all standard. As this book is intended for a diverse audience, however, this chapter is included in order to make the balance of the book comprehensible to those readers who have never seen, for example, the z-transform. In particular, a thorough comprehension of the material in this chapter is necessary to understand the presentation of digital filter banks in Chapter 11, but it will also prove useful elsewhere.

3.1 Linear Time-Invariant Systems

This section presents a very important class of systems for all areas of signal processing, namely, *linear time-invariant systems* (LTI). Such systems may not fall into the most general class of systems, but are, nonetheless, important inasmuch as their simplicity

Distant Speech Recognition   Matthias Wölfel and John McDonough
© 2009 John Wiley & Sons, Ltd

conduces to their tractability for analysis, and hence enables the development of a detailed theory governing their operation and design. We consider the class of discrete-time or digital linear time-invariant systems, as digital filters offer much greater flexibility along with many possibilities and advantages over their analog counterparts. We also briefly consider, however, the class of continuous-time systems, as this development will be required for our initial analysis of array processing algorithms in Chapter 13. We will initially present the properties of such systems in the time domain, then move to the frequency and z-transform domains, which will prove in many cases to be more useful for analysis.

3.1.1   Time Domain Analysis

A *discrete-time system* (DTS) is defined as a transform operator $\mathcal{T}$ that maps an input sequence $x[n]$ onto an output sequence $y[n]$ with the sample index n, such that

$$y[n] = \mathcal{T}\{x[n]\}. \tag{3.1}$$

The class of systems that can be represented through an operation such as (3.1) is very broad. Two simple examples are:

- *time delay*,

$$y[n] = x[n - n_d] \tag{3.2}$$

  where n_d is an integer delay factor; and
- *moving average*,

$$y[n] = \frac{1}{M_2 - M_1 + 1} \sum_{m=M_1}^{M_2} x[n - m]$$

  where M_1 and M_2 determine the average position and length.

While (3.1) characterizes the most general class of discrete-time systems, the analysis of such systems would be difficult or impossible without some further restrictions. We now introduce two assumptions that will result in a much more tractable class of systems.

A DTS is said to be *linear* if

$$\mathcal{T}\{x_1[n] + x_2[n]\} = \mathcal{T}\{x_1[n]\} + \mathcal{T}\{x_2[n]\} = y_1[n] + y_2[n], \tag{3.3}$$

and

$$\mathcal{T}\{ax_1[n]\} = a\mathcal{T}\{x_1[n]\} = a\,y_1[n]. \tag{3.4}$$

Equation (3.3) implies that transforming the sum of the two input sequences $x_1[n]$ and $x_2[n]$ produces the same output as would be obtained by summing the two individual outputs $y_1[n]$ and $y_2[n]$, while (3.4) implies that transforming a scaled input sequence

$a\,x_1[n]$ produces the same sequence as scaling the original output $y_1[n]$ by the same scalar factor a. Both of these properties can be combined into the *principle of superposition*:

$$\mathcal{T}\{ax_1[n] + bx_2[n]\} = a\mathcal{T}\{x_1[n]\} + b\mathcal{T}\{x_2[n]\} = a\,y_1[n] + b\,y_2[n],$$

which is understood to hold true for all a and b, and all $x_1[n]$ and $x_2[n]$. Linearity will prove to be a property of paramount importance when analyzing discrete-time systems. We now consider a second important property. Let

$$y_d[n] = \mathcal{T}\{x[n - n_d]\},$$

where n_d is an integer delay factor. A system is *time-invariant* if

$$y_d[n] = y[n - n_d],$$

which implies that transforming a delayed version $x[n - n_d]$ of the input produces the same sequence as delaying the output of the original sequence to obtain $y[n - n_d]$. As we now show, LTI systems are very tractable for analysis. Moreover, they have a wide range of applications.

The *unit impulse* sequence $\delta[n]$ is defined as

$$\delta[n] \triangleq \begin{cases} 1, & n = 0, \\ 0, & \text{otherwise.} \end{cases}$$

The *shifting property* of the unit impulse allows any sequence $x[n]$ to be expressed as

$$x[n] = \sum_{m=-\infty}^{\infty} x[m]\,\delta[n - m],$$

which follows directly from the fact that $\delta[n - m]$ is nonzero only for $n = m$. This property is useful in characterizing the response of a LTI system to arbitrary inputs, as we now discuss.

Let us define the *impulse response* $h_m[n]$ of a general system $\mathcal{T}$ as

$$h_m[n] \triangleq \mathcal{T}\{\delta[n - m]\}. \tag{3.5}$$

If $y[n] = \mathcal{T}\{x[n]\}$, then we can use the shifting property to write

$$y[n] = \mathcal{T}\left\{ \sum_{m=-\infty}^{\infty} x[m]\,\delta[n - m] \right\}.$$

If $\mathcal{T}$ is linear, then the operator $\mathcal{T}\{\}$ works exclusively on the time index n, which implies that the coefficients $x[m]$ are effectively constants and are *not* modified by the system. Hence, we can write

$$y[n] = \sum_{m=-\infty}^{\infty} x[m]\,\mathcal{T}\{\delta[n - m]\} = \sum_{m=-\infty}^{\infty} x[m]\,h_m[n], \tag{3.6}$$

where the final equality follows from (3.5). If T is also time-invariant, then

$$h_m[n] \triangleq h[n - m],\tag{3.7}$$

and substituting (3.7) into (3.6) yields

$$y[n] = \sum_{m=-\infty}^{\infty} x[m]\,h[n - m] = \sum_{m=-\infty}^{\infty} h[m]\,x[n - m].\tag{3.8}$$

Equation (3.8) is known as the *convolution sum*, which is such a useful and frequently occurring operation that it is denoted with the symbol $*$ and typically express (3.8) with the shorthand notation,

$$y[n] = x[n] * h[n].\tag{3.9}$$

From (3.8) or (3.9) it follows that the response of a LTI system T to any input $x[n]$ is completely determined by its impulse response $h[n]$.

In addition to linearity and time-invariance, the most desirable feature a system may possess is that of *stability*. A system is said to be *bounded input–bounded output* (BIBO) stable, if every bounded input sequence $x[n]$ produces a bounded output sequence $y[n]$. For LTI systems, BIBO stability requires that $h[n]$ is absolutely summable, such that,

$$S = \sum_{m=-\infty}^{\infty} |h[m]| < \infty,$$

which we now prove. Consider that

$$|y[n]| = \left| \sum_{m=-\infty}^{\infty} h[m]\,x[n - m] \right| \le \sum_{m=-\infty}^{\infty} |h[m]|\,|x[n - m]|,\tag{3.10}$$

where the final inequality in (3.10) follows from the triangle inequality (Churchill and Brown 1990, sect. 4). If $x[n]$ is bounded, then for some $B_x > 0$,

$$|x[n]| \le B_x \; \forall \; -\infty < n < \infty.\tag{3.11}$$

Substituting (3.11) into (3.10), we find

$$|y[n]| \le B_x \sum_{m=-\infty}^{\infty} |h[m]| = B_x\, S < \infty,$$

from which the claim follows.

The *complex exponential sequence* $e^{j\omega n} \; \forall \; -\infty < n < \infty$ is an *eigenfunction* of any LTI system. This implies that if $e^{j\omega n}$ is taken as an input to a LTI system, the output is a scaled version of $e^{j\omega n}$, as we now demonstrate. Define $x[n] = e^{j\omega n}$ and substitute this input into (3.8) to obtain

$$y[n] = \sum_{m=-\infty}^{\infty} h[m]\, e^{j\omega(n-m)} = e^{j\omega n} \sum_{m=-\infty}^{\infty} h[m]\, e^{-j\omega m}. \tag{3.12}$$

Defining the *frequency response* of a LTI system as

$$H(e^{j\omega}) \triangleq \sum_{m=-\infty}^{\infty} h[m]\, e^{-j\omega m}, \tag{3.13}$$

enables (3.12) to be rewritten as

$$y[n] = H(e^{j\omega})\, e^{j\omega n},$$

whereupon it is apparent that the output of the LTI system differs from its input only through the *complex scale factor* $H(e^{j\omega})$. As a complex scale factor can introduce both a magnitude scaling and a phase shift, but nothing more, we immediately realize that these operations are the only possible modifications that a LTI system can perform on a complex exponential signal. Moreover, as *all* signals can be represented as a sum of complex exponential sequences, it becomes apparent that a LTI system can only apply a magnitude scaling and a phase shift to *any* signal, although both terms may be frequency dependent.

3.1.2 Frequency Domain Analysis

The LTI eigenfunction $e^{j\omega n}$ forms the link between the time and frequency domain analysis of LTI systems, inasmuch as this sequence is equivalent to the *Fourier kernel*. For any sequence $x[n]$, the *discrete-time Fourier transform* is defined as

$$X(e^{j\omega}) \triangleq \sum_{n=-\infty}^{\infty} x[n]\, e^{-j\omega n}. \tag{3.14}$$

In light of (3.13) and (3.14), it is apparent that the frequency response of a LTI system is nothing more than the Fourier transform of its impulse response. The samples of the original sequence can be recovered from the *inverse Fourier transform*,

$$x[n] \triangleq \frac{1}{2\pi} \int_{-\pi}^{\pi} X(e^{j\omega})\, e^{j\omega n}\, d\omega. \tag{3.15}$$

In order to demonstrate the validity of (3.15), we need only consider that

$$\frac{1}{2\pi} \int_{-\pi}^{\pi} e^{j\omega(n-m)}\, d\omega = \begin{cases} 1, & \text{for } n = m, \\ 0, & \text{otherwise,} \end{cases} \tag{3.16}$$

a relationship which is easily proven. When $x[n]$ and $X(e^{j\omega})$ satisfy (3.14–3.15), we will say they form a *transform pair*, which we denote as

$$x[n] \leftrightarrow X(e^{j\omega}).$$

We will adopt the same notation for other transform pairs, but not specifically indicate this in the text for the sake of brevity.

To see the effect of time delay in the frequency domain, let us express the Fourier transform of a time delay (3.2) as

$$Y(e^{j\omega}) = \sum_{n=-\infty}^{\infty} y[n] e^{-j\omega n} = \sum_{n=-\infty}^{\infty} x[n - n_{\mathrm{d}}] e^{-j\omega n}. \tag{3.17}$$

Introducing the change of variables $n' = n - n_{\mathrm{d}}$ in (3.17) provides

$$Y(e^{j\omega}) = \sum_{n'=-\infty}^{\infty} x[n'] e^{-j\omega(n'+n_{\mathrm{d}})} = e^{-j\omega n_{\mathrm{d}}} \sum_{n'=-\infty}^{\infty} x[n'] e^{-j\omega n'},$$

which is equivalent to the transform pair

$$x[n - n_{\mathrm{d}}] \leftrightarrow e^{-j\omega n_{\mathrm{d}}} X(e^{j\omega}). \tag{3.18}$$

As indicated by (3.18), the effect of a time delay in the frequency domain is to induce a *linear phase shift* in the Fourier transform of the original signal. In Chapter 13, we will use this property to perform beamforming in the subband domain by combining the subband samples from each sensor in an array using a phase shift that compensates for the propagation delay between a desired source and a given sensor.

To analyze the effect of the convolution (3.8) in the frequency domain, we can take the Fourier transform of $y[n]$ and write

$$Y(e^{j\omega}) = \sum_{n=-\infty}^{\infty} y[n] e^{-j\omega n} = \sum_{n=-\infty}^{\infty} \left\{ \sum_{m=-\infty}^{\infty} x[m] h[n - m] \right\} e^{-j\omega n}.$$

Changing the order of summation and re-indexing with $n' = n - m$ provides

$$Y(e^{j\omega}) = \sum_{m=-\infty}^{\infty} x[m] \sum_{n=-\infty}^{\infty} h[n - m] e^{-j\omega n} = \sum_{m=-\infty}^{\infty} x[m] \sum_{n'=-\infty}^{\infty} h[n'] e^{-j\omega(n'+m)}$$

$$= \sum_{m=-\infty}^{\infty} x[m] e^{-j\omega m} \sum_{n'=-\infty}^{\infty} h[n'] e^{-j\omega n'}. \tag{3.19}$$

Equation (3.19) is then clearly equivalent to

$$Y(e^{j\omega}) = X(e^{j\omega}) H(e^{j\omega}). \tag{3.20}$$

This simple but important result indicates that time domain *convolution* is equivalent to frequency domain *multiplication*, which is one of the primary reasons that frequency domain operations are to be preferred over their time domain counterparts. In addition to its inherent simplicity, we will learn in Section 3.2 that frequency domain implementations of LTI systems are often more efficient than time domain implementations.

The most general LTI system can be specified with a linear constant coefficient difference equation of the form

$$y[n] = -\sum_{l=1}^{L} a_l y[n-l] + \sum_{m=0}^{M} b_m x[n-m]. \tag{3.21}$$

Equation (3.21) specifies the relation between the output signal $y[n]$ and the input signal $x[n]$ in the time domain. Transforming (3.21) into the frequency domain and making use of the linearity of the Fourier transform along with the time delay property (3.18) provides the input–output relation

$$Y(e^{j\omega}) = -\sum_{l=1}^{L} a_l e^{-j\omega l} Y(e^{j\omega}) + \sum_{m=0}^{M} b_m e^{-j\omega m} X(e^{j\omega}). \tag{3.22}$$

Based on (3.20), we can then express the frequency response of such a LTI system as

$$H(e^{j\omega}) = \frac{Y(e^{j\omega})}{X(e^{j\omega})} = \frac{\displaystyle\sum_{l=0}^{L} b_l e^{-j\omega l}}{1 + \displaystyle\sum_{m=1}^{M} a_m e^{-j\omega m}}. \tag{3.23}$$

Windowing and Modulation

If we multiply the signal x with a *windowing function* w in the time domain we can write

$$y[n] = x[n]\, w[n], \tag{3.24}$$

which is equivalent to

$$Y(e^{j\omega}) = \frac{1}{2\pi} \int_{-\pi}^{\pi} X(e^{j\theta})\, W(e^{j(\omega-\theta)})\, d\theta \tag{3.25}$$

in the frequency domain. Equation (3.25) represents a periodic convolution of $X(e^{j\omega})$ and $W(e^{j\omega})$. This implies that $X(e^{j\omega})$ and $W(e^{j\omega})$ are convolved, but as both are periodic functions of ω, the convolution extends only over a single period. The operation defined by (3.24) is known as *windowing* when $w[n]$ has a generally lowpass frequency response, such as those windows discussed in Section 5.1. In the case of windowing, (3.25) implies that the spectrum $X(e^{j\omega})$ will be *smeared* through convolution with $W(e^{j\omega})$. This effect will become important in Section 3.3 during the presentation of the short-time Fourier transform. If $W(e^{j\omega})$ has large sidelobes, it implies that some of the frequency resolution of $X(e^{j\omega})$ will be lost.

On the other hand, the operation (3.24) is known as *modulation* when $w[n] = e^{j\omega_c n}$ for some angular frequency $0 < \omega_c \leq \pi$. In this case, (3.25) implies that the spectrum will be *shifted* to the right by ω_c, such that

$$Y(e^{j\omega}) = X\left(e^{j(\omega-\omega_c)}\right). \tag{3.26}$$

Equation (3.26) follows from

$$H_c(e^{j\omega}) = \sum_{n=-\infty}^{\infty} h_c[n]\, e^{-j\omega n} = \sum_{n=-\infty}^{\infty} e^{j\omega_c n}\, h[n]\, e^{-j\omega n}$$

$$= \sum_{n=-\infty}^{\infty} h[n]\, e^{-j(\omega-\omega_c)n} = H\left(e^{j(\omega-\omega_c)}\right).$$

In Chapter 11 we will use (3.26) to design a set of filters or a digital filter bank from a single lowpass prototype filter.

Cross-correlation

There is one more property of the Fourier transform, which we derive here, that will prove useful in Chapter 10. Let us define the *cross-correlation* x_{12} of two sequences $x_1[n]$ and $x_2[n]$ as

$$x_{12}[n] \triangleq \sum_{m=-\infty}^{\infty} x_1[m]\, x_2[n+m]. \tag{3.27}$$

Then through manipulations analogous to those leading to (3.20), it is straightforward to demonstrate that

$$X_{12}(e^{j\omega}) = X_1^*(e^{j\omega})\, X_2(e^{j\omega}), \tag{3.28}$$

where $x_{12}[n] \leftrightarrow X_{12}(e^{j\omega})$.

The definition of the inverse Fourier transform (3.15) together with (3.28) imply that

$$x_{12}[n] = \frac{1}{2\pi} \int_{-\pi}^{\pi} X_1^*(e^{j\omega})\, X_2(e^{j\omega})\, e^{j\omega n}\, d\omega. \tag{3.29}$$

3.1.3 z-Transform Analysis

The z-transform can be viewed as an *analytic continuation* (Churchill and Brown 1990, sect. 102) of the Fourier transform into the complex or z-plane. It is readily obtained by replacing $e^{j\omega}$ in (3.14) with the complex variable z, such that

$$X(z) \triangleq \sum_{n=-\infty}^{\infty} x[n] z^{-n}. \tag{3.30}$$

When (3.30) holds, we will say, just as in the case of the Fourier transform, that $x[n]$ and $X(z)$ constitute a transform pair, which is denoted as $x[n] \leftrightarrow X(z)$. It is readily verified that the convolution theorem also holds in the z-transform domain, such that when the output $y[n]$ of a system with input $x[n]$ and impulse response $h[n]$ is given by (3.8), then

$$Y(z) = X(z) H(z). \tag{3.31}$$

The term $H(z)$ in (3.31) is known as the *system* or *transfer function*, and is analogous to the frequency response in that it specifies the relation between input and output in the z-transform domain. Similarly, a time delay has a simple manifestation in the z-transform domain, inasmuch as it follows that

$$x[n - n_d] \leftrightarrow z^{-n_d} X(z).$$

Finally, the equivalent of (3.26) in the z-transform domain is

$$e^{j\omega_c n} h[n] \leftrightarrow H(z e^{-j\omega_c}). \tag{3.32}$$

The inverse z-transform is formally specified through the *contour integral* (Churchill and Brown 1990, sect. 32),

$$x[n] \triangleq \frac{1}{2\pi j} \oint_C X(z) z^{n-1} dz, \tag{3.33}$$

where C is the *contour of integration*. Parameterizing the unit circle as the contour of integration in (3.33) through the substitution $z = e^{j\omega} \forall -\pi \leq \omega \leq \pi$ leads immediately to the inverse Fourier transform (3.15).

While the impulse response of a LTI system uniquely specifies the z-transform of such a system, the converse is *not* true. This follows from the fact that (3.30) represents a *Laurent series expansion* (Churchill and Brown 1990, sect. 47) of a function $X(z)$ that is analytic in some annular region, which implies it possesses continuous derivatives of *all* orders. The bounds of this annular region, which is known as the *region of convergence* (ROC), will be determined by the locations of the poles of $X(z)$. Moreover, the coefficients in the series expansion of $X(z)$, which is to say the sample values in the impulse response $x[n]$, will be different for different annular ROCs. Hence, in order to uniquely specify the impulse response $x[n]$ corresponding to a given $X(z)$, we must also specify the ROC of $X(z)$. For reasons which will shortly become apparent, we will uniformly assume that the ROC includes the unit circle as well as all points exterior to the unit circle.

For systems specified through linear constant coefficient difference equations such as (3.21), it holds that

$$H(z) = \frac{Y(z)}{X(z)} = \frac{\sum_{l=0}^{L} b_l z^{-l}}{1 + \sum_{m=1}^{M} a_m z^{-m}}. \tag{3.34}$$

This equation is the z-transform equivalent of (3.23).

While (3.33) is correct, the contour integral can be difficult to calculate directly. Hence, the inverse z-transform is typically evaluated with less formal methods, which we now illustrate with several examples.

Example 3.1 Consider the geometric sequence

$$x[n] = a^n u[n], \tag{3.35}$$

for some $|a| < 1$, where $u[n]$ is the *unit step function*,

$$u[n] \triangleq \begin{cases} 1, & \text{for } n \geq 0, \\ 0, & \text{otherwise.} \end{cases}$$

Substituting (3.35) into (3.30) and making use of the identity

$$\sum_{n=0}^{\infty} \beta^n = \frac{1}{1-\beta} \; \forall \, |\beta| < 1,$$

where $\beta = a z^{-1}$, yields

$$a^n u[n] \leftrightarrow \frac{1}{1 - a z^{-1}}. \tag{3.36}$$

The requirement $|\beta| = |a z^{-1}| < 1$ implies the ROC for (3.35) is specified by $|z| > |a|$. Note that (3.36) is also valid for complex a. □

Example 3.2 Consider now the *decaying sinusoid*,

$$x[n] = u[n] \rho^n \cos \omega_c n, \tag{3.37}$$

for some real $0 < \rho < 1$ and $0 \leq \omega_c \leq \pi$. Using Euler's formula, $e^{j\theta} = \cos \theta + j \sin \theta$, to rewrite (3.37) provides

$$x[n] = u[n] \frac{\rho^n}{2} \left(e^{j\omega_c n} + e^{-j\omega_c n} \right). \tag{3.38}$$

Applying (3.36) to (3.38) with $a = \rho e^{\pm j\omega}$ then yields

$$u[n]\, a^n \, \cos \omega_c n \leftrightarrow \frac{1}{2} \left(\frac{1}{1 - \rho\, z^{-1}\, e^{j\omega_c}} + \frac{1}{1 - \rho\, z^{-1}\, e^{-j\omega_c}} \right)$$

$$= \frac{1 - \rho\, z^{-1}\, \cos \omega_c}{1 - 2\rho\, z^{-1}\, \cos \omega_c + \rho^2\, z^{-2}}.$$

Moreover, the requirement $|\beta| = |\rho z^{-1} e^{j\omega_c}| < 1$ implies that the ROC of (3.37) is $|z| > \rho$. $\qquad\qquad\qquad\qquad\qquad\qquad\qquad\qquad\qquad\qquad\qquad\qquad\qquad\qquad\quad\square$

Examples 3.1 and 3.2 treated the calculation of the z-transform from the specification of a time series. It is often more useful, however, to perform calculations or filter design in the z-transform domain, then to transform the resulting system output or transfer function back into the time domain, as is done in the next example. Before considering this example, however, we need two definitions (Churchill and Brown 1990, sect. 56 and sect. 57) from the theory of complex analysis.

Definition 3.1.1 (**simple zero**) *A function $H(z)$ is said to have a* simple zero *at $z = z_0$ if $H(z_0) = 0$ but*

$$\left. \frac{d\, H(z)}{d\, z} \right|_{z=z_0} \neq 0.$$

Before stating the next definition, we recall that a function $H(z)$ is said to be *analytic* at a point $z = z_0$ if it possesses continuous derivatives of all orders there.

Definition 3.1.2 (**simple pole**) *A function $H(z)$ is said to have a* simple pole *at z_0 if it can be expressed in the form*

$$H(z) = \frac{\phi(z)}{z - z_0},$$

where $\phi(z)$ is analytic at $z = z_0$ and $\phi(z_0) \neq 0$.

Example 3.3 Consider the rational system function as defined in (3.34) which, in order to find the impulse response $h[n]$ that pairs with $H(z)$, has to be expressed in factored form as

$$H(z) = K \frac{\prod_{l=1}^{L}(1 - c_l\, z^{-1})}{\prod_{m=1}^{M}(1 - d_m\, z^{-1})}, \qquad\qquad (3.39)$$

where $\{c_l\}$ and $\{d_m\}$ are respectively, the sets of *zeros* and *poles* of $H(z)$, and K is a real constant. The representation (3.39) is always possible, inasmuch as the *fundamental theorem of algebra* (Churchill and Brown 1990, sect. 43) states that any polynomial of order P can be factored into P zeros, provided that all zeros are *simple*. It follows that

(3.39) can be represented with the *partial fraction expansion*,

$$H(z) = \sum_{m=1}^{M} \frac{A_m}{1 - d_m z^{-1}}, \tag{3.40}$$

where the constants A_m can be determined from

$$A_m = (1 - d_m z^{-1}) H(z)\big|_{z=d_m}. \tag{3.41}$$

Equation (3.41) can be readily verified by combining the individual terms of (3.40) over a common denominator. Upon comparing (3.36) and (3.40) and making use of the linearity of the z-transform, we realize that

$$h[n] = u[n] \sum_{m=1}^{M} A_m d_m^n. \tag{3.42}$$

With arguments analogous to those used in the last two examples, the ROC for (3.42) is readily found to be

$$|z| > \max_m |d_m|.$$

Clearly for real $h[n]$ any complex poles d_m must occur in complex conjugate pairs, which is also true for complex zeros c_m.   □

By definition, a *minimum phase system* has all of its zeros and poles within the unit circle. Hence, assuming that $|c_l| < 1 \, \forall l = 1, \ldots, L$ and $|d_m| < 1 \, \forall m = 1, \ldots, M$ is tantamount to assuming that $H(z)$ as given in (3.39) is a minimum phase system. Minimum phase systems are in many cases tractable because they have stable *inverse systems*. The inverse system of $H(z)$ is by definition that system $H^{-1}(z)$ achieving (Oppenheim and Schafer 1989, sect. 5.2.2)

$$H^{-1}(z) H(z) = z^{-D},$$

for some integer $D \geq 0$. Hence, the inverse of (3.39) can be expressed as

$$H^{-1}(z) = \frac{1}{H(z)} = K^{-1} \frac{\prod_{m=1}^{M}(1 - d_m z^{-m})}{\prod_{l=1}^{L}(1 - c_l z^{-l})}.$$

Clearly, $H^{-1}(z)$ is minimum phase, just as $H(z)$, which in turn implies that both are stable. We will investigate a further implication of the minimum phase property in Section 5.4 when discussing cepstral coefficients.

Equations (3.23) and (3.34) represent a so-called *auto-regressive, moving average* (ARMA) model. From the last example it is clear that the z-transform of such a model contains both pole and zero locations. We will also see that its impulse response is, in general, infinite in duration, which is why such systems are known as *infinite impulse response* (IIR) systems. Two simplifications of the general ARMA model are possible, both of which are frequently used in signal processing and adaptive filtering, wherein the parameters of a LTI system are iteratively updated to optimize some criterion (Haykin 2002). The first such simplification is the *moving average* model

$$y[n] = \sum_{m=0}^{M} b_m x[n-m]. \tag{3.43}$$

Systems described by (3.43) have impulse responses with finite duration, and hence are known as *finite impulse response* (FIR) systems. The z-transforms of such systems contain only zero locations, and hence they are also known as *all-zero* filters. As FIR systems with bounded coefficients are always stable, they are often used in adaptive filtering algorithms. We will use such FIR systems for the beamforming applications discussed in Chapter 13.

The second simplification of (3.21) is the *auto-regressive* (AR) model, which is characterized by the difference equation

$$y[n] = -\sum_{m=1}^{M} a_m y[n-m] + x[n]. \tag{3.44}$$

Based on Example 3.3, it is clear that such AR systems are IIR just as ARMA systems, but their z-transforms contain only poles, and hence are also known as *all-pole* filters. AR systems find frequent application in speech processing, and are particularly useful for spectral estimation based on linear prediction, as described in Section 5.3.3, as well as the minimum variance distortionless response, as described in Section 5.3.4.

From (3.42), it is clear that all poles $\{d_k\}$ must lie within the unit circle if the system is to be BIBO stable. This holds because poles within the unit circle correspond to exponentially *decaying* terms in (3.42), while poles outside the unit circle would correspond to exponentially *growing* terms. The same is true of both AR and ARMA models. Stability, on the other hand, is not problematic for FIR systems, which is why they are more often used in adaptive filtering applications. It is, however, possible to build such adaptive filters using an IIR system (Haykin 2002, sect. 15).

Once the system function has been expressed in factored form as in (3.39), it can be represented graphically as the pole-zero plot (Oppenheim and Schafer 1989, sect. 4.1) shown on the left side of Figure 3.1, wherein the pole and zero locations in the complex plane are marked with $\times$ and $\circ$ respectively. To see the relation between the pole-zero plot and the Fourier transform shown on the right side of Figure 3.1, it is necessary to associate the unit circle in the z-plane with the frequency axis of the Fourier transform through the parameterization $z = e^{j\omega}$ for $-\pi \leq \omega \leq \pi$. For a simple example in which there is a simple pole at $d_1 = 0.8$ and a simple zero at $c_1 = -0.6$, the magnitude of the frequency response can be expressed as

$$H(e^{j\omega}) = \left| \frac{z - c_1}{z - d_1} \right|_{z=e^{j\omega}} = \left| \frac{e^{j\omega} + 0.6}{e^{j\omega} - 0.8} \right|. \tag{3.45}$$

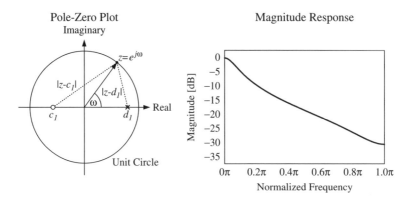

Figure 3.1 Simple example of the pole-zero plot in the complex z-plane and the corresponding frequency domain representation

The quantities $|e^{j\omega} + 0.6|$ and $|e^{j\omega} - 0.8|$ appearing on the right-hand side of (3.45) are depicted with dotted lines on the left side of Figure 3.1. Clearly, the point $z = 1$ corresponds to $\omega = 0$, which is much closer to the pole $z = 0.8$ than to the zero $z = -0.6$. Hence, the magnitude $|H(e^{j\omega})|$ of the frequency response is a maximum at $\omega = 0$. As ω increases from 0 to π, the test point $z = e^{j\omega}$ sweeps along the upper half of the unit circle, and the distance $|e^{j\omega} - 0.8|$ becomes ever larger, while the distance $|e^{j\omega} + 0.6|$ becomes ever smaller. Hence, $|H(e^{j\omega})|$ decreases with increasing ω, as is apparent from the right side of Figure 3.1. A filter with such a frequency response is known as a *lowpass filter*, because low-frequency components are passed (nearly) without attenuation, while high-frequency components are suppressed.

While the simple filter discussed above is undoubtedly lowpass, it would be poorly suited for most applications requiring a lowpass filter. This lack of suitability stems from the fact that the transition from the *passband*, wherein all frequency components are passed without attenuation, to the *stopband*, wherein all frequency components are suppressed, is very gradual rather than *sharp*; i.e., the *transition band* from passband to stopband is very wide. Moreover, depending on the application, the stopband suppression provided by such a filter may be inadequate. The science of digital filter design through pole-zero placement in the z-plane is, however, very advanced at this point. A great many possible designs have been proposed in the literature that are distinguished from one another by, for example, their stopband suppression, passband ripple, phase linearity, width of the transition band, etc. Figure 3.2 shows the pole-zero locations and magnitude response of a lowpass filter based on a tenth-order Chebychev Type II design. As compared with the simple design depicted in Figure 3.1, the Chebychev Type II design provides a much sharper transition from passband to stopband, as well as much higher stopband suppression. Oppenheim and Schafer (1989, sect. 7) describe several other well-known digital filter designs. In Chapter 11, we will consider the design of a filter that serves as a prototype for all filters in a digital filter bank. In such a design, considerations such as stopband suppression, phase linearity, and total response error will play a decisive role.

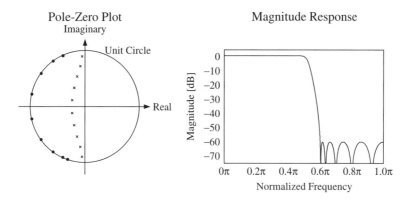

Figure 3.2 Pole-zero plot in the z-plane of a tenth-order Chebychev Type II filter and the corresponding frequency response magnitude

Parseval's Theorem

Parseval's theorem concerns the equivalence of calculating the energy of a signal in the time or transform domain. In the z-transform domain, Parseval's theorem can be expressed as

$$\sum_{n=-\infty}^{\infty} x^2[n] = \frac{1}{2\pi j} \oint_C X(v)\,X(v^{-1})\,v^{-1}\,dv, \tag{3.46}$$

where the contour of integration is most often taken as the unit circle. In the Fourier transform domain, this becomes

$$\sum_{n=-\infty}^{\infty} x^2[n] = \frac{1}{2\pi} \oint_{-\pi}^{\pi} \left| X\left(e^{j\omega}\right) \right|^2 \, d\omega. \tag{3.47}$$

3.1.4 *Sampling Continuous-Time Signals*

While discrete-time signals are a useful abstraction inasmuch as they can be readily calculated and manipulated with digital computers, it must be borne in mind that such signals do not occur in nature. Hence, we consider here how a real continuous-time signal may be converted to the digital domain or *sampled*, then converted back to the continuous-time domain or *reconstructed* after some digital processing. In particular, we will discuss the well-known Nyquist–Shannon sampling theorem.

The *continuous-time Fourier transform* is defined as

$$X(\omega) \triangleq \int_{-\infty}^{\infty} x(t)\,e^{-\omega t}\,dt, \tag{3.48}$$

for real $-\infty < \omega < \infty$. This transform is defined over the entire real line. Unlike its discrete-time counterpart, however, the continuous-time Fourier transform is *not* periodic.

We adopt the notation $X(\omega)$ with the intention of emphasizing this lack of periodicity. The continuous-time Fourier transform possesses the same useful properties as its discrete-time counterpart (3.14). In particular, it has the inverse transform,

$$x(t) \triangleq \frac{1}{2\pi} \int_{-\infty}^{\infty} X(\omega) e^{\omega t} \, d\omega \; \forall \; -\infty < t < \infty. \tag{3.49}$$

It also satisfies the convolution theorem,

$$y(t) = \int_{-\infty}^{\infty} h(\tau) x(t - \tau) \, d\tau \leftrightarrow Y(\omega) = H(\omega) X(\omega).$$

The continuous-time Fourier transform also possesses the time delay property,

$$x(t - t_d) \leftrightarrow e^{-j\omega t_d} X(\omega), \tag{3.50}$$

where t_d is a real-valued time delay.

We will now use (3.48–3.49) to analyze the effects of sampling as well as determine which conditions are necessary to perfectly reconstruct the original continuous-time signal. Let us define a continuous-time *impulse train* as

$$s(t) = \sum_{n=-\infty}^{\infty} \delta(t - nT),$$

where T is the *sampling interval*. The continuous-time Fourier transform of $s(t)$ can be shown to be

$$S(\omega) = \frac{2\pi}{T} \sum_{m=-\infty}^{\infty} \delta(\omega - m\omega_s),$$

where $\omega_s = 2\pi/T$ is the *sampling frequency* or *rate* in radians/second.

Consider the continuous-time signal $x_c(t)$ which is to be sampled through multiplication with the impulse train according to

$$x_s(t) = x_c(t) s(t) = x_c(t) \sum_{n=-\infty}^{\infty} \delta(t - nT) = \sum_{n=-\infty}^{\infty} x_c(nT) \delta(t - nT).$$

Then the spectrum $X_s(\omega)$ of the sampled signal x_s consists of a series of scaled and shifted replicas of the original continuous-time Fourier transform $X_c(\omega)$, such that,

$$X_s(\omega) = \frac{1}{2\pi} X_c(\omega) * S(\omega) = \frac{1}{T} \sum_{m=-\infty}^{\infty} X_c(\omega - m\omega_s).$$

The last equation is proven rigorously in Section B.13. Figure 3.3 (Original Signal) shows the original spectrum $X_c(\omega)$, which is assumed to be bandlimited such that

$$X_c(\omega) = 0 \; \forall \; |\omega| > \omega_N,$$

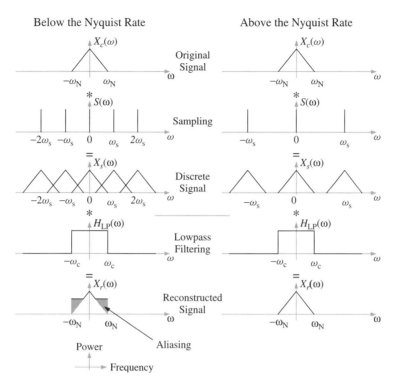

Figure 3.3 Effect of sampling and reconstruction in the frequency domain. Perfect reconstruction requires that $\omega_N < \omega_c < \omega_s - \omega_N$

for some real $\omega_N > 0$. Figure 3.3 (Sampling) shows the trains $S(\omega)$ of frequency-domain impulses resulting from the sampling operation for two cases: The Nyquist sampling criterion is *not* satisfied (left) and it is satisfied (right). Shown in Figure 3.3 (Discrete Signal) are the spectra $X_s(\omega)$ for the undesirable and desirable cases, whereby the continuous-time signal $x_c(t)$ was sampled insufficiently and sufficiently often to enable recovery of the original spectrum with a lowpass filter. In the first case the original spectrum overlaps with its replicas. In the second case – where the Nyquist sampling theorem is satisfied – the original spectrum and its images do not overlap, and $x_c(t)$ can be uniquely determined from its samples

$$x_s[n] = x_s(nT) \, \forall \, n = 0, \pm 1, \pm 2, \ldots. \tag{3.51}$$

Reconstructing $x_c(t)$ from its samples requires that the sampling rate satisfy the *Nyquist criterion*, which can be expressed as

$$\omega_s = \frac{2\pi}{T} > 2\,\omega_N. \tag{3.52}$$

This inequality is a statement of the famous Nyquist sampling theorem. The bandwidth ω_N of the continuous-time signal $x_c(t)$ is known as the *Nyquist frequency*, and $2\omega_N$, the lower bound on the allowable sampling rate, is known as the *Nyquist rate*. The reconstructed spectrum $X_r(\omega)$ is obtained by filtering according to

$$X_r(\omega) = H_{\mathrm{LP}}(\omega)\, X_s(\omega),$$

where $H_{\mathrm{LP}}(\omega)$ is the frequency response of the lowpass filter.

Figure 3.3 (Reconstructed Signal, left side) shows the spectral overlap that results in $X_r(\omega)$ when the Nyquist criterion is *not* satisfied. In this case, high-frequency components are mapped into low-frequency regions, a phenomenon known as *aliasing*, and it is no longer possible to isolate the original spectrum from its images with $H_{\mathrm{LP}}(\omega)$. Hence, it is no longer possible to perfectly reconstruct $x_c(t)$ from its samples in (3.51). On the right side of Figure 3.3 (Reconstructed Signal) is shown the perfectly reconstructed spectrum $X_r(\omega)$ obtained when the Nyquist criterion is satisfied. In this case, the original spectrum X_c can be isolated from its images with the lowpass filter $H_{\mathrm{LP}}(\omega)$, and perfect reconstruction is possible based on the samples (3.51) of the original signal $x_c(t)$.

The first component of a complete digital filtering system is invariably an analog anti-aliasing filter, which serves to bandlimit the input signal (Oppenheim and Schafer 1989, sect. 3.7.1). As implied from the foregoing discussion, such bandlimiting is necessary to prevent aliasing. The bandlimiting block is then followed by a sampler, then by the digital filter itself, and finally a digital-to-analog conversion block. Ideally the last of these is a lowpass filter $H_{\mathrm{LP}}(\omega)$, as described above. Quite often, however, $H_{\mathrm{LP}}(\omega)$ is replaced by a simpler *zero-order hold* (Oppenheim and Schafer 1989, sect. 3.7.4.).

While filters can be implemented in the continuous-time or analog domain, working in the digital domain has numerous advantages in terms of flexibility and adaptability. In particular, a digital filter can easily be adapted to changing acoustic environments. Moreover, digital filters can be implemented in software, and hence offer far greater flexibility in terms of changing the behavior of the filter during its operation. In Chapter 13, we will consider the implementation of several adaptive beamformers in the digital domain, but will begin the analysis of the spatial filtering effects of a microphone array in the continuous-time domain, based on relations (3.48) through (3.50).

3.2   The Discrete Fourier Transform

While the Fourier and z-transforms are very useful conceptual devices and possess several interesting properties, their utility for implementing real LTI systems is limited at best. This follows from the fact that both are defined for continuous-valued variables. In practice, real signal processing algorithms are typically based either on difference equations in the case of IIR systems, or the *discrete Fourier transform* (DFT) and its efficient implementation through the *fast Fourier transform* (FFT) in the case of FIR systems. The FFT was originally discovered by Carl Friedrich Gauss around 1805. Its widespread popularity, however, is due to the publication of Cooley and Tukey (1965), who are credited with having independently re-invented the algorithm. It can be calculated with any of a number of efficient algorithms (Oppenheim and Schafer 1989, sect. 9), implementations of which are commonly available. The presentation of such algorithms, however, lies outside of our present scope. Here we consider instead the properties of the DFT, and, in particular, how the DFT may be used to implement LTI systems.

Let us begin by defining

- the *analysis equation*,

$$\tilde{X}[m] \triangleq \sum_{n=0}^{N-1} \tilde{x}[n] \, W_N^{mn} \tag{3.53}$$

- and the *synthesis equation*,

$$\tilde{x}[n] \triangleq \frac{1}{N} \sum_{m=0}^{N-1} \tilde{X}[m] \, W_N^{-mn}, \tag{3.54}$$

of the *discrete Fourier series* (DFS), where $W_N = e^{-j(2\pi/N)}$ is the Nth root of unity. As is clear from (3.53–3.54), both $\tilde{X}[m]$ and $\tilde{x}[n]$ are periodic sequences with a period of N, which is the reason behind their designation as discrete Fourier series. In this section, we first show that $\tilde{X}[m]$ represents a sampled version of the discrete-time Fourier transform $X(e^{j\omega})$ of some sequence $x[n]$, as introduced in Section 3.1.2. We will then demonstrate that $\tilde{x}[n]$ as given by (3.54) is equivalent to a time-aliased version of $x[n]$. Consider then the finite length sequence $x[n]$ that is equivalent to the periodic sequence $\tilde{x}[n]$ over one period of N samples, such that

$$x[n] \triangleq \begin{cases} \tilde{x}[n], & \forall \, 0 \leq n \leq N-1, \\ 0, & \text{otherwise.} \end{cases} \tag{3.55}$$

The Fourier transform of $x[n]$ can then be expressed as

$$X(e^{j\omega}) = \sum_{-\infty}^{\infty} x[n] \, e^{-j\omega n} = \sum_{n=0}^{N-1} \tilde{x}[n] \, e^{-j\omega n}. \tag{3.56}$$

Upon comparing (3.53) and (3.56), it is clear that

$$\tilde{X}[m] = X(e^{j\omega})\big|_{\omega = 2\pi m/N} \quad \forall \, m \in \mathbb{N}. \tag{3.57}$$

Equation (3.57) indicates that $\tilde{X}[m]$ represents the *periodic* sequence obtained by sampling $X(e^{j\omega})$ at N equally spaced frequencies over the range $0 \leq \omega < 2\pi$. The following simple example illustrates how a periodic sequence may be represented in terms of its DFS coefficients $\tilde{X}[m]$ according to (3.54).

Example 3.4 Consider the impulse train with period N defined by

$$\tilde{x}[n] = \sum_{l=-\infty}^{\infty} \delta[n + lN].$$

Given that $\tilde{x}[n] = \delta[n] \, \forall \, 0 \leq n \leq N - 1$, we can calculate the DFS coefficients according to (3.53) to obtain

$$\tilde{X}[m] = \sum_{n=0}^{N-1} \delta[n] \, W_N^{mn} = 1 \, \forall \, m = 0, \ldots, N - 1. \tag{3.58}$$

Substituting (3.58) into (3.54) provides the representation

$$\tilde{x}[n] = \sum_{l=-\infty}^{\infty} \delta[n + lN] = \frac{1}{N} \sum_{m=0}^{N-1} W_N^{-mn} = \frac{1}{N} \sum_{m=0}^{N-1} e^{j(2\pi/N)mn}, \tag{3.59}$$

which is the desired result. $\qquad\qquad\qquad\qquad\qquad\qquad\qquad\qquad\qquad\qquad\qquad\qquad\qquad\quad$ □

We next investigate what effect the sampling of $X(e^{j\omega})$ mentioned above has on the sequence returned by (3.54). To begin, we substitute the definition (3.14) of $X(e^{j\omega})$ into (3.57), then the resulting sequence $\tilde{X}[m]$ into the DFS synthesis equation (3.54), to obtain

$$\tilde{x}[n] = \frac{1}{N} \sum_{l=0}^{N-1} \left[\sum_{m=-\infty}^{\infty} x[m] \, W_N^{lm} \right] W_N^{-ln}.$$

Upon changing the order of summation, the last equation becomes

$$\tilde{x}[n] = \sum_{m=-\infty}^{\infty} x[m] \left[\frac{1}{N} \sum_{l=0}^{N-1} W_N^{-l(n-m)} \right],$$

which, in light of (3.59), can be rewritten as

$$\tilde{x}[n] = \sum_{m=-\infty}^{\infty} x[m] \sum_{r=-\infty}^{\infty} \delta[n - m + rN] = x[n] * \sum_{r=-\infty}^{\infty} \delta[n + rN].$$

Equivalently,

$$\tilde{x}[n] = \sum_{r=-\infty}^{\infty} x[n + rN]. \tag{3.60}$$

From (3.60) it is clear that $\tilde{x}[n]$ is the periodic sequence that results when $x[n]$ is repeated every N samples, and the repetitions are summed together. If $x[n]$ is of length N or less, then a single period of $\tilde{x}[n]$ will be equivalent to the original sequence $x[n] \, \forall \, 0 \leq n \leq N - 1$. Otherwise, $\tilde{x}[n]$ will be a *time-aliased* version of $x[n]$. This fact gives some indication of the duality that exists between the time and transform domains for discrete Fourier series; i.e., sampling in one domain introduces a periodicity in the other (Oppenheim and Schafer 1989, sect. 8.2.3). Barring some limitation on the extent of either $x[n]$ or $X(e^{j\omega})$, this periodicity can lead to aliasing, which precludes the possibility of recovering the original signal perfectly. These considerations lead us to introduce the discrete Fourier transform and the *inverse discrete Fourier transform*, which are respectively defined as

$$X[m] \triangleq \sum_{n=0}^{N-1} x[n] \, W_N^{mn}, \tag{3.61}$$

$$x[n] \triangleq \frac{1}{N} \sum_{m=0}^{N-1} X[m] \, W_N^{-mn}. \tag{3.62}$$

3.2.1   Realizing LTI Systems with the DFT

In prior sections, we established that time domain convolution is equivalent to frequency or transform domain multiplication. This fact, when combined with the computational efficiency afforded by a fast implementation of the DFT, provides for the efficient realization of LTI systems with FIR. Such implementations entail the use of the FFT to transform input sequences as well as the impulse response of a LTI system into the frequency domain, where they are multiplied. The output of the system is then reconstructed through the inverse FFT. Two problems may arise with this approach, however:

- The first problem is due to the time domain aliasing induced by the frequency domain sampling inherent in the DFT; i.e., unless the final system output is limited in length, it will be time-aliased by the DFT.
- The second problem is that if the input sequence is arbitrarily long, it becomes increasingly inefficient to apply longer and longer DFTs to calculate the system output.

Both of these problems can be solved by applying either the *overlap-add* or *overlap-save* methods. To develop these methods, first consider that the convolution of a sequence $x_1[n]$ of length L with another sequence $x_2[n]$ of length P produces a sequence $x_3[n]$ of length $L + P - 1$. This is apparent from the specialization of the convolution sum,

$$x_3[n] = \sum_{m=-\infty}^{\infty} x_1[m] \, x_2[n-m]$$

or in the Fourier domain

$$X_3(e^{j\omega}) = X_1(e^{j\omega}) \, X_2(e^{j\omega}). \tag{3.63}$$

Moreover, based on the exposition of the DFT above, we can readily obtain a sampled version of $X_3(e^{j\omega})$ according to

$$\tilde{X}_3[m] = \tilde{X}_1[m] \, \tilde{X}_2[m], \tag{3.64}$$

where

$$\tilde{X}_i[m] = X_i(e^{j\omega})\big|_{\omega=2\pi m/N} \quad \forall \, m = 0, \dots, N-1, \, i = 1, 2, 3.$$

The final sequence $x_3[n]$ corresponding to the convolution of $x_1[n]$ and $x_2[n]$ can be obtained by applying the DFT synthesis equation (3.54) to $\tilde{X}_3[m]$, in order to obtain the *periodic* sequence $\tilde{x}_3[n]$, and then setting all but the first period to zero. The problem, however, is one of determining the number of points N at which $X_3(e^{j\omega})$ must be sampled in order to ensure that the first period of $\tilde{x}_3[n]$ is a faithful representation of $x_3[n]$. Based on the realization that the non-zero length of $x_3[n]$ will be $L + P - 1$, clearly we must require

$$N \geq L + P - 1. \tag{3.65}$$

This relation can always be satisfied for FIR systems. If the DFT is to be used to realize a LTI system, however, the input sequence is for all intents and purposes infinitely long. Hence, it must be segmented prior to processing to avoid an arbitrarily long processing delay and unacceptable computational expense. For these reasons, we consider here the overlap-add and overlap-save methods.

3.2.2 Overlap-Add Method

Initially, let us assume we wish to efficiently implement a LTI system with impulse response $h[n]$ of length P. The input sequence $x[n]$ of the system has a length, in general, much greater than P. We begin by partitioning $x[n]$ into frame segments k of length L according to

$$x[n] = \sum_{k=0}^{\infty} x_k[n - kL], \tag{3.66}$$

where each $x_k[n]$ is specified by

$$x_k[n] = \begin{cases} x[n + kL], & \forall\, 0 \leq n \leq L - 1, \\ 0, & \text{otherwise.} \end{cases} \tag{3.67}$$

Note that (3.66–3.67) imply that the zeroth sample of $x_k[n]$ is in fact the kLth sample of $x[n]$. As convolution is a linear, time-invariant operation, the output of the system can be *patched together* according to

$$y[n] = x[n] * h[n] = \sum_{k=0}^{\infty} y_k[n - kL], \tag{3.68}$$

where each segment $y_k[n]$ is obtained from

$$y_k[n] = x_k[n] * h[n]. \tag{3.69}$$

While the time domain convolution in (3.69) is conceptually correct, we will in fact calculate each segment $y_k[n]$ by applying the synthesis equation (3.54) to the sampled spectrum

$$\tilde{X}_k[m] = \tilde{Y}_k[m]\, \tilde{H}[m], \tag{3.70}$$

then isolating the first period of the resulting sequence $\tilde{x}_k[n]$, which will be equivalent to $x_k[n]$ provided (3.65) is satisfied. This provides computational efficiency. It is accompanied, however, by two drawbacks: Firstly, as we will learn in Section 11.1, the rectangular window implicit in (3.67) has poor stopband characteristics, which, as implied by the discussion in Section 3.1.2, will cause a smearing in the frequency domain. Secondly, it is computationally awkward, in that, when applied to subband domain adaptive filtering as discussed in Chapter 13, the estimated impulse response must be time limited to one-half the length of the DFT. This implies that the frequency samples of the impulse must be inverse transformed to the time domain, then time windowed to limit their duration, then transformed again into the frequency domain to be multiplied with the DFT of the input sequence. These considerations are behind our presentation of digital filter banks in Chapter 11, which do not suffer from the aforementioned shortcomings.

3.2.3 Overlap-Save Method

An alternative method for implementing LTI systems is the overlap-save method, whereby the L-point circular convolution of a P-point impulse response $h[n]$ and a L-point segment $x_k[n]$ is implemented. Thereafter, that portion of the output segment $\tilde{y}_k[n]$ corresponding to the linearly convolved segment $y_k[n]$ is identified, and these segments are patched together to form the final output:

$$y[n] = \sum_{r=0}^{\infty} y_k[n - r(L - P + 1) + P - 1], \qquad (3.71)$$

where

$$y_k[n] = \begin{cases} \tilde{y}_k[n], & \forall\, P - 1 \leq n \leq L - 1, \\ 0, & \text{otherwise.} \end{cases} \qquad (3.72)$$

The computational efficiency of both the overlap-add and overlap-save methods is a direct result of that of the FFT; that is, with the aid of the FFT, the N-point convolutions required by both methods can be performed in $\mathcal{O}(N \log N)$ time. Were these operations implemented naively in the time domain, the computational expense would be $\mathcal{O}(N^2)$; see Oppenheim and Schafer (1989 sect. 9).

3.3 Short-Time Fourier Transform

It is often the case that the characteristics of the signal $x[n]$, which is to be spectrally analyzed or digitally processed, change with time. In such cases, it is useful to isolate a portion of the entire sequence, and perform spectral analysis only on this portion. The *time-dependent Fourier transform* is defined as

$$\overline{X}[n, e^{j\lambda}) \triangleq \sum_{m=-\infty}^{\infty} x[n + m]\, w[m]\, e^{-j\lambda m}, \qquad (3.73)$$

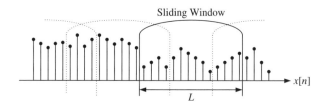

Figure 3.4   Application of the window sequence $w[n]$ to isolate a segment of $x[n]$ for spectral analysis

where $w[n]$ is a *window sequence*. Following Oppenheim and Schafer (1989 sect. 11.3), we use the mixed notation $\overline{X}[n, e^{j\lambda})$ to indicate that functional dependence on both the discrete variable n and the continuous variable λ. As defined in (3.73), the time-dependent Fourier transform is obtained by using the window $w[n]$ to isolate a segment of the sequence $x[n]$, then taking the discrete-time Fourier transform of this windowed segment. Note that in the definition (3.73), the window is held stationary and the $x[n]$ is shifted n samples to the left. This process is depicted in Figure 3.4, wherein $w[n]$ has non-zero support of length L. Inasmuch as $\overline{X}[n, e^{j\lambda})$ is the discrete-time Fourier transform of $x[n + m]\,w[m]$, the time-dependent Fourier transform is invertible if $w[0] \neq 0$. In particular, based on the inverse Fourier transform (3.15), we can write,

$$x[n + m]\,w[m] = \frac{1}{2\pi} \int_0^{2\pi} \overline{X}[n, e^{j\lambda})\, e^{j\lambda m}\, d\lambda \,\forall\, -\infty < m < \infty,$$

which implies,

$$x[n] = \frac{1}{2\pi\, w[0]} \int_0^{2\pi} \overline{X}[n, e^{j\lambda})\, e^{j\lambda m}\, d\lambda.$$

Rearranging the sum (3.73) provides another useful interpretation of the time-dependent Fourier transform. In particular, substituting $m' = n + m$ in (3.73) enables $\overline{X}[n, e^{j\lambda})$ to be rewritten as

$$\overline{X}[n, e^{j\lambda}) = \sum_{m'=-\infty}^{\infty} x[m']\,w[-(n - m')]\, e^{j\lambda(n - m')}. \tag{3.74}$$

Equation (3.74) can be interpreted as the convolution

$$\overline{X}[n, e^{j\lambda}) = x[n] * h_\lambda[n], \tag{3.75}$$

where

$$h_\lambda = w[-n]\, e^{j\lambda n}. \tag{3.76}$$

This leads to the representation

$$H_\lambda(e^{j\omega}) = W\left(e^{j(\lambda - \omega)}\right). \tag{3.77}$$

A slightly different interpretation of the time-dependent Fourier transform can be defined as

$$\hat{X}[n, e^{j\lambda}] \triangleq \sum_{m=-\infty}^{\infty} x[m]\, w[m-n]\, e^{-j\lambda m}. \qquad (3.78)$$

Referring to the windowed time sequence depicted in Figure 3.4, we can interpret the difference between (3.73) and (3.78) as follows. Whereas in (3.73) it is the sequence $x[m]$ that is shifted to the *left* by n samples past a stationary window $w[m]$, in (3.78), the time sequence $x[m]$ remains stationary while the window is shifted to the *right* by n samples. The exact relationship between (3.73) and (3.78) is readily found to be

$$\hat{X}[n, e^{j\lambda}] = e^{-j\lambda n}\overline{X}[n, e^{j\lambda}]. \qquad (3.79)$$

Representing the time-dependent Fourier transform in terms of the continuous frequency variable λ is useful conceptually, but not conducive to efficient implementation. The latter requires computations based on discrete values. Hence, as in the case of the DFT, let us sample λ in (3.73) at M equally spaced frequencies $\lambda_m = 2\pi m/M$ for some $M \geq L$. This leads to the *short-time Fourier transform*,

$$\overline{X}_m[n] \triangleq \overline{X}[n, e^{j2\pi m/M}] = \sum_{l=0}^{L-1} x[n+l]\, w[l]\, e^{-j(2\pi/M)ml} \quad \forall\, 0 \leq m \leq M-1 \qquad (3.80)$$

where m denotes the subband index. Upon comparing (3.53) and (3.80) it is evident that $\overline{X}_m[n]$ is the DFT of the windowed sequence $x[n+l]\,w[l]$. From the inverse DFT (3.54), it then follows that

$$x[n+l]\,w[l] = \frac{1}{M}\sum_{m=0}^{M-1} \overline{X}_m[n]\, e^{j(2\pi/M)lm} \quad \forall\, 0 \leq l \leq L-1. \qquad (3.81)$$

Assuming that the window $w[l] \neq 0 \,\forall\, 0 \leq l \leq L-1$, it is possible to recover $x[n], \ldots, x[n+L-1]$ from

$$x[n+l] = \frac{1}{M\,w[l]}\sum_{m=0}^{M-1} \overline{X}_m[n]\, e^{j(2\pi/M)lm} \quad \forall\, 0 \leq l \leq L-1. \qquad (3.82)$$

Equation (3.82) implies that perfect recovery of $x[n]$ is possible provided at least as many samples of λ are taken as there are nonzero samples in the window; i.e., such that $M \geq L$.

Inasmuch as (3.80) implies a sampling of $\overline{X}[n, \lambda]$ in λ, it also implies a sampling of (3.74–3.76) in λ. In particular, we can rewrite (3.75) and (3.76) as

$$\overline{X}_m[n] = x[n] * h_m[n] \,\forall\, 0 \leq m \leq M-1, \qquad (3.83)$$

where

$$h_m[n] = w[-n]\, e^{j(2\pi/M)mn}. \qquad (3.84)$$

Analysis Filter Bank

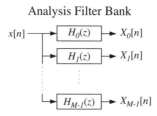

Figure 3.5 Analysis filter banks with individual filter functions $\{H_m(z)\}$

Equations (3.83–3.84) can be viewed as the bank of M filters depicted in Figure 3.5, where the mth filter has frequency response

$$H_m(e^{j\omega}) = W\left(e^{j(2\pi m/M - \omega)}\right). \tag{3.85}$$

This interpretation of the short-time Fourier transform indicates its utility for short-time spectral analysis, which will prove useful in Section 5.1, where we will discuss the segmentation and analysis of speech for the purpose of extracting acoustic features for automatic recognition. For the purpose of automatic speech recognition, the analysis window is fairly short, typically between 16 and 32 ms, because it is only over such relatively small time durations that speech is stationary. Moreover, the requirements for speech recognition in regard to frequency resolution are modest. More problematic is the application of such a short analysis window for beamforming. This follows from the fact that a short window $w[n]$ implies the introduction of large sidelobes in the frequency domain, and thereby a loss of frequency resolution, as implied by (3.25).

As we will learn in Chapter 11, provided that the window sequence $w[n]$ is carefully designed, the m filters in the filter bank shown in Figure 3.5 divide the entire frequency range into M largely nonoverlapping *subbands*. The outputs of the M subband filters, which we refer to as *subband samples*, can be appropriately filtered and combined in order to resynthesize the original signal to machine precision. To obtain nonoverlapping subbands requires high-frequency resolution. Achieving high resolution in the frequency domain requires using a longer analysis window, and thus sacrificing some time resolution. This illustrates a general rule: Frequency resolution can, in general, only be obtained at the cost of time resolution, and vice versa. Adaptive filtering and beamforming, unlike speech recognition, typically require high-frequency resolution for optimal performance.

We will refer to any operations whereby the subband samples are modified prior to resynthesis as being conducted in the *subband domain*. As is well known (Haykin 2002, sect. 7), frequency or subband domain implementations of adaptive filters can be far more computationally efficient than their time domain counterparts. The subband domain also offers other advantages with respect to frequency domain discussed in Section 3.2.1.

3.4 Summary and Further Reading

We began this chapter with a review of the basics of DSP, including a short introduction of LTI systems, the Fourier transform, and the z-transform. Next there was a brief discussion of how filters can be designed through pole-zero placement in the complex z-plane in order to provide some desired frequency response. We then discussed the effects of sampling a continuous time signal to obtain a digital representation, as well as the efficient implementation of LTI systems with the discrete Fourier transform. Next we provided a brief presentation of the short-time Fourier transform, which will prove useful for the development in subsequent chapters.

The rapid development of the theory of DSP in the 1960s, like so many other developments in science and engineering, was made possible because the prerequisite mathematical theory was already in place. In this case, the necessary theory was that of *complex analysis* due largely to Cauchy and his contemporaries in the 1700s. Those readers interested in this crown jewel of mathematics are referred to any of a number of excellent texts, including Churchill and Brown (1990) and Greene and Krantz (1997). A review of these texts will provide a deeper understanding of such concepts as analyticity, poles and zeros, and the Cauchy integral theorem, among others.

A classic work on the basics of DSP, FFT algorithms and filter design techniques is Oppenheim and Schafer (1989). Those readers with little background who are seeking very gentle introductions to the concepts of DSP are advised to consult Rosen and Howell (1990) and Lyons (2004). The former work is almost completely nonmathematical. The latter goes more into the mathematical details, but makes use of more images and supporting material than the standard DSP textbooks. Other useful references for beginners are Broesch (1997) and Stein (2000). Proakis and Manolakis (2007) is a very good and up-to-date reference on all current techniques. Haykin (2002) is a very illuminating, well-written guide on all issues pertaining to adaptive filtering.

3.5 Principal Symbols

Symbol	Description
ω	angle frequency
Ω_p	passband
D	decimation factor
$G_m(e^{j\omega})$	frequency response of mth synthesis filter
$h[n]$	impulse response
$H_d(e^{j\omega})$	desired passband frequency response
$H_m(e^{j\omega})$	frequency response of mth analysis filter
$\mathbf{h}$	uniform DFT filter bank analysis prototype
k	frame index
m	subband index
m_d	processing delay
M	number of subbands
n	sample index
T	sampling interval
$\mathcal{T}\{\cdot\}$	transformation operator

Symbol	Description
$u[n]$	step sequence
$w[n]$	window function
$x, x[n]$	input signal, input sequence
$y, y[n]$	output signal, output sequence
$W_N = e^{-j2\pi/N}$	Nth root of unity

4

Bayesian Filters

Many problems in science and engineering can be formulated as the estimation of the *state* of a physical system based on some evidence or observations. Quite often, the state itself is not directly observable, but a mathematical model of the system is available indicating how the internal state of the system interacts with the external world in order to produce observations. In such cases, the problem can be posed as that of forming a probabilistic estimate of the internal state of the system conditioned on the available sequence of observations, which are typically corrupted by noise or other distortions. Forming such an estimate is the function of any one of the several variants of *Bayesian filters*. In the context of distant speech recognition, Bayesian filters have two primary applications:

- tracking nonstationary distortions in order to estimate the original speech from the mouth of a speaker;
- tracking the physical position of one or more active speakers to facilitate beamforming.

Choosing a particular state-space determines which aspects of the physical system are modeled. For example, to track a person in a room it may be sufficient to consider the position only. For a more sophisticated model, the speaker's velocity and acceleration might also be included. Typically, not all aspects of a physical system can be modeled. To account for the unmodeled portions of a system, a Bayesian filter includes random or stochastic components, which allow the unknown elements to be modeled statistically. This implies, however, that the state itself can only be determined statistically. The statistical characterization of the state returned by a Bayesian filter is the so-called filtering density.

The best known Bayesian filter is undoubtedly the *Kalman filter* (Kalman 1960), which – under the twin assumptions of Gaussianity and linearity – is the optimal minimum mean square error estimator for the hidden state of a system. As we will discuss in the present chapter, when a system is nonlinear or non-Gaussian, as is very often the case, a variety of other Bayesian filters can be used instead of the Kalman filter, but the optimality of the estimate is no longer guaranteed. The performance of the estimator must then be determined through simulations.

Local filters overcome the linearity assumption by linearizing the nonlinearities in the system through a local first-order Taylor series approximation or Stirling's polynomial interpolation. Thus local filters are valid only within a small neighborhood or working region of the state space. Well-known Bayesian filters based on the Taylor series are the *extended Kalman filter* and the *iterated extended filter* (Jazwinski 1970). The *unscented Kalman filter* (Julier and Uhlmann 2004), on the other hand, does not rely on a linearization, but instead propagates a set of test points through the system in order to determine how the several mean vectors and covariance matrices required for updating the state estimate are transformed by the system and observation nonlinearities.

Global filters provide an estimate which is valid in nearly the entire state space by solving for nonlinear or non-Gaussian system components through analytical or numerical methods. The analytical approaches could be based on approximating the filtering density with, for example, a mixture of Gaussian distributions (Sorenson and Alspach 1971) or by a spline function (de Figueiredo and Jan 1971). Numerical approaches evaluate the conditional pdf only at isolated grid points, which is known as the *point-mass method* (Bucy 1969). Alternatively, the filtering density can be approximated by a set of weighted samples in the state space as with the *particle filter* (Doucet 1998; Gordon *et al.* 1993), which is based on the concept of a *sequential Monte Carlo estimation*. A summary of the Kalman filter along with two popular nonlinear filters – namely, the extended Kalman filter and the particle filter – is presented in Table 4.1.

The balance of this chapter is organized as follows. We begin by presenting the basic concepts of sequential Bayesian estimation in Section 4.1, where we show that the filtering density can be updated through two operations, namely, *prediction* and *correction*. Estimation of the filtering density is of paramount importance inasmuch as all other estimates – such as the minimum mean square error estimate – can be readily calculated from the filtering density. In Section 4.2 we have a brief presentation of the Wiener filter. Section 4.3 begins with a discussion of the conventional Kalman filter, then considers several variations. As mentioned previously, the *extended Kalman filter* and *iterated extended Kalman filter* are both based on linearizing the nonlinear elements of a system and observation functional about the current state estimate. The difference between the two stems from the fact that the iterated extended Kalman filter will potentially refine the current state estimate by linearizing the nonlinear elements several times at each time step. Section 4.3.4 discusses the numerical stability of the conventional Kalman filter and how it can be improved through the update formulae based on the Cholesky decomposition. As discussed in Section 4.3.5, the *probabilistic data association filter* is a variation of the Kalman filter that first estimates which of several observations is likely to be associated

Table 4.1 Comparison between the Kalman, the extended Kalman and particle filter

Item	Kalman filter	Extended Kalman filter	Particle filter
Conditional density	Gaussian	Gaussian	Non-Gaussian
Dynamical model	Linear	Nonlinear	Nonlinear
Measurement model	Linear	Nonlinear	Nonlinear
Measurement noise	Gaussian	Gaussian	Non-Gaussian
Propagation of stat.	Mean & covar.	Mean & covar.	Probability density

with the state or track, as opposed to noise or other spurious acoustic events. The *joint probabilistic data association filter*, as discussed in Section 4.3.6, in addition to being able to use multiple observations can also maintain state vectors for several tracks simultaneously. Finally, in Section 4.4 we present the particle filter, which is based on a completely different set of assumptions than the Kalman filter and its variants. Namely, the particle filter approximates the filtering density with a set of discrete points or *particles* and their corresponding weights.

While this chapter is intended to present elements of the theory of Bayesian estimation, subsequent chapters will discuss practical applications. Whereas in Chapter 6, we will discuss how the methods presented here can be used to enhance speech which has been corrupted by noise or reverberation, in Chapter 10, we will see how such filters can be used to track a moving speaker.

4.1   Sequential Bayesian Estimation

Here we will present the fundamental concepts of Bayesian filtering. A schematic illustrating the operation of a Bayesian filter is shown in Figure 4.1, in which $\mathbf{x}_k$ denotes an unobservable state at time k, and $\mathbf{y}_k$ denotes an observation. As mentioned at the outset, the function of a Bayesian filter is to provide a statistical model of $\mathbf{x}_k$ conditioned on the sequence of observations $\mathbf{y}_{1:k}$. To provide such a statistical characterization, a Bayesian filter posits a state-space model, which is doubly stochastic in that its operation is governed by two equations, both of which include stochastic components. For the discrete-time case, these are

- the *state equation* or *system model*

$$\mathbf{x}_k = \mathbf{f}_k(\mathbf{x}_{k-1}, \mathbf{u}_{k-1}) \tag{4.1}$$

- and the *observation equation* or *measurement model*

$$\mathbf{y}_k = \mathbf{h}_k(\mathbf{x}_k, \mathbf{v}_k) \tag{4.2}$$

where $\mathbf{f}_k$ and $\mathbf{h}_k$ represent time-varying, nonlinear transition and observation functions, and $\mathbf{u}_k$ and $\mathbf{v}_k$ denote the *process* and *observation noise*, respectively.

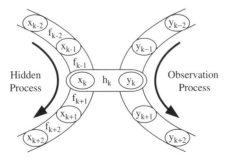

Figure 4.1   State-space model

In the most general formulation, the objective of the state estimation problem is to track the time evolution of the *filtering density* $p(\mathbf{x}_k|\mathbf{y}_{1:m})$. Depending on which observations $\mathbf{y}_{1:m}$ are used as evidence to determine $p(\mathbf{x}_k|\mathbf{y}_{1:m})$, state estimation falls into one of three categories:

- *smoothing*, whereby $k < m$;
- *filtering*, whereby $k = m$;
- *prediction*, whereby $k > m$.

Here we will be concerned exclusively with the *filtering* problem.

The system model (4.1) represents a *first-order Markov process*, which implies that the current state of the system depends solely on the state immediately prior, such that

$$p(\mathbf{x}_k|\mathbf{x}_{0:k-1}) = p(\mathbf{x}_k|\mathbf{x}_{k-1}) \, \forall \, k \in \mathbb{N}. \qquad (4.3)$$

The observation model (4.2) implies that the observations are dependent only on the current system state and are independent of both prior observations and prior states once the current state has been specified, such that,

$$p(\mathbf{y}_k|\mathbf{x}_{0:k}, \mathbf{y}_{1:k-1}) = p(\mathbf{y}_k|\mathbf{x}_k) \, \forall \, k \in \mathbb{N}. \qquad (4.4)$$

In Chapter 7, we will refer to the latter property as *conditional independence*. As previously mentioned, our intention is to investigate methods for tracking the filtering density $p(\mathbf{x}_k|\mathbf{y}_{1:k})$ as it evolves in discrete time k. A general solution can be obtained by applying *Bayes' rule*, which can be expressed as,

$$p(\mathbf{x}_k|\mathbf{y}_{1:k}) = \frac{p(\mathbf{y}_{1:k}|\mathbf{x}_k) \, p(\mathbf{x}_k)}{p(\mathbf{y}_{1:k})}. \qquad (4.5)$$

This filtering density can be tracked throughout time by sequentially performing two operations:

- *Prediction*
  The prior pdf of the state at time k can be obtained from the *Chapman–Kolmogorov equation* (Papoulis 1984),

$$p(\mathbf{x}_k|\mathbf{y}_{1:k-1}) = \int p(\mathbf{x}_k|\mathbf{x}_{k-1}) \, p(\mathbf{x}_{k-1}|\mathbf{y}_{1:k-1}) \, d\mathbf{x}_{k-1}, \qquad (4.6)$$

  where the evolution of the state $p(\mathbf{x}_k|\mathbf{x}_{k-1})$ is defined by the state equation (4.1).
- *Correction or update*
  The current observation is "folded" into the estimate of the filtering density through an invocation of Bayes' rule:

$$p(\mathbf{x}_k|\mathbf{y}_{1:k}) = \frac{p(\mathbf{y}_k|\mathbf{x}_k) \, p(\mathbf{x}_k|\mathbf{y}_{1:k-1})}{p(\mathbf{y}_k|\mathbf{y}_{1:k-1})} = \frac{p(\mathbf{x}_k, \mathbf{y}_k|\mathbf{y}_{1:k-1})}{p(\mathbf{y}_k|\mathbf{y}_{1:k-1})}, \qquad (4.7)$$

where the normalization constant in the denominator of (4.7) is given by

$$p(\mathbf{y}_k|\mathbf{y}_{1:k-1}) = \int p(\mathbf{y}_k|\mathbf{x}_k)\, p(\mathbf{x}_k|\mathbf{y}_{1:k-1})\, d\mathbf{x}_k.$$

That (4.7) is correct can be seen through the following chain of equalities:

$$
\begin{aligned}
p(\mathbf{x}_k|\mathbf{y}_{1:k}) &= \frac{p(\mathbf{y}_{1:k}|\mathbf{x}_k)p(\mathbf{x}_k)}{p(\mathbf{y}_{1:k})} \\
&= \frac{p(\mathbf{y}_k, \mathbf{y}_{1:k-1}|\mathbf{x}_k)p(\mathbf{x}_k)}{p(\mathbf{y}_k, \mathbf{y}_{1:k-1})} \\
&= \frac{p(\mathbf{y}_k|\mathbf{y}_{1:k-1}, \mathbf{x}_k)p(\mathbf{y}_{1:k-1}|\mathbf{x}_k)p(\mathbf{x}_k)}{p(\mathbf{y}_k|\mathbf{y}_{1:k-1})p(\mathbf{y}_{1:k-1})} \\
&= \frac{p(\mathbf{y}_k|\mathbf{y}_{1:k-1}, \mathbf{x}_k)\, p(\mathbf{y}_{1:k-1}, \mathbf{x}_k)\, p(\mathbf{x}_k)}{p(\mathbf{y}_k|\mathbf{y}_{1:k-1})\, p(\mathbf{y}_{1:k-1})\, p(\mathbf{x}_k)} \\
&= \frac{p(\mathbf{y}_k|\mathbf{x}_k)p(\mathbf{x}_k|\mathbf{y}_{1:k-1})}{p(\mathbf{y}_k|\mathbf{y}_{1:k-1})} \\
&= \frac{p(\mathbf{x}_k, \mathbf{y}_k|\mathbf{y}_{1:k-1})}{p(\mathbf{y}_k|\mathbf{y}_{1:k-1})}.
\end{aligned}
$$

The decisive elements in (4.7) are $p(\mathbf{x}_k|\mathbf{y}_{1:k-1})$, which is determined by the system model (4.1) together with (4.6), and $p(\mathbf{y}_k|\mathbf{x}_k)$, which is determined by the observation model (4.2).

The formulation of the tracking problem as one of density estimation is actually very powerful inasmuch as it subsumes many other well-known forms of parameter estimation. All point estimates of $\mathbf{x}_k$, such as maximum likelihood, the median, or the conditional mean, can be obtained as soon as $p(\mathbf{x}_k|\mathbf{y}_{1:k})$ is known. Consider the *minimum mean square error* (MMSE) estimate of $\mathbf{x}_k$, which can be formulated as follows. First we define the mean square error as

$$
\xi(\hat{\mathbf{x}}_k) = \mathcal{E}_{p(\mathbf{x}_k|\mathbf{y}_{1:k})}\left\{|\hat{\mathbf{x}}_k - \mathbf{x}_k|^2 \,|\, \mathbf{y}_{1:k}\right\} = \int |\hat{\mathbf{x}}_k - \mathbf{x}_k|^2\, p(\mathbf{x}_k|\mathbf{y}_{1:k})\, d\mathbf{x}_k
$$

$$
= \underbrace{|\hat{\mathbf{x}}_k|^2 \int p(\mathbf{x}_k|\mathbf{y}_{1:k})\, d\mathbf{x}_k}_{\hat{\mathbf{x}}_k^2} + \underbrace{\int |\mathbf{x}_k|^2\, p(\mathbf{x}_k|\mathbf{y}_{1:k})\, d\mathbf{x}_k}_{\mathcal{E}_{p(\mathbf{x}_k|\mathbf{y}_{1:k})}\left\{\mathbf{x}_k^2|\mathbf{y}_{1:k}\right\}} - \underbrace{2\hat{\mathbf{x}}_k^T \int \mathbf{x}_k\, p(\mathbf{x}_k|\mathbf{y}_{1:k})\, d\mathbf{x}_k}_{2\hat{\mathbf{x}}_k\mathcal{E}_{p(\mathbf{x}_k|\mathbf{y}_{1:k})}\{\mathbf{x}_k|\mathbf{y}_{1:k}\}}
$$

where $|\cdot|$ denotes the Euclidean norm. To calculate the optimal estimate, it is necessary to take the partial derivative of $\xi(\hat{\mathbf{x}}_k)$ with respect to $\hat{\mathbf{x}}_k$ on both sides of the prior equation and equate it to zero, whereby we find

$$
\frac{\partial \xi(\mathbf{x}_k)}{\partial \hat{x}_k} = 2\hat{x}_k - 2\mathcal{E}_{p(\mathbf{x}_k|\mathbf{y}_{1:k})}\{\mathbf{x}_k|\mathbf{y}_{1:k}\} = 0.
$$

Hence, it is apparent that the optimal MMSE estimator is equivalent to the *conditional mean*,

$$\mathcal{E}_{p(\mathbf{x}_k|\mathbf{y}_{1:k})}\{\mathbf{x}_k|\mathbf{y}_{1:k}\} = \int \mathbf{x}_k\, p(\mathbf{x}_k|\mathbf{y}_{1:k})\, d\mathbf{x}_k. \tag{4.8}$$

Similarly, it follows that knowledge of the filtering density $p(\mathbf{x}_k|\mathbf{y}_{1:k-1})$ enables all other less general estimates of $\mathbf{x}_k$ to be readily calculated.

4.2 Wiener Filter

Stochastic filter theory was established by the pioneering work of Norbert Wiener (1949), Wiener and Hopf (1931), and Andrey Kolmogorov (1941a, b). A Wiener filter provides the optimal static, linear, MMSE solution, where the mean square error is calculated between the output of the filter and some desired signal. We discuss the Wiener filter in this section, because such a filter is equivalent to the Kalman filter described in Section 4.3 without any process noise. Hence, the Wiener filter is in fact a Bayesian estimator (Simon 2006, sect. 8.5.2). We will derive both the time and frequency domain solutions for the *finite impulse response* (FIR) filter.

4.2.1 Time Domain Solution

Let $x[n]$ denote the desired signal and let $d[n]$ represent some additive distortion. The primary assumptions inherent in the Wiener filter are that the second-order statistics of both $x[n]$ and $d[n]$ are stationary. The corrupted signal is then defined as

$$y[n] \triangleq x[n] + d[n].$$

The time domain output of the FIR Wiener filter, which is the estimate $\hat{x}[n]$ of the desired signal $x[n]$, is by definition obtained from the convolution

$$\hat{x}[n] \triangleq \sum_{l=0}^{L-1} h[l]\, y[n-l], \tag{4.9}$$

where $h[n]$ is the filter impulse response of length L. Upon defining

$$\mathbf{h} \triangleq \begin{bmatrix} h[0] & h[1] & \cdots & h[L-1] \end{bmatrix}^T,$$

$$\mathbf{y}[n] \triangleq \begin{bmatrix} y[n] & y[n-1] & \cdots & y[n-L+1] \end{bmatrix}^T,$$

the output of the filter can be expressed as

$$\hat{x}[n] = \mathbf{h}^T\, \mathbf{y}[n].$$

The estimation error is $\epsilon[n] \triangleq x[n] - \hat{x}[n]$, and the squared-estimation error is given by

$$\zeta \triangleq \mathcal{E}\{\epsilon^T[n]\,\epsilon[n]\} = \mathcal{E}\{(x[n] - \mathbf{h}^T\,\mathbf{y}[n])^T (x[n] - \mathbf{h}^T\,\mathbf{y}[n])\}. \tag{4.10}$$

which must be minimized. Equation (4.10) can be rewritten as

$$\zeta = \mathcal{E}\{x^T[n]\,x[n]\} - 2\,\mathbf{h}^T\,\mathbf{r}_{xy} + \mathbf{h}^T\,\mathbf{R}_y\,\mathbf{h},$$

where

$$\mathbf{R}_y \triangleq \mathcal{E}\{\mathbf{y}[n]\,\mathbf{y}^T[n]\},$$

$$\mathbf{r}_{xy} \triangleq \mathcal{E}\{\mathbf{y}[n]\,x[n]\}.$$

The Wiener filter is based on the assumption that the components $\mathbf{R}_y$ and $\mathbf{r}_{xy}$ are stationary. In order to solve for the optimal filter coefficients, we set

$$\frac{\partial \zeta}{\partial \mathbf{h}} = -2\,\mathbf{r}_{xy} + 2\,\mathbf{h}^T\,\mathbf{R}_y = \mathbf{0}, \tag{4.11}$$

which leads immediately to the famous *Wiener–Hopf equation*

$$\mathbf{R}_y\,\mathbf{h} = \mathbf{r}_{xy}. \tag{4.12}$$

The solution for the optimal coefficients is then

$$\mathbf{h}_{\mathrm{o}} = \mathbf{R}_y^{-1}\mathbf{r}_{xy}.$$

Note that the optimal solution can also be found through the well-known *orthogonality principle* (Stark and Woods 1994), which can be stated as

$$\mathcal{E}\{\mathbf{y}[n-i]\,\epsilon[n]\} = \mathbf{0}\,\forall i = 0, \ldots, L-1. \tag{4.13}$$

In other words, the orthogonality principle requires that the estimation error $\epsilon[n]$ is orthogonal to all of the inputs $y[n-i]$ for $i = 0, \ldots, L-1$ used to form the estimate $\hat{x}[n]$.

4.2.2 *Frequency Domain Solution*

In order to derive the Wiener filter in the frequency domain, let us express (4.13) as

$$\mathcal{E}\left\{y[n-i]\left[x[n] - \sum_{l=0}^{L-1} h_{\mathrm{opt}}[l]\,y[n-l]\right]\right\} = 0\,\forall i = 0, \ldots, L-1.$$

Equivalently, we can write

$$r_{xy}[n] - h_{\mathrm{opt}}[n] * r_y[n] = 0, \tag{4.14}$$

where the cross-correlation sequence of $x[n]$ and $y[n]$ as well as the autocorrelation sequence of $y[n]$ are, respectively,

$$r_{xy}[l] \triangleq \begin{cases} \mathcal{E}\{y[n-l]\,x[n]\}, & \forall l = 0, \ldots, L-1, \\ 0, & \text{otherwise}, \end{cases}$$

$$r_y[l] \triangleq \begin{cases} \mathcal{E}\{y[n-l]\,y[n]\}, & \forall l = -L+1, \ldots, L-1, \\ 0, & \text{otherwise}. \end{cases}$$

Taking the Fourier transform of (4.14) provides

$$\Sigma_{XY}(\omega) - H_{\text{opt}}(\omega)\, \Sigma_Y(\omega) = 0,$$

where[1] $r_{xy}[n] \leftrightarrow \Sigma_{XY}(\omega)$, $h[n] \leftrightarrow H_{\text{opt}}(\omega)$, and $r_y[n] \leftrightarrow \Sigma_Y(\omega)$. This leads immediately to the solution

$$H_{\text{opt}}(\omega) = \frac{\Sigma_{XY}(\omega)}{\Sigma_Y(\omega)}. \tag{4.15}$$

Given that $X(\omega)$ and $D(\omega)$ are statistically independent by assumption, it follows that

$$\Sigma_Y(\omega) = \Sigma_X(\omega) + \Sigma_D(\omega),$$

$$\Sigma_{XY}(\omega) = \Sigma_X(\omega).$$

Hence, we can rewrite (4.15) as

$$H_{\text{opt}}(\omega) = \frac{\Sigma_X(\omega)}{\Sigma_X(\omega) + \Sigma_D(\omega)}, \tag{4.16}$$

the form in which the Wiener filter is most often seen. Alternatively, the frequency response of the filter can be expressed as

$$H_{\text{opt}}(\omega) = \frac{1}{1 + \dfrac{\Sigma_D(\omega)}{\Sigma_X(\omega)}},$$

from which it is apparent that when the spectral power of the disturbance comes to dominate that of the signal, the gain of the filter is reduced. When the signal dominates the disturbance, on the other hand, the gain increases. In all cases it holds that

$$0 \leq |H_{\text{opt}}(\omega)| \leq 1.$$

As presented here, the classical Wiener filter presents something of a paradox in that it requires that the desired signal $x[n]$ or its power spectrum $\Sigma_X(\omega)$ is known before the

---

[1] The notation $r_y[n] \leftrightarrow \Sigma_Y(\omega)$ indicates that $r_y[n]$ and $\Sigma_Y(\omega)$ comprise a Fourier transform pair; see Section 3.1.2 for details.

filter coefficients can be designed. Were this information available, there would be no need of a Wiener filter. The art of practical Wiener filter design consists of nothing more than the robust estimation of the desired signal $\Sigma_X(\omega)$ and noise $\Sigma_D(\omega)$ components appearing in (4.15). References indicating how this can be achieved are presented at the ends of Sections 6.3.1 and 13.3.5.

4.3 Kalman Filter and Variations

In this section, we present the best known set of solutions for estimating the filtering density, namely the *Kalman filter* (KF) (Kalman 1960) and its several variations.

4.3.1 Kalman Filter

The Kalman filter provides a closed form means of sequentially updating $p(\mathbf{x}_k|\mathbf{y}_{1:k})$ under two critical assumptions:

- The transition and observation models $\mathbf{f}_k$ and $\mathbf{h}_k$ are linear.
- The process and observation noises $\mathbf{u}_k$ and $\mathbf{v}_k$ are Gaussian.

As the linear combination of Gaussian r.v.s is also Gaussian, these assumptions taken together imply that both $\mathbf{x}_k$ and $\mathbf{y}_k$ will remain Gaussian for all time k. Note that the combination of Gaussians in the nonlinear domain, such as the logarithmic domain, results in a non-Gaussian distribution, as described in Section 9.3.1. As mentioned previously, under these conditions, the KF is the optimal MMSE estimator.

In keeping with the aforementioned linearity assumption, the state model (4.1–4.2) can be expressed as

$$\mathbf{x}_k = \mathbf{F}_{k|k-1}\,\mathbf{x}_{k-1} + \mathbf{u}_{k-1}, \tag{4.17}$$

$$\mathbf{y}_k = \mathbf{H}_k\mathbf{x}_k + \mathbf{v}_k, \tag{4.18}$$

where $\mathbf{F}_{k|k-1}$ and $\mathbf{H}_k$ are the known *transition* and *observation* matrices. The noise terms $\mathbf{u}_k$ and $\mathbf{v}_k$ in (4.17–4.18) are by assumption zero mean, white Gaussian random vector processes with covariance matrices

$$\mathbf{U}_k = \mathcal{E}\{\mathbf{u}_k\mathbf{u}_k^T\}, \qquad \mathbf{V}_k = \mathcal{E}\{\mathbf{v}_k\mathbf{v}_k^T\},$$

respectively. Moreover, by assumption $\mathbf{u}_k$ and $\mathbf{v}_k$ are statistically independent.

By definition, the transition matrix $\mathbf{F}_{k|k-1}$ has two important properties:

- *product rule*

$$\mathbf{F}_{k|m}\,\mathbf{F}_{m|n} = \mathbf{F}_{k|n}, \tag{4.19}$$

- *inverse rule*

$$\mathbf{F}_{k|k-1}\mathbf{F}_{k-1|k} = \mathbf{F}_{k-1|k}\mathbf{F}_{k|k-1} = \mathbf{I}, \tag{4.20}$$

which implies that $\mathbf{F}_{k|k-1}^{-1} = \mathbf{F}_{k-1|k}$.

Once more let $\mathbf{y}_{1:k-1}$ denote all past observations up to time $k-1$, and let $\hat{\mathbf{y}}_{k|k-1}$ denote the MMSE estimate of the next observation $\mathbf{y}_k$ given all prior observations, such that,

$$\hat{\mathbf{y}}_{k|k-1} = \mathcal{E}\{\mathbf{y}_k | \mathbf{y}_{1:k-1}\}.$$

By definition, the *innovation* is the difference

$$\mathbf{s}_k \triangleq \mathbf{y}_k - \hat{\mathbf{y}}_{k|k-1} \tag{4.21}$$

between the actual and the predicted observations. This quantity is given the name innovation, because it contains all the "new information" required for sequentially updating the filtering density $p(\mathbf{x}_k|\mathbf{y}_{1:k})$; i.e., the innovation contains that information about the time evolution of the system that cannot be predicted from the state-space model.

The innovations process has three important properties:

- *Orthogonality*
  The innovation process $\mathbf{s}_K$ at time K is *orthogonal* to all past observations $\mathbf{y}_1, \ldots, \mathbf{y}_{k-1}$, such that

$$\mathcal{E}\{\mathbf{s}_K \mathbf{y}_k^T\} = \mathbf{0} \,\forall\, 1 \leq k \leq K - 1.$$

- *Whiteness*
  The innovations are orthogonal to each other, such that

$$\mathcal{E}\{\mathbf{s}_K \mathbf{s}_k^T\} = \mathbf{0} \,\forall\, 1 \leq k \leq K - 1.$$

- *Reversibility*
  There is a one-to-one correspondence between the observed data $\mathbf{y}_{1:k} = \{\mathbf{y}_1, \ldots, \mathbf{y}_k\}$ and the sequence of innovations $\mathbf{s}_{1:k} = \{\mathbf{s}_1, \ldots, \mathbf{s}_k\}$, such that one can always be uniquely recover from the other (Haykin 2002, sect. 10).

We will now present the principal quantities and relations in the operation of the KF. Our presentation is intended to convey intuition rather than provide a rigorous derivation. Those seeking the latter are referred to any one of a number of excellent texts, such as Haykin (2002, sect. 10), Grewal and Andrews (1993), or Simon (2006, sect. 5.1). Our presentation of the KF has, roughly speaking, four phases:

1. Provide an expression for calculating the correlation matrix of the innovations process.
2. Obtain an expression for the sequential update of the MMSE state estimate.

3. Define the Kalman gain, which plays the pivotal role in the sequential state update considered in phase two.
4. State the Riccati equation, which provides the means to update the state estimation error covariance matrices required to calculate the Kalman gain.

We begin by stating how the predicted observation may be calculated based on the current state estimate, according to,

$$\hat{\mathbf{y}}_{k|k-1} = \mathbf{H}_k \hat{\mathbf{x}}_{k|k-1}. \tag{4.22}$$

In light of (4.21) and (4.22), we may write

$$\mathbf{s}_k = \mathbf{y}_k - \mathbf{H}_k \hat{\mathbf{x}}_{k|k-1}. \tag{4.23}$$

Substituting (4.18) into (4.23), we find

$$\mathbf{s}_k = \mathbf{H}_k \boldsymbol{\epsilon}_{k|k-1} + \mathbf{v}_k, \tag{4.24}$$

where

$$\boldsymbol{\epsilon}_{k|k-1} \triangleq \mathbf{x}_k - \hat{\mathbf{x}}_{k|k-1} \tag{4.25}$$

is the *predicted state estimation error* at time k, using all data up to time $k-1$. It can be readily shown that $\boldsymbol{\epsilon}_{k|k-1}$ is orthogonal to $\mathbf{u}_k$ and $\mathbf{v}_k$ (Haykin 2002, sect. 10.1). Using (4.24) and exploiting the statistical independence of $\mathbf{u}_k$ and $\mathbf{v}_k$, the correlation matrix of the innovations sequence can be expressed as

$$\mathbf{S}_k \triangleq \mathcal{E}\left\{\mathbf{s}_k \mathbf{s}_k^T\right\} = \mathbf{H}_k \mathbf{K}_{k|k-1} \mathbf{H}_k^T + \mathbf{V}_k, \tag{4.26}$$

where the *predicted state estimation error covariance matrix* is defined as

$$\mathbf{K}_{k|k-1} \triangleq \mathcal{E}\left\{\boldsymbol{\epsilon}_{k|k-1} \boldsymbol{\epsilon}_{k|k-1}^T\right\}. \tag{4.27}$$

As described previously, the sequential update of the filtering density can be partitioned into two steps:

- First, there is a *prediction*, which can be expressed as

$$\hat{\mathbf{x}}_{k|k-1} = \mathbf{F}_{k|k-1} \hat{\mathbf{x}}_{k-1|k-1}, \tag{4.28}$$

a direct specialization of the Chapman–Kolmogorov equation (4.6). Clearly the prediction is so-called because it is made without the advantage of any information derived from the current observation $\mathbf{y}_k$.

- The latter information is instead folded into the current estimate through the *update* or *correction*, according to

$$\hat{\mathbf{x}}_{k|k} = \hat{\mathbf{x}}_{k|k-1} + \mathbf{G}_k \mathbf{s}_k, \tag{4.29}$$

where the *Kalman gain* is defined as

$$\mathbf{G}_k \triangleq \mathcal{E}\{\mathbf{x}_k \mathbf{s}_k^T\}\mathbf{S}_k^{-1}, \tag{4.30}$$

for $\mathbf{x}_k$, $\mathbf{s}_k$, and $\mathbf{S}_k$ given by (4.17), (4.23), and (4.26), respectively. Note that (4.29) is of paramount importance, as it shows how the MMSE or Bayesian state estimate can be recursively updated – that is, it is only necessary to premultiply the prior estimate $\hat{\mathbf{x}}_{k|k-1}$ by the transition matrix $\mathbf{F}_{k|k-1}$, then to add a correction factor consisting of the Kalman gain $\mathbf{G}_k$ multiplied by the innovation $\mathbf{s}_k$. Hence, the entire problem of recursive MMSE estimation under the assumptions of linearity and Gaussianity reduces to the calculation of the Kalman gain (4.30), whereupon the state estimate can be updated according to (4.29). From (4.28) and (4.29), we deduce that the KF has the predictor–corrector structure shown in Figure 4.2.

The Kalman gain (4.30) can be efficiently calculated according to

$$\mathbf{G}_k = \mathbf{K}_{k|k-1}\mathbf{H}_k^T\mathbf{S}_k^{-1}, \tag{4.31}$$

where the correlation matrix $\mathbf{S}_k$ of the innovations sequence is defined in (4.26). The *Riccati equation* then specifies how $\mathbf{K}_{k|k-1}$ can be sequentially updated, namely as,

$$\mathbf{K}_{k|k-1} = \mathbf{F}_{k|k-1}\,\mathbf{K}_{k-1}\,\mathbf{F}_{k|k-1}^T + \mathbf{U}_{k-1}. \tag{4.32}$$

The matrix $\mathbf{K}_k$ in (4.32) is, in turn, obtained through the recursion,

$$\mathbf{K}_k = \mathbf{K}_{k|k-1} - \mathbf{G}_k \mathbf{H}_k \mathbf{K}_{k|k-1} = (\mathbf{I} - \mathbf{G}_k \mathbf{H}_k)\mathbf{K}_{k|k-1}. \tag{4.33}$$

This matrix $\mathbf{K}_k$ can be interpreted as the covariance matrix of the *filtered state estimation error* (Haykin 2002, sect. 10), such that,

$$\mathbf{K}_k \triangleq \left\{\boldsymbol{\epsilon}_k \boldsymbol{\epsilon}_k^T\right\},$$

where

$$\boldsymbol{\epsilon}_k \triangleq \mathbf{x}_k - \hat{x}_{k|k}.$$

Note the critical difference between $\boldsymbol{\epsilon}_{k|k-1}$ and $\boldsymbol{\epsilon}_k$, namely, $\boldsymbol{\epsilon}_{k|k-1}$ is the error in the state estimate made *without* knowledge of the current observation $\mathbf{y}_k$, while $\boldsymbol{\epsilon}_k$ is the error in the state estimate made *with* knowledge of $\mathbf{y}_k$.

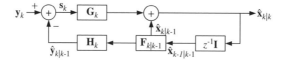

Figure 4.2 Predictor–corrector structure of the Kalman filter

Algorithm 4.1 Calculations for Kalman filter based on the Riccati equation

Input vector process: $\mathbf{y}_1, \mathbf{y}_2, \ldots, \mathbf{y}_k$
Known parameters:

- state transition matrix: $\mathbf{F}_{k|k-1}$
- observation matrix: $\mathbf{H}_k$
- covariance matrix of process noise: $\mathbf{U}_k$
- covariance matrix of measurement noise: $\mathbf{V}_k$
- initial diagonal loading: σ_D^2

Initial conditions:

$$\hat{\mathbf{x}}_{1|0} = \mathbf{x}_0; \quad \mathbf{K}_{1|0} = \frac{1}{\sigma_D^2}\mathbf{I}$$

Computation: $k = 1, 2, 3, \ldots$

$$\hat{\mathbf{x}}_{k|k-1} = \mathbf{F}_{k|k-1}\,\hat{\mathbf{x}}_{k-1|k-1} \text{ (prediction)} \tag{4.34}$$

$$\mathbf{K}_{k|k-1} = \mathbf{F}_{k|k-1}\,\mathbf{K}_{k-1}\,\mathbf{F}_{k|k-1}^T + \mathbf{U}_{k-1} \tag{4.35}$$

$$\mathbf{S}_k = \mathbf{H}_k\,\mathbf{K}_{k|k-1}\,\mathbf{H}_k^T + \mathbf{V}_k \tag{4.36}$$

$$\mathbf{G}_k = \mathbf{K}_{k|k-1}\,\mathbf{H}_k^T\,\mathbf{S}_k^{-1} \tag{4.37}$$

$$\mathbf{s}_k = \mathbf{y}_k - \mathbf{H}_k\,\hat{\mathbf{x}}_{k|k-1} \tag{4.38}$$

$$\hat{\mathbf{x}}_{k|k} = \hat{\mathbf{x}}_{k|k-1} + \mathbf{G}_k\,\mathbf{s}_k \text{ (correction)} \tag{4.39}$$

$$\mathbf{K}_k = (\mathbf{I} - \mathbf{G}_k\,\mathbf{H}_k)\,\mathbf{K}_{k|k-1} \tag{4.40}$$

The calculation of the KF based on the Riccati equation is summarized in Algorithm 4.1. A schematic of the KF is shown in Figure 4.3. The predicted state estimate $\hat{\mathbf{x}}_{k|k-1}$ is projected into the observation space as in (4.22) to obtain the predicted observation $\hat{\mathbf{y}}_{k|k-1}$ and the innovation $\mathbf{s}_k$ as in (4.23). The predicted state-error covariance matrix $\mathbf{K}_{k|k-1}$ is obtained from $\mathbf{K}_{k-1}$ according to (4.32). As $\mathbf{K}_{k|k-1}$ includes the contribution of the positive-definite[2] covariance matrix $\mathbf{U}_k$ of the process noise, it will typically be larger than $\mathbf{K}_{k-1}$. The covariance matrix $\mathbf{K}_{k|k-1}$ is also projected into the observation space according to (4.26) to obtain the covariance matrix $\mathbf{S}_k$ of the innovation $\mathbf{s}_k$. Both $\mathbf{K}_{k|k-1}$ and $\mathbf{S}_k$ are then used to calculate the Kalman gain $\mathbf{G}_k$ as in (4.31), whereupon the state estimate update proceeds according to (4.29).

Having defined all necessary quantities, the connection between the KF and the sequential Bayesian filtering algorithm (4.6–4.7) can now be made explicit by writing

---

[2] A real symmetric matrix M is positive definite if $\mathbf{z}^T\mathbf{Mz} > 0\,\forall$ nonzero vectors $\mathbf{z}$.

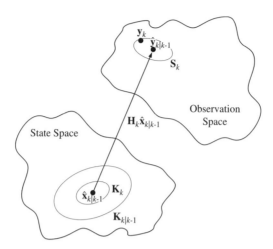

Figure 4.3   Schematic illustrating the operation of the Kalman filter

- the prediction as

$$p(\mathbf{x}_k|\mathbf{y}_{1:k-1}) = \mathcal{N}(\mathbf{x}_k; \hat{\mathbf{x}}_{k|k-1}, \mathbf{K}_{k|k-1}), \tag{4.41}$$

- and the correction as

$$p(\mathbf{x}_k|\mathbf{y}_{1:k}) = \mathcal{N}(\mathbf{x}_k; \hat{\mathbf{x}}_{k|k}, \mathbf{K}_k), \tag{4.42}$$

where $\mathcal{N}(\mathbf{x}; \boldsymbol{\mu}, \boldsymbol{\Sigma})$ is the multidimensional Gaussian pdf with mean vector $\boldsymbol{\mu}$ and covariance matrix $\boldsymbol{\Sigma}$. In particular, it is now obvious that (4.41–4.42) are specializations of (4.6–4.7), respectively, for the case wherein all pdfs are Gaussian and transition and observation functions are linear in $\mathbf{x}_k$.

4.3.2   Extended Kalman Filter

While the KF is optimal under the conditions discussed at the beginning of Section 4.3.1, these conditions seldom hold in most practical applications. In particular, it often happens that the state and observation equations (4.1–4.2) are *not* linear. One of the first attempts to generalize the KF to handle such nonlinearities was the *extended Kalman filter* (EKF). To formulate the EKF, we first posit a less restrictive state-space model, namely,

$$\mathbf{x}_k = \mathbf{F}_{k|k-1}\mathbf{x}_{k-1} + \mathbf{u}_{k-1}, \tag{4.43}$$

$$\mathbf{y}_k = \mathbf{H}_k(\mathbf{x}_k) + \mathbf{v}_k, \tag{4.44}$$

where the observation *functional*[3] $\mathbf{H}_k(\mathbf{x}_k)$ is in general nonlinear and time-varying. The main idea behind the EKF is then to linearize this functional about the most recent state

---

[3] Most authors formulate the extended KF with a nonlinear process functional $\mathbf{F}_{k|k-1}(\mathbf{x}_k)$ in addition to the observation functional $\mathbf{H}_k(\mathbf{x}_k)$; see, for example, Haykin (2002, sect. 10.10). This more general formulation is not required for the description of the speaker tracking system in Section 10.2.

estimate $\hat{\mathbf{x}}_{k|k-1}$. The corresponding linearization can be written as

$$\overline{\mathbf{H}}_k(\hat{\mathbf{x}}_{k|k-1}) \triangleq \left. \frac{\partial \mathbf{H}_k(\mathbf{x})}{\partial \mathbf{x}} \right|_{\mathbf{x} = \hat{\mathbf{x}}_{k|k-1}}, \tag{4.45}$$

where entry [4] (i, j) of $\overline{\mathbf{H}}_k(\mathbf{x})$ is the partial derivative of the ith component of $\mathbf{H}_k(\mathbf{x})$ with respect to the jth component of $\mathbf{x}$.

Based on (4.45), we can express the first-order Taylor series of $\mathbf{H}_k(\mathbf{x}_k)$ as

$$\mathbf{H}_k(\mathbf{x}_k) \approx \mathbf{H}_k(\hat{\mathbf{x}}_{k|k-1}) + \overline{\mathbf{H}}_k(\hat{\mathbf{x}}_{k|k-1}) \left(\mathbf{x}_k - \hat{\mathbf{x}}_{k|k-1} \right). \tag{4.46}$$

Using this linearization, the nonlinear state-space equations (4.43–4.44) can be written as

$$\mathbf{x}_k = \mathbf{F}_{k|k-1} \mathbf{x}_{k-1} + \mathbf{u}_{k-1}, \tag{4.47}$$

$$\overline{\mathbf{y}}_k \approx \overline{\mathbf{H}}_k(\hat{\mathbf{x}}_{k|k-1}) \, \mathbf{x}_k + \mathbf{v}_k, \tag{4.48}$$

where we have defined

$$\overline{\mathbf{y}}_k \triangleq \mathbf{y}_k - \left[\mathbf{H}_k(\hat{\mathbf{x}}_{k|k-1}) - \overline{\mathbf{H}}_k(\hat{\mathbf{x}}_{k|k-1}) \, \hat{\mathbf{x}}_{k|k-1} \right]. \tag{4.49}$$

As everything on the right-hand side of (4.49) is known at time k, the term $\overline{\mathbf{y}}_k$ can be regarded as an observation.

The EKF is obtained by applying the computations in (4.34–4.40) to the linearized model in (4.47–4.49), whereupon we find

$$\hat{\mathbf{x}}_{k|k-1} = \mathbf{F}_{k|k-1} \hat{\mathbf{x}}_{k-1|k-1} \tag{4.50}$$

$$\hat{\mathbf{x}}_{k|k} = \hat{\mathbf{x}}_{k|k-1} + \mathbf{G}_k \, \mathbf{s}_k \tag{4.51}$$

$$\mathbf{s}_k = \overline{\mathbf{y}}_k - \overline{\mathbf{H}}_k(\hat{\mathbf{x}}_{k|k-1}) \hat{\mathbf{x}}_{k|k-1} = \mathbf{y}_k - \mathbf{H}_k(\hat{\mathbf{x}}_{k|k-1}). \tag{4.52}$$

The use of such a linearized model can be equated with the *Gauss–Newton method*, wherein higher order terms in the series expansion (4.46) are neglected. The connection between the KF and the Gauss–Newton method is well known, as is the fact that the convergence rate of the latter is superlinear if the error $\mathbf{y} - \mathbf{H}_k(\mathbf{x})$ is small near the optimal solution. Further details are given by Bertsekas (1995, sect. 1.5).

A linear approximation such as that used in the EKF can produce large approximation errors for severely nonlinear functions. Such considerations led to the development of a further refinement of the KF, which we consider next.

4.3.3 *Iterated Extended Kalman Filter*

As the EKF uses a first-order Taylor series expansion (4.46), its performance can be poor if the current state estimate $\hat{\mathbf{x}}_{k|k-1}$ is far from the true state $\mathbf{x}$. A further refinement of the

---

[4] It should always be borne in mind that $\overline{\mathbf{H}}_k(\mathbf{x})$ is a *matrix*.

EKF, dubbed the *iterated extended Kalman filter* (IEKF), uses several *local iterations* at each time step k to move the state estimate closer to the true state.

In order to develop the IEKF, we note that the update in the EKF can be expressed as,

$$\mathbf{S}_k(\hat{\mathbf{x}}_{k|k-1}) = \overline{\mathbf{H}}_k(\hat{\mathbf{x}}_{k|k-1})\mathbf{K}_{k|k-1}\overline{\mathbf{H}}_k^T(\hat{\mathbf{x}}_{k|k-1}) + \mathbf{V}_k, \tag{4.53}$$

$$\mathbf{G}_k(\hat{\mathbf{x}}_{k|k-1}) = \mathbf{K}_{k|k-1}\overline{\mathbf{H}}_k^T(\hat{\mathbf{x}}_{k|k-1})\mathbf{S}_k^{-1}(\hat{\mathbf{x}}_{k|k-1}), \tag{4.54}$$

$$\mathbf{s}_k(\hat{\mathbf{x}}_{k|k-1}) = \mathbf{y}_k - \mathbf{H}_k(\hat{\mathbf{x}}_{k|k-1}), \tag{4.55}$$

$$\hat{\mathbf{x}}_{k|k} = \hat{\mathbf{x}}_{k|k-1} + \mathbf{G}_k(\hat{\mathbf{x}}_{k|k-1})\,\mathbf{s}_k(\hat{\mathbf{x}}_{k|k-1}), \tag{4.56}$$

where we have explicitly indicated the dependence of the relevant quantities on $\hat{\mathbf{x}}_{k|k-1}$. As described by Jazwinski (1970, sect. 8.3), in the IEKF, (4.53–4.56) are replaced with the local iteration,

$$\mathbf{S}_k(\boldsymbol{\eta}_i) = \overline{\mathbf{H}}_k(\boldsymbol{\eta}_i)\,\mathbf{K}_{k|k-1}\overline{\mathbf{H}}^T(\boldsymbol{\eta}_i) + \mathbf{V}_k, \tag{4.57}$$

$$\mathbf{G}_k(\boldsymbol{\eta}_i) = \mathbf{K}_{k|k-1}\overline{\mathbf{H}}_k^T(\boldsymbol{\eta}_i)\mathbf{S}_k^{-1}(\boldsymbol{\eta}_i), \tag{4.58}$$

$$\mathbf{s}_k(\boldsymbol{\eta}_i) = \mathbf{y}_k - \mathbf{H}_k(\boldsymbol{\eta}_i), \tag{4.59}$$

$$\boldsymbol{\zeta}_k(\boldsymbol{\eta}_i) \triangleq \mathbf{s}_k(\boldsymbol{\eta}_i) - \overline{\mathbf{H}}_k(\boldsymbol{\eta}_i)\left(\hat{\mathbf{x}}_{k|k-1} - \boldsymbol{\eta}_i\right), \tag{4.60}$$

$$\boldsymbol{\eta}_{i+1} \triangleq \hat{\mathbf{x}}_{k|k-1} + \mathbf{G}_k(\boldsymbol{\eta}_i)\boldsymbol{\zeta}_k(\boldsymbol{\eta}_i), \tag{4.61}$$

where $\overline{\mathbf{H}}_k(\boldsymbol{\eta}_i)$ is the linearization of $\mathbf{H}_k(\boldsymbol{\eta}_i)$ about $\boldsymbol{\eta}_i$. The local iteration is initialized at $i = 1$ by setting

$$\boldsymbol{\eta}_1 = \hat{\mathbf{x}}_{k|k-1}.$$

Note that $\boldsymbol{\eta}_2 = \hat{\mathbf{x}}_{k|k}$ as defined in (4.56). Hence, if the local iteration is run only once, the IEKF reduces to the EKF. Normally (4.57–4.61) are repeated, however, until there are no substantial changes between $\boldsymbol{\eta}_i$ and $\boldsymbol{\eta}_{i+1}$. Both $\mathbf{G}_k(\boldsymbol{\eta}_i)$ and $\overline{\mathbf{H}}_k(\boldsymbol{\eta}_i)$ are updated for each local iteration. After the last iteration I, we set

$$\hat{\mathbf{x}}_{k|k} = \boldsymbol{\eta}_I$$

and this value is used to update $\mathbf{K}_k$ and $\mathbf{K}_{k+1|k}$. Jazwinski (1970, sect. 8.3) reports that the IEKF provides faster convergence in the presence of significant nonlinearities in the observation equation, especially when the initial state estimate $\boldsymbol{\eta}_1 = \hat{\mathbf{x}}_{k|k-1}$ is far from the optimal value.

4.3.4 Numerical Stability

All variants of the KF discussed in Sections 4.3.1 through 4.3.3 are based on the Riccati equation (4.32–4.33). Unfortunately, the Riccati equation possesses poor numerical stability properties (Haykin 2002, sect. 11) as can be seen from the following: Moving

(4.32) one step forward in time, we have

$$\mathbf{K}_{k+1|k} = \mathbf{F}_{k+1|k} \, \mathbf{K}_k \, \mathbf{F}_{k+1|k}^T + \mathbf{U}_k. \tag{4.62}$$

Substituting (4.33) into (4.62) provides

$$\mathbf{K}_{k+1|k} = \mathbf{F}_{k+1|k} \, \mathbf{K}_{k|k-1} \, \mathbf{F}_{k+1|k}^T - \mathbf{F}_{k+1|k} \, \mathbf{G}_k \, \mathbf{H}_k \, \mathbf{K}_{k|k-1} \, \mathbf{F}_{k+1|k}^T + \mathbf{U}_k. \tag{4.63}$$

Manipulating (4.31), we can write

$$\mathbf{S}_k \, \mathbf{G}_k^T = \mathbf{H}_k \, \mathbf{K}_{k|k-1}. \tag{4.64}$$

Then upon substituting (4.64) for the matrix product $\mathbf{H}_k \mathbf{K}_{k|k-1}$ appearing in (4.63), we find

$$\mathbf{K}_{k+1|k} = \mathbf{F}_{k+1|k} \, \mathbf{K}_{k|k-1} \, \mathbf{F}_{k+1|k}^T - \mathbf{F}_{k+1|k} \, \mathbf{G}_k \, \mathbf{S}_k \, \mathbf{G}_k^T \, \mathbf{F}_{k+1|k}^T + \mathbf{U}_k \tag{4.65}$$

which illustrates the problem inherent in the Riccati equation: As $\mathbf{K}_{k+1|k}$ is the covariance matrix of the predicted state error $\boldsymbol{\epsilon}_{k+1|k}$, it must be positive definite. Similarly, $\mathbf{S}_k$ is the covariance matrix of the innovation $\mathbf{s}_k$ and must also be positive definite. Moreover, if $\mathbf{F}_{k+1|k}$ and $\mathbf{G}_k$ are full rank, then the terms $\mathbf{F}_{k+1|k} \mathbf{K}_{k|k-1} \mathbf{F}_{k+1|k}^T$ and $\mathbf{F}_{k+1|k} \mathbf{G}_k \mathbf{S}_k \mathbf{G}_k^T \mathbf{F}_{k+1|k}^T$ are also positive definite. Therefore, (4.65) implies that a positive-definite matrix $\mathbf{K}_{k+1|k}$ must be calculated as the *difference* of the positive-definite matrix $\mathbf{F}_{k+1|k} \mathbf{K}_{k|k-1} \mathbf{F}_{k+1|k}^T + \mathbf{U}_k$ and positive-definite matrix $\mathbf{F}_{k+1|k} \mathbf{G}_k \mathbf{S}_k \mathbf{G}_k^T \mathbf{F}_{k+1|k}^T$. Due to finite precision errors, the resulting matrix $\mathbf{K}_{k+1|k}$ can become indefinite after a sufficient number of iterations, at which point the KF exhibits a behavior known as *explosive divergence* (Haykin 2002, sect. 11).

As discussed in Section 10.2.1, a more stable implementation of the KF can be developed based on the *Cholesky decomposition* (see Section B.3) or *square-root* of $\mathbf{K}_{k+1|k}$, which is by definition that unique lower triangular matrix $\mathbf{K}_{k+1|k}^{1/2}$ achieving

$$\mathbf{K}_{k+1|k} \triangleq \mathbf{K}_{k+1|k}^{1/2} \, \mathbf{K}_{k+1|k}^{T/2}.$$

The Cholesky decomposition of a matrix exists if and *only* if that matrix is symmetric and positive definite (Golub and Van Loan 1996a, sect. 4.2.3). The basic idea behind the square-root implementation of the KF is to update $\mathbf{K}_{k+1|k}^{1/2}$ instead of $\mathbf{K}_{k+1|k}$ directly. By updating or *propagating* $\mathbf{K}_{k+1|k}^{1/2}$ forward in time, it can be assured that $\mathbf{K}_{k+1|k}$ remains positive definite. Thereby, a numerically stable algorithm is obtained *regardless* of the precision of the machine on which it runs. Moreover, a square-root implementation effectively doubles the numerical precision of the direction form implementation (Simon 2006, sect. 6.3–6.4). This added precision comes at the price, however, of somewhat more computation. Section 10.2.1 presents a procedure whereby $\mathbf{K}_{k+1|k}^{1/2}$ can be efficiently propagated in time using a series of *Givens rotations*; the latter are described in Section B.15.

4.3.5   Probabilistic Data Association Filter

The *probabilistic data association filter* (PDAF) is a generalization of the KF wherein
the Gaussian pdf associated with the location of a speaker or a *track* is supplemented
with a pdf for spurious observations or *clutter* (Bar-Shalom and Fortmann 1988, sect.
6.4). Through the inclusion of the clutter model, the PDAF is able to make use of seve-
ral observations $\mathbf{Y}_k = \{\mathbf{y}_k^{(i)}\}_{i=1}^{m_k}$ for each time instant, where m_k is the total number of
validated observations for time k; the process of validation will be described shortly. The
basic assumption inherent in the PDAF is that the current state is normally distributed
with mean $\hat{\mathbf{x}}_{k|k-1}$ and covariance matrix $\mathbf{K}_{k|k-1}$, such that

$$p(\mathbf{x}_k|\mathbf{Y}_{1:k-1}) = \mathcal{N}(\mathbf{x}_k; \hat{\mathbf{x}}_{k|k-1}, \mathbf{K}_{k|k-1}),$$

where $\hat{\mathbf{x}}_{k|k-1}$ and $\mathbf{K}_{k|k-1}$ are the predicted state estimate and state estimation error covari-
ance matrix given by (4.28) and (4.32), respectively.

The first step in the update is the *validation* of observations, which occurs as follows.
We assume that the true observation $\mathbf{y}_k$ at time k conditioned on all prior observations
$\mathbf{Y}_k$ is normally distributed according to

$$p(\mathbf{y}_k|\mathbf{Y}_k) = \mathcal{N}(\mathbf{y}_k; \hat{\mathbf{y}}_{k|k-1}, \mathbf{S}_k)$$

where $\hat{\mathbf{y}}_{k|k-1}$ is the predicted observation (4.22), and $\mathbf{S}_k$ is the correlation matrix (4.26)
of the innovations sequence $\mathbf{s}_k$. Now define the *volume of the validation region* $V_k(\gamma)$ as
that region of the observation space where the true observation will be found with high
probability:

$$V_k(\gamma) \triangleq \{\mathbf{y} : (\mathbf{y} - \mathbf{H}_k\hat{\mathbf{x}}_{k|k-1})^T \mathbf{S}_k^{-1}(\mathbf{y} - \mathbf{H}_k\hat{\mathbf{x}}_{k|k-1}) \leq \gamma\}$$
$$= \{\mathbf{y} : \mathbf{s}_k^T \mathbf{S}_k^{-1} \mathbf{s}_k \leq \gamma\}, \tag{4.66}$$

where $\mathbf{s}_k$ is the innovation defined in (4.21) and γ is a gating parameter whose function is
explained next. The validation region is an ellipsoid in the observation space that contains
a given amount of probability mass within the smallest possible volume. Observations
falling within this validation region are treated as valid, those falling outside it are ignored.
The process of validation is depicted graphically in Figure 4.4. The basic problem solved

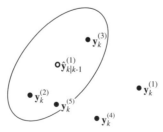

Figure 4.4   Validation of observations. Observations $\mathbf{y}_k^{(1)}$ and $\mathbf{y}_k^{(4)}$ are ignored as they fall outside
the validation region

by a data association filter is firstly either to associate a validated observation with a track, or to discard it as arising from the clutter.

The *gate probability* is defined as the probability that the correct observation falls within the gate,

$$P_G \triangleq P\{\mathbf{y}_k \in V_k(\gamma)\}. \tag{4.67}$$

At most one observation can be attributed to the track itself. The remaining observations are associated with a *background* or *clutter model*, which implies they will be ignored for the purpose of updating the individual state estimates. In order to determine which observations are attributed to actual tracks and which to the clutter model, let us begin by defining the *association events*

$$\theta_k^{(i)} \triangleq \{y_k^{(i)} \text{ is the correct observation for track } i \text{ at time } k\}, \tag{4.68}$$

$$\theta_k^{(0)} \triangleq \{\text{all observations are clutter at time } k\}. \tag{4.69}$$

The posterior probability of each association event can be expressed as

$$\beta_k^{(i)} = P(\theta_k^{(i)}|\mathbf{Y}_{1:k}) \,\forall\, i = 0, \dots, m_k.$$

As the events $\{\theta_k^{(i)}\}_{i=0}^{m_k}$ are exhaustive and mutually exclusive, we have

$$\sum_{i=0}^{m_k} \beta_k^{(i)} = 1. \tag{4.70}$$

Conditioned on each association event $\theta_k^{(i)}$, let us define the state estimate

$$\hat{\mathbf{x}}_{k|k}^{(i)} \triangleq \mathcal{E}\{\mathbf{x}_k|\theta_k^{(i)}, \mathbf{Y}_{1:k}\} = \hat{\mathbf{x}}_{k|k-1} + \mathbf{G}_k \mathbf{s}_k^{(i)}, \tag{4.71}$$

where

$$\hat{\mathbf{x}}_{k|k-1} = \mathbf{F}_{k|k-1} \hat{\mathbf{x}}_{k-1|k-1},$$

and

$$\mathbf{s}_k^{(i)} = \mathbf{y}_k^{(i)} - \mathbf{H}_k \hat{\mathbf{x}}_{k|k-1} \tag{4.72}$$

is the innovation for observation $\mathbf{y}_k^{(i)}$. Under the null association event $\theta_k^{(0)}$, no observation is associated with the track, and the state update reduces to

$$\hat{\mathbf{x}}_{k|k}^{(0)} = \hat{\mathbf{x}}_{k|k-1}. \tag{4.73}$$

Moreover, invoking the total probability theorem, the filtered state estimate can be expressed as

$$\hat{\mathbf{x}}_{k|k} = \sum_{i=0}^{m_k} \hat{\mathbf{x}}_{k|k}^{(i)} \beta_k^{(i)}. \tag{4.74}$$

Substituting (4.72) and (4.73) into (4.74), we find that the combined update can be expressed as

$$\hat{\mathbf{x}}_{k|k} = \hat{\mathbf{x}}_{k|k-1} + \mathbf{G}_k \, \mathbf{s}_k, \tag{4.75}$$

where the *combined innovation* is

$$\mathbf{s}_k = \sum_{i=1}^{m_k} \mathbf{s}_k^{(i)} \, \beta_k^{(i)}. \tag{4.76}$$

While it may appear otherwise, (4.75) is actually highly nonlinear due to the dependence of the association events $\{\theta_k^{(i)}\}$ on the innovations.

The Riccati equation must be suitably modified to account for the additional uncertainty associated with the multiple innovations $\{\mathbf{s}_k^{(i)}\}$, as well as the possibility of the null event $\theta_k^{(0)}$. In particular, it is necessary to replace (4.33) with,

$$\mathbf{K}_k = \beta_k^{(0)} \, \mathbf{K}_{k|k-1} + (1 - \beta_k^{(0)}) \, \mathbf{K}_k^c + \tilde{\mathbf{K}}_k, \tag{4.77}$$

where

$$\mathbf{K}_k^c \triangleq (\mathbf{I} - \mathbf{G}_k \mathbf{H}_k) \mathbf{K}_{k|k-1} \tag{4.78}$$

is the covariance of the state in the absence of uncertainty as to the correct observation, and

$$\tilde{\mathbf{K}}_k \triangleq \mathbf{G}_k \left[\left(\sum_{i=1}^{m_k} \beta_k^{(i)} \, \mathbf{s}_k^{(i)} \mathbf{s}_k^{(i)T} \right) - \mathbf{s}_k \mathbf{s}_k^T \right] \mathbf{G}_k^T. \tag{4.79}$$

Comparing (4.33) and (4.78), it is clear that $\mathbf{K}_k^c$ is equivalent to the filtered state estimation-error correlation matrix from the conventional KF. A proof of (4.77–4.79) can be found in Bar-Shalom and Fortmann (1988, Appendix D.3). As it is not known which of the m_k validated measurements is correct, the term $\tilde{\mathbf{K}}_k$ in (4.77), which is positive semidefinite, increases the covariance of the updated state to reflect the uncertainty of the origin of each observation.

We next consider how the association posterior probabilities $\beta_k^{(i)}$ required in (4.74) can be evaluated. First, we express the probabilities explicitly as

$$\beta_k^{(i)} \triangleq P(\theta_k^{(i)}|\mathbf{Y}_{1:k}) = P(\theta_k^{(i)}|\mathbf{Y}_k, m_k, \mathbf{Y}_{1:k-1}) \, \forall \, i = 1, \ldots, m_k. \tag{4.80}$$

The additional conditioning on m_k in (4.80) affects nothing, as this information is already contained in $\mathbf{Y}_k$. Applying Bayes' rule enables (4.80) to be rewritten as

$$\beta_k^{(i)} = \frac{p(\mathbf{Y}_k|\theta_k^{(i)}, m_k, \mathbf{Y}_{1:k-1}) \, P(\theta_k^{(i)}|m_k, \mathbf{Y}_{1:k-1})}{p(\mathbf{Y}_k|m_k, \mathbf{Y}_{1:k-1})} \quad \forall i = 0, \dots, m_k, \tag{4.81}$$

where

$$p(\mathbf{Y}_k|m_k, \mathbf{Y}_{1:k-1}) = \sum_{j=0}^{m_k} p(\mathbf{Y}_k|\theta_k^{(j)}, m_k, \mathbf{Y}_{1:k-1}) \, P(\theta_k^{(j)}|m_k, \mathbf{Y}_{1:k-1}).$$

The pdf of the correct observation $\mathbf{y}_k^{(i)}$ is

$$p(\mathbf{y}_k^{(i)}|\theta_k^{(i)}, m_k, \mathbf{Y}_{1:k-1}) = P_G^{-1} \mathcal{N}(\mathbf{y}_k^{(i)}; \hat{\mathbf{y}}_{k|k-1}, \mathbf{S}_k) = P_G^{-1} \mathcal{N}(\mathbf{s}_k^{(i)}; \mathbf{0}, \mathbf{S}_k)$$

$$= P_G^{-1} |2\pi \mathbf{S}_k|^{1/2} \exp\left[-\frac{1}{2} \mathbf{s}_k^{(i)T} \mathbf{S}_k^{-1} \mathbf{s}_k^{(i)}\right], \tag{4.82}$$

where P_G is the gate probability (4.67); i.e., the probability that the correct observation falls within the validation gate. The term P_G^{-1} appears in (4.82) to correct for having restricted the normal density to the validation gate. The pdf in (4.81) can thus be rewritten as

$$p(\mathbf{Y}_k|\theta_k^{(i)}, m_k, \mathbf{Y}_{1:k-1}) = \begin{cases} V_k^{-m_k+1} \, P_G^{-1} \mathcal{N}(\mathbf{s}_k^{(i)}; \mathbf{0}, \mathbf{S}_k), & \forall i = 1, \dots, m_k, \\ V_k^{-m_k}, & \text{for } i = 0, \end{cases} \tag{4.83}$$

where V_k is the volume of the validation region, which we will shortly define. Let $P_F(m_k)$ denote the *probability mass function* (PMF) of the number of false measurements, and let P_D denote the track detection probability; i.e., the probability that the correct observation is detected at all. Bar-Shalom and Fortmann (1988, Appendix D.4) prove that the *a priori* probability of $\beta_k^{(i)}$ given only the number of validated observations is $P(\beta_k^{(i)}|m_k, \mathbf{Y}_{1:k-1}) = P(\beta_k^{(i)}|m_k)$ which can be calculated as

$$P(\beta_k^{(i)}|m_k) = \begin{cases} \frac{1}{m_k} P_{DG} \left[P_{DG} + (1 - P_{DG})\frac{P_F(m_k)}{P_F(m_k-1)}\right]^{-1}, & \forall i = 1, \dots, m_k \\ (1 - P_{DG})\frac{P_F(m_k)}{P_F(m_k-1)} \left[P_{DG} + (1 - P_{DG})\frac{P_F(m_k)}{P_F(m_k-1)}\right]^{-1}, & \text{for } i = 0, \end{cases}$$
$$\tag{4.84}$$

where $P_{DG} = P_D P_G$. There are two possible models for the PMF:

1. *Parametric model*
   The parametric model is a Poisson density with parameter λV_k,

$$P_F(m_k) = e^{-\lambda V_k} \frac{(\lambda V_k)^{m_k}}{m_k!} \quad \forall m_k = 0, 1, 2, \dots, \tag{4.85}$$

where λ is the average number of false observations per unit volume. As V_k is the volume of the validation region, λV_k is the number of false observations expected to be observed within the track gate.

2. *Nonparametric model*

The nonparametric model is a "diffuse" prior,

$$P_F(m_k) = \frac{1}{N} \; \forall \, m_k = 0, 1 \ldots, N-1, \tag{4.86}$$

where N may be as large as is necessary, inasmuch as this factor will cancel out of (4.84).

Substituting the parametric model (4.85) in (4.84), we find the *Poisson model*

$$P(\beta_k^{(i)}|m_k) = \begin{cases} \dfrac{P_{DG}}{P_{DG}\, m_k + (1 - P_{DG})\, \lambda\, V_k}, & \forall \, i = 1, \ldots, m_k, \\[3ex] \dfrac{(1 - P_{DG})\, \lambda\, V_k}{P_{DG}\, m_k + (1 - P_{DG})\, \lambda\, V_k}, & \text{for } i = 0. \end{cases} \tag{4.87}$$

Substituting the nonparametric or "diffuse" prior (4.86) in (4.84), on the other hand, provides

$$P(\beta_k^{(i)}|m_k) = \begin{cases} P_{DG}/m_k, & \forall \, i = 1, \ldots, m_k, \\ 1 - P_{DG}, & \text{for } i = 0. \end{cases} \tag{4.88}$$

Note that the nonparametric model (4.88) can be obtained from the Poisson model (4.87) by simply substituting $\lambda = m_k/V_k$, which is equivalent to replacing the Poisson parameter with the spatial density of validated observations.

The volume of the elliptical validation region (4.66) is

$$V_k = c_{n_y} |\gamma\, \mathbf{S}_k|^{1/2} = c_{n_y} \gamma^{n_y/2} |\mathbf{S}_k|^{1/2}, \tag{4.89}$$

where γ is the gate parameter in (4.66), n_y is the dimension of the observation $\mathbf{y}$, and c_{n_y} is the volume of the n_y-dimensional unit hypersphere.

Substituting (4.83) and (4.89) into (4.81) and manipulating provides

$$\beta_k^{(i)} = \frac{e^{(i)}}{b + \sum_{j=1}^{m_k} e^{(j)}} \; \forall \, i = 1, \ldots, m_k, \tag{4.90}$$

$$\beta_k^{(0)} = \frac{b}{b + \sum_{j=1}^{m_k} e^{(j)}}, \tag{4.91}$$

where

$$e^{(i)} \triangleq \exp\left(-\frac{1}{2} \mathbf{s}_k^{(i)T} \mathbf{S}_k^{-1} \mathbf{s}_k^{(i)} \right),$$

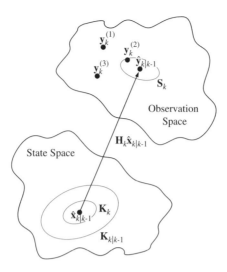

Figure 4.5   Schematic illustrating the operation of the probabilistic data association filter

and

$$b \triangleq \lambda |2\pi \mathbf{S}_k|^{1/2} (1 - P_{DG})/P_D$$

$$= (2\pi/\gamma)^{n_y/2} \lambda \, V_k \, c_{n_y} (1 - P_{DG})/P_D. \tag{4.92}$$

The nonparametric model is the same as the above except that λV_k must be replaced with m_k in (4.92).

Figure 4.5 illustrates schematically the operation of the PDAF. Comparing Figure 4.3 with Figure 4.5, it is clear that the state estimation error covariance matrix $\mathbf{K}_{k|k-1}$ is projected onto the observation space to obtain the innovations covariance matrix $\mathbf{S}_k$. Once $\hat{\mathbf{y}}_{k|k-1}$ and $\mathbf{S}_k$ are known, the multiple observations can be probabilistically associated with either the actual track or the clutter model, as in (4.81).

4.3.6   Joint Probabilistic Data Association Filter

The *joint probabilistic data association filter* (JPDAF) extends the PDAF in order to handle the case of multiple active tracks (Bar-Shalom and Fortmann 1988, sect. 9.3), in addition to multiple observations for each time instant. This capacity will prove useful for the development in Section 10.3, where the acoustic tracking of multiple simultaneous speakers is considered.

Consider the set $\mathbf{Y}_k = \{\mathbf{y}_k^{(j)}\}_{j=1}^{m_k}$ of all observations occuring at time instant k and let $\mathbf{Y}_{1:K-1} = \{\mathbf{Y}_k\}_{k=1}^{K-1}$ denote the set of all past observations. The first step in the JPDAF algorithm is the evaluation of the conditional probabilities of the *joint association events*

$$\boldsymbol{\theta} = \bigcap_{i=1}^{m_k} \theta^{(j,t_j)},$$

where the atomic events are defined as

$$\theta^{(j,t)} \triangleq \{\text{observation } j \text{ originated from track } t\} \, \forall \, j = 1, \ldots, m_k \text{ and } t = 0, 1, \ldots, T.$$

Here, t_j denotes the index of the track with which the jth observation is associated in the current joint association event. As before, $t = 0$ denotes the clutter model which generates only spurious observations.

Shortly we must derive the posterior probabilities of the joint association events. For this purpose, validation gates on the individual tracks will not be used. Rather, it will be assumed that each observation lies within the validation region of each track, which implies that $P_G = 1$.

To avoid undue computational burden, validation gates will be used for the selection of *feasible joint events*. Through the mechanism we now present, we will avoid considering those joint events with negligible probabilities. Let us define the *validation matrix* as

$$\Omega \triangleq [\omega^{(j,t)}] \, \forall \, j = 1, \ldots, m_k, \, t = 0, 1, \ldots, T,$$

with binary elements $\omega^{(j,t)}$ indicating whether measurement j lies in the validation gate of track t. Inasmuch as the track $t = 0$ corresponds to the clutter model, from which all observations might have originated, the corresponding column in Ω contains only 1's. Consider the scenario with three tracks and five observations depicted in Figure 4.6. The corresponding validation matrix is

$$\Omega = \begin{bmatrix} 1 & 0 & 1 & 0 \\ 1 & 1 & 0 & 0 \\ 1 & 1 & 1 & 1 \\ 1 & 0 & 1 & 1 \\ 1 & 1 & 0 & 1 \end{bmatrix}.$$

Considering the validation matrix, the first observation may, as stated previously, have come from the clutter model; hence $\omega^{(1,0)} = 1$. Examining Figure 4.6, it is clear that $\mathbf{y}_k^{(1)}$

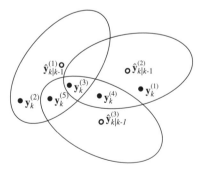

Figure 4.6 Distribution of five observations $\{\mathbf{y}_k^{(j)}\}_j$ with respect to the validation regions for three tracks centered about the predicted observations $\{\hat{\mathbf{y}}_{k|k-1}^{(t)}\}_t$.

is well outside the validation region for Track 1, which is centered around $\hat{\mathbf{y}}_{k|k-1}^{(1)}$; hence, $\omega^{(1,1)} = 0$. We observe, however, that $\mathbf{y}_k^{(1)}$ falls within the validation region of Track 2, centered at $\hat{\mathbf{y}}_{k|k-1}^{(2)}$, but outside the validation region of Track 3, centered at $\hat{\mathbf{y}}_{k|k-1}^{(3)}$; hence, $\omega^{(1,2)} = 1$ and $\omega^{(1,3)} = 0$. The validation of the other elements in Ω proceeds similarly.

The joint association event $\boldsymbol{\theta}$ can be expressed as the matrix

$$\hat{\Omega}(\boldsymbol{\theta}) = [\hat{\omega}^{(j,t)}(\boldsymbol{\theta})]$$

which comprises the components in Ω corresponding to the associations in $\boldsymbol{\theta}$, such that

$$\hat{\omega}^{(j,t)}(\boldsymbol{\theta}) \triangleq \begin{cases} 1, & \text{if } \theta^{(j,t)} \subset \boldsymbol{\theta}, \\ 0, & \text{otherwise.} \end{cases} \tag{4.93}$$

A *feasible event* is defined as an event wherein:

- an observation has exactly one source, which may be the clutter model, such that

$$\sum_{t=0}^{T} \hat{\omega}^{(j,t)}(\boldsymbol{\theta}) = 1 \; \forall \; j = 1, 2, \ldots, m_k;$$

- no more than one observation can originate from any track, such that the *track detection indicator*,

$$\delta^{(t)}(\boldsymbol{\theta}) \triangleq \sum_{j=1}^{m_k} \hat{\omega}^{(j,t)}(\boldsymbol{\theta}), \tag{4.94}$$

satisfies

$$\delta^{(t)}(\boldsymbol{\theta}) \leq 1 \; \forall \; t = 1, 2, \ldots, T.$$

Given this definition, the matrices $\hat{\Omega}$ can be constructed by scanning Ω and picking one nonzero element for each row and one nonzero element for each column, except for column $t = 0$, which corresponds to a false alarm and thus may have an unrestricted number of nonzero elements.

The track detection indicator (4.94) specifies whether track t has been detected in $\boldsymbol{\theta}$, which implies that an observation has been associated with it. Let us also define an *observation association indicator* as

$$\tau^{(j)}(\boldsymbol{\theta}) \triangleq \sum_{t=1}^{T} \hat{\omega}^{(j,t)}(\boldsymbol{\theta}) \; \forall \; j = 1, 2, \ldots, m_k \tag{4.95}$$

to indicate whether an observation j has been associated with an actual track in $\boldsymbol{\theta}$. Then the number of unassociated or *false* observations in $\boldsymbol{\theta}$ is

$$\phi(\boldsymbol{\theta}) \triangleq \sum_{j=1}^{m_k} \left[1 - \tau^{(j)}(\boldsymbol{\theta}) \right]. \tag{4.96}$$

We are now ready to derive the conditional probability of an association event $\boldsymbol{\theta}_k$ at time k. We begin by applying Bayes' rule as

$$P(\boldsymbol{\theta}|\mathbf{Y}_{1:k}) = P(\boldsymbol{\theta}_k|\mathbf{Y}_k, \mathbf{Y}_{1:k-1}) = \frac{1}{c} p(\mathbf{Y}_k|\boldsymbol{\theta}_k, \mathbf{Y}_{1:k-1}) P(\boldsymbol{\theta}_k|\mathbf{Y}_{1:k-1})$$

$$= \frac{1}{c} p(\mathbf{Y}_k|\boldsymbol{\theta}_k, \mathbf{Y}_{1:k-1}) P(\boldsymbol{\theta}_k), \tag{4.97}$$

where c is the normalization constant required to ensure that $P(\boldsymbol{\theta}|\mathbf{Y}_{1:k})$ is a valid discrete probability distribution. Note that the unneeded conditioning on $\mathbf{Y}_{1:k-1}$ in the term $P(\boldsymbol{\theta}_k|\mathbf{Y}_{1:k-1})$ has been eliminated in the last line of (4.97). The pdf on the right-hand side of (4.97) can be expressed as

$$p(\mathbf{Y}_k|\boldsymbol{\theta}_k, \mathbf{Y}_{1:k-1}) = \prod_{j=1}^{m_k} p\left(\mathbf{y}_k^{(j)}|\theta_k^{(j,t_j)}, \mathbf{Y}_{1:k-1} \right). \tag{4.98}$$

The conditional pdf of a single observation conditioned on its track of origin is then assumed to be

$$p\left(\mathbf{y}_k^{(j)}|\theta_k^{(j,t_j)}, \mathbf{Y}_{1:k-1} \right) = \begin{cases} \mathcal{N}^{(t_j)}\left(\mathbf{y}_k^{(j)} \right), & \text{if } \tau^{(j)}(\boldsymbol{\theta}_k) = 1, \\ V^{-1}, & \text{if } \tau^{(j)}(\boldsymbol{\theta}_k) = 0, \end{cases} \tag{4.99}$$

where V is the volume of the observation space. The observation associated with track t_j has the pdf

$$\mathcal{N}^{(t_j)}\left(\mathbf{y}_k^{(j)} \right) = \mathcal{N}\left(\mathbf{y}_k^{(j)}; \hat{\mathbf{y}}_k^{(t_j)}, \mathbf{S}_k^{(t_j)} \right), \tag{4.100}$$

where $\hat{\mathbf{y}}_k^{(t_j)}$ and $\mathbf{S}_k^{(t_j)}$ denote, respectively, the predicted observation and innovation covariance matrix associated with track t_j. As implied by the second condition in (4.99), observations unassociated with any track are assumed to be uniformly distributed in the entire volume V of the observation space.

Using (4.99), the pdf (4.98) can be written as

$$p(\mathbf{Y}_k|\boldsymbol{\theta}_k, \mathbf{Y}_{1:k-1}) = V^{-\phi(\boldsymbol{\theta}_k)} \prod_{j=1}^{m_k} \left[\mathcal{N}^{(t_j)}\left(\mathbf{y}_k^{(j)} \right) \right]^{\tau^{(j)}(\boldsymbol{\theta}_k)}. \tag{4.101}$$

In (4.101), V^{-1} is raised to the power $\phi(\boldsymbol{\theta})$, the total number of false observations in event $\boldsymbol{\theta}_k$, and the observation association indicator $\tau^{(j)}(\boldsymbol{\theta}_k)$ selects the pdfs of the observations associated with actual tracks in $\boldsymbol{\theta}_k$.

Next, it is necessary to calculate the prior probability $P(\boldsymbol{\theta}_k)$ of the association event $\boldsymbol{\theta}_k$, which is the last factor in (4.97). Let $\boldsymbol{\delta}(\boldsymbol{\theta}_k)$ denote the vector of track detection indicators (4.94) corresponding to $\boldsymbol{\theta}_k$, and let $\phi(\boldsymbol{\theta}_k)$ denote the total number (4.96) of false observations. Note that both $\boldsymbol{\delta}(\boldsymbol{\theta}_k)$ and $\phi(\boldsymbol{\theta}_k)$ are completely determined given $\boldsymbol{\theta}_k$. Hence, we can write

$$P(\boldsymbol{\theta}_k) = P(\boldsymbol{\theta}_k, \boldsymbol{\delta}(\boldsymbol{\theta}_k), \phi(\boldsymbol{\theta}_k)). \tag{4.102}$$

This joint probability can be rewritten as

$$P(\boldsymbol{\theta}_k) = P(\boldsymbol{\theta}_k | \boldsymbol{\delta}(\boldsymbol{\theta}_k), \phi(\boldsymbol{\theta}_k)) \, P(\boldsymbol{\delta}(\boldsymbol{\theta}_k), \phi(\boldsymbol{\theta}_k)). \tag{4.103}$$

The first term in (4.103) can be calculated beginning with the assumption that in association event $\boldsymbol{\theta}_k$, the set of tracks that are detected is equivalent to $m_k - \phi(\boldsymbol{\theta}_k)$, the number of observations actually associated. The number of measurement-to-track assignment events $\boldsymbol{\theta}_k$ in which the same set of tracks is detected, is given by the number of permutations of the m_k measurements taken $m_k - \phi(\boldsymbol{\theta}_k)$ at a time, where $m_k - \phi(\boldsymbol{\theta}_k)$ is the number of tracks to which a measurement may be assigned under the same detection event. Therefore, assuming that each such event is equally likely *a priori*, we find

$$P(\boldsymbol{\theta}_k | \boldsymbol{\delta}(\boldsymbol{\theta}_k), \phi(\boldsymbol{\theta}_k)) = \left(\frac{m_k}{m_k - \phi(\boldsymbol{\theta}_k)} \right)^{-1} = \frac{\phi(\boldsymbol{\theta}_k)! \, (m_k - \phi(\boldsymbol{\theta}_k))!}{m_k!}. \tag{4.104}$$

The last factor in (4.103) can be expressed as

$$P(\boldsymbol{\delta}(\boldsymbol{\theta}_k), \phi(\boldsymbol{\theta}_k)) = \prod_{t=1}^{T} \left(P_D^{(t)} \right)^{\delta^{(t)}(\boldsymbol{\theta}_k)} \left(1 - P_D^{(t)} \right)^{1 - \delta^{(t)}(\boldsymbol{\theta}_k)} P_F(\phi(\boldsymbol{\theta}_k)), \tag{4.105}$$

where $P_D^{(t)}$ is the detection probability of track t, and $P_F(\phi(\boldsymbol{\theta}_k))$ is the prior PMF of the number of false observations. The indicators $\delta^{(t)}(\boldsymbol{\theta}_k)$ have been used in (4.105) to select probabilities of detection and non-detection events according to the association event $\boldsymbol{\theta}_k$ under consideration.

Substituting (4.104) and (4.105) into (4.103) enables the prior probability of $\boldsymbol{\theta}_k$ to be expressed as

$$P(\boldsymbol{\theta}_k) = \frac{\phi(\boldsymbol{\theta}_k)!}{m_k!} P_F(\phi(\boldsymbol{\theta}_k)) \prod_{t=1}^{T} \left(P_D^{(t)} \right)^{\delta^{(t)}(\boldsymbol{\theta}_k)} \left(1 - P_D^{(t)} \right)^{1 - \delta^{(t)}(\boldsymbol{\theta}_k)}. \tag{4.106}$$

Then, substituting (4.101) and (4.106) into (4.97), we obtain the *a posteriori* probability
of $\boldsymbol{\theta}_k$ as

$$
P(\boldsymbol{\theta}_k|\mathbf{Y}_{1:k}) = \frac{1}{c}\frac{\phi!}{m_k!}P_F(\phi)V^{-\phi}\prod_{j=1}^{m_k}\left[\mathcal{N}^{(t_j)}(\mathbf{y}_k^{(j)})\right]^{\tau^{(j)}}\prod_{t=1}^{T}\left(P_D^{(t)}\right)^{\delta^{(t)}}\left(1-P_D^{(t)}\right)^{1-\delta^{(t)}},
$$

(4.107)

where the dependence of ϕ, $\delta^{(t)}$ and $\tau^{(j)}$ on the association event $\boldsymbol{\theta}_k$ has been suppressed
out of notational convenience.

Similar to the PDAF, the JPDAF has two versions which are distinguished by the
definition of the PMF $P_F(\phi)$ over the number of false observations. The *parametric
JPDAF* uses the Poisson PMF (4.85) in (4.107), which leads immediately to the cancellation of the terms $\phi!$ and V^ϕ. Moreover, the terms $e^{-\lambda V}$ and $m_k!$ also cancel, as they
appear in all numerator terms as well as the denominator c of (4.107). Hence, under the
Poisson prior, the joint association posterior probabilities can be expressed as

$$
P(\boldsymbol{\theta}_k|\mathbf{Y}_{1:k}) = \frac{\lambda^\phi}{c'}\prod_{j=1}^{m_k}\left[\mathcal{N}^{(t_j)}(\mathbf{y}_k^{(j)})\right]^{\tau^{(j)}}\prod_{t=1}^{T}\left(P_D^{(t)}\right)^{\delta^{(t)}}\left(1-P_D^{(t)}\right)^{1-\delta^{(t)}},
$$

(4.108)

where c' is the new normalization constant. The *nonparametric JPDAF* uses the diffuse
prior (4.86), which we repeat as

$$
P_F(\phi) = \epsilon \,\forall\, \phi.
$$

(4.109)

Substituting (4.109) into (4.107) and canceling the constants ϵ and $m_k!$ yields

$$
P(\boldsymbol{\theta}_k|\mathbf{Y}_{1:k}) = \frac{1}{c}\frac{\phi!}{V^\phi}\prod_{j=1}^{m_k}\left[\mathcal{N}^{(t_j)}\left(\mathbf{y}_k^{(j)}\right)\right]^{\tau_j}\prod_{t=1}^{T}\left(P_D^{(t)}\right)^{\delta^{(t)}}\left(1-P_D^{(t)}\right)^{1-\delta^{(t)}}.
$$

The marginal association probabilities corresponding to (4.90–4.91) are obtained from
the joint probabilities by summing over all joint events in which the marginal event of
interest occurs. Using the definition (4.93), this summation can be written as

$$
\beta^{(j,t)} \triangleq P(\theta^{(j,t)}|\mathbf{Y}_{1:k}) = \sum_{\boldsymbol{\theta}} P(\boldsymbol{\theta}|\mathbf{Y}_{1:k})\,\hat{\omega}^{(j,t)}(\boldsymbol{\theta}) \,\forall\, j = 1, \ldots, m_k \text{ and } t = 0, 1, \ldots, T.
$$

(4.110)

For any given track, it is only necessary to marginalize out the effect of all other tracks
to obtain the required *a posteriori* probabilities. Thereafter, the state update for each
track can be made separately according to (4.72–4.76). We will, however, omit these
straightforward details.

Figure 4.7 illustrates schematically the operation of the JPDAF. Comparing Figure 4.5
with Figure 4.7, it is clear that the principal extension of the JPDAF with respect to the
PDAF is the inclusion of the capacity to handle multiple tracks in addition to multiple
observations. Observe that the JPDAF maintains separate state spaces for each track, but
there is a single observation space for all tracks. Once, the probabilistic data association
has been performed at each time step, the state estimates are then updated independently.

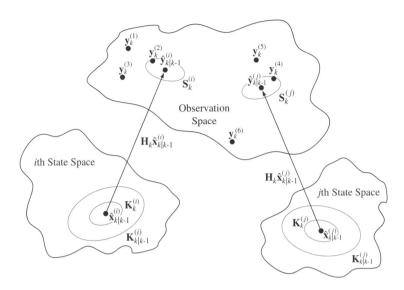

Figure 4.7   Schematic illustrating the operation of the joint probabilistic data association filter

4.4   Particle Filters

We now begin our exposition of the second class of filters following from the Bayesian formulation (4.6–4.7). Substantial development of numerical approaches based on sequential Monte Carlo methods within nonlinear state-space estimation began in the nineties even though first applications had been presented 20 years before (Handshin and Mayne 1969). The elegance of these methods lies in the combination of the powerful Monte Carlo sampling technique – which will be discussed in Section 4.4.1 – with Bayesian inference, which forms the basis of all methods discussed in this chapter. Such methods, have, in particular, been used for parameter and state estimation. For the latter it is commonly referred to as *particle filtering*. We will postpone discussing the reason behind the designation "particle" until the principal characteristics of the filter have been presented.

In Section 4.3, we were able to develop an algorithm for sequentially updating the filtering density $p(\mathbf{x}_k|\mathbf{y}_{1:k})$ by assuming all relevant pdfs appearing in (4.6–4.7) to be Gaussian, which is a very tractable, if limited, assumption. The *particle filter* (PF) is founded on a different set of methods enabling $p(\mathbf{x}_k|\mathbf{y}_{1:k})$ to be sequentially updated when the Gaussian assumption is relaxed.

4.4.1   Approximation of Probabilistic Expectations

In this section we will consider different ways to approximate pdfs $p(\mathbf{x})$ which will be used to evaluate expectations of the form

$$\mathcal{E}_{p(\mathbf{x})}\{\mathbf{f}(\mathbf{x})\} = \int \mathbf{f}(\mathbf{x})\, p(\mathbf{x})\, d\mathbf{x}, \qquad (4.111)$$

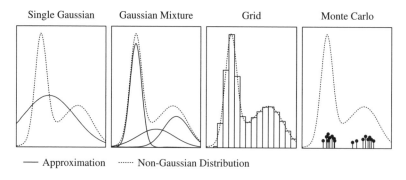

Figure 4.8   Different approximations of a non-Gaussian distribution

where $\mathbf{f}(\mathbf{x})$ is some arbitrary vector-valued function. This capacity is useful in that forming the MMSE state estimate as in (4.8) requires evaluating an integral of the form (4.111). The vector function $\mathbf{f}(\mathbf{x})$ can, however, have other forms such as that required to estimate higher order moments of $\mathbf{x}$. Eventually, we will replace $p(\mathbf{x})$ in (4.111) with the filtering density $p(\mathbf{x}_k|\mathbf{y}_{1:k})$, which is potentially highly non-Gaussian. Several different approximations of non-Gaussian pdfs are shown in Figure 4.8. We will discuss the merits and drawbacks of each.

Single Gaussian

Due to its analytic tractability, the simplest method for approximating the filtering density is with a single Gaussian pdf. The KF described in Section 4.3.1 assumes that the filtering density $p(\mathbf{x}_k|\mathbf{y}_{1:k})$ is represented by a single Gaussian pdf as in (4.42). The expectation (4.111) can then be approximated as

$$\int_{\mathbf{x}} \mathbf{f}(\mathbf{x})\, p(\mathbf{x})\, d\mathbf{x} \approx \int_{\mathbf{x}} \mathbf{f}(\mathbf{x})\, \mathcal{N}(\mathbf{x}; \boldsymbol{\mu}, \boldsymbol{\Sigma})\, d\mathbf{x}, \tag{4.112}$$

where $\boldsymbol{\mu}$ and $\boldsymbol{\Sigma}$ are, respectively, the mean vector and covariance matrix of the approximate pdf. From Figure 4.8 it is apparent, however, that a single Gaussian cannot provide a good approximation of general non-Gaussian pdfs.

Gaussian Mixture Model

A mixture model uses a weighted combination of pdfs of a particular form to model an arbitrary pdf. The *Gaussian mixture model* (GMM) is defined as the weighted sum of M Gaussian pdfs according to

$$p(\mathbf{x}) \triangleq \sum_{m=1}^{M} w_m \mathcal{N}(\mathbf{x}; \boldsymbol{\mu}_m, \boldsymbol{\Sigma}_m), \tag{4.113}$$

where the weights satisfy $0 \leq w_m \leq 1 \, \forall m = 1, \ldots, M$ and $\sum_{m=1}^{M} w_m = 1$. Through the use of a sufficient number of Gaussian components, any pdf can be approximated to arbitrary accuracy.

Substituting (4.113) into (4.111), we find

$$\int_{\mathbf{x}} \mathbf{f}(\mathbf{x}) \, p(\mathbf{x}) \, d\mathbf{x} \approx \sum_{m=1}^{M} w_m \int_{\mathbf{x}} \mathbf{f}(\mathbf{x}) \, \mathcal{N}(\mathbf{x}; \boldsymbol{\mu}_m, \boldsymbol{\Sigma}_m) \, d\mathbf{x}. \tag{4.114}$$

As will be discussed in Chapter 8, the parameters of the pdfs for both single and multidimensional Gaussians as well as the GMM in (4.113) can be readily estimated. Moreover, solving for the conditional mean in (4.8) using either (4.112) or (4.113) is straightforward. This does not imply, however, that (4.112) and (4.114) will admit closed-form solutions for any conceivable $\mathbf{f}(\mathbf{x})$. Moreover, their solution by numerical means may well prove cumbersome or intractable depending on the form of $\mathbf{f}(\mathbf{x})$, especially given that the numerical optimization may need to be conducted recursively in a high-dimensional space.

Grid-Based Approximation

Let $\mathcal{G} = \{\mathbf{x}^{(m)}, m = 1, \ldots, M\}$ denote a set of support points which are aligned to form an equidistant grid, and let V denote the volume spanned by $\mathcal{G}$. Then $\mathcal{E}_{p(x)}\{\mathbf{f}(\mathbf{x})\}$ can be approximated as

$$\int \mathbf{f}(\mathbf{x}) \, p(\mathbf{x}) \, d\mathbf{x} \approx \frac{V}{M} \sum_{m=1}^{M} \mathbf{f}(\mathbf{x}^{(m)}) \, p(\mathbf{x}^{(m)}). \tag{4.115}$$

The number of support points where $\mathbf{f}(\mathbf{x}) \, p(\mathbf{x})$ must be evaluated grows exponentially with the dimensionality of the state space. What makes grid-based methods particularly ineffective for problems of high dimension is that they must maintain many points in regions of the state space that are relatively unimportant; i.e., many points $\{\mathbf{x}^{(m)}\}$ must be maintained where $\mathbf{f}(\mathbf{x}^{(m)}) p(\mathbf{x}^{(m)})$ is effectively zero.

Monte Carlo Integration

Monte Carlo[5] *integration* or *sampling* is a stochastic numerical integration method that uses a number of randomly chosen samples to approximate an integral. A sequence $\{\mathbf{x}^{(1)}, \mathbf{x}^{(2)}, \ldots, \mathbf{x}^{(M)}\}$ of i.i.d. *random samples* or *support points* are drawn from the pdf $p(\mathbf{x})$ in order to ensure that the samples are located primarily in regions with high probability mass. Thereafter $p(\mathbf{x})$ is approximated by the *empirical density function*

$$\hat{p}(\mathbf{x}) \triangleq \frac{1}{M} \sum_{m=1}^{M} \delta(\mathbf{x} - \mathbf{x}^{(m)}). \tag{4.116}$$

---

[5] Monte Carlo estimation was named after the principality of Monaco which is famous for gambling. The name was suggested by Stanislaw Ulam.

Arbitrary expectations of the form $\mathcal{E}_{p(\mathbf{x})}\{\mathbf{f}(\mathbf{x})\}$ can then be approximated by replacing the continuous pdf $p(\mathbf{x})$ with the empirical density $\hat{p}(\mathbf{x})$, such that

$$
\mathcal{E}_{p(\mathbf{x})}\{\mathbf{f}(\mathbf{x})\} \approx \int \mathbf{f}(\mathbf{x})\,\hat{p}(\mathbf{x})\,d\mathbf{x}
$$

$$
= \int \mathbf{f}(\mathbf{x}) \cdot \frac{1}{M} \sum_{m=1}^{M} \delta(\mathbf{x} - \mathbf{x}^{(m)}) \cdot d\mathbf{x}
$$

$$
= \frac{1}{M} \sum_{m=1}^{M} \mathbf{f}(\mathbf{x}^{(m)}) = \hat{\mathbf{f}}_M. \tag{4.117}
$$

It may not seem obvious that this is a reasonable approximation. *Kolmogorov's strong law of large numbers*, however, states that $\hat{\mathbf{f}}_M$ *converges almost surely* to $\mathcal{E}_{p(\mathbf{x})}\{\mathbf{f}(\mathbf{x})\}$ as $M \to \infty$ (Robert and Casella 2004). The convergence rate is determined by the *central limit theorem*, which for scalar $\hat{f}_M$ states

$$
\sqrt{M}\left(\hat{f}_M - \mathcal{E}_{p(\mathbf{x})}\{f(\mathbf{x})\}\right) \sim \mathcal{N}(0, \sigma^2),
$$

where σ^2 denotes the variance of $\hat{f}_M$.

One crucial advantage of Monte Carlo sampling over nearly all deterministic numerical methods is that its estimation accuracy is nearly independent of the dimensionality of the state space. Hence, for high-dimensional spaces it is often the algorithm of first choice. In low-dimensional spaces other methods, such as grid-based integration, may outperform Monte Carlo sampling.

Before Monte Carlo sampling can be applied to a given problem, two fundamental questions must be answered:

- How can we draw random samples $\mathbf{x}^{(m)}$ from the probability distribution $p(\mathbf{x})$?
- How can we estimate the expectation $\mathcal{E}_{p(\mathbf{x})}\{\mathbf{f}(\mathbf{x})\}$ of a function $\mathbf{f}(\mathbf{x})$ with respect to the pdf $p(\mathbf{x})$?

These questions will be addressed in the coming sections during our brief introduction to Monte Carlo methods, and these methods will be applied to the task of speech feature enhancement and speaker tracking in Chapters 5 and 10, respectively. For a more detailed discussion of Monte Carlo methods the reader should consult Robert and Casella (2004).

Importance Sampling

The idea of *importance sampling* is to draw samples from a *proposal* or *importance* pdf $q(\mathbf{x})$ instead of the true distribution $p(\mathbf{x})$, as in practice it is often hard to obtain samples from $p(\mathbf{x})$ directly. The choice of the proposal pdf is a critical aspect in importance sampling. The name importance sampling stems from the view that the samples are drawn from regions of *importance*. Rewriting the integral (4.111) that is required to evaluate

$\mathcal{E}_{p(\mathbf{x})}\mathbf{f}(\mathbf{x})$, we obtain the *importance sampling fundamental identity* (Robert and Casella 2004), which can be expressed as,

$$\mathcal{E}_{p(\mathbf{x})}\{\mathbf{f}(\mathbf{x})\} = \int \mathbf{f}(\mathbf{x}) \, p(\mathbf{x}) \, d\mathbf{x} = \int \mathbf{f}(\mathbf{x}) \, w(\mathbf{x}) \, q(\mathbf{x}) \, d\mathbf{x} = \mathcal{E}_{q(\mathbf{x})}\{\mathbf{f}(\mathbf{x})w(\mathbf{x})\}, \qquad (4.118)$$

where

$$w(\mathbf{x}) \triangleq \frac{p(\mathbf{x})}{q(\mathbf{x})} \qquad (4.119)$$

denotes the *importance weight* or *importance ratio*. In order that the weights can be calculated as in (4.119), it is necessary that $q(\mathbf{x}) \neq 0 \, \forall \, \mathbf{x}$, where $p(\mathbf{x}) \neq 0$, and that the ratio $p(\mathbf{x})/q(\mathbf{x})$ is otherwise well defined.

By drawing samples from $q(\mathbf{x})$ we obtain the empirical density $\hat{q}(\mathbf{x})$ and hence can approximate $\mathcal{E}_{q(\mathbf{x})}\{\mathbf{f}(\mathbf{x})w(\mathbf{x})\}$ by Monte Carlo integration. Thus, $\mathcal{E}_{p(\mathbf{x})}\{\mathbf{f}(\mathbf{x})\}$ can be approximated as the weighted summation

$$\mathcal{E}_{p(\mathbf{x})}\{\mathbf{f}(\mathbf{x})\} \approx \frac{1}{M} \sum_{m=1}^{M} w\left(\mathbf{x}^{(m)}\right) \mathbf{f}\left(\mathbf{x}^{(m)}\right), \qquad (4.120)$$

where the samples $\left\{\mathbf{x}^{(m)}\right\}$ are drawn from $q(\mathbf{x})$. If the normalization factor of $p(\mathbf{x})$ is unknown, the importance weights can only be evaluated up to a normalization constant. To ensure that $\sum_{m=1}^{M} w\left(\mathbf{x}^{(m)}\right) \equiv 1$, it is then necessary to calculate the *normalized importance weights*,

$$\tilde{w}\left(\mathbf{x}^{(m)}\right) = \frac{w\left(\mathbf{x}^{(m)}\right)}{\sum_{n=1}^{M} w\left(\mathbf{x}^{(n)}\right)} \, \forall \, m = 1, \ldots, M. \qquad (4.121)$$

This normalization will be more rigorously justified in Section 4.4.2. Note that importance sampling as given in (4.120) is biased but consistent, where the latter implies that the bias vanishes as $M \to \infty$. The approximation in (4.120) is equivalent to replacing $p(\mathbf{x})$ with the *weighted empirical density*

$$\hat{p}(\mathbf{x}) \triangleq \sum_{m=1}^{M} \tilde{w}\left(\mathbf{x}^{(m)}\right) \delta\left(\mathbf{x} - \mathbf{x}^{(m)}\right), \qquad (4.122)$$

where the normalized weights $\tilde{w}\left(\mathbf{x}^{(m)}\right)$ are given by (4.121).

4.4.2 Sequential Monte Carlo Methods

Evaluating (4.111) using either a single Gaussian pdf or GMM to approximate $p(\mathbf{x})$ represents a potentially intractable integration. As mentioned above, the basic idea of Monte Carlo estimation is to avoid such an intractable operation replacing $p(\mathbf{x})$ with a discrete approximation $\hat{p}(\mathbf{x})$ as in (4.116).

A *particle* is the name given to a sample point of a pdf and its associated weight. The more likely is a given region of the state space, the higher the density of particles or the higher their individual weights, or both, as indicated in Figure 4.8 (Monte Carlo). The pdf can then be approximated through such a distribution of sample points and their associated weights. In sequential Monte Carlo methods, the distribution of particles evolves over time as the *a posteriori* probabilities are updated by new observations, the weights are recalculated, and the particles recursively propagated according to Bayes' rule.

To apply Monte Carlo integration and importance sampling to the problem of sequentially estimating the filtering density discussed in Section 4.1, it is necessary to replace the current estimate $p(\mathbf{x}_{k-1}|\mathbf{y}_{1:k-1})$ with its empirical counterpart. This is achieved by drawing samples from $p(\mathbf{x}_k|\mathbf{y}_{1:k-1})$, which then allow $p(\mathbf{x}_k, \mathbf{y}_k|\mathbf{y}_{1:k-1})$ and thereafter $p(\mathbf{y}_k|\mathbf{y}_{1:k-1})$ and $p(\mathbf{x}_k|\mathbf{y}_{1:k})$ in (4.7) to be *sequentially* updated with Monte Carlo integration. Hence, a possibly difficult or intractable numerical integration will be replaced with a relatively straightforward Monte Carlo integration.

Sequential Importance Sampling

In this section, we develop the means to sequentially update the filtering density $p(\mathbf{x}_k|\mathbf{y}_{1:k})$. In order to obtain an algorithm with tractable complexity, we will adopt an approach based on the importance sampling described in Section 4.4.1. To begin, let us assume that we have a weighted empirical density $\hat{p}(\mathbf{x}_{k-1}|\mathbf{y}_{1:k-1})$ of the form (4.116). Then we can estimate $p(\mathbf{x}_k, \mathbf{y}_k|\mathbf{y}_{1:k-1})$ through Monte Carlo integration according to

$$p(\mathbf{x}_k, \mathbf{y}_k|\mathbf{y}_{1:k-1}) = p(\mathbf{y}_k|\mathbf{x}_k) \int p(\mathbf{x}_k|\mathbf{x}_{k-1}) \cdot \hat{p}(\mathbf{x}_{k-1}|\mathbf{y}_{1:k-1}) \, d\mathbf{x}_{k-1}$$

$$= p(\mathbf{y}_k|\mathbf{x}_k) \int p(\mathbf{x}_k|\mathbf{x}_{k-1}) \cdot \frac{1}{M} \sum_{m=1}^{M} \delta\left(\mathbf{x}_{k-1} - \mathbf{x}_{k-1}^{(m)}\right) \cdot d\mathbf{x}_{k-1}$$

$$= \frac{1}{M} \sum_{m=1}^{M} p(\mathbf{y}_k|\mathbf{x}_k) \int p(\mathbf{x}_k|\mathbf{x}_{k-1}) \cdot \delta\left(\mathbf{x}_{k-1} - \mathbf{x}_{k-1}^{(m)}\right) \cdot d\mathbf{x}_{k-1}$$

$$= \frac{1}{M} \sum_{m=1}^{M} p(\mathbf{y}_k|\mathbf{x}_k) \cdot p\left(\mathbf{x}_k|\mathbf{x}_{k-1}^{(m)}\right). \tag{4.123}$$

Clearly (4.123) would be equivalent to the prediction step (4.6) if $\mathbf{y}_k$ were marginalized out of the likelihood.

We will, however, proceed somewhat differently here. That is, let us replace the prediction and correction steps in the "standard" formulation of the Bayesian filter in (4.6) and (4.7) by firstly approximating $p(\mathbf{x}_k, \mathbf{y}_k|\mathbf{y}_{1:k-1})$ with the weighted empirical density

$$\hat{p}(\mathbf{x}_k, \mathbf{y}_k|\mathbf{y}_{1:k-1}) = \frac{1}{M} \sum_{m=1}^{M} p\left(\mathbf{y}_k|\mathbf{x}_k^{(m)}\right) \delta\left(\mathbf{x}_k - \mathbf{x}_k^{(m)}\right), \tag{4.124}$$

where the sample points $\mathbf{x}_k^{(m)} \, \forall m$ are obtained by drawing samples from the importance density $q(\mathbf{x}_k | \mathbf{x}_{k-1}^{(m)}, \mathbf{y}_k)$. As the optimal importance density

$$q_{\text{opt}}(\mathbf{x}_k | \mathbf{x}_{k-1}^{(m)}, \mathbf{y}_k) = p(\mathbf{x}_k | \mathbf{x}_{k-1}^{(m)}, \mathbf{y}_k) = \frac{p(\mathbf{y}_k | \mathbf{x}_k, \mathbf{x}_{k-1}^{(m)}) p(\mathbf{x}_k | \mathbf{x}_{k-1}^{(m)})}{p(\mathbf{y}_k | \mathbf{x}_{k-1}^{(m)})}$$

is often difficult to obtain, a suboptimal choice of the importance density is frequently used. The most popular suboptimal choice is the transitional prior

$$q_{\text{sub}}(\mathbf{x}_k | \mathbf{x}_{k-1}^{(m)}, \mathbf{y}_k) = p(\mathbf{x}_k | \mathbf{x}_{k-1}^{(m)}).$$

Secondly, we must calculate the normalization constant $p(\mathbf{y}_k | \mathbf{y}_{1:k-1})$ required to form the new state estimate according to

$$p(\mathbf{x}_k | \mathbf{y}_{1:k}) = \frac{p(\mathbf{x}_k, \mathbf{y}_k | \mathbf{y}_{1:k-1})}{p(\mathbf{y}_k | \mathbf{y}_{1:k-1})}. \tag{4.125}$$

This is readily achieved by replacing the marginalization

$$p(\mathbf{y}_k | \mathbf{y}_{1:k-1}) = \int p(\mathbf{x}_k, \mathbf{y}_k | \mathbf{y}_{1:k-1}) \, d\mathbf{x}_k$$

with a Monte Carlo integration of the form

$$p(\mathbf{y}_k | \mathbf{y}_{1:k-1}) \approx \int \hat{p}(\mathbf{x}_k, \mathbf{y}_k | \mathbf{y}_{1:k-1}) \, d\mathbf{x}_k$$

$$= \int \frac{1}{M} \sum_{m=1}^{M} p\left(\mathbf{y}_k | \mathbf{x}_k^{(m)}\right) \delta\left(\mathbf{x}_k - \mathbf{x}_k^{(m)}\right) d\mathbf{x}_k$$

$$= \frac{1}{M} \sum_{m=1}^{M} p\left(\mathbf{y}_k | \mathbf{x}_k^{(m)}\right). \tag{4.126}$$

Substituting (4.124) and (4.126) into (4.125) and setting

$$w_k^{(m)} \triangleq w(\mathbf{x}_k^{(m)}) = p\left(\mathbf{y}_k | \mathbf{x}_k^{(m)}\right)$$

then provides the discrete density

$$\hat{p}(\mathbf{x}_k | \mathbf{y}_{1:k}) = \frac{1}{M} \sum_{m=1}^{M} \tilde{w}_k^{(m)} \delta\left(\mathbf{x}_k - \mathbf{x}_k^{(m)}\right), \tag{4.127}$$

where the normalized weights are given by

$$\tilde{w}_k^{(m)} = \frac{w_k^{(m)}}{\sum_{n=1}^{M} w_k^{(n)}}. \tag{4.128}$$

Hence, applying the normalization constant $p(\mathbf{y}_k|\mathbf{y}_{1:k-1})$ in (4.125) is equivalent to normalizing the weights of the new weight empirical density as in (4.128).

Once $\hat{p}(\mathbf{x}_k|\mathbf{y}_{1:k})$ has been evaluated, the MMSE estimate which, as mentioned in Section 4.1, is equivalent to the conditional mean, can be readily determined from

$$\hat{\mathbf{x}}_k = \int \mathbf{x}_k \, \hat{p}(\mathbf{x}_k|\mathbf{y}_{1:k}) \, d\mathbf{x}_k = \sum_{m=1}^{M} \tilde{w}_k^{(m)} \mathbf{x}_k^{(m)}. \tag{4.129}$$

With this formulation, the PF fits exactly into the prediction–correction form for Bayesian filters in (4.41) and (4.42). The prediction step (4.41) corresponds to drawing new samples from $p\left(\mathbf{x}_k|\mathbf{x}_{k-1}^{(m)}\right)$, and the correction step corresponds to calculating $p(\mathbf{x}_k|\mathbf{y}_k)$ according to (4.127) and (4.128). There remain, however, several other issues that must be resolved in order for the PF to function effectively.

Degeneracy and Effective Sample Size

Degeneracy, a well-known problem in sequential Monte Carlo methods, implies that the vast majority of the probability mass is concentrated on one or, at most, a few particles. Indeed, it can be shown theoretically that the variance of the particle weights can only increase with time; hence, degeneracy is unavoidable (Doucet *et al.* 2000). The detrimental effect of degeneracy is that a great deal of computation is expended updating particles whose contribution to the approximation of the filtering density $p(\mathbf{x}_k|\mathbf{y}_{1:k})$ is effectively zero. A suitable measure of degeneracy is the *effective sample size* (Kong *et al.* 1994), defined as,

$$M_{\mathrm{eff}} \triangleq \frac{1}{\sum_{m=1}^{M} \left(\tilde{w}_k^{(m)}\right)^2},$$

where $\{\tilde{w}_k^{(m)}\}$ are the normalized weights given by (4.121). The effective sample size must lie between 1 and M, inclusive. The limiting cases are:

- If only a single sample has nonzero weight, such that $\tilde{w}_k^{(m)} = 1$ for some m and $\tilde{w}_k^{(n)} = 0 \,\forall\, n \neq m$, then $M_{\mathrm{eff}} = 1$.
- If all samples are equally weighted such that $\tilde{w}_k^{(m)} = 1/M \,\forall\, m = 1, 2, \ldots, M$, then

$$M_{\mathrm{eff}} = \frac{1}{M \cdot \left(\frac{1}{M}\right)^2} = M.$$

Hence, M_{eff} is an effective indicator for determining whether or not the probability mass has "collapsed" into a few particles. Whenever M_{eff} falls below a predefined threshold, a *resampling* operation should be performed, as we now describe.

Sequential Importance Resampling

The basic idea of importance resampling is to maintain as many particles as possible in regions with high probability. This is accomplished by the replication or selection of samples with high importance weights and the elimination of samples with low importance weights. Resampling is usually applied between two importance sampling steps, either at every step or only if found to be necessary based on the calculation of M_{eff}.

Resampling is, by definition, a mapping of the random measure $\{\mathbf{x}_k^{(m)}, w_k^{(m)}\}$ into the random measure $\{\tilde{\mathbf{x}}_k^{(m)}, 1/M\}$; i.e., the initial set of weighted particles is replaced with a new subset of the initial particles, all of which have the same weight. This is achieved by sampling with replacement from the approximate discrete representation of $p(\mathbf{x}_k|\mathbf{y}_{1:k})$ given by

$$p(\mathbf{x}_k|\mathbf{y}_{1:k}) \approx \sum_{m=1}^{M} w_k^{(m)} \delta\left(\mathbf{x}_k - \mathbf{x}_k^{(m)}\right).$$

The process of resampling can be divided into two stages, where, for convenience, we assume that the weights have already been normalized:

1. *Replication factor calculation*
   The number $N_k^{(m)}$ of *children* or *replication factor* is determined for each particle.
2. *Resampling*
   The particles are split, based on the replication factors $\left\{N_k^{(m)}\right\}$ calculated in the prior step.

Finding the replication factors is conceptually simple: Firstly, the *cumulative sum of weights* is calculated according to

$$W_k^{(m)} = \sum_{n=1}^{m-1} w_k^{(n)}, \tag{4.130}$$

where, by assumption, $W_k^{(M)} = 1$. Secondly, M random samples $u_n \in [0,1] \,\forall\, n = 1, \ldots, M$ are generated, and for each u_n that particle $\mathbf{x}_k^{(n)*} = \mathbf{x}_k^{(m)}$ is chosen such that $u_n \in \left[W_k^{(m)}, W_k^{(m+1)}\right]$. This results in a new set of particles in which some particles may appear multiple times, but all particles have uniform weight. A particle filter with resampling is summarized in Algorithm 4.2. This algorithm is known as *multinomial resampling*, because the replication factor $N_k^{(m)}$ for each particle is effectively drawn from a multinomial distribution with probabilities $\{w_k^{(m)}\}$ (Douc and Cappe 2005). Note that resampling should be performed only after updating the filtering density, as resampling induces some additional random variation in the current particle set.

As the particles with high weights are chosen many times in the basic multinomial resampling algorithm, a loss in diversity among the particles can occur. Moreover, in the case of small processing noise, all particles can easily collapse to a single point. To restore the diversity of the samples after resampling, Berzuini and Gilks (2001) proposed the *resample-move* algorithm.

Algorithm 4.2   Particle filter with resampling

1. Draw the samples $\mathbf{x}_k^{(m)} \propto p\left(\mathbf{x}_k|\mathbf{x}_{k-1}^{(m)}\right) \forall m$.

2. Calculate the importance weights $w_k^{(m)} \propto p\left(\mathbf{y}_k|\mathbf{x}_k^{(m)}\right) \forall m$.

3. Normalize the importance weights

$$\tilde{w}_k^{(m)} = w_k^{(m)} / \sum_{n=1}^{N} w_k^{(n)} \, \forall \, m = 1, \ldots, M.$$

4. Calculate the new filtering density $p(\mathbf{x}_k|\mathbf{y}_{1:k}) \approx \sum_{m=1}^{M} \tilde{w}_k^{(m)} \delta\left(\mathbf{x}_k - \mathbf{x}_k^{(m)}\right)$

5. Perform resampling by sampling $\mathbf{x}_k^{(m)}$ from $p(\mathbf{x}_k|\mathbf{y}_{1:k}) \forall m$ to obtain a new set of particles $\{\tilde{\mathbf{x}}^{(m)}\}$ with uniform weights.

Semi-deterministic Resampling

Unfortunately, real random resampling, as described in the previous section, has a high variability. Therefore, a large number of samples are required in order to obtain a reliable approximation of the filtering density. Naive algorithms for resampling have a complexity of $\mathcal{O}(M^2)$. This complexity can be reduced to $\mathcal{O}(M \log M)$, however, by applying a *binary search* to the cumulative density $\left\{W_k^{(m)}\right\}$ in (4.130) to find the particle $\mathbf{x}_k^{(n)*} = \mathbf{x}_k^{(m)}$ such that $u_n \in \left[W_k^{(m)}, W_k^{(m+1)}\right]$. But even with this reduction in computation, the resampling algorithm is still relatively inefficient.

   Kitagawa (1996) argued that it is unnecessary to perform random resampling, as the purpose of resampling is solely to obtain a uniformly-weighted empirical density that mimics the filtering pdf. Based on this argument, Kitagawa proposed two novel semi-deterministic resampling algorithms, which later become known as *systematic resampling* (SR) and *stratified resampling*. The steps in the systematic resampling algorithm are illustrated in Listing 4.1. As is clear from the pseudocode, the drawing of M random variables $\{u_n\}$ from $u_n \in [0, 1]$, which was performed in the multinomial resampling algorithm, is replaced by the drawing of a single $u \in [0, 1/M]$ in Line 02, where M is the total number of particles. Thereafter, the replication factor $N_k^{(m)}$ of every particle is determined by the while loop in Lines 07 through 09. Incrementing u in Line 09 has the effect of replacing the selection of $u_n \in [0, 1]$ in the multinomial resampling algorithm with the choice

$$u_n \leftarrow \begin{cases} u, & \text{for } n = 1, \\ u + (n-1)S, & \text{otherwise}, \end{cases}$$

where the step size $S = 1/M$ is set in Line 01. Hence, only a single random variable need be drawn in order to determine the replication factors $\left\{N_k^{(m)}\right\}$ for all particles $\left\{\mathbf{x}_k^{(m)}\right\}$.

Listing 4.1 Systematic resampling algorithm for calculating the replication factor $N_k^{(m)}$ for each particle $\mathbf{x}_k^{(m)}$

```
00   def systematicResample():
01       S ← 1/M
02       draw u ∈ U[0, S]
03       W ← 0
04       for  m = 1, . . . , M :
05           W ← W + w_k^(m)
06           N_k^(m) ← 0
07           while u ≤ W :
08               N_k^(m) ← N_k^(m) + 1
09               u ← u + S
```

Listing 4.2 Residual systematic resampling algorithm for calculating the replication factor $N_k^{(m)}$ for each particle $\mathbf{x}_k^{(m)}$

```
00   def residualSystematicResample():
01       S ← 1/M
02       draw u ∈ U[0, S]
03       for  m = 1, . . . , M :
04           N_k^(m) ← ⌊(w_k^(m) − u)M⌋ + 1
05           u ← u + N_k^(m) S − w_k^(m)
```

As shown in Listing 4.2, a more efficient implementation of systematic resampling, which is known as *residual systematic resampling* (RSR), was proposed by Bolic *et al.* (2003). In the listing, the notation $\lfloor \bullet \rfloor$ indicates the largest integer less than the argument. As is apparent upon comparing Listings 4.1 and 4.2, the difference between the SR and RSR algorithms lies in the fact that u is calculated with respect to the origin of the cumulative sum of weights in SR. In RSR, on the other hand, u is updated with respect to the origin of the weight currently under consideration. Hence, it is necessary to subtract the weight $w_k^{(m)}$ when updating u in Line 06.

As noted by Bolic *et al.*, the RSR algorithm provides the following advantages with respect to the SR procedure:

1. The RSR algorithm contains only a single loop.
2. The complexity of the RSR algorithm is $\mathcal{O}(M)$.
3. The RSR algorithm is suitable for pipeline implementations, as it contains no conditional branches.

For these reasons the RSR algorithm is a popular choice for the practical implementation of particle filters.

Schematics of the Particle Filter

A schematic of the operation of the particle filter is shown in Figure 4.9. The operation of the particle filter can be summarized as follows: At the top of the figure, we begin with a

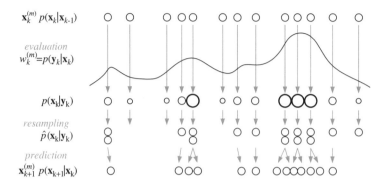

Figure 4.9 Illustration of a particle filter with importance sampling and resampling. The size of the circles represent the weights of the samples

uniformly weighted set of particles $\{w_k^{(m)}, 1/M\}$. The particle locations are then combined with the new observation $\mathbf{y}_k$ to give each particle its correct weight, resulting in a random measure $\{\mathbf{x}_k^{(m)}, w_k^{(m)}\}$ approximating $p(\mathbf{x}_k|\mathbf{y}_{1:k})$. Then resampling as described above is conducted resulting in a uniformly weighted measure $\{\tilde{\mathbf{x}}_k^{(m)}, 1/M\}$ that still approximates $p(\mathbf{x}_k|\mathbf{y}_{1:k})$. Finally, new particles $\{\mathbf{x}_{k+1}^{(m)}, 1/M\}$ are generated for the next step.

4.5 Summary and Further Reading

In this chapter we have described general techniques for Bayesian parameter estimation, which is a very flexible approach to estimating the state of a system, which is not directly observable, based on a sequence of observations. After formulating the general problem, we considered three classes of specific filters: the Wiener filter, then the KF and its variants, and finally the class of particle filters that have only recently appeared in the literature. As we will learn in subsequent chapters, all three classes of filters are very useful for accomplishing several tasks related to distant speech recognition.

We have presented a brief summary about Bayesian filters. More extensive treatments can be found in the literature, a few of which we summarize here. Good introductions to the properties and characteristics of Wiener filters are available in Hänsler and Schmidt (2004, sect. 5) and Haykin (2002, sect. 2). An extremely accessible presentation of the basic properties of the KF as well as a discussion of robust implementations of square-root adaptive filters can be found in Haykin (2002, sect. 10). Further details about KFs along with an extensive set of exercises and computer simulations can be found in Grewal and Andrews (1993). Simon (2006) provides an extensive treatment of the Kalman filter and many variations, as well as an extensive bibliography. That work also presents many of the historically significant milestones in the development of Kalman filter. Bar-Shalom and Fortmann (1988) describes all relevant details of the PDAF and JPDAF, including their use for target tracking applications.

A very interesting and readable presentation of the Bayesian tracking framework, as well as a tutorial on particle filters, is given by Ristic *et al.* (2004). Robert and Casella (2004) presents the theoretical underpinnings of particle filters.

Simon (2006 sect. 11–12) presents the theory of the H_∞ filter, which is more robust than the conventional KF when the characteristics of the physical plant are not precisely known or modeled. Simon (2006, sect. 14) also discusses another relatively new formulation of the Bayesian filter, namely the *unscented KF*. The latter uses a series of test points to deduce the characteristics of the system nonlinearity, and can provide better convergence in the presence of extreme nonlinearities in the system model. Another alternative to the conventional particle filter based on *Fourier densities* has been proposed in Brunn *et al.* (2006a, 2006b). Similary, an approach based on Dirac densities was proposed by Schrempf *et al.* (2006).

4.6 Principal Symbols

Symbol	Description
ϵ_k	filter state estimation error at time k
$\epsilon_{k\|k-1}$	predicted state estimation error at time k
γ	gating parameter
$\mathbf{f}_k$	transition function at time k
$\mathbf{F}_{k\|k-1}$	transition matrix from time $k-1$ to k
$\mathcal{G}$	set of support points
$\mathbf{G}_k$	Kalman gain at time k
$\mathbf{h}_k$	observation function at time k
$\mathbf{H}_k$	observation matrix
$\mathbf{H}_k(\mathbf{x})$	nonlinear observation functional
$\overline{\mathbf{H}}_k(\mathbf{x}_{k\|k-1})$	linearized observation functional
i	iteration index
m	samples
M	number of samples
$p(\mathbf{x})$	prior distribution (*a priori* knowledge before the observation)
$p(\mathbf{x}_k\|\mathbf{y}_{1:k})$	filtering density
$p(\mathbf{x}_{k+1}\|\mathbf{x}_k)$	(state) transition probability, evolution
$p(\mathbf{y}_k\|\mathbf{x}_k)$	output probability, likelihood function
$q(\mathbf{x})$	proposal (or importance) distribution
S	step width
$\mathbf{s}_k$	innovation at time k
$\mathbf{S}_k$	innovation covariance matrix at time k
t	track index
T	number of tracks
$\mathbf{U}_k$	transition noise covariance matrix at time k
V	volume

Symbol	Description
$\mathbf{V}_k$	observation noise covariance matrix at time k
$\mathbf{u}_k$	process noise at time k
$\mathbf{v}_k$	observation noise at time k
w	weight
$\tilde{w}$	normalized weight
W	cumulative weight
$\mathbf{x}_k$	state at time k
$\mathbf{x}_{0:k}$	state sequence from time 0 to k
$\mathbf{y}_k$	observation at time k
$\mathbf{y}_{1:k-1}$	observation sequence from time 1 to $k-1$

5

Speech Feature Extraction

Acoustic modeling requires that the speech waveform $s(t)$ is processed in such a way that it produces a sequence of feature vectors of a relative small number of dimensions. This reduction is necessary to not waste resources of a model to represent irrelevant portions of the space. The transformation of the input data into a set of dimension-reduced features is called *speech feature extraction*, *acoustic preprocessing* or *front-end processing*. The set of transforms must be carefully chosen such that the resulting features will contain only relevant information to perform the desired task. Feature extraction as applied in *automatic speech recognition* (ASR) systems aims to preserve the information needed to determine the phonetic class while being invariant to other factors including speaker differences such as accent, emotions or speaking rate or other distortion such as background noise, channel distortion or reverberation. This step is critical, because if useful information is lost in the feature extraction step it cannot be recovered in later processing.

Over the years many different speech feature extraction methods have been proposed. The variety of methods are distinguished by the extent to which they incorporate information about the human auditory processing and perception, robustness to distortions and length of the observation window.

Since the 1940s short-time frequency analysis (Koenig *et al.* 1946) has been used to carry out speech analysis and became the fundamental approach underlying any speech processing front-end. The nonlinear frequency resolution of the ear is implemented into the front-end by a nonlinear scaling prior to spectral analysis, by the bilinear transform, or, possibly, by nonlinear scaled filter banks. The application of the cepstrum marks a milestone in speech feature extraction. Already introduced to speech processing by Noll (1964), it took more than a decade to become widely accepted in speech recognition and adopted by the two most widely used front-ends, namely, *mel frequency cepstral coefficients* (MFCC) (Davis and Mermelstein 1980), and *perceptual linear prediction* (PLP) (Hermansky 1990). After the cepstral transform both front-ends are traditionally augmented by either *dynamic features*, which were introduced into speech feature extraction by Furui (1986), or a stacking of neighboring frames. The dimension of the augmented features might be reduced by linear discriminant analysis (Häb-Umbach and Ney 1992) or neural networks.

Distant Speech Recognition Matthias Wölfel and John McDonough
© 2009 John Wiley & Sons, Ltd

5.1 Short-Time Spectral Analysis

Speech is a *quasi-stationary* signal, which implies that the vocal tract shape, and thus its transfer function, remain nearly unchanged over time intervals of 5 to 25 ms duration. ASR systems, and especially the front-ends of such systems, typically assume that a signal is stationary for the duration of an analysis window. Hence, it is necessary to split an utterance that is to be recognized into short segments. Section 5.1.1 discusses methods used to accomplish this. Section 5.1.2 introduces the spectrogram, which is a means of displaying the spectral content of speech in the time–frequency axis.

5.1.1 Speech Windowing and Segmentation

Selecting the length of an analysis segment for ASR involves tradeoff between conflicting requirements. On the one hand, the segment must be short enough to provide the required time resolution. On the other, it must be long enough to ensure adequate frequency resolution in the *power spectrum*, which describes a signal's power as a function of frequency. In addition, during voiced speech the segment must be long enough to be insensitive to its exact position relative to the glottal cycle. The advantage of a long observation segment is that it smooths out some of the temporal variations of unvoiced speech. The disadvantage is that it blurs rapid events, such as the release of stop consonants.

The choice of frame shift and size is dependent on the velocity of the articulators, which determines how quickly the vocal tract changes shape. Some speech sounds, such as stop consonants or diphthongs, have sharp spectral transitions with a spectral peak shift of up to 80 Hz/ms (Markel and Gray 1980). It is common to adjust the frame shift and analysis window size together; as a shorter frame shift can track more rapid variations of the shape of the vocal tract, the analysis window size should also be shortened to achieve better localization in time of short-lived movements of the articulators.

To apply the segmentation, the entire speech signal must be windowed. This implies it is multiplied componentwise with an analysis window of a duration of between 16 to 32 ms, which is known as the *frame size*. For every new frame, the window is shifted 5 to 15 ms, which is known as the *frame shift*. Choosing the right window shape is very important, as this shape determines the properties – in particular, frequency resolution – of the speech segment in the frequency domain. This is clear from (3.24–3.25) and the related discussion; i.e., the windowing theorem states that the Fourier transform of the time window is convolved with the short-term spectrum of the actual signal. This means that true spectral characteristics of the signal will be "smeared" with the Fourier transform of the window.

The simplest analysis window of length $N_w + 1$ is the *rectangular window*, shown in Figure 5.1, which can be expressed as

$$w[n] = \begin{cases} 1, & \forall \ 0 \le n \le N_w, \\ 0, & \text{otherwise.} \end{cases} \qquad (5.1)$$

Because of the abrupt discontinuities at its edges, the rectangular window introduces large *sidelobes* in the frequency domain, as shown in Figure 5.2. As mentioned above, these large sidelobes lead to a smearing of spectral energy. Through this smearing, energy at a given frequency appears to leak into adjacent frequency regions. Hence, this effect is also known as *spectral leakage*. The amount of spectral leakage is directly related to the size of the sidelobes in the frequency domain.

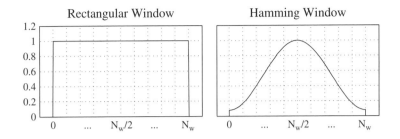

Figure 5.1 Rectangular and Hamming window

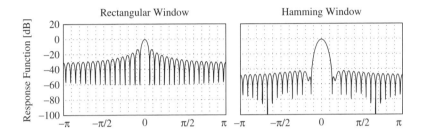

Figure 5.2 Fourier transforms of the rectangular and Hamming window sequences

To reduce the smearing effect, the height and width of the sidelobes of the windowing function in the frequency domain must be as small as possible. Windowing functions without abrupt discontinuities at their edges in the time domain are known to have smaller sidelobes. Many types of windowing functions of this kind have been proposed in the literature, including the Hann, Blackman, Kaiser, and Bartlett windows. A detailed description of the different windowing functions can be found, for example, in Oppenheim and Schafer (1989, sect. 7.4.1). In speech recognition, the *Hamming window*,

$$w[n] = \begin{cases} 0.54 - 0.46\cos\left(\frac{2\pi n}{N_w}\right), & \forall\, 0 \leq n \leq N_w, \\ 0, & \text{otherwise,} \end{cases} \tag{5.2}$$

is used almost exclusively. The Hamming window is illustrated on the right side of Figure 5.1. From frequency responses of the rectangular and Hamming windows shown in Figure 5.2, it is clear that although the rectangular window has a narrower main lobe, its sidelobes are much higher than those of the Hamming window. In fact, the first sidelobe of the rectangular window is only 13 dB below the main lobe. Hence, significant spectral leakage is to be expected when the rectangular window is used.

5.1.2 The Spectrogram

The *spectrogram* is a graphical representation of the energy density as a function of angular frequency ω and discrete time frame k,

$$\text{spectrogram}_k(e^{j\omega}) \triangleq \left| X[k, e^{j\omega}] \right|^2$$

where $X[k, e^{j\omega})$ is the time-dependent Fourier transform defined in (3.73). A spectrogram is typically displayed in gray scale, such that the higher the energy at a specific frequency and a given time, the darker this region appears in the time–frequency plane. Hence, spectral peaks are shown in black, while spectral valleys are shown in white. Values in between have a gray shade. Due to the large dynamic range of human speech, spectrograms are alternatively displayed in a logarithmic scale

$$\text{logarithmic spectrogram}_k(e^{j\omega}) = 20 \log_{10} \left| X[k, e^{j\omega}) \right|.$$

Depending on the window size used, we differentiate between:

- The *wide-band* spectrogram – In this case a short duration window of less than a pitch period, typically 10 ms, is used. This provides good time resolution, but smears the harmonic structure, thereby yielding spectra similar to those of spectral envelopes.
- The *narrow-band* spectrogram – In this case a long duration window of at least the length of two pitch periods is used. The narrow-band spectrogram provides good frequency resolution but poor time resolution. Due to the increased frequency resolution, the harmonics of f_0 can be observed as horizontal striations during segments of voiced speech.

Figure 5.3 shows plots of wide-band and narrow-band spectrograms. It also presents spectrograms with additive and reverberant distortions. While the additive noise fills up the regions of the time–frequency plane with low speech energy, reverberation smears the spectral energy along the time axis.

A spectrogram is sometimes also referred to as a *sonogram* or *voiceprint*. Spectrograms of speech signals are often used to analyze phonemes and their transitions. As mentioned above, spectrograms are based on the time-dependent or short-time Fourier transform, which serves as an intermediate step in nearly all current speech feature extraction techniques. Note that any of the spectral analysis techniques presented in Section 5.3 can be used to calculate the spectrogram.

5.2 Perceptually Motivated Representation

Experience has proven that feature extraction techniques based on characteristics of the human auditory system are likely to provide ASR performance that is superior to naive or ad hoc techniques. This stems from the fact that the human auditory system evolved concurrently over millions of years with the human speech production apparatus, and hence is highly "tuned" to the perception and recognition of human speech. This section describes front-end implementations motivated by one or more aspects of the human auditory system.

5.2.1 Spectral Shaping

To model the sensitivity of the human ear, some feature extraction schemes apply a finite impulse response filter with a single coefficient. This is known as a *pre-emphasis filter*,

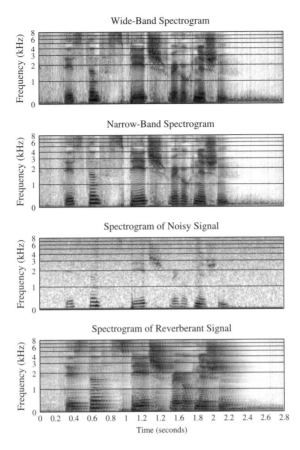

Figure 5.3 Narrow-band, mel-scaled, logarithmic spectrogram of clean speech and wide-band, mel-scaled, logarithmic spectrograms of clean speech, noisy speech and reverberant speech. All spectrograms were produced from the phrase "distant speech recognition" spoken by a male speaker

whose transfer function can be expressed as

$$H_{\text{pre-emphasis}}(z) = 1 + \alpha_{\text{pre-emphasis}} z^{-1},$$

where $\alpha_{\text{pre-emphasis}}$ typically assumes values in the range $-1.0 \leq \alpha_{\text{pre-emphasis}} \leq -0.95$. The pre-emphasis filter, however, also emphasizes frequencies above 5 kHz, where the human auditory system becomes less sensitive. To overcome this limitation, more sophisticated pre-emphasis filters have been proposed in the literature (Markel and Gray 1980). Many current speech recognition systems, however, do not apply a pre-emphasis stage and let the acoustic model compensate for the shape of the spectral slope.

5.2.2   Bark and Mel Filter Banks

The use of filter banks in the front-end of an ASR system is intended to model the operation of the *cochlea* in the inner ear, which behaves as if it were composed of overlapping bandpass filters. The passband of each filter is known as a *critical band*

Table 5.1 Critical bands which define the frequency bandwidth in which the ear integrates the excitation

f_{Bark}	f_c	Δf	f_{Bark}	f_c	Δf	f_{Bark}	f_c	Δf
0.5	50	100	8.5	1000	160	16.5	3400	550
1.5	150	100	9.5	1170	190	17.5	4000	700
2.5	250	100	10.5	1370	210	18.5	4800	900
3.5	350	100	11.5	1600	240	19.5	5800	1100
4.5	450	110	12.5	1850	280	20.5	7000	1300
5.5	570	120	13.5	2150	320	21.5	8500	1800
6.5	700	140	14.5	2500	380	22.5	10500	2500
7.5	840	150	15.5	2900	450	23.5	13500	3500

f_{Bark} critical band rate; f_c center frequency; Δf bandwidth

(Fletcher 1940). Two pure tones are said to lie in the same critical band if their frequencies are so close together that there is a considerable overlap in their amplitude envelopes in the basilar membrane. The *Bark scale*, named after Heinrich Barkhausen, who proposed the first subjective measurements of loudness, was among the first attempts to describe the effect of these critical bands. The center frequencies and bandwidths of the Bark scale are given in Table 5.1. The spacing of the critical bands is nonlinear, but corresponds to a psychoacoustic scale proposed by Zwicker (1961),

$$f_{Bark}(f) = 13 \arctan(0.00076 f) + 3.5 \arctan\left(\left(\frac{f}{7500}\right)^2\right)$$

where the frequency f is in Hertz. The bandwidth of the critical bands can be approximated as

$$\Delta f = 25 + 75\left(1 + 1.4 f^2\right)^{0.69}.$$

An alternative expression of the Bark scale, due to Schroeder (Hermansky 1990), is given by

$$f_{Bark}(f) = 6 \log\left(\frac{f}{600} + \sqrt{\left(\frac{f}{600}\right)^2 + 1}\right). \tag{5.3}$$

The *mel scale* proposed by Stevens *et al.* (1937) is an alternative nonlinear scaling of the frequency axis which models the nonlinear pitch perception characteristics of the human ear. The mel scale is based on experiments wherein human subjects were asked to divide given frequency ranges into four perceptually equal intervals. Alternatively, the subjects were asked to adjust a frequency to be perceptually equivalent to one half of a given frequency. Its name has been abbreviated from the word *melody* to indicate that the scale is based on pitch comparisons. The mel frequency can be approximated by

$$f_{mel}(f) = 1127.01048 \log\left(1 + \frac{f}{700}\right). \tag{5.4}$$

The use of nonlinear scales, such as the Bark or mel scale, is very popular in automatic speaker and speech recognition. These scales can be applied either by a nonlinear filter bank as discussed next for the mel scale, or approximated by a bilinear transform as described in Section 5.2.3.

The mel filter bank is defined by M triangular filters ($m = 1, 2, \ldots, M$) averaging the spectral energy around each center frequency

$$
H_m[k] = \begin{cases}
0, & k < f[m-1], \\[2mm]
\dfrac{2(k-f[m-1])}{(f[m+1]-f[m-1])(f[m]-f[m-1])}, & f[m-1] \le k \le f[m], \\[2mm]
\dfrac{2(f[m+1]-k)}{(f[m+1]-f[m-1])(f[m+1]-f[m])}, & f[m] \le k \le f[m+1], \\[2mm]
0, & k > f[m+1]
\end{cases}
\tag{5.5}
$$

where $f[\cdot]$ is a function of the lowest f_{lowest} and highest f_{highest} frequencies of the filter bank, as well as the sampling frequency f_{sampling} and the number of bins in the linear frequency domain N according to

$$
f[m] = \frac{N}{f_{\text{sampling}}} f_{\text{mel}}^{-1}\left(f_{\text{lowest}} + m\,\frac{f_{\text{lowest}} - f_{\text{highest}}}{M+1} \right).
\tag{5.6}
$$

The inverse of the mel frequency can be calculated from (5.4) as

$$
f_{\text{mel}}^{-1}(f) = 700\left(\exp\frac{f}{1127.01048} - 1 \right).
\tag{5.7}
$$

The bandwidths of the triangular filters are assigned such that the three dB points (i.e., where the spectral power falls to one-half its maximum) are exactly half way between the center frequencies of the filters. The triangular filters generally increase in spacing and decrease in height for higher frequencies, although some implementations use an equal height. Such a mel filter bank is depicted in Figure 5.4.

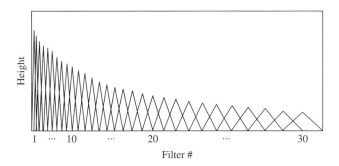

Figure 5.4   Mel filter bank

5.2.3 Warping by Bilinear Transform – Time vs Frequency Domain

Instead of approximating the mel or Bark scale through critical band filtering of the power spectrum using a nonlinearly scaled filter bank, it is possible to directly map the linear frequency axis ω to a nonlinear frequency axis $\tilde{\omega}$. This mapping process is called *frequency warping*. A convenient way to implement frequency warping is through a conformal map, such as a first-order all-pass filter (Oppenheim and Schafer 1989, sect. 5.5), which is also known as the *bilinear transform* (BLT) (Braccini and Oppenheim 1974; Oppenheim *et al.* February 1971), or a Blaschke factor (Greene and Krantz 1997, sect. 9.1). It is defined in the z-domain as

$$\tilde{z}^{-1} = Q(z) = \frac{z^{-1} - \alpha}{1 - \alpha \cdot z^{-1}} \forall -1 < \alpha < +1, \tag{5.8}$$

where α is the *warp factor*. A particular characteristic of the BLT is that it preserves the unit circle, such that

$$\left| Q\left(e^{j\omega}\right) \right| = 1 \forall -\pi < \omega \leq \pi.$$

Indeed, this latter property is the reason behind the designation *all-pass*. In Section 9.2.2, we will consider how such an all-pass transform can be used to formulate an effective means of adapting the cepstral means of a hidden Markov model to the characteristics of a particular speaker. Here we will be exclusively concerned with the time and frequency domain. The relationship between $\tilde{\omega}$ and ω is nonlinear as indicated by the phase function of the all-pass filter (Matsumoto and Moroto 2001)

$$\tilde{\omega} = \arg\left(e^{-j\omega}\right) = \omega + 2\arctan\left(\frac{\alpha \sin \omega}{1 - \alpha \cos \omega}\right). \tag{5.9}$$

A good approximation of the mel and Bark scale by the BLT is possible if the warp factor is set accordingly. The optimal warp factor depends on the sampling frequency and can be found by different optimization methods (Smith and Abel 1999). Figure 5.5

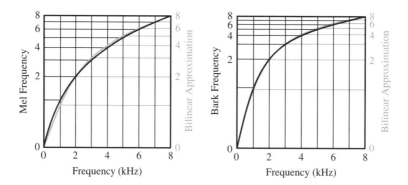

Figure 5.5 Mel frequency (scale shown along left edge of left image) and Bark frequency (scale shown along left edge of right image) can be approximated by a BLT (scale shown along right edges) for a sampling rate of 16 kHz, $\alpha_{\mathrm{mel}} = 0.4595$, $\alpha_{\mathrm{Bark}} = 0.6254$

compares the mel scale and the Bark scale with their approximations by the BLT for a sampling frequency of 16 kHz.

Frequency warping through BLTs can be applied in the *time domain*, the *frequency domain* or in the *cepstral domain*, which is discussed in Section 5.4. In all cases, the frequency axis is nonlinearly scaled. The effect on the spectral resolution, however, varies for the three different domains. This effect can be explained as follows:

- *Warping in the time domain* modifies the values in the autocorrelation matrix. Hence, in the case of autoregressive models (3.44), more coefficients are used to model lower frequencies and fewer coefficients to model higher frequencies. This effect is described for spectral envelope estimation in Sections 5.3.3 and 5.3.6.
- *Warping in the frequency or cepstral domain* does not change the spectral resolution as the transform is applied after spectral analysis.

  As indicated by Nocerino *et al.* (1985), a general warping transform in the same domain, such as the BLT, is equivalent to a matrix multiplication

$$\mathbf{f}_{\text{warp}} = \mathbf{L}(\alpha)\,\mathbf{f},$$

where α denotes the warp factor as before, and $\mathbf{L}(\alpha)$ the transform matrix. It follows that the values $\mathbf{f}_{\text{warp}}$ on the warped scale are a linear interpolation of the values $\mathbf{f}$ on the linear scale. When spectral estimation is based on linear prediction or the minimum variance distortionless response, which are discussed in Sections 5.3.3 and 5.3.4, respectively, the prediction coefficients are not altered as they are calculated before the BLT is applied.

Figure 5.6 demonstrates the effect of warping applied either in the time or in the frequency domain on the spectral envelope and compares the warped spectral envelopes with the unwarped spectral envelope.

As an example, we will briefly investigate the change of spectral resolution for the most interesting case, where the BLT is applied in the time domain with a warp factor

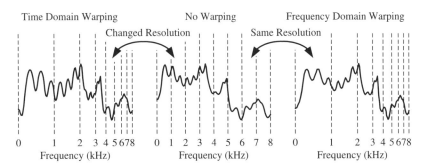

Figure 5.6 Spectral envelopes without warping and warping applied in the time or frequency domain. While warping in the time domain changes the spectral resolution and frequency axis, warping in frequency domain does not alter the spectral resolution but still changes the frequency axis

$\alpha > 0$. In this case, we observe that spectral resolution decreases as frequency increases. In comparison to the resolution provided by the linear frequency scale, corresponding to $\alpha = 0$, the warped frequency resolution increases for low frequencies up to the *turning point* (TP) frequency (Härmä and Laine 2001)

$$f_{\text{tp}}(\alpha) = \pm \frac{f_s}{2\pi} \arccos(\alpha), \tag{5.10}$$

where f_s represents the sampling frequency. At the TP frequency, the spectral resolution is not affected. Above the TP frequency, the frequency resolution decreases in comparison to the resolution provided by the linear frequency scale. For $\alpha < 0$, spectral resolution increases as frequency increases.

As observed by Strube (1980), prediction error minimization of the predictors $\tilde{a}_m$ in the warped domain is equivalent to the minimization of the output power of the warped inverse filter,

$$\tilde{A}(z) = 1 + \sum_{m=1}^{M} \tilde{a}_m \tilde{z}^{-m}(z), \tag{5.11}$$

in the linear domain, where each unit delay element z^{-1} is replaced by a BLT $\tilde{z}^{-1}$. The prediction error is therefore given by

$$E\left(e^{j\omega}\right) = \left|\tilde{A}\left(e^{j\omega}\right)\right|^2 P\left(e^{j\omega}\right), \tag{5.12}$$

where $P\left(e^{j\omega}\right)$ is the power spectrum of the signal. From Parseval's theorem as discussed in Section 3.1.3, it then follows that the total squared prediction error can be expressed as

$$\sigma^2 = \int_{-\pi}^{\pi} E\left(e^{j\tilde{\omega}}\right) d\tilde{\omega} = \int_{-\pi}^{\pi} E\left(e^{j\omega}\right) W^2\left(e^{j\omega}\right) d\omega, \tag{5.13}$$

where $W(z)$ denotes the weighting filter

$$W(z) \triangleq \frac{\sqrt{1-\alpha^2}}{1-\alpha z^{-1}}. \tag{5.14}$$

The minimization of the squared prediction error σ^2, however, does *not* lead to minimization of the power, but the power of the error signal filtered by the weighting filter $W(z)$, which is apparent from the presence of this factor in (5.13). Thus, the BLT introduces an unwanted spectral tilt. To compensate for this negative effect, the inverse weighting function

$$\left|\tilde{W}(\tilde{z}) \cdot \tilde{W}(\tilde{z}^{-1})\right|^{-1} = \frac{\left|1 + \alpha \cdot \tilde{z}^{-1}\right|^2}{1-\alpha^2} \tag{5.15}$$

can be applied. The effect of the spectral tilt introduced by the BLT and its correction (5.15) are depicted in Figure 5.7.

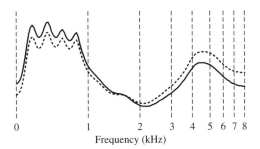

Frequency (kHz)

Figure 5.7   The plot of two warped spectral envelopes $\alpha = \alpha_{\mathrm{mel}}$ demonstrates the effect of spectral tilt. While the spectral tilt is not compensated for the dashed line, it is compensated for the solid line. It is clear to see that high frequencies are emphasized for $\alpha > 0$ if no compensation is applied

5.3   Spectral Estimation and Analysis

Spectral estimation and analysis are fundamental components of speech feature extraction for automatic recognition and many other speech-processing algorithms, including compression, coding, and voice conversion. These applications impose a variety of requirements on the spectral estimate, including

- spectral resolution,
- nonlinear modeling of the frequency axis,
- variance of the estimated spectra, and
- capacity to model the frequency response function of the vocal tract during voiced speech.

To satisfy these requirements, a broad variety of solutions have been proposed in the literature, all of which can be classified into either

- *nonparametric methods* based on periodograms (e.g., the power spectrum) or
- *parametric methods*, using a small number of parameters estimated from the data (e.g., linear prediction).

In this section, we will concentrate on spectral estimation techniques which are useful in extracting the features to be used by an ASR system. An overview of the different approaches is given in Table 5.2. Moreover, an introduction to spectral analysis, covering many alternative methods not treated here, can be found in Stoica and Moses (2005).

5.3.1   Power Spectrum

A very simple approach to the spectral analysis of a signal $x[n]$ for $n = 0, \ldots, M$ is to calculate its *power spectrum*. The power spectrum can be obtained through the calculation of the *discrete circular autocorrelation*

$$\phi[l] = \sum_{n=0}^{M-1-l} x[n]x[(n+l)\%M]. \tag{5.16}$$

Table 5.2 Overview of spectral estimation methods

Spectrum	Properties		
	Detail	Resolution	Sensitive to pitch
Fourier	exact	linear, static	very high
mel filter bank	smooth	nonlinear, static	high
LP	approx.	linear, static	medium
perceptual LP	approx.	nonlinear, static	medium
warped LP	approx.	nonlinear, static	medium
warped-twice LP	approx.	nonlinear, adaptive	medium
MVDR	approx.	linear, static	low
warped MVDR	approx.	nonlinear, static	low
warped-twice MVDR	approx.	nonlinear, adaptive	low

LP = linear prediction; MVDR = minimum variance distortionless response

Thereafter, the discrete Fourier transform of the autocorrelation coefficients is calculated, resulting in the *discrete power spectrum*

$$S[m] = \sum_{l=0}^{M-1} \phi[l] e^{-j2\pi lm/M} \forall m = 0, 1, \ldots, M-1$$

where m is the discrete angle frequency.

The power spectrum is widely used in speech processing because it can be quickly calculated via the *fast Fourier transform* (FFT). Nonetheless, it is poorly suited to the estimation of speech spectra intended for automatic recognition, because it models spectral peaks and valleys equally well. This characteristic is bad for two reasons:

- *The effect of the fundamental frequency* – The power spectrum cannot suppress the effect of the fundamental frequency and its harmonics in voiced speech, and therefore provides a poor estimate of the response function of the vocal tract. In all Western languages, it is only the response of the vocal tract that serves to distinguish between words.
- *The effect of ambient noise* – Noise in the logarithmic power domain is most evident in spectral valleys; hence, an exact representation of these regions is less useful than an approximation of the spectral power. The spectral peaks, on the other hand, should be faithfully represented as they contain the most relevant information and, as we will see later, are less distorted by additive distortions.

5.3.2 Spectral Envelopes

The *spectral envelope* is a plot of power vs frequency representing the resonances of the vocal tract. The spectral envelope is typically subject to certain smoothness criteria, such that the spectral effects of the periodic or noisy excitations, which are provided by the vocal cords or by the turbulent flow of air through a constriction of the vocal tract, respectively, are excluded. The spectral envelope faithfully represents the spectral peaks of the power spectra but may devote less precision to modeling the spectral valleys. Such

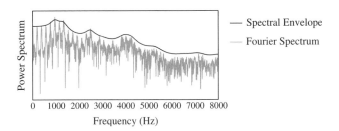

Figure 5.8   The spectral envelope and original Fourier spectrum

a spectral envelope is shown in Figure 5.8. It may impossible or undesirable to model every individual peak; e.g., if a group of peaks is close together. In such cases, the spectral envelope should provide a reasonable approximation. In addition, the method used for estimating the spectral envelope should be stable and applicable to a wide range of signals with very different characteristics. To provide robustness in the presence of distortion, it is desirable that a local change in signal frequency does not affect the intensity of the spectral estimate at frequencies well-apart from this point. Moreover, spectral envelope representation should be resilient to distortions in the data.

   We begin our discussion of the spectral envelope estimation with the most popular method, namely, linear prediction. In the next section we describe more advanced methods that overcome the limitations of linear prediction.

5.3.3   LP Envelope

The estimation of an all-pole model (discussed briefly in Section 3.1.3) via *linear prediction* (LP) is a well-known technique. As with the FFT, LP was invented by Carl Friedrich Gauss as a means to predict the reappearance of the asteroid Ceres after it had been lost in the glare of the sun. Atal and Schroeder (1967) developed an entire family of predictors for speech coding. Probably the best-known speech coder is *code-excited LP* (Atal and Schroeder 1984). Nowadays, LP is a widely-used method in various speech applications such as speech modeling, coding, synthesis, and extraction of speech features for automatic recognition.

   The idea behind LP is to predict the signal $x[n]$ at time n by a weighted linear combination of M immediately preceding samples and some input $u[n]$, such that,

$$\hat{x}_M[n] = -\sum_{m=1}^{M} a_m \, x[n-m] + G \cdot u[n],$$

where M is known as the *model order* and G as the *gain*. Hence, it is necessary to determine the values of the LP coefficients $\{a_m\}_{m=1}^{M}$. Assuming that $u[n]$ is unknown and thus that $x[n]$ must be predicted solely from a weighted combination of prior samples, the error between $x[n]$ and the prediction $\hat{x}_M[n]$ is given by the *error term*

$$e_M[n] = x[n] + \sum_{m=1}^{M} a_m \, x[n-m].$$

The higher $e_M[n]$, the worse the quality of the model defined by the linear prediction coefficients

$$\mathbf{a} = \begin{bmatrix} a_1 & a_2 & \cdots & a_m \end{bmatrix}^T.$$

The vector of prediction coefficients $\mathbf{a}$, can be estimated by minimizing the total power of the prediction error:

$$\hat{\mathbf{a}} = \underset{\mathbf{a}=[a_1,a_2,\cdots,a_M]}{\mathrm{argmin}} \sum_{n=-\infty}^{\infty} \left(x[n] + \sum_{m=1}^{M} a_m\, x[n-m] \right)^2. \tag{5.17}$$

The prediction coefficients can be obtained by solving the matrix equation

$$\mathbf{a} = \mathbf{\Phi}^{-1}\phi, \tag{5.18}$$

where

$$\mathbf{\Phi} = \begin{bmatrix} \Phi[1,1] & \Phi[1,2] & \cdots & \Phi[1,M] \\ \Phi[2,1] & \Phi[2,2] & \cdots & \Phi[2,M] \\ \vdots & \vdots & \ddots & \vdots \\ \Phi[M,1] & \Phi[M,2] & \cdots & \Phi[M,M] \end{bmatrix}$$

and

$$\phi = \begin{bmatrix} \Phi[1,0], \Phi[2,0], \ldots, \Phi[M,0] \end{bmatrix}.$$

The autocorrelation coefficients can be calculated by

$$\Phi[m,n] = \mathcal{E}\{x[m]\,x[n]\}.$$

Three principal methods exist for minimizing (5.18):

- the *autocorrelation method*,
- the *covariance method* which is based on the covariance matrix, and
- the *lattice method*,

All methods yield slightly different LP coefficients (Makhoul 1975). A detailed introduction to LP can be found in Strobach (1990). Due to space limitations, here we will discuss only the autocorrelation method which is used almost exclusively in speech recognition. This is, firstly, due to the stability of the estimate, which stems from the fact that the predictor filter of the autocorrelation method has zeros only inside the unit circle in the z-plane. Secondly, there exists an efficient algorithm for the calculation of the prediction coefficients known as the *Levinson–Durbin recursion* (Makhoul 1975). The latter is summarized in Algorithm 5.1.

The autocorrelation method constrains the evaluation integral to the range from 0 to $M-1$ and assumes that all values outside this integral to be zero. Under the previous constraints, the autocorrelation matrix can be simplified to

$$\Phi[m,n] = \Phi[0, |m-n|].$$

Algorithm 5.1 The Levinson–Durbin recursion

1. Initialize with $a_{0,0} = 1$ and $e_0 = \phi[0]$
2. For $m = 1, 2, \cdots, M$

$$k_m = \frac{-1}{\epsilon_{m-1}} \sum_{i=0}^{m-1} \phi[i-m] \, a_{i,m-1}$$

with

$$a_{i,m} = \begin{cases} 1, & i = 0, \\ a_{i,m-1} + k_m \, a^*_{m-i,m-1}, & i = 1, 2, \cdots, m-1, \\ k_m, & i = m \end{cases}$$

and

$$e_m = e_{m-1}(1 - |k_m|^2)$$

3. The final set of linear prediction coefficients are given by $\{a_i = a_{i,M}\}_i$.

Frequency Domain Formulation

So far we have introduced the basic concept of LP from a time domain formulation. By applying the z-transform to (5.17), we obtain the formulation in the transform domain:

$$\hat{\mathbf{a}} = \underset{\mathbf{a}=[a_1,a_2,\cdots,a_M]}{\operatorname{argmin}} \sum_{n=-\infty}^{\infty} \left(\left(z^n + \sum_{m=1}^{M} a_m \, z^{n-m} \right) X(z) \right)^2.$$

Assuming that $x[k]$ is deterministic, we can set $z = e^{j\omega}$ and apply Parseval's theorem (3.47) to replace the infinite summation by a finite integral, as

$$\hat{\mathbf{a}} = \underset{\mathbf{a}=[a_1,a_2,\cdots,a_M]}{\operatorname{argmin}} \frac{1}{2\pi} \int_{-\pi}^{\pi} \left| A\left(e^{j\omega}\right) \cdot X\left(e^{j\omega}\right) \right|^2 d\omega, \qquad (5.19)$$

where

$$A\left(e^{j\omega}\right) = 1 + \sum_{m=1}^{M} a_m \, e^{-jm\omega}. \qquad (5.20)$$

Once the LP coefficients $\mathbf{a}$ and the squared prediction error $e_M = G^2$ have been obtained from the Levinson–Durbin recursion, the transfer function of the discrete all-pole model can be expressed as

$$H(z) = \frac{G}{A(z)} = \frac{G}{1 + \sum_{m=1}^{M} a_m \, z^{-m}}, \qquad (5.21)$$

where the *gain* G matches the scale of the LP model to the spectrum of the original signal. The all-pole spectral estimate $\hat{S}\left(e^{j\omega}\right)$, henceforth known as the *LP envelope*, is then given by

$$\hat{S}\left(e^{j\omega}\right) = \left|H\left(e^{j\omega}\right)\right|^2 = \frac{e_M}{\left|1 + \sum_{m=1}^{M} a_m \, e^{-jm\omega}\right|^2}. \qquad (5.22)$$

Limitation of LP Envelopes

To understand the limitation of LP envelopes for modeling voiced speech, we need only follow Murthi and Rao (2000) and represent the short-time spectrum of a segment of voiced speech as an overtone series. Let $\omega_0 = 2\pi f_0$ where f_0 denotes the fundamental frequency, and let $L = f_s/2f_0$ denote the number of harmonics, where f_s is the sampling frequency. Then the model for the short-term spectrum can be expressed as

$$S_{\text{harmonic}}\left(e^{j\omega}\right) = \sum_{l=1}^{L} 2\pi \frac{|b_l|^2}{4} \left[\delta(\omega + \omega_0 l) + \delta(\omega - \omega_0 l)\right], \qquad (5.23)$$

where b_l is the amplitude of the lth harmonic. We can now set $\left|X\left(e^{j\omega}\right)\right|^2 = S_{\text{harmonic}}\left(e^{j\omega}\right)$ and substitute (5.23) into (5.19) to obtain

$$\hat{\mathbf{a}} = \underset{\mathbf{a}=[a_1,a_2,\cdots,a_M]}{\text{argmin}} \frac{1}{2\pi} \int_{-\omega}^{\omega} \left|A\left(e^{j\omega}\right)\right|^2 \cdot S_{\text{harmonic}}\left(e^{j\omega}\right) d\omega,$$

or, equivalently,

$$\underset{\mathbf{a}=[a_1,a_2,\cdots,a_M]}{\text{argmin}} \sum_{l=1}^{L} \frac{|b_l|^2}{2} \left|A(e^{jl\omega_0})\right|^2.$$

To achieve the desired minimization of the prediction error, the LP filter (5.20) attempts to null out the harmonics $k\omega_0$ present in the original spectrum. With increasing model order M, the ability of the LP filter to null out these harmonics increases. But in the process, the zeros of the LP filter move ever closer to the unit circle, thereby causing sharper contours in the spectral envelope (5.22) and an overestimation of the spectral power at the harmonics (Murthi and Rao 2000). Such effects are particularly problematic for medium- and high-pitched voices. As such, the LP method does not provide spectral envelopes which reliably estimate the power at the harmonic frequencies in voiced speech.

5.3.4 MVDR Envelope

Here we briefly review the *minimum variance distortionless response* (MVDR)[1] as originally introduced by Capon (1969). It has been adopted by Lacoss (1971) who demonstrated that this method provides an unbiased minimum variance estimate of the spectral components. He has also shown that the MVDR spectral estimate yields spectral peaks which are proportional to the power at that frequency. This is in contrast to the LP spectral

---

[1] Also known as Capon's method or the maximum-likelihood method (Musicus 1985).

estimate which yields spectral peaks which are proportional to the square of the power at that frequency. In order to overcome the problems associated with LP, Murthi and Rao (1997) proposed the use of the MVDR for all-pole modeling of speech signals. A detailed discussion of speech spectral estimation using the MVDR can be found in Murthi and Rao (2000).

MVDR spectral estimation can be posed as a problem in filter bank design, wherein the final filter bank is subject to the *distortionless constraint* (Haykin 2002, sect. 2.8):

The signal at the frequency of interest (FOI) ω_{foi} *must pass undistorted with unity gain.*

This condition can be expressed as

$$H\left(e^{j\omega_{\text{foi}}}\right) = \sum_{m=0}^{M} h[m] e^{-jm\omega_{\text{foi}}} = 1.$$

This constraint can be rewritten in vector form as

$$\mathbf{v}^H\left(e^{j\omega_{\text{foi}}}\right)\mathbf{h} = 1,$$

where $\mathbf{v}\left(e^{j\omega_{\text{foi}}}\right)$ is the *fixed frequency vector*,

$$\mathbf{v}\left(e^{j\omega}\right) \triangleq \begin{bmatrix} 1 & e^{-j\omega} & e^{-j2\omega} & \cdots & e^{-jM\omega} \end{bmatrix}^T,$$

and $\mathbf{h}$ is the *stacked impulse response*,

$$\mathbf{h} \triangleq \begin{bmatrix} h[0] & h[1] & \cdots & h[M] \end{bmatrix}^T.$$

The distortionless filter $\mathbf{h}$ can now be obtained by solving for the constrained minimization problem:

$$\min_{\mathbf{h}} \mathbf{h}^H \mathbf{\Phi} \mathbf{h} \text{ subject to } \mathbf{v}^H\left(e^{j\omega_{\text{foi}}}\right)\mathbf{h} = 1 \tag{5.24}$$

where $\mathbf{\Phi}$ is the $(M+1) \times (M+1)$ Toeplitz autocorrelation matrix with (m, n)th element $\Phi[m, n] = \phi[m-n]$ of the input signal

$$\phi[n] = \sum_{m=0}^{M} x[m] x[m-n].$$

The solution of the constrained minimization problem can be expressed as (Haykin 2002, sect. 2.8)

$$\mathbf{h} = \frac{\mathbf{\Phi}^{-1} \mathbf{v}(e^{j\omega_{\text{foi}}})}{\mathbf{v}^H(e^{j\omega_{\text{foi}}}) \mathbf{\Phi}^{-1} \mathbf{v}(e^{j\omega_{\text{foi}}})}.$$

This implies that $\mathbf{h}$ is the impulse response of the distortionless filter for the frequency ω_{foi}. The MVDR envelope of the spectrum $S(e^{-j\omega})$ at frequency ω_{foi} is then obtained as the output of the optimized constrained filter

$$S_{\text{MVDR}}(e^{j\omega_{\text{foi}}}) = \frac{1}{2\pi} \int_{-\pi}^{\pi} \left| H(e^{j\omega_{\text{foi}}}) \right|^2 S(e^{-j\omega})\, d\omega. \tag{5.25}$$

Although MVDR spectral estimation was posed as a problem of designing a distortionless filter for a given frequency ω_{foi}, this was only a conceptual device. The MVDR spectral envelope can in fact be represented in parametric form for all frequencies and computed as

$$S_{\text{MVDR}}(e^{j\omega}) = \frac{1}{\mathbf{v}^H(e^{j\omega})\,\boldsymbol{\Phi}^{-1}\mathbf{v}(e^{j\omega})}.$$

Under the assumption that the $(M+1) \times (M+1)$ Hermitian Toeplitz correlation matrix $\boldsymbol{\Phi}$ is positive definite and thus invertible, Musicus (1985) derived a fast algorithm to calculate the MVDR spectral envelope from a set of *linear prediction coefficients* (LPCs), as given in Algorithm 5.2.

The MVDR envelope copes well with the problem of overestimation of the spectral power at the harmonics of voiced speech. To show this, we once more model voiced speech as the sum of harmonics (5.23). Using the frequency form of the MVDR envelope given by (5.25), the spectral estimate at $\omega_l = \omega_0 l \ \forall l = 1, 2, \ldots$ is given by

$$S_{\text{MVDR}}(e^{j\omega_0 l}) = \sum_{l=1}^{L} \frac{|b_l|^2}{4} \left\{ |H(e^{j\omega_l})|^2 + |H(e^{-j\omega_l})|^2 \right\},$$

where b_l is the amplitude of the lth harmonic. Thus the MVDR distortionless filter $\mathbf{h}$ faithfully preserves the input power at $\omega_0 l$ while treating the other $(2L - 1)$ exponentials as interference and attempting to minimize their influence on the output of the filter. Hence, the MVDR envelope models the perceptually important speech harmonics very well. Unlike warped envelopes, however, it does not mimic the human auditory system and does not model the different frequency bands with varying accuracy.

Spectral Relationship between LP and MVDR Spectral Envelopes

Burg (1972) showed that the MVDR spectral envelope of model order M can also be expressed as the harmonic mean of the LP spectra $S_{\text{LP}}^{(M)}(e^{j\omega})$ of orders 0 through M:

$$S_{\text{MVDR}}^{(M)}(e^{j\omega}) = \left[\sum_{m=0}^{M} \frac{1}{S_{\text{LP}}^{(m)}(e^{j\omega})} \right]^{-1}.$$

This relationship also holds for warped LP and warped MVDR spectral envelopes as discussed in the following sections. The given relation explains why the (warped) MVDR spectral envelope exhibits a smoother frequency response with decreased variance than the corresponding (warped) LP spectrum (Murthi and Rao 2000) if compared for the same model order.

Algorithm 5.2   Fast MVDR spectral envelope calculation

1. Compute the LPCs $a_{0\cdots M}^{(M)}$ of order M and the prediction error e_M
2. Correlate the LPCs, as

$$
\mu_m = \begin{cases} \dfrac{1}{e_M} \displaystyle\sum_{i=0}^{M-m} (M+1-m-2i)\, a_i^{(M)}\, a_{i+m}^{*(M)}, & m = 0, 1, 2, \cdots, M, \\[2em] \mu_{-m}^*, & m = -M, \cdots, -1 \end{cases}
$$

3. Compute the *MVDR envelope*

$$
S_{\mathrm{MVDR}}(e^{j\omega}) = \frac{1}{\displaystyle\sum_{m=-M}^{M} \mu_m e^{-j\omega m}}
$$

5.3.5 *Perceptual LP Envelope*

The LP and MVDR all-pole models approximate speech spectra equally well at all frequency bands. To eliminate this inconsistency between LP- or MVDR-based spectral estimation and the human auditory analysis, two widely used modifications exist, both of which will be described in this and the following section.

The *perceptual linear prediction* (PLP) method, as proposed by Hermansky (1990), is outlined in Algorithm 5.3. It modifies LP spectral analysis through the introduction of the Bark scale (5.3) and logarithmic amplitude compression prior to the Levinson–Durbin recursion which is described in Algorithm 5.1. The logarithmic amplitude is implemented by raising the magnitude of the spectral components to a power of 0.33 in order to simulate the power law of human hearing. Due to the previously mentioned modifications, which are performed in the frequency domain, the autocorrelation coefficients cannot be computed directly. Hence, additional Fourier transforms are required.

5.3.6 *Warped LP Envelope*

An alternative to PLP, for which there is no need to convert between time and frequency domains, is to perform LP analysis on a *warped* frequency axis. This is accomplished by replacing the unit delay element $e^{-jk\omega}$ with a cascade of first-order all-pass filters, such as were presented in Section 5.2.3. The application of the BLT prior to LP analysis was proposed by Strube (1980).

The inverse filter on the warped frequency axis,

$$
\tilde{A}(e^{j\tilde{\omega}}) = 1 + \sum_{m=1}^{M} \tilde{a}_m \frac{e^{-jm\omega} - \alpha}{1 - \alpha \cdot e^{-jm\omega}},
$$

can then be estimated with the Levinson–Durbin recursion, Algorithm 5.1, using the warped autocorrelation coefficients. Note that applying the BLT to the spectrum of a

Algorithm 5.3 Perceptual linear prediction

1. Calculation of the windowed power spectrum
2. Critical band integration (5.3) realized by a filter-bank defined as

$$
C_k(\omega) = \begin{cases}
10^{f_{\text{Bark}} - f_{\text{Bark}}^{(k)}}, & f_{\text{Bark}} \leq f_{\text{Bark}}^{(k)} - 0.5, \\
1, & f_{\text{Bark}}^{(k)} - 0.5 < f_{\text{Bark}} < f_{\text{Bark}}^{(k)} + 0.5, \\
10^{-2.5(f_{\text{Bark}} - f_{\text{Bark}}^{(k)} + 0.5)}, & f_{\text{Bark}} \geq f_{\text{Bark}}^{(k)} - 0.5
\end{cases}
$$

where the center frequency $f_{\text{Bark}}^{(k)}$ of filter k is given by $f_{\text{Bark}}^{(k)} = 0.994k$.
3. Equally loudness pre-emphasis

$$
E(\omega) = 1.151 \sqrt{\frac{\left(\omega^2 + 1.44 \cdot 10^6\right)\omega^2}{\left(\omega^2 + 1.6 \cdot 10^5\right)\left(\omega^2 + 9.61 \cdot 10^6\right)}}
$$

4. Intensity to loudness compensation

$$
Q(\omega) = F^{1/3}(\omega) = \left(E(\omega) \int_0^\pi C_k(\omega) S(\omega) d\omega \right)^{1/3}
$$

5. Inverse Fourier transform
6. Calculation of the PLP coefficients by the Levinson–Durbin recursion as outlined in Algorithm 5.1.

finite sequence produces a spectrum corresponding to an infinite sequence,

$$
\tilde{X}(\tilde{z}) = \sum_{n=0}^{\infty} \tilde{x}[n]\,\tilde{z}^{-n} = X(z) = \sum_{n=0}^{N-1} x[n]\,z^{-n}.
$$

Thus the direct calculation of the warped autocorrelation coefficients,

$$
\tilde{\phi}[m] = \sum_{n=0}^{\infty} \tilde{x}[n]\,\tilde{x}[n-m], \tag{5.26}
$$

is not feasible. To overcome this problem, a variety of solutions have been proposed (Edler and Schuller 2000; Strube 1980; Tokuda *et al.* 1995). Here we give the algorithm proposed by Matsumoto *et al.* (1998). To obtain the warped predictors, we must solve the normal equations

$$
\sum_{y=1}^{p} \tilde{\Phi}[m, n]\,\tilde{a}_{m,n} = -\tilde{\Phi}[m, 0], \; \forall\, m = 1, 2, \cdots, p, \tag{5.27}
$$

where

$$\tilde{\Phi}[m, n] \triangleq \sum_{l=0}^{\infty} y_m[l]\, y_n[l],$$

and $y_m[n]$ is the output of the mth-order all-pass filter excited by $y_0[n] = x[n]$. The last equation implies that $\tilde{\Phi}[m, n]$ is a component of the warped autocorrelation function

$$\tilde{\Phi}[m, n] = \tilde{\phi}[|m - n|]. \tag{5.28}$$

Thus, (5.27) is revealed to be an autocorrelation function, exactly like the autocorrelation equation found in standard LP analysis. Furthermore, as $\tilde{\Phi}[m, n]$ depends only on the difference $|m - n|$, we can replace (5.26) by

$$\tilde{\phi}[|m - n|] = \sum_{l=0}^{N-1-|m-n|} x[l]\, y_{|m-n|}[l], \tag{5.29}$$

where $y_i[n]$ is the output sequence given by

$$y_i[n] = \alpha \cdot (y_i[n - 1] - y_{i-1}[n]) - y_{i-1}[n - 1].$$

Hence, the warped autocorrelation coefficients $\tilde{\Phi}[m, n]$ can be calculated with a finite sum.

Given the warped LP coefficients, we can now obtain the transfer function $H_{\text{warped LP}}(z)$. Thereby, we derive an all-pole spectral estimate in the warped frequency domain, henceforth referred to as the *warped LP envelope*:

$$S_{\text{warped LP}}(e^{j\omega}) = \left| H_{\text{warped LP}}\left(e^{j\omega}\right) \right|^2 = \frac{\tilde{e}_M}{\left| 1 + \sum_{m=1}^{M} \tilde{a}_m e^{-jm\omega} \right|^2}. \tag{5.30}$$

Note that if α is set appropriately, the spectrum (5.30) is already in the mel warped frequency domain and therefore it is necessary to either

- eliminate the mel spaced triangular filter bank traditionally used in the extraction of mel frequency cepstral coefficients, or
- replace it by a filter bank of uniform half-overlapping triangular filters to provide feature reduction or additional spectral smoothing.

If we are interested in an envelope which is in the linear frequency domain, we can calculate the spectral estimate as

$$\tilde{S}(e^{j\omega}) = \frac{\tilde{e}_M}{\left| 1 + \sum_{k=1}^{M} \tilde{a}_k \frac{e^{-jk\omega} - \alpha}{1 - \alpha \cdot e^{-jk\omega}} \right|^2},$$

which, in comparison with the conventional LP envelope, uses more parameters to describe the lower frequencies and fewer parameters to describe the higher frequencies.

The warping of the LP envelope addresses the inconsistency between LP spectral estimation and that performed by the human auditory system. Unfortunately, for high-pitched
voiced speech the lower harmonics become so sparse that single harmonics appear as spectral poles, which is highly undesirable in all-pole modeling. One proposed approach to
overcome this drawback is to weight the warped autocorrelation coefficient $\tilde{\phi}[m]$ with a
lag window (Matsumoto and Moroto 2001). An alternative is to use the warped MVDR
envelope as described in the next section.

5.3.7 Warped MVDR Envelope

To overcome the problems inherent in LP while emphasizing the perceptually relevant
portions of the spectrum, the BLT must be applied prior to MVDR spectral envelope
estimation (Wölfel and McDonough 2005). The derivation of the so-called *warped MVDR*
will be presented in this section. Let us define the *warped frequency vector* $\tilde{\mathbf{v}}$ as

$$\tilde{\mathbf{v}}(e^{j\omega}) \triangleq \left[1 \quad \frac{e^{-j\omega}-\alpha}{1-\alpha \cdot e^{-j\omega}} \quad \frac{e^{-j2\omega}-\alpha}{1-\alpha \cdot e^{-j2\omega}} \quad \cdots \quad \frac{e^{-jM\omega}-\alpha}{1-\alpha \cdot e^{-jM\omega}} \right]^T .$$

In order to calculate the distortionless filter $\tilde{\mathbf{h}}$ in the warped domain, we must once more
solve the constrained minimization problem

$$\min_{\tilde{\mathbf{h}}} \tilde{\mathbf{h}}^H \tilde{\mathbf{\Phi}} \tilde{\mathbf{h}} \quad \text{subject to} \quad \tilde{\mathbf{v}}^H (e^{j\omega_{\text{foi}}})\tilde{\mathbf{h}} = 1, \tag{5.31}$$

where $\tilde{\mathbf{\Phi}}$ is the Toeplitz autocorrelation matrix as defined by (5.28). Clearly, this solution
is different from MVDR on the linear frequency scale. The way to solve for the warped
constrained minimization problem, however, is very similar to its unwarped counterpart.
The warped MVDR envelope of the spectrum $S(e^{-j\omega})$ at frequency ω_{foi} can be obtained
as the output of the optimal filter,

$$S_{\text{warpedMVDR}}(e^{j\omega_{\text{foi}}}) = \frac{1}{2\pi} \int_{-\pi}^{\pi} \left| \tilde{H}(e^{j\omega_{\text{foi}}}) \right|^2 S(e^{-j\omega})d\omega, \tag{5.32}$$

under the constraint

$$\tilde{H}(e^{j\omega_{\text{foi}}}) = \sum_{m=0}^{M} \tilde{h}(m) \frac{e^{-jm\omega_{\text{foi}}} - \alpha}{1 - \alpha \cdot e^{-jm\omega_{\text{foi}}}} = 1.$$

Assuming that the Hermitian Toeplitz correlation matrix $\tilde{\mathbf{\Phi}}$ is positive definite and thus
invertible, Musicus' (1985) algorithm, as given in Algorithm 5.2, can be readily applied to
compute the warped MVDR spectral envelope. The LPCs and the error term in Step 1 of
Algorithm 5.2, however, must be replaced by their warped counterparts from Section 5.3.6.
Note that the spectrum (5.32) derived by the modified, fast algorithm has a warped
frequency axis and should be handled as suggested in Section 5.3.6.

If we are interested in a warped envelope estimate on the linear frequency axis, we can replace Step 3 of Algorithm 5.2 by

$$\tilde{S}_{\text{MVDR}}(e^{j\omega}) = \frac{1}{\sum_{m=-M}^{M} \tilde{\mu}_m \frac{e^{-jm\omega} - \alpha}{1 - \alpha \cdot e^{-jm\omega}}} \cdot$$

5.3.8 Warped-Twice MVDR Envelope

From Section 5.2.3, we know that the BLT, when applied in the time domain prior to spectral analysis, enables the frequency axis to be warped while simultaneously altering the spectral resolution. Alternatively, the spectral resolution is unaltered if the BLT is applied in the frequency domain. Moreover, it is possible to compensate for the warping of the frequency axis due to the BLT in the time domain through a second BLT in the frequency domain. Thus it is possible to move spectral resolution to higher or lower frequencies while keeping the frequency axis fixed (Nakatoh *et al.* 2004; Wölfel 2006). Due to the application of two warping stages in MVDR spectral estimation, this approach is dubbed *warped-twice MVDR*.

Compensation

The warped-twice MVDR envelope must be applied with special care to compensate for unwanted distortions. To fit the final frequency axis to a particular, but fixed, frequency axis (e.g., the mel-scale α_{mel}) the compensation warp factor must be calculated as

$$\beta = \frac{\alpha - \alpha_{\text{mel}}}{1 - \alpha \cdot \alpha_{\text{mel}}} \cdot$$

The spectral tilt introduced from both BLTs with warp factors α and β can be expressed with a single warp factor as

$$\chi = \frac{\alpha + \beta}{1 + \alpha \cdot \beta} \cdot \tag{5.33}$$

A derivation of (5.33) is provided in Section B.9. A compensation of the spectral tilt is now possible by applying the inverted weighting function

$$\left| \tilde{W}(\tilde{z}) \cdot \tilde{W}(\tilde{z}^{-1}) \right|^{-1} \tag{5.34}$$

to the warped autocorrelation coefficients. This weighting function can easily be realized as a second-order finite impulse response filter

$$\widehat{\phi}[m] = \frac{1 + \chi^2 + \chi \cdot \tilde{\phi}[m-1] + \chi \cdot \tilde{\phi}[m+1]}{1 - \chi^2}. \tag{5.35}$$

Algorithm 5.4   Warped-twice MVDR spectral envelope calculation

1. Compute the warped autocorrelation coefficients $\tilde{\phi}[0], \cdots, \tilde{\phi}[M+1]$ as given in (5.29)
2. Calculate the compensation warp factor

$$\beta = \frac{\alpha - \alpha_{\mathrm{mel}}}{1 - \alpha \cdot \alpha_{\mathrm{mel}}}$$

3. Compensate spectral tilt on warped autocorrelation coefficients

$$\widehat{\phi}[m] = \frac{1 + \chi^2 + \chi \cdot \tilde{\phi}[m-1] + \chi \cdot \tilde{\phi}[m+1]}{1 - \chi^2}$$

   with the warp factor

$$\chi = \frac{\alpha + \beta}{1 + \alpha \cdot \beta}$$

4. Compute the warped LPCs $\tilde{a}_{0\cdots M}^{(M)}$ of order M and the prediction error power $\tilde{e}_M$
5. Correlate the warped LPCs, as

$$\widehat{\mu}_k = \begin{cases} \dfrac{1}{\tilde{e}_M} \displaystyle\sum_{m=0}^{M-k} (M+1-k-2m)\, \widehat{a}_m^{(M)} \widehat{a}_{m+k}^{*(M)}, & k = 0, 1, \cdots, M, \\[6mm] \widehat{\mu}_{-k}^*, & k = -M, \cdots, -1 \end{cases}$$

6. Compute the *warped-twice MVDR envelope*

$$S_{\mathrm{W2MVDR}}(e^{j\omega}) = \frac{1}{\displaystyle\sum_{m=-M}^{M} \widehat{\mu}_m \dfrac{e^{j\omega} - \beta}{1 - \beta \cdot e^{j\omega}}}$$

Note that (5.35) requires the calculation of $M+1$ autocorrelation coefficients $\tilde{\phi}[m]$ as defined in (5.29).

A fast computation of the warped-twice MVDR envelope of model order M is possible by extending Musicus' algorithm as outlined in Algorithm 5.4. The required compensation steps are described in more detail in the next section. A flowchart of the individual processing steps, including a steering function as defined in the next section, is given in Figure 5.9.

Steering Function

In order to achieve the best ASR performance, the free parameters of the warped-twice MVDR envelope must be adapted in such a way that characteristics relevant for classification are emphasized while less relevant information is suppressed. A function providing

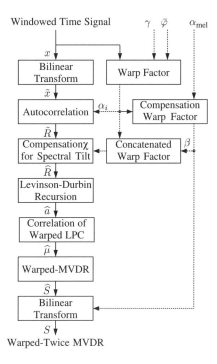

Figure 5.9 Overview of warped-twice minimum variance distortionless response. Symbols are defined as in the text

this property is called a *steering function*. Nakatoh *et al.* (2004) proposed a method for steering the spectral resolution to lower or higher frequencies whereby, for every frame k, the ratio of the first two autocorrelation coefficients were used as a steering function

$$\varphi_k \triangleq \frac{\phi_k[1]}{\phi_k[0]}. \qquad (5.36)$$

The factor γ is introduced in order to adjust the sensitivity of the steering function. Moreover, the bias $\overline{\varphi}$, which is obtained by averaging over all values in the training set, keeps the average of α close to α_{mel}. This leads to

$$\alpha_k = \gamma \cdot (\varphi_k - \overline{\varphi}) + \alpha_{\text{mel}}. \qquad (5.37)$$

Figure 5.10 gives the different values of the normalized first autocorrelation coefficient φ averaged over all samples for each individual phoneme. A clear separation between the fricatives, in particular the sibilants, and nonfricatives can be observed.

5.3.9 Comparison of Spectral Estimates

Figure 5.11 displays plots of the spectral envelopes derived from the power spectrum as well as the LP and MVDR models on both a linear and nonlinear frequency scale. The warp factor for the warped LP and warped MVDR was set to 0.4595 so as to simulate the

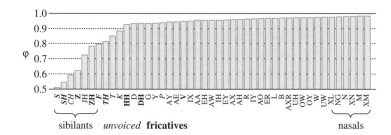

Figure 5.10 Values of the normalized first autocorrelation coefficient for different phonemes. Different phone classes are distinguished by different ranges of values: The sibilants have in general small values, the unvoiced phones fall in the middle of the range, and the nasals have the highest values

mel frequency for a signal sampled at 16 kHz. Due to its stronger smoothing properties, the model order of the MVDR-envelope was set to 30, while that of the LP envelope was set to 15.

The spectral estimates on the nonlinear frequency scale differ from those in the linear frequency scale inasmuch as more parameters are apportioned to describe the lower as compared to the higher frequency regions. Thus the warped estimates provide a higher spectral resolution in low frequencies and lower spectral resolution in higher frequency regions. Therefore, warping prior to spectral analysis provides properties which cannot be achieved when the spectral analysis is *followed* by frequency warping.

The warp factors, applied in the time or frequency domain, warp the frequency axis. The effect can be used to apply the mel scale or, when done on a speaker-dependent basis, to implement vocal tract length normalization. Better results with piecewise linear warping as described in Section 9.1.1 may be achieved (Wölfel 2003), although this effect is strongly dependent on the sampling rate.

The MVDR envelope prevents the unwanted overestimation of the harmonic peaks in medium- and high-pitched voiced speech that is seen in the LP envelope. As is apparent from Figure 5.11, the LP envelope overestimates the spectral peak at 4 kHz, which is apparent upon comparing the LP envelope with the Fourier spectrum. Unlike the LP spectral envelope, the MVDR envelope provides a broad peak which matches the true spectrum better.

Figure 5.12 compares the influence of the model order and the warp factor on the warped-twice MVDR spectral envelope estimate. While the model order varies the overall spectral resolution of the estimate, the warp factor moves spectral resolution to higher or lower frequencies.

5.3.10 Scaling of Envelopes

In this section, we investigate the influence of additive noise on the spectral peaks of the power spectrum and those of spectral envelopes. The peaks in the logarithmic domain are known to be particularly robust to additive noise, as $\log(a + b) \approx \log(\max\{a, b\})$ (Barker and Cooke 1997). This fact is best illustrated by plotting energies in the logarithmic power frequency domain before and after additive distortion on the x- and y-axis, respectively. The gray line in Figure 5.13, shows the ideal case of a noise free speech signal; here all

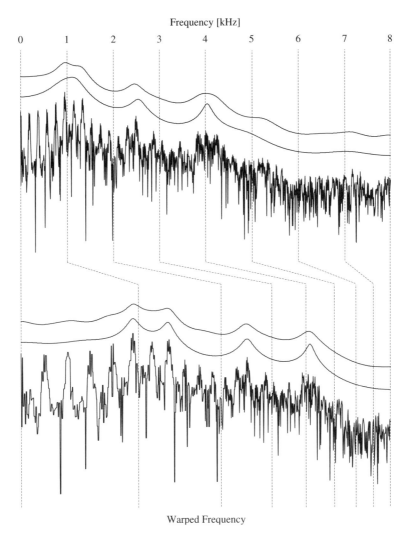

Figure 5.11 Different spectral estimations of voiced speech. From top to bottom: minimum variance distortionless response envelope with model order 30, linear prediction envelope with model order 15 and Fourier spectrum, and their mel warped, $\alpha = 0.4595$, counterparts with same model order

points fall on the line $x = y$. In the case of additive noise, the lower values of the power spectrum are lifted to higher energies; i.e., the low-energy components are masked by noise and their information is *missing*. This effect is more apparent on the power spectrum. The MVDR envelope, however, shows a broad band instead of a narrow ribbon even in the high-energy regions which is due to the high variance of the maximum amplitude in spectral envelope estimation techniques. The spectral peaks of the envelope are not as robust to additive noise as the spectral peaks of the logarithmic power spectrum. Scaling the spectral envelope to the highest peak of the Fourier spectrum provides more robust

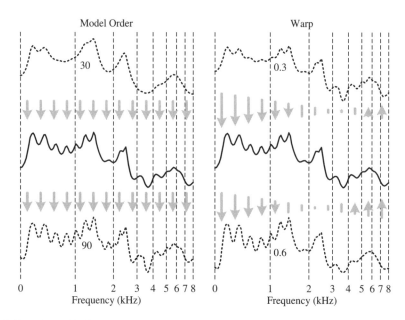

Figure 5.12 The solid lines show warped-twice MVDR spectral envelopes with model order 60, $\alpha = \alpha_{\text{mel}} = 0.4595$. Its counterparts with lower and higher model order and warp factor α are given by dashed lines. The arrows point in the direction of higher resolution. While the model order changes the overall spectral resolution at all frequencies, the warp factor moves spectral resolution to lower or higher frequencies. At the turning point frequency ($f_{tp,\alpha=0.3} = 3.23\text{kHz}$, $f_{tp,\alpha=\text{mel}} = 2.78\,\text{kHz}$, $f_{tp,\alpha=0.6} = 2.38\,\text{kHz}$), the resolution is not affected and the direction of the arrows changes

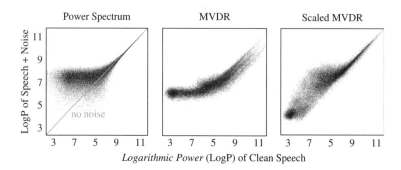

Figure 5.13 Influence of noise on captured speech in the logarithmic power domain. The average signal-to-noise ratio is 8 dB

features than both the conventional envelope, as is clear upon comparing the features of the envelope with those of the scaled envelope in Figure 5.13, and the power spectrum, as can be seen by comparing the features of the power spectrum with those of the scaled envelope.

5.4 Cepstral Processing

Cepstral features were originally invented by Bogert *et al.* (1963) to distinguish between earthquakes and underground nuclear explosions. Noll (1964) introduced them into speech processing for vocal pitch detection. The name *cepstrum*, which stems from the reversal of the first four letters of *spectrum*, was adopted because its inventors realized they were performing operations in a transform domain that were more often performed in the time domain. The cepstrum was introduced to ASR by Davis and Mermelstein (1980). One year later its successful use for the purpose of speaker verification was reported by Furui (1981). Nowadays, the cepstral features are widely used in a broad variety of speech applications. Our interest in cepstral features here stems from their overwhelming prevalence as the feature of first choice for ASR.

5.4.1 Definition and Characteristics of Cepstral Sequences

Consider a stable sequence $x[n]$ with z-transform $X(z)$. By definition the *complex cepstrum* is that stable sequence $\hat{x}[n]$ whose z-transform is

$$\hat{X}(z) \triangleq \log X(z),$$

where $\log(\cdot)$ in this case is the complex-valued logarithm (Churchill and Brown 1990, sect. 26). It is the use of the complex-valued logarithm in its definition that gives the complex cepstrum its name, for, as we will shortly see, $\hat{x}[n]$ is real-valued for all real $x[n]$. Based on the definition (3.33) of the inverse z-transform, we can write

$$\hat{x}[n] \triangleq \frac{1}{2\pi j} \oint_C \hat{X}(z) \, z^{n-1} dz = \frac{1}{2\pi j} \oint_C \log X(z) \, z^{n-1} dz,$$

where the contour of integration C must lie in the region of convergence of $\hat{X}(z) = \log X(z)$. As we require $\hat{x}[n]$ to be stable, this region of convergence must contain the unit circle. Hence, the contour C can be parameterized as $C = \{z = e^{j\omega} \, \forall \, \omega \in (-\pi, \pi]\}$, whereupon we find

$$\hat{x}[n] = \frac{1}{2\pi} \int_{-\pi}^{\pi} \log X\left(e^{j\omega}\right) e^{j\omega n} d\omega, \tag{5.38}$$

which is equivalent to the inverse Fourier transform of $\log X\left(e^{j\omega}\right)$.

The *real cepstrum*, the computation of which is illustrated in Figure 5.14, is that sequence $c_x[n]$ defined as

$$c_x[n] \triangleq \frac{1}{2\pi} \int_{-\pi}^{\pi} \log \left|X\left(e^{j\omega}\right)\right| e^{j\omega n} d\omega. \tag{5.39}$$

Note that it is common to first apply a set of mel filters, as described in Section 5.2.2, to the output of the block $\log |\cdot|$ in the figure before applying the inverse *discrete Fourier transform* (DFT). Expressing $X(e^{j\omega})$ in polar coordinates as

$$X\left(e^{j\omega}\right) = \left|X\left(e^{j\omega}\right)\right| \exp\left(j \arg X\left(e^{j\omega}\right)\right),$$

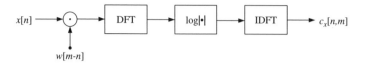

Figure 5.14 Computation of the short-time, real cepstrum

it follows that

$$\hat{X}\left(e^{j\omega}\right) = \log X\left(e^{j\omega}\right) = \log\left|X\left(e^{j\omega}\right)\right| + j \arg X\left(e^{j\omega}\right).$$

Hence, it is clear upon comparing (5.38) and (5.39) that $c_x[n]$ is the inverse transform of the real part of $\log X\left(e^{j\omega}\right)$. Hence, $c_x[n]$ must be equivalent to the conjugate-symmetric portion of $\hat{x}[n]$, such that

$$c_x[n] = \frac{\hat{x}[n] + \hat{x}^*[-n]}{2}. \tag{5.40}$$

In order to further investigate the characteristics of cepstral sequences, we must redefine the system transfer function (3.39) as

$$H(z) \triangleq K \frac{\prod_{m=1}^{M_o}(1 - a_m z) \prod_{m=1}^{M_i}(1 - c_m z^{-1})}{\prod_{m=1}^{N_o}(1 - b_m z) \prod_{m=1}^{N_i}(1 - d_m z^{-1})}, \tag{5.41}$$

where we will uniformly assume $|a_m|, |b_m|, |c_m|, |d_m| < 1$. In (5.41) M_i and N_i are, respectively, the numbers of zeros and poles *inside* the unit circle, while M_o and N_o are the numbers of zeros and poles *outside* the unit circle. This transfer function describes a mixed phased system in that it has zeros and poles both *inside* and *outside* the unit circle, the latter being $\{a_k\}$ and $\{b_k\}$, respectively.

In order to determine the time-series representations of that $\hat{h}[n]$ whose transform pair is $\hat{H}(z)$, we can make use of the series expansions

$$\log\left(1 - \alpha z^{-1}\right) = -\sum_{n=1}^{\infty} \frac{\alpha^n}{n} z^{-n} \; \forall \, |z| > |\alpha|, \tag{5.42}$$

$$\log\left(1 - \beta z\right) = -\sum_{n=1}^{\infty} \frac{\beta^n}{n} z^n \; \forall \, |z| < |\beta|^{-1}. \tag{5.43}$$

Hence, we can express $\hat{h}[n]$ as

$$\hat{h}[n] = \begin{cases} \log|K|, & n = 0, \\[2ex] -\displaystyle\sum_{m=1}^{M_i} \frac{c_m^n}{n} + \sum_{m=1}^{N_i} \frac{d_m^n}{n}, & \forall\, n > 0, \\[3ex] \displaystyle\sum_{m=1}^{M_o} \frac{a_m^{-n}}{n} - \sum_{m=1}^{N_o} \frac{b_m^{-n}}{n}, & \forall\, n < 0. \end{cases} \tag{5.44}$$

Several characteristics of cepstral sequences emerge from (5.44). Firstly, it is clear that a minimum phase system will have a *causal* sequence of cepstral coefficients, which implies $\hat{h}[n] = 0 \forall n < 0$. Secondly, for real $h[n]$, which implies that the complex poles and zeros of $H(z)$ occur in complex-conjugate pairs, $\hat{h}[n]$ will also be real as stated at the outset. A third implication of (5.44) is that the cepstral coefficients $\hat{h}[n]$ decay at *least* as fast as $1/n$. Hence, the lower order coefficients will contain *most* of the information about the overall spectral shape of $H\left(e^{j\omega}\right)$.

Another consequence of (5.40) and (5.44) is that for any transform pair $x[n] \leftrightarrow X\left(e^{j\omega}\right)$ with complex cepstrum $\hat{x}[n]$, it is possible to define a second cepstral sequence $\hat{x}_{\min}[n]$ that corresponds to the *minimum phase* transform pair $x_{\min}[n] \leftrightarrow X_{\min}\left(e^{j\omega}\right)$ whereby

$$\left| X\left(e^{j\omega}\right) \right| = \left| X_{\min}\left(e^{j\omega}\right) \right|.$$

In other words, on the unit circle the spectra corresponding to $\hat{x}[n]$ and $\hat{x}_{\min}[n]$ have the same magnitude and differ only in phase.

In order to derive an expression for $\hat{x}_{\min}[n]$ in terms of $\hat{x}[n]$, we first assume $\hat{x}_{\min}[n] = 0$ $\forall n < 0$ in (5.40), and write

$$\hat{x}_{\min}[n] = \begin{cases} 0, & \forall n < 0, \\ c_x[0] & n = 0, \\ 2c_x[n] & \forall n > 0. \end{cases} \qquad (5.45)$$

Then upon substituting (5.40) into (5.45) for the general $\hat{x}[n]$, we obtain

$$\hat{x}_{\min}[n] = \begin{cases} 0, & \forall n < 0, \\ \hat{x}[0], & n = 0, \\ 2\hat{x}[n], & \forall n > 0, \end{cases} \qquad (5.46)$$

where, in writing the equality for the case $n > 0$, we have made use of the fact that $\hat{x}[n]$ is real-valued.

The low-order cepstral coefficients, especially $\hat{x}_{\min}[0]$ and $\hat{x}_{\min}[1]$, can be given a particular intuitive meaning. The initial value $\hat{x}_{\min}[0]$ represents the average power of the input signal, although it is often replaced by more robust measurements of signal power for purposes of ASR. The next value $\hat{x}_{\min}[1]$ indicates the distribution of spectral energy between low and high frequencies. A positive value indicates a sonorant sound, as the preponderance of the spectral energy will be concentrated in the low-frequency regions. A negative value, on the other hand, indicates a fricative, inasmuch as most of the spectral energy will be concentrated at high frequencies (Deng and O'Shaughnessy 2003). Higher order cepstral coefficients represent ever increasing levels of spectral detail. Note that a finite input sequence results in an infinite number of cepstral coefficients. It is a well-established fact, however, that a finite number of coefficients, typically ranging between 12 and 20 depending on the sampling rate, is sufficient for accurate ASR (Huang 2001). This is confirmed in Section 5.7.2, which discusses the fact that cepstral coefficients with low order contribute more to the class separability than the cepstral coefficients with higher order. The ideal number of cepstral coefficients also depends on the smoothness of the spectral estimate which is determined by whether or not a filter bank is used and the type and model order of the spectral envelope used.

5.4.2 Homomorphic Deconvolution

We now develop another characteristic of cepstral sequences, which will prove very useful for ASR. The utility of this characteristic stems from the fact that it is possible to remove the effect of the periodic excitation produced by the vocal cords from a sequence of cepstral coefficients by simply *discarding* the higher order coefficients. In order to model the periodic excitation in the source-filter model of speech production discussed in Section 2.2.1 let us define the transform pairs,

$$x[n] \leftrightarrow X\left(e^{j\omega}\right), \qquad h[n] \leftrightarrow H\left(e^{j\omega}\right), \qquad p[n] \leftrightarrow P\left(e^{j\omega}\right).$$

and assume that the sequence $x[n]$ is given by the convolution,

$$x[n] = h[n] * p[n],$$

where $h[n]$ is the impulse response of a *linear time-invariant* (LTI) system and $p[n]$ is a periodic excitation of that system with period T_0. From (3.20) and the preceding discussion, it then follows that

$$\log X\left(e^{j\omega}\right) = \log H\left(e^{j\omega}\right) + \log P\left(e^{j\omega}\right).$$

Upon taking the inverse Fourier transform of the last equation, we arrive at

$$\hat{x}[n] = \hat{h}[n] + \hat{p}[n], \tag{5.47}$$

which implies that if two sequences are *convolved* in the time domain, their corresponding complex cepstra are *added*. Moreover, in light of (5.46), we can rewrite (5.47) as

$$\hat{x}_{\min}[n] = \hat{h}_{\min}[n] + \hat{p}_{\min}[n]. \tag{5.48}$$

That $h[n]$ is the impulse response of a LTI system implies that $\hat{h}[n]$ will have the form (5.44). Moreover, it is evident that $\hat{h}_{\min}[n]$ corresponds to a system function as in (5.41), where all terms $(1 - a_m z)$ and $(1 - b_m z)$ contributing zeros and poles outside the unit circle have been replaced by the terms $(1 - a_m z^{-1})$ and $(1 - b_m z^{-1})$, respectively; i.e., the zeros and poles outside of the unit circle have been replaced by their (conjugate) reciprocals, which fall inside the unit circle.

Oppenheim and Schafer (1989, sect. 12.8.2) have shown when $p[n]$ is a periodic excitation with period T_0, it follows that $\hat{p}[0] = 0$ and that $\hat{p}[n]$ will also be periodic with a period of $N_0 = T_0/T_s$ samples, where T_s is the sampling interval discussed in Section 3.1.4. This implies that $\hat{p}[n]$ is nonzero only at $\hat{p}[kN_0]$. In consequence of these facts it is clear that $\hat{h}_{\min}[n]$ can be recovered nearly perfectly from the so-called *liftering* operation

$$\hat{h}_{\min}[n] \approx \hat{x}_{\min}[n]\, w[n], \tag{5.49}$$

where

$$w[n] = \begin{cases} 1, & \forall\, 0 \le n < N_0, \\ 0, & \text{otherwise.} \end{cases}$$

The significance of the liftering operation for purposes of ASR can be immediately seen by assuming that $h[n]$ is the impulse response of a speaker's vocal tract, and $p[n]$ is the periodic excitation provided by the vocal cords during segments of voiced speech. Then (5.49) implies that, in the cepstral domain, the spectral envelope determined by the shape of the vocal tract can be separated from the periodic excitation of the vocal cords by discarding all but the lowest order cepstral coefficients.

5.4.3   Calculating Cepstral Coefficients

For the purposes of ASR, the minimum phase equivalent $\hat{x}_{\min}[n]$ of the cepstral sequence $\hat{x}[n]$ is almost invariably used to generate acoustic features. Such features can be calculated by using the inverse DFT to calculate $c_x[n]$ as in (5.39), then using this intermediate result to calculate $\hat{x}_{\min}[n]$ as in (5.45).

Another common alternative to the use of the inverse DFT is to apply the Type 2 *discrete cosine transform* (DCT) directly to the log-power spectral density $\log \left| X\left(e^{j\omega}\right)\right|$, such that

$$\hat{x}_{\min}[n] = \sum_{m=0}^{M-1} \log \left| X\left(e^{j\omega_m}\right)\right| T_{n,m}^{(2)}, \tag{5.50}$$

where $T_{n,m}^{(2)}$ are the components of the Type 2 DCT given in equation (B.1).

Yet another alternative is to calculate the cepstral coefficients $\hat{x}_{\min}[n]$ from a set of linear prediction coefficients $\{a_n\}$, as described in Sections 5.3.3, 5.3.5, and 5.3.6. All of these models result in an all-pole estimate of the spectral envelope of the form given in (5.21) and (5.22). Once the LPCs $\{a_n\}$ have been calculated, it is straightforward to extract the corresponding cepstral coefficients through the recursion,

$$\hat{x}_n = -a_n - \frac{1}{n} \sum_{m=1}^{n-1} m\, a_{n-m}\, \hat{x}_m \ \forall\, n = 1, \dots, N. \tag{5.51}$$

To demonstrate (5.51), we begin by writing

$$\hat{X}(z) = -\log A(z),$$

where

$$A(z) = 1 + \sum_{n=1}^{} a_n z^{-1}.$$

As the all-pole model (5.20) is minimum phase, the Laurent series expansion of $\hat{X}(z)$ will involve only negative powers of z. Upon replacing z by z^{-1} in both $A(z)$ and $\hat{X}(z)$, it then follows that

$$\hat{X}(z^{-1}) = \sum_{m=1}^{\infty} \hat{x}_m z^m = -\log \sum_{k=0}^{M} a_k z^k = -\log A(z^{-1}), \tag{5.52}$$

where M is the order of the LP model. Differentiating both sides of (5.52) by z provides

$$\sum_{m=1}^{\infty} m \, \hat{x}_m \, z^{m-1} = -\frac{\displaystyle\sum_{n=1}^{M} n \, a_n \, z^{n-1}}{\displaystyle\sum_{k=0}^{M} a_k \, z^k}.$$

Rewriting the last equation, we arrive at

$$-\sum_{m=1}^{\infty}\sum_{k=0}^{M} m \, a_k \, \hat{x}_m \, z^{k+m-1} = \sum_{n=1}^{M} n \, a_n \, z^{n-1}. \tag{5.53}$$

It is now possible to equate the coefficients of equivalent powers of z by requiring $n = k + m$, from which it follows

$$n \, a_n = -\sum_{m=1}^{n} m \, a_{n-m} \, \hat{x}_m = -n \, \hat{x}_n - \sum_{m=1}^{n-1} m \, a_{n-m} \, \hat{x}_m, \tag{5.54}$$

where the final equality follows from $a_0 = 1$. A minor rearrangement of (5.54) is then sufficient to demonstrate (5.51).

It is also possible to obtain a nonrecursive relation which allows linear prediction coefficients to be calculated from cepstral coefficients. A more detailed discussion together with the relevant derivation can be found in Schroeder (1981, Appendix B.1).

5.5 Comparison between Mel Frequency, Perceptual LP and warped MVDR Cepstral Coefficient Front-Ends

In the previous sections different components in the feature extraction process have been introduced. In this section we will assemble the different components into three different front-ends and discuss their advantages and disadvantages. Undoubtedly the most popular features extraction methods used in ASR are MFCCs by Davis and Mermelstein (1980) and PLP cepstral coefficients by Hermansky (1990). It depends on the task which of the two methods leads to better recognition accuracy. It is reported in the literature that MFCCs give better results under clean conditions without significant mismatch between the training data. PLP cepstral coefficients, on the other hand, provide better results in noisy or mismatched conditions. This is probably due to the unequal modeling of spectral energy by the spectral envelopes as discussed in a previous section.

More recently, novel feature extraction schemes based on the minimum variance distortionless response providing reliable and robust estimates of the spectral envelopes have been proposed; see, Wölfel and McDonough (2005); Wölfel et al. (2003) and Dharanipragada et al. (2007). The warped or perceptual MVDR spectral envelope front-ends have demonstrated consistent improvements over the two widely used methods for noise free as well as noisy recordings.

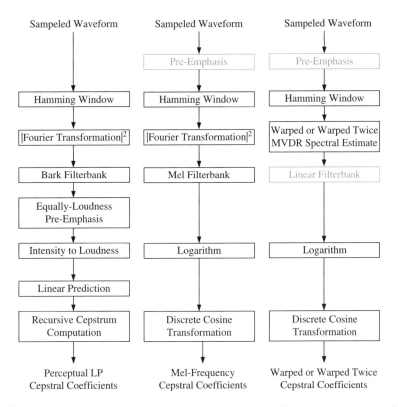

Figure 5.15 Three front-ends are compared, namely perceptual linear prediction, mel frequency and warped or warped-twice MVDR cepstral coefficients. The components in gray color are compulsive and might not be used

Figure 5.15 shows flowcharts of the components of the two traditional ASR front-ends, as well as the MVDR front-end, which was more recently proposed in the literature. The cepstral features extracted by any of the front-ends shown are typically processed with mean and variance normalization as described in Section 6.9.2. Thereafter, the normalized features are further augmented through the calculation of delta and delta–delta coefficients or frame stacking followed by discriminant analysis, as discussed in the next section.

5.6 Feature Augmentation

For time segments of 100 ms or less, human phonetic categorization and discrimination is poor. As used in ASR, a typical analysis window has a length of no more than 32 ms. This suggests that the observation context of the input feature vector as provided by a single window of short-time spectral analysis should be extended. Indeed, analysis of longer time spans seems to be essential to speech feature extraction (Yang *et al.* 2000).

5.6.1 Static and Dynamic Parameter Augmentation

It has proven useful to augment the current speech frame either by static or by dynamic features. Extended static features are easily obtained by concatenating to the current frame as many as seven consecutive frames to the left and right. Furui (1986) described how dynamic information could be helpful for ASR. To obtain dynamic information the difference between consecutive frames might be taken. For a more reliable estimate, to minimize the harmful effects of random interframe variations, dynamic information is commonly estimated over five or seven frames. In addition one can also include acceleration, however, for a reliable estimate of those second-order dynamics an even longer timespan is required. Yang *et al.* (2005) have investigated the effect of static and dynamic features. They found that dynamic features are more resilient to additive noise than their static counterparts. Also note that dynamic features are immune to short-time convolutional distortion which is associated with a constant offset in the logarithmic spectrum or cepstrum domain.

The absolute measure can be thought of a zeroth-order derivation. A first-order derivation can be approximated by the linear phase filter

$$\dot{s}[k] \approx \sum_{m=-M}^{+M} m \, s[k+m]. \tag{5.55}$$

Higher orders can be derived by consecutively reapplying (5.55) to the output of the previous order. The signal output of the first-order derivation is referred to as *delta coefficients* and the second-order *delta–delta coefficients* respectively. As differentiation filters tend to amplify noise in the measurement it is suggested to compensate for this negative effect by, e.g., spline interpolation or band-limited differentiation.

An alternative is a simple *stacking* or *concatenation* of neighboring frames

$$\mathbf{c}_{\text{stacked}}[k] = \begin{bmatrix} \mathbf{c}_{k-M} \\ \vdots \\ \mathbf{c}_k \\ \vdots \\ \mathbf{c}_{k+M} \end{bmatrix}.$$

With the introduction of the weighting matrix $\mathbf{W}$ we can write

$$\mathbf{c}_{\text{final}} \triangleq \mathbf{W} \mathbf{c}_{\text{stacked}}[k], \tag{5.56}$$

which represents a single parameter vector that contains all desired information about the signal for each frame. Note that (5.56), if the values of $\mathbf{W}$ are set accordingly, allows all kind of linear filter operation such as differentiation, averaging, and weighting. Stacking static or dynamic features increase the dimensionality of the speech feature vector. In Section 5.7 we discuss data-driven methods to determine the values of $\mathbf{W}$ and the reduction of the dimension of the augmented speech feature vector.

5.6.2   Feature Augmentation by Temporal Patterns

Hermansky and Sharma (1998) introduced a method which learns temporal patterns based on between 50 and 100 consecutive frames of speech features, which amounts to between 0.5 and 1 second. The features are derived from *logarithmic critical band energies* (LCBEs) covering at least two syllables. The raw representation of such features can be transformed into posterior probabilities of phonetic classes (Morgan and Bourlard 1995).

Hermansky and Sharma used a two-staged multi-layer perception architecture, dubbed TRAPS, which stands for *temporal patterns*. The first stage learns *critical band* phone probabilities conditioned on the input features. The second stage merges the output of each of the individual outputs of the first stage. This way of learning temporal information from the time–frequency plane is constrained to emphasize temporal trajectories of narrowband components by modeling correlation among long-term LCBE trajectories from different frequency bands. TRAPS features perform about as well as conventional short-term features. They significantly reduce *word error rates* (WERs), however, when used in combination with the conventional features, as they provide different information which can be combined to good effect.

A successor of TRAPS is dubbed *hidden activation TRAPS* (HATS) and differs inasmuch as it uses the hidden activations of the critical bands instead of their outputs as inputs to the second stage (Chen *et al.* 2003a). In Chen *et al.* (2004a) several different approaches to using long-term temporal information are compared. In that study, it was confirmed that the constraint on learning from the time–frequency plane is important.

5.7   Feature Reduction

Feature reduction is commonly applied as a preprocessing step to overcome the *curse of dimensionality* through a reduction in the dimensionality of the feature space. The term coined by Richard Bellman (1961) describes the problem whereby volume increases exponentially as additional dimensions are added to a space. For example, consider a subcube embedded within a cube in an arbitrarily high-dimensional space. In order to capture 10% of the volume in three dimensions, the subcube would need to be 46.42% as large on a side as the cube, while in 10 dimensions the subcube would need to be 79.43% as large. This effect is demonstrated for different dimensions in Figure 5.16.

The implication of the curse of dimensionality for feature extraction is that dimensionality reduction is typically required in order to perform robust parameter estimation with a limited amount of training data. An exact relationship between the number of parameters in the feature space and the expected error, the number of available training samples or imposed constrains cannot be established. A rule of thumb, however, has been suggested to prevent *overfitting*, whereby estimation errors are introduced by choosing a high-dimensional observation space when only a limited number of training samples are available. The rule of thumb suggests using at least 10 times as many training samples per class as the number of features. This ratio, however, should be increased for complex classifiers as used in ASR.

Different objective functions and procedures for feature transform and reduction exist to satisfy for different requirements:

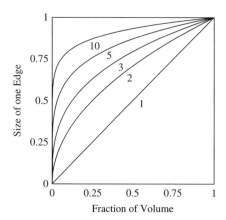

Figure 5.16 Illustration of the curse of dimensionality for different feature dimensions as indicated by the number next to each curve

- minimizing reconstruction error, e.g., principal component analysis (Dunteman 1989),
- retaining interesting directions, e.g., projection pursuit (Friedman and Tukey 1974),
- making features as independent as possible, e.g., independent component analysis (Section 12.2),
- maximizing class separability, e.g., linear discriminant analysis (Section 5.7.2),
- minimizing classification error, e.g., discriminative parameter estimation (Section 8.2).

For feature reduction focusing on speech recognition, however, only the latter two are of interest.

5.7.1 Class Separability Measures

In this and the coming sections we consider the problem of estimating a projection matrix $\mathbf{W}$ that provides maximal separability between classes in the projected feature space. For each of M classes, let us assume that there exists a set $\{\mathbf{y}_{m,k}\}$ of labeled training samples. Let the number of samples in the mth class be denoted by K_m, the mean of all samples in the mth class as $\boldsymbol{\mu}_m$, and let the mean of all samples regardless of class be denoted by $\boldsymbol{\mu}$. Class separability is a classical concept in pattern recognition, and can be expressed as a function of two out of three scatter matrices, which are defined as

- the *within-class scatter matrix*

$$\mathbf{S}_{\mathrm{w}} = \frac{1}{K} \sum_{m=1}^{M} \left[\sum_{k=1}^{K_m} (\mathbf{y}_{m,k} - \boldsymbol{\mu}_m)(\mathbf{y}_{m,k} - \boldsymbol{\mu}_m)^T \right], \tag{5.57}$$

- the *between-class scatter matrix*

$$\mathbf{S}_{\mathrm{b}} = \frac{1}{K} \sum_{m=1}^{M} K_m (\boldsymbol{\mu}_m - \boldsymbol{\mu})(\boldsymbol{\mu}_m - \boldsymbol{\mu})^T, \tag{5.58}$$

- the *total scatter matrix*

$$\mathbf{S}_{\mathrm{t}} = \mathbf{S}_{\mathrm{w}} + \mathbf{S}_{\mathrm{b}} = \frac{1}{K} \sum_{m=1}^{M} \left[\sum_{k=1}^{K_m} (\mathbf{y}_{m,k} - \boldsymbol{\mu})(\mathbf{y}_{m,k} - \boldsymbol{\mu})^T \right],$$

where K_m is the number of samples in the mth class, and K is the total number of samples. It is interesting to note that the given matrices are invariant under a coordinate shift (Fukunaga 1990). Moreover, from the above relations, it is clear that any of the scatter matrices can always be derived from the other two.

For a high-class separability all vectors belonging to the same class must be close together and well separated from the feature vectors of other classes. This implies that the final transformed feature will have a relatively small within-class and a relatively large between-class scatter matrix. The most widely used class separability measure is likely to be

$$d = \mathrm{trace}\left(\mathbf{S}_{\mathrm{w}}^{-1}\mathbf{S}_{\mathrm{b}}\right). \tag{5.59}$$

If $\mathbf{S}_{\mathrm{w}}$ in nonsingular, the class separability can be approximated by

$$d = \mathrm{trace}\left(\mathbf{S}_{\mathrm{b}}\right)/\mathrm{trace}\left(\mathbf{S}_{\mathrm{w}}\right). \tag{5.60}$$

An alternative representation of class separability replaces the trace by the determinant

$$d = \det\left(\mathbf{S}_{\mathrm{w}}^{-1}\mathbf{S}_{\mathrm{b}}\right). \tag{5.61}$$

Class separability is well correlated with WER and thus can serve as an analysis tool to compare different feature extraction or enhancement algorithms, or to compare the quality of individual channels as described in Section 12.1. In the next section, we will use class separability as an objective function in order to rank and select dimensions for feature transformation.

5.7.2 Linear Discriminant Analysis

The basic idea of discriminant analysis is to find a mapping function represented by a transform $\mathbf{W}$ such that the class separability as defined in Section 5.7.1 is maximized. *Linear discriminant analysis* (LDA) assumes that all classes have a common covariance matrix $\boldsymbol{\Sigma}_m = \boldsymbol{\Sigma} \; \forall \, m$. Linear combinations are particularly attractive because of their computational simplicity.

LDA can be used as a restricted Gaussian classifier (Hastie *et al.* 2001). The popularity of LDA for ASR, however, stems from its capacity to project feature vectors into a space of lower dimensionality while retaining all or nearly all information relevant for classification (Häb-Umbach and Ney 1992). This is possible inasmuch as the centroids of M classes can be represented in an affine subspace of dimension $\leq M - 1$, and because, under a transform obtained through LDA, the dimensions are ordered according to their relevance for classification. The latter point is evident from Figure 5.17. The 195-dimensional feature vector was obtained by concatenating 15 consecutive frames of cepstral coefficients each comprising 13 normalized coefficients. From the figure, it is clear

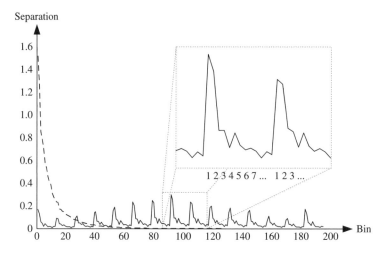

Figure 5.17 Class separation, after (5.59), for each of the 195 individual feature bins before (solid line) and after linear discriminant analysis (dashed line) processing. The initial 195 features are a concatenation of 15 frames each containing 13 cepstral coefficients

that the lower order cepstral coefficients contribute more to the class separability than the higher order coefficients. Moreover, it is apparent that the center frames contribute more to the class separability than the neighboring frames on the left and right. The dashed line in Figure 5.17 indicates the class separability after LDA, which decreases monotonically with increasing dimension. We will shortly describe how such a set of projections can be determined based on the scatter matrices defined in the last section, and how each such projection can be ranked in terms of class separability. After such a ranking has been made, only the most important projections need be retained. The elimination of detail that is superfluous for the purpose of classification through linear discriminant analysis has been shown to improve ASR performance.

The LDA problem, as originally formulated by Fisher (1936), is to find a linear projection $\mathbf{W}$ such that the between-class variance of the projected features, which are defined as,

$$\mathbf{Z} \triangleq \mathbf{W}^T \mathbf{X},$$

is maximized relative to the within-class variance. From Figure 5.18 it is obvious why this criterion makes sense. Although the projection along axis A achieves maximal separation between the means of the two classes, there is still considerable overlap between the classes in the projected space. The projection along axis B, on the other hand, produces less separation between the projected means, but greater separability overall. Hence, it is necessary to consider the structure of the class covariance matrices in determining the optimal projection.

In order to determine the optimal linear project, let us begin by determining the best single linear projection $\mathbf{w}$. This can be achieved by maximizing the *Rayleigh quotient*,

$$Q(\mathbf{S}_b, \mathbf{S}_w) \triangleq \max_{\mathbf{w}} \frac{\mathbf{w}^T \mathbf{S}_b \mathbf{w}}{\mathbf{w}^T \mathbf{S}_w \mathbf{w}}, \tag{5.62}$$

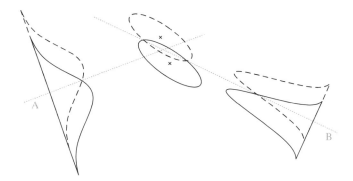

Figure 5.18 The discriminant direction without considering the variance (A) and the discriminant direction as derived by linear discriminant analysis (B). The crosses represent the class centroides and the oval are the contours of the covariance.

where S_w and S_b are defined in (5.57) and (5.58), respectively. Alternatively, the problem can be formulated as the constrained minimization problem

$$\min_{\mathbf{w}} -\frac{1}{2}\mathbf{w}^T\mathbf{S}_b\mathbf{w} \text{ subject to } \mathbf{w}^T\mathbf{S}_w\mathbf{w} = 1. \tag{5.63}$$

This constrained optimization problem is readily solved with the method of Lagrange multipliers. We need only formulate the objective function

$$\mathcal{L}(\mathbf{w}) = -\frac{1}{2}\mathbf{w}^T\mathbf{S}_b\mathbf{w} + \frac{1}{2}\lambda(\mathbf{w}^T\mathbf{S}_w\mathbf{w} - 1). \tag{5.64}$$

Taking the partial derivative with respect to $\mathbf{w}$ on both sides of (5.64) and equating to zero then yields

$$\mathbf{S}_b\mathbf{w} = \lambda\mathbf{S}_w\mathbf{w},$$

or, as $\mathbf{S}_w$ is positive definite and hence invertible,

$$\mathbf{S}_w^{-1}\mathbf{S}_b\mathbf{w} = \lambda\mathbf{w}. \tag{5.65}$$

Equation (5.65) is known as a a *generalized eigenvalue problem* (Crawford 1976). It differs from the standard eigenvalue problem in that $\mathbf{S}_w^{-1}\mathbf{S}_b$ will not, in general, be symmetric. It is possible, however, to convert (5.65) to a standard eigenvalue problem by first performing an eigenvalue decomposition on $\mathbf{S}_b$ such that

$$\mathbf{S}_b = \mathbf{U}\boldsymbol{\Lambda}\mathbf{U}^T, \tag{5.66}$$

which implies that the square root[2] of $\mathbf{S}_b$ can be defined as

$$\mathbf{S}_b^{1/2} \triangleq \mathbf{U}\boldsymbol{\Lambda}^{1/2}\mathbf{U}^T. \tag{5.67}$$

---

[2] Note that this is *not* the same square root as that based on the Cholesky decomposition, as described in Sections 10.2.1 and 13.4.4.

Thereupon, we can define

$$\mathbf{v} \triangleq \mathbf{S}_b^{1/2} \mathbf{w}. \tag{5.68}$$

Substituting (5.68) into (5.65) then provides,

$$\mathbf{S}_b^{1/2} \mathbf{S}_w \mathbf{S}_b^{1/2} \mathbf{v} = \lambda \mathbf{v},$$

which is a standard eigenvalue problem for the symmetric positive-definite matrix $\mathbf{S}_b^{1/2} \mathbf{S}_w \mathbf{S}_b^{1/2}$. It is then readily verified that the eigenvector $\mathbf{v}_{\max}$ sought is that corresponding to the maximum eigenvalue $\lambda_{\max}$ of $\mathbf{S}_b^{1/2} \mathbf{S}_w \mathbf{S}_b^{1/2}$. The corresponding projection $\mathbf{w}_{\max}$ is then given by

$$\mathbf{w}_{\max} = \mathbf{S}_b^{-1/2} \mathbf{v}_{\max}.$$

Assume now that a projection of total length M is sought, and let $\mathbf{w}_m$ denote the mth column of $\mathbf{W}$. To obain the entire linear projection matrix $\mathbf{W}$, the eigenvectors $\{\mathbf{v}_m\}$ corresponding to the M highest eigenvalues are retained and used to calculate the projections $\{\mathbf{w}_m\}$ from

$$\mathbf{w}_m = \mathbf{S}_b^{-1/2} \mathbf{v}_m \ \forall \, m = 1, \ldots, M.$$

The projection vectors $\{\mathbf{w}_m\}$ then comprise the columns of $\mathbf{W}$. It is also possible to solve for the optimal LDA projection matrix through simultaneous diagonalization (Fukunaga 1990, sect. 10.2).

LDA is a simple and powerful technique that is able to compute time derivatives implicitly (Eisele *et al.* 1996), often with better effect than explicit calculation. The disadvantage of the LDA is that it is data dependent, and thus suffers performance degradations whenever there is a mismatch between training and test data.

5.7.3 Heteroscedastic Linear Discriminant Analysis

In the Bayes' sense, LDA is the optimum solution for normal distributions with a common covariance matrix $\mathbf{\Sigma}_m = \mathbf{\Sigma} \, \forall m$. This assumption, however, does not hold for the classification of speech signals and the optimal solution is somewhat more difficult to achieve.

Let us define the mean vector $\boldsymbol{\mu}_m$ of the mth class as

$$\boldsymbol{\mu}_m \triangleq \frac{1}{N_m} \sum_{k=1}^{N_m} \mathbf{y}_{m,k},$$

where N_m is the number of features assigned to the mth class, and $\mathbf{y}_{m,k}$ is the kth feature assigned to class m. Similarly, let the scatter matrix for the mth class be defined as

$$\mathbf{S}_m \triangleq \frac{1}{N_m} \sum_{k=1}^{N_m} \left(\mathbf{x}_{m,k} - \boldsymbol{\mu}_m \right) \left(\mathbf{x}_{m,k} - \boldsymbol{\mu}_m \right)^T.$$

The objective function used in heteroscedastic linear discriminant analysis can then be defined as

$$Q(\mathbf{S}_b, \{\mathbf{S}_m\}) \triangleq \max_{\mathbf{W}} \prod_{m=1}^{M} \left(\frac{\left|\mathbf{W}^T \mathbf{S}_b \mathbf{W}\right|}{\left|\mathbf{W}^T \mathbf{S}_m \mathbf{W}\right|} \right)^{N_m}$$

or, upon taking the logarithm,

$$\log Q(\mathbf{S}_b, \{\mathbf{S}_m\}) = \underset{\mathbf{W}}{\operatorname{argmax}} \, N \log \left|\mathbf{W}^T \mathbf{S}_b \mathbf{W}\right| - \sum_{m=1}^{M} N_m \log \left|\mathbf{W}^T \mathbf{S}_m \mathbf{W}\right|, \qquad (5.69)$$

where the total number of features in the training set is given by

$$N = \sum_m N_m.$$

Unfortunately, (5.69) has no analytic solution, which implies that a numerical optimization routine, such as the method of conjugate gradients (Bertsekas 1995, sect. 1.6), must be used to find the optimal transformation matrix $\mathbf{W}$. Like LDA, heteroscedastic LDA is invariant to linear transformations of the data in the original feature space. Figure 5.19 illustrates an example where the conventional and heteroscedastic LDA solutions for the optimal transformation matrix differ. While the solution of the heteroscedastic LDA is able to handle different covariance matrices, the LDA optimizes for a global covariance matrix (not shown in the image). A detailed introduction to heteroscedastic LDA is provided in Hastie *et al.* (2001) and with regard to speech recognition in Kumar and Andreou (1998).

Further extensions to LDA are maximum likelihood-based heteroscedastic LDA (Kumar and Andreou 1996) which puts the optimization function inside the ML estimation framework. This approach can be readily extended to adopt the minimum phone error framework, dubbed MPE-LDA (Zhang and Matsoukas 2005), to incorporate *a priori* knowledge of confusable hypotheses. An alternative approach using the same optimization criteria as MPE-LDA is presented next.

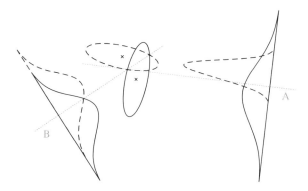

Figure 5.19 The discriminant direction as derived by conventional (B) and heteroscedastic linear discriminant analysis (A), respectively

5.8 Feature-Space Minimum Phone Error

The main idea of *feature-space minimum phone error* (fMPE) (Povey *et al.* 2004) is to adjust a feature vector that improves the minimum phone error objective function as defined in Section 8.2.3. To allow for a modification of the feature vector $\mathbf{x}_k$ at time k a linear transform matrix $\mathbf{M}$ is applied to a high-dimensional conditioning feature vector $\mathbf{g}_k$:

$$\mathbf{y}_k = \mathbf{x}_k + \mathbf{M}\mathbf{g}_k,$$

where $\mathbf{g}$ is an intermediate high-dimensional feature vector

$$\mathbf{g}_k = (p(1|\mathbf{x}_k), p(2|\mathbf{x}_k), \ldots, p(q|\mathbf{x}_k))^T$$

composed of the posterior probabilities $p(l|\mathbf{x}_k)$ which have to be evaluated on each frame. As there are approximately 100 000 Gaussians, Povey proposed to cluster them and evaluate only the most likely cluster centers. $\mathbf{M}$ represents a transform matrix that project $\mathbf{g}_k$ into the dimension of the feature $\mathbf{x}_k$. The posterior tells which of the Gaussians is close to the current frame and thus fMPE can be viewed as a region dependent feature correction function.

With the optimization of the transform matrix $\mathbf{M}$ by gradient descent the model parameters λ are also updated by iterating between updating the fMPE features and a retraining of the hidden Markov model parameters by maximum likelihood. Therefore the fMPE estimation procedure is to find a transform that maximizes the minimum phone error objective function

$$\hat{\mathbf{M}} = \underset{\mathbf{M}}{\arg\max}\ F_{\text{MPE}}(\mathbf{y}, \lambda).$$

5.9 Summary and Further Reading

This chapter has covered fundamentals of acoustic feature extraction. We have learned that efficient feature extraction techniques are based on characteristics of the human auditory system. A major contribution in acoustic feature extraction is the estimation of the spectral representation. Our focus on spectral estimation has been limited to the estimation of the Fourier transform and various spectral envelope techniques. A general introduction to spectral analysis can be found in Stoica and Moses (2005). The bilinear transform has been introduced as an alternative to nonlinearly scaled filter banks to represent the nonlinear resolution of the human ear. The cepstral representation is another important technique which can be found in nearly every acoustic front-end. We have briefly introduced the cepstrum. The interested reader who is seeking for a more rigorous treatment should refer to Oppenheim and Schafer (1989) where an entire chapter is devoted to cepstrum processing. Human, as well as automatic, phonetic categorization and discrimination is poor for short observation windows, which suggests extending the observation context. Feature reduction is another important step in obtaining robust features for ASR. We have introduced LDA and heteroscedastic LDA which is frequently used in acoustic front-ends. Hastie *et al.* (2001) give a detailed introduction to discriminant analysis and present both linear and quadratic discriminant analyses, as well as a compromise between those two dubbed regularized discriminant analyses.

A good, however dated, overview of feature extraction is given in the paper by Picone (1993). Two landmark books which are even older than the mentioned overview, but well written – covering the mathematical theory as well as applications of digital speech signal processing – are Rabiner and Schafer (1978) and the reissue of Deller *et al.* (1999). A more recent source is the chapter on speech signal representation by Huang *et al.* (2001, sect. 6) which covers some aspects that are relevant for speech feature extraction.

5.10 Principal Symbols

Symbol	Description
ϵ	excitation signal
ϕ	autocorrelation vector
Φ	autocorrelation matrix
ω	angular frequency, $\omega = 2\pi f$
a	linear prediction coefficient
b	amplitude
c	cepstral coefficient
e	error term
f	frequency
h	impulse response
H	transfer function
k	frame index
L	number of harmonics
m	discrete angle frequency
n	discrete-time index
s	fixed frequency vector
s	speech signal
S	spectrum
$\mathbf{S}$	scatter matrix
t	continuous time
$\mathbf{W}$	projection matrix
$W(z)$	weighting filter
x	clean signal
y	noisy signal

6

Speech Feature Enhancement

In *automatic speech recognition* (ASR) the distortion of the acoustic features can be compensated for either in the model domain or in the feature domain. The former techniques adapt the model on the distorted test data in such a way as if the model were *trained* on distorted data. Feature domain techniques, on the other hand, attempt to remove or suppress the distortion itself. It has been shown in various publications, such as Deng *et al.* (2000); Sehr and Kellermann (2007), that feature domain techniques provide better system performance than simply matching the training and testing conditions. The problem is especially severe for speech corrupted with reverberation. In particular, for reverberation times above 500 ms, ASR performance with respect to a model trained on clean speech does not improve significantly even when the acoustic model of the recognizer has been trained on data from the same acoustic environment (Baba *et al.* 2002).

The term *enhancement* indicates an improvement in speech quality. For speech observations, enhancement can be expressed either in terms of *intelligibility*, which is an indicator of how well the speech can be understood by a human, or *signal quality*, which is an indicator of how badly the speech is corrupted, or it can include both of these measures. For the purpose of automatic classification, features must be manipulated to provide a higher class separability. It is possible to perform speech feature enhancement in an independent preprocessing step, or within the front-end of the ASR system during feature extraction. In both cases it is not necessary to modify the decoding stage and it might not require any changes to the acoustic models of the ASR system, except for methods that change the means or variances of the features, such as cepstral mean and variance normalization. If the training data, however, is distorted itself, it might be helpful to enhance the training features as well.

In general the speech enhancement problem can be formulated as the estimation of cleaned speech coefficients by maximizing or minimizing certain objective criteria using additional knowledge, which could represent prior knowledge about the characteristics of the desired speech signal or unwanted distortion, for example. A common and widely accepted distortion measure was introduced in Chapter 4, namely, the squared error distortion,

$$d(\hat{x}, x) = |f(\hat{x}) - f(x)|^2$$

Distant Speech Recognition   Matthias Wölfel and John McDonough
© 2009 John Wiley & Sons, Ltd

where the function $f(x)$ – which could be anyone of x, $|x|$, x^2, or $\log x$ – determines the *fidelity criterion* of the estimator.

As the term *speech enhancement* is very broad and can potentially cover a wide variety of techniques, including:

- additive noise reduction,
- dereverberation,
- blind source separation,
- beamforming,
- reconstruction of lost speech packets in digital networks, or
- bandwidth extension of narrowband speech,

it is useful to provide some more specificity. An obvious classification criteria is provided by the number and type of sensors used. *Single-channel methods*, as described in this section, obtain the input from just a single microphone while *multi-channel methods* rely on observations from an array of sensors. These methods can be further categorized by the type of sensors. An example of the fusion of audio and visual features in order to improve recognition performance is given by Almajai *et al.* (2007). As discussed in Chapters 12 and 13, respectively, blind source separation and beamforming combine acoustic signals captured only with microphones. These techniques differ inasmuch beamforming assumes more prior information – namely, the geometry of the sensor array and position of the speaker – is available. Single and multi-channel approaches can be combined to further improve the signal or feature in terms of the objective function used, such as *signal-to-noise ratio* (SNR), class separability, or word error rate.

In this book we want to use the term *speech feature enhancement* exclusively to describe algorithms or devices whose purpose is to improve the speech features, where a single corrupted waveform or single corrupted feature stream is available. The goal is an improved classification accuracy which may not necessarily result in an improved or pleasing sound quality if reconstruction is at all possible. As seen in previous sections, additive noise and reverberation are the most frequently encountered problems in *distant speech recognition* (DSR) and our investigations are limited to methods of removing the effects of these distortions.

Work on speech enhancement addressing noise reduction has been a research topic since the early 1960s when Manfred Schröder at Bell Labs began working in the field. Schröder's analog implementation of *spectral subtraction*, however, is not well known inasmuch as it was only published in patents (Schröder 1965, 1968). In 1974 Weiss *et al.* (1974) proposed an algorithm in the autocorrelation domain. Five years later Boll (1979) proposed a similar algorithm which, however, worked in the spectra domain. Boll's algorithm became one of the earliest and most popular approaches to speech enhancement. A broad variety of variations to Boll's basic spectral subtraction approach followed.

Cepstral mean normalization (CMN), another popular approach, which in contrast to the aforementioned methods is designed to compensate for channel distortion, was proposed by Atal (1974) already in 1974. CMN came into wide use, however, only in the early 1990s. The effects of additive noise on cepstral coefficients as well as various remedies were investigated in the PhD dissertations by Acero (1990a), Gales (1995), and Moreno (1996).

Considering speech feature enhancement as a Bayesian filtering problem leads to the application of a series of statistical algorithms intended to estimate the state of a dynamical system. Such Bayesian filters are described in Chapter 4. Pioneering work in that direction was presented by Lim and Oppenheim (1978) where an autoregressive model was used for a speech signal distorted by additive white Gaussian noise. Lim's algorithm estimates the autoregressive parameters by solving the Yule–Walker equation with the current estimate of the speech signal and obtains an improved speech signal by applying a Wiener filter to the observed signal. Paliwal and Basu (1987) extended this idea by replacing the Wiener filter with a *Kalman filter* (KF). That work was likely the first application of the KF to speech feature enhancement. In the years following different sequential speech enhancement methods were proposed and the single Gaussian model was replaced by a Gaussian mixture (Lee *et al.* 1997). Several extensions intended to overcome the strict assumptions of the KF have appeared in the literature. The *interacting multiple model*, wherein several KFs in different stages interact with each other, was proposed by Kim (1998). Just recently very powerful methods based on partice filters have been proposed to enhance the speech features in the logarithmic spectral domain (Singh and Raj 2003; Yao and Nakamura 2002). This idea has been adopted and augmented by Wölfel (2008a) to jointly track, estimate and compensate for additive and reverberant distortions.

6.1   Noise and Reverberation in Various Domains

We begin our exposition by defining a signal model. Let $\mathbf{x} = [x_1, x_2, \cdots, x_M]$ denote the original speech sequence, let $\mathbf{h} = [h_1, h_2, \cdots, h_M]$ denote convolutional distortions such as the room impulse response, and let $\mathbf{n} = [n_1, n_2, \cdots, n_M]$ denote the additive noise sequence. The signal model can then be expressed as

$$\mathbf{y}^{(t)} = \mathbf{h}^{(t)} * \mathbf{x}^{(t)} + \mathbf{n}^{(t)}, \tag{6.1}$$

in the *discrete-time domain*, which we indicate with the superscript (t). Next we develop equivalent representations of the signal model in alternative domains, which will be indicated with suitable superscripts. The relationship, however, between additive and convolution distortion as well as the clean signal might become nontrivial after the transformation into different domains. In particular, ignoring the phase will lead to approximate solutions, which are frequently used due to their relative simplicity. An overview of the relationship between the original and clean signal is presented in Table 6.1.

The advantage of time domain techniques is that they can be applied on a sample-by-sample basis, while all alternative domains presented here require windowing the signals and processing an entire block of data at once.

6.1.1   Frequency Domain

Representing the waveform as a sum of sinusoids by the application of the Fourier transform leads to the *spectral domain* representation,

$$\mathbf{y}^{(f)} = \mathbf{h}^{(f)}\mathbf{x}^{(f)} + \mathbf{n}^{(f)}, \tag{6.2}$$

Table 6.1 Relation and approximation between the clean signal distorted by additive and convolutional distortions in different domains

Domain	Relationship and approximation
time	$\mathbf{y}^{(t)} = \mathbf{h}^{(t)} * \mathbf{x}^{(t)} + \mathbf{n}^{(t)}$
spectra	$\mathbf{y}^{(f)} = \mathbf{h}^{(f)}\mathbf{x}^{(f)} + \mathbf{n}^{(f)}$
power spectra	$\begin{aligned} \mathbf{y}^{(p)} &= \left\|\mathbf{h}^{(f)}\mathbf{x}^{(f)} + \mathbf{n}^{(f)}\right\|^2 \\ &\approx \left\|\mathbf{h}^{(f)}\mathbf{x}^{(f)}\right\|^2 + \left\|\mathbf{n}^{(f)}\right\|^2 \end{aligned}$
logarithmic spectra	$\begin{aligned} \mathbf{y}^{(l)} &= \log\left\{\left\|\mathbf{h}^{(f)}\mathbf{x}^{(f)} + \mathbf{n}^{(f)}\right\|^2\right\} \\ &\approx \mathbf{h}^{(l)} + \mathbf{x}^{(l)} + \log\left\{1 + e^{\mathbf{n}^{(l)} - \mathbf{h}^{(l)} - \mathbf{x}^{(l)}}\right\} \end{aligned}$
cepstra	$\begin{aligned} \mathbf{y}^{(c)} &= \mathbf{T}\log\left\{\left\|\mathbf{h}^{(f)}\mathbf{x}^{(f)} + \mathbf{n}^{(f)}\right\|^2\right\} \\ &\approx \mathbf{h}^{(c)} + \mathbf{x}^{(c)} + \mathbf{T}\log\left(1 + e^{\mathbf{T}^{-1}(\mathbf{n}^{(c)} - \mathbf{h}^{(c)} - \mathbf{x}^{(c)})}\right) \end{aligned}$

where, in this chapter, we will adopt the convention that the vector–vector product $\mathbf{h}^{(f)}\mathbf{x}^{(f)}$ is calculated component-by-component. Two obvious advantages of the spectral domain over the time domain are that the convolutional term is now represented by a multiplication and that the several frequency components can now be treated independently. This relationship holds also for all other spectral domain representations. Another advantage is the easier integration of nonlinearities derived from psychoacoustic models into the signal model.

Block Convolution

Common speech feature extraction front-ends, as described in Chapter 5, introduce a segmentation of the observation sequence and thus (6.2) no longer correctly describes the convolution term. This disparity becomes more severe if the convolution sequence is longer than the length of the segmentation window. This is commonly encountered in DSR where room impulse responses are much longer than the 10 to 32 ms of the segmentation window.

In order to implement linear convolution in the spectra domain, either the overlap-add or overlap-save method, both of which are described in Sections 3.2.2 and 3.2.3, respectively, can be used (Stockham 1966). Otherwise a digital filter bank, such as described in Chapter 11, can be used to implement this task. The latter solution has the advantage in that the subbands are more sharply separated due to the better transition from pass- to stopband and the superior stopband suppression. The short-time spectra domain speech signal can be expressed with definition (3.73) as

$$X_k[m] = \overline{X}[kF + n, e^{j2\pi m/M})$$

where k represents the frame, M the window length and F the frame shift. Similarly, the short-time spectral domain representation of the impulse response can be written as

$$H_k[m] = \overline{H}[kF + n, e^{j2\pi m/M}).$$

For the sake of simplicity, the windowing of the convolved sequence is neglected in the following development. We should always bear the above in mind, however, and if applied extend the given equations appropriately.

6.1.2 Power Spectral Domain

The power carried by the wave per frequency is calculated by squaring the frequency domain components. This squaring operation leads to the *power spectral domain* representation,

$$\mathbf{y}^{(p)} = \left|\mathbf{y}^{(f)}\right|^2 = \left|\mathbf{h}^{(f)}\mathbf{x}^{(f)} + \mathbf{n}^{(f)}\right|^2 = \left|\mathbf{h}^{(f)}\mathbf{x}^{(f)}\right|^2 + \left|\mathbf{n}^{(f)}\right|^2 + \mathbf{e}^{(p)}, \qquad (6.3)$$

where

$$\mathbf{e}^{(p)} = (\mathbf{h}^{(f)}\mathbf{x}^{*(f)})\mathbf{n}^{(f)} + \mathbf{h}^{(f)}\mathbf{x}^{(f)}\mathbf{n}^{*(f)} = 2\left|\mathbf{h}^{(f)}\mathbf{x}^{(f)}\right|\left|\mathbf{n}^{(f)}\right|\cos\theta, \qquad (6.4)$$

and θ denotes the frequency-dependent phase difference between the clean speech and noise signal. Under the assumption that the speech and noise signal are uncorrelated stationary random processes (6.3) can be approximated as

$$\mathbf{y}^{(p)} \approx \left|\mathbf{h}^{(f)}\mathbf{x}^{(f)}\right|^2 + \left|\mathbf{n}^{(f)}\right|^2. \qquad (6.5)$$

Upon comparing (6.3) and (6.5), it becomes clear that (6.4) represents the error between the exact and approximate representation.

If the clean speech signal and the additive noise term are considered to be uncorrelated, θ is uncorrelated and has a uniform distribution between $-\pi$ and $+\pi$. As explained subsequently in Section 12.2.2, the *central limit theorem* states, however, that the sum of i.i.d. r.v.s will be Gaussian distributed as the number of such r.v.s approaches infinity. This fact prompted Deng *et al.* (2004a) to assume Gaussian distributions to model the phase after the application of the mel filterbank, as it contains contributions of many frequency components. The output of the low-frequency filters, however, are obtained from the combination of very few or even a single spectral bin. Hence, we would not reasonably expect these outputs to be Gaussian. The distributions of the phase error in the power spectral domain for different mel-frequency bins are shown in Figure 6.1. The figure, in fact, confirms our expectation that the Gaussian approximation is poor for the lowest spectral bins. In particular, low frequencies are nearly uniformly distributed, while higher frequencies are indeed well approximated by a Gaussian distribution. The phase is mostly uncorrelated between the different mel power spectral bins.

Filter Banks

If not already implemented with the bilinear transform in the time domain, the nonlinear scaling of the frequency axis can now be applied by a transfer matrix $\mathbf{W}$ to produce a reduced number of nonlinear scaled energy bins

$$\mathbf{W}\mathbf{y}^{(p)} = \mathbf{W}\left|\mathbf{h}^{(f)}\mathbf{x}^{(f)}\right|^2 + \mathbf{W}\left|\mathbf{n}^{(f)}\right|^2 + 2\mathbf{W}\left|\mathbf{h}^{(f)}\mathbf{x}^{(f)}\right|\left|\mathbf{n}^{(f)}\right|\cos\theta.$$

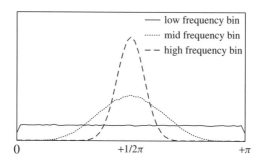

Figure 6.1 Distribution of the phase error in the power spectral domain for different mel-scaled frequency bins

As this step does not alter our principal assumptions, it will be neglected during further analysis. It should be, however, kept in mind that some simplifications cannot be performed due to the application of **W**.

6.1.3 Logarithmic Spectral Domain

To obtain a more meaningful representation of the speech signal, as discussed in Section 2.3.3 or by Acero (1990a), the use of the *logarithmic spectral domain*

$$y^{(l)} = \log\left\{\left|\mathbf{h}^{(f)}\mathbf{x}^{(f)} + \mathbf{n}^{(f)}\right|^2\right\}, \tag{6.6}$$

has been proposed. To develop the relationship between **y**, **h**, **x** and **n** in the logarithmic spectral domain, we define

$$\mathbf{n}^{(l)} = \log\left|\mathbf{n}^{(f)}\right|^2, \ \mathbf{h}^{(l)} = \log\left|\mathbf{h}^{(f)}\right|^2, \ \mathbf{x}^{(l)} = \log\left|\mathbf{x}^{(f)}\right|^2.$$

By rewriting (6.3) as

$$e^{\mathbf{y}^{(l)}} = e^{\mathbf{h}^{(l)}}e^{\mathbf{x}^{(l)}} + e^{\mathbf{n}^{(l)}} + \frac{e^{(p)}}{\left|\mathbf{h}^{(f)}\right|\left|\mathbf{x}^{(f)}\right|\left|\mathbf{n}^{(f)}\right|}e^{(\mathbf{h}^{(l)}+\mathbf{x}^{(l)}+\mathbf{n}^{(l)})/2},$$

and by applying the chain of transformations, we obtain

$$
\begin{aligned}
\mathbf{y}^{(l)} &= \log\left(e^{\mathbf{h}^{(l)}+\mathbf{x}^{(l)}} + e^{\mathbf{n}^{(l)}} + \frac{e^{(p)}}{\left|\mathbf{h}^{(f)}\right|\left|\mathbf{x}^{(f)}\right|\left|\mathbf{n}^{(f)}\right|}e^{(\mathbf{h}^{(l)}+\mathbf{x}^{(l)}+\mathbf{n}^{(l)})/2}\right) \\
&= \log\left(e^{\mathbf{h}^{(l)}+\mathbf{x}^{(l)}}\left(1 + e^{\mathbf{n}^{(l)}-\mathbf{h}^{(l)}-\mathbf{x}^{(l)}} + \frac{e^{(p)}}{\left|\mathbf{h}^{(f)}\right|\left|\mathbf{x}^{(f)}\right|\left|\mathbf{n}^{(f)}\right|}e^{(\mathbf{n}^{(l)}-\mathbf{h}^{(l)}-\mathbf{x}^{(l)})/2}\right)\right) \\
&= \mathbf{h}^{(l)} + \mathbf{x}^{(l)} + \log\left(1 + e^{\mathbf{n}^{(l)}-\mathbf{h}^{(l)}-\mathbf{x}^{(l)}} + \frac{e^{(p)}}{\left|\mathbf{h}^{(f)}\right|\left|\mathbf{x}^{(f)}\right|\left|\mathbf{n}^{(f)}\right|}e^{(\mathbf{n}^{(l)}-\mathbf{h}^{(l)}-\mathbf{x}^{(l)})/2}\right)
\end{aligned}
$$

$$= \mathbf{h}^{(l)} + \mathbf{x}^{(l)} + \log\left(1 + e^{\mathbf{n}^{(l)} - \mathbf{h}^{(l)} - \mathbf{x}^{(l)}}\right)$$

$$+ \log\left(1 + \frac{\mathbf{e}^{(p)} e^{(\mathbf{n}^{(l)} - \mathbf{h}^{(l)} - \mathbf{x}^{(l)})/2}}{\left|\mathbf{h}^{(f)}\right| \left|\mathbf{x}^{(f)}\right| \left|\mathbf{n}^{(f)}\right| \left(1 + e^{\mathbf{n}^{(l)} - \mathbf{h}^{(l)} - \mathbf{x}^{(l)}}\right)}\right)$$

$$= \mathbf{h}^{(l)} + \mathbf{x}^{(l)} + \log\left(1 + e^{\mathbf{n}^{(l)} - \mathbf{h}^{(l)} - \mathbf{x}^{(l)}}\right)$$

$$+ \log\left(1 + \frac{\mathbf{e}^{(p)}}{\left|\mathbf{h}^{(f)}\right| \left|\mathbf{x}^{(f)}\right| \left|\mathbf{n}^{(f)}\right| \cosh[(\mathbf{n}^{(l)} - \mathbf{h}^{(l)} - \mathbf{x}^{(l)})/2]}\right)$$

$$= \mathbf{h}^{(l)} + \mathbf{x}^{(l)} + \log\left(1 + e^{\mathbf{n}^{(l)} - \mathbf{h}^{(l)} - \mathbf{x}^{(l)}}\right) + \mathbf{e}^{(l)}$$

where $\mathbf{1} = [1, 1, \cdots, 1]^T$. The final equation can be approximated as

$$\mathbf{y}^{(l)} \approx \mathbf{h}^{(l)} + \mathbf{x}^{(l)} + \log\left(1 + e^{\mathbf{n}^{(l)} - \mathbf{h}^{(l)} - \mathbf{x}^{(l)}}\right), \tag{6.7}$$

which is, in fact, exact if the speech and noise signal are uncorrelated stationary random processes. The *error term*

$$\mathbf{e}^{(l)} = \log\left(1 + \frac{\mathbf{h}^{(f)} \mathbf{x}^{(f)} \mathbf{n}^{(f)} \cos\theta}{\left|\mathbf{h}^{(f)}\right| \left|\mathbf{x}^{(f)}\right| \left|\mathbf{n}^{(f)}\right| \cosh\left\{\log\left|\mathbf{n}^{(f)}\right| - \log\left|\mathbf{h}^{(f)}\right| - \log\left|\mathbf{x}^{(f)}\right|\right\}}\right)$$

is difficult to evaluate, but Deng *et al.* (2002) have empirically verified that the average value of $\mathbf{e}^{(l)}$ is close to zero and that θ is approximately Gaussian distributed.

One advantage of working in the logarithmic spectral domain is that convolution becomes an additive term known as *spectral tilt*. Furthermore, it has been shown that the application of feature enhancement techniques (Ephraim, Y. and Malah, D. 1984; Hu and Loizou 2007) in the logarithmic spectral domain provides better distortion attenuation than their application in the power spectral domain. In contrast to the power spectral domain, the logarithmic spectral domain has a linear relationship to the cepstral domain. This implies that all operations are performed on the (very nearly) final features of the recognition system, which is a distinct advantage.

6.1.4 Cepstral Domain

The features of most ASR systems are represented in the *cepstral domain*, in which the signal model can be expressed as

$$\mathbf{y}^{(c)} = \mathbf{T}\log\mathbf{y}^{(p)} = \mathbf{T}\log\left\{\left|\mathbf{h}^{(f)} \mathbf{x}^{(f)} + \mathbf{n}^{(f)}\right|^2\right\}, \tag{6.8}$$

where $\mathbf{T}$ denotes the Type 2 cosine transform matrix as defined in Section B.1. Because the cepstral domain can be expressed by a simple matrix multiplication with the logarithmic spectral domain, it can be approximated by

$$\mathbf{y}^{(c)} \approx \mathbf{T}\mathbf{h}^{(l)} + \mathbf{T}\mathbf{x}^{(l)} + \mathbf{T}\log\left(\mathbf{1} + e^{\mathbf{n}^{(l)} - \mathbf{h}^{(l)} - \mathbf{x}^{(l)}}\right)$$
$$= \mathbf{h}^{(c)} + \mathbf{x}^{(c)} + \mathbf{T}\log\left(\mathbf{1} + e^{\mathbf{T}^{-1}(\mathbf{n}^{(c)} - \mathbf{h}^{(c)} - \mathbf{x}^{(c)})}\right). \tag{6.9}$$

The evaluation of the error term

$$\mathbf{e}^{(c)} = \mathbf{T}\,\mathbf{e}^{(l)},$$

is further complicated in the cepstral domain through the required matrix multiplication with $\mathbf{T}$.

The advantage of the cepstral coefficients is a decorrelation of the bins. This, however, comes at the cost that the corruption is not independent across the feature dimensions as in the spectral domain. Thus, some processing steps, such as model combination, require the inverse transformation into the logarithmic spectral domain prior to their application.

6.2   Two Principal Approaches

Various approaches to single channel speech feature enhancement have been proposed in the literature. A broad variety of techniques segment the time signal prior to processing. We will focus exclusively on such techniques in this chapter, as the ASR system requires the segmentation of speech in any event. Other algorithms, such as subspace approaches (Ephraim and Van Trees 1995; Hermus *et al.* 2007), are covered in various publications. A good overview is presented in Loizou (2007).

Two principal classes of methods for speech feature enhancement exist. In the first class, the enhanced features are estimated *directly*. In the second class of methods, the distortions induced in the signal are initially estimated for subsequent removal; hence, such techniques are dubbed *indirect*, inasmuch as most approaches belonging to this class compensate for the distortion in either the power spectral, logarithmic spectral or cepstral domain.

These two approaches can, for example, be expressed in the *minimum mean squared error* (MMSE) Bayesian filter framework,[1] as discussed in Chapter 4 and will be investigated in the following. Other estimators, such as maximum likelihood,[2] differ primarily in the operating assumptions made and optimization criteria used. The solution for the *direct MMSE Bayesian estimate* is given by the conditional mean (4.8) and repeated here for convenience

$$\mathcal{E}\{\mathbf{x}_k | \mathbf{y}_{1:k}\} = \int \mathbf{x}_k \, p(\mathbf{x}_k | \mathbf{y}_{1:k}) \, d\mathbf{x}_k.$$

---

[1] The Wiener filter is optimal in the MMSE sense for complex spectrum estimation as it assumes a linear relationship between the observed data and the estimator. This is not the case, however, in the magnitude spectrum, logarithmic magnitude spectrum or cepstral domains all of which are considered here.

[2] As the maximum likelihood approach does not provide enough attenuation it is typically not used by itself, but in conjunction with other techniques.

The solution of the *indirect MMSE Bayesian estimate*, however, is obtained by first estimating the additive $\mathbf{n}_k$ and convolutional $\mathbf{h}_k$ distortions, respectively,

$$p(\mathbf{x}_k|\mathbf{y}_{1:k}) = \int \int p(\mathbf{x}_k, \mathbf{n}_k, \mathbf{h}_k|\mathbf{y}_{1:k}) \, d\mathbf{n}_k \, d\mathbf{h}_k.$$

Thereafter the calculation of the conditional mean (4.8) can be expressed as

$$\mathcal{E}\{\mathbf{x}_k|\mathbf{y}_{1:k}\} = \int \mathbf{x}_k \int \int p(\mathbf{x}_k, \mathbf{n}_k, \mathbf{h}_k|\mathbf{y}_{1:k}) \, d\mathbf{n}_k \, d\mathbf{h}_k \, d\mathbf{x}_k.$$

Using the relation $p(\mathbf{x}_k, \mathbf{n}_k, \mathbf{h}_k|\mathbf{y}_{1:k}) = p(\mathbf{x}_k|\mathbf{y}_{1:k}, \mathbf{n}_k, \mathbf{h}_k) \, p(\mathbf{n}_k, \mathbf{h}_k|\mathbf{y}_{1:k})$ and interchanging the order of integration, we obtain

$$\mathcal{E}\{\mathbf{x}_k|\mathbf{y}_{1:k}\} = \int \int f_k(\mathbf{y}_{1:k}, \mathbf{n}_k, \mathbf{h}_k) \, p(\mathbf{n}_k, \mathbf{h}_k|\mathbf{y}_{1:k}) \, d\mathbf{n}_k \, d\mathbf{h}_k, \qquad (6.10)$$

where

$$f_k(\mathbf{y}_{1:k}, \mathbf{n}_k, \mathbf{h}_k) = \int \mathbf{x}_k \, p(\mathbf{x}_k|\mathbf{y}_{1:k}, \mathbf{n}_k, \mathbf{h}_k) \, d\mathbf{x}_k \qquad (6.11)$$

is a linear or nonlinear function mapping the sequence of observations $\mathbf{y}_{1:k}$ and distortions $\mathbf{n}_k, \mathbf{h}_k$ to the clean speech $\mathbf{x}_k$. In the indirect approach, the distortions $\mathbf{n}_k$ and $\mathbf{h}_k$ are treated as the *state* in the Bayesian framework. Hence, it is assumed that neither $\mathbf{n}_k$ nor $\mathbf{h}_k$ can be observed directly. Rather, they can only be inferred, and once inferred their effects can be removed to obtain an estimate of clean speech $\mathbf{x}_k$. Note that both approaches, the direct and the indirect, can account for speech and non-speech regions, or even for different phoneme classes, through, for example, soft-decision gain modifications (McAulay and Malpass 1980).

6.3   Direct Speech Feature Enhancement

This section describes four popular direct approaches which estimate the features of clean speech from the distorted observation. While the approach presented first is widely used for speech enhancement in general, the last three approaches have been developed particularly for ASR. The second and third are applied in the logarithmic spectral domain while the fourth works directly in the cepstral domain.

6.3.1   Wiener Filter

The direct estimate of the original speech signal can be obtained through the application of the Wiener filter described in Section 4.2, as

$$\hat{X}(\omega) = \hat{G}(\omega)Y(\omega).$$

Under the assumptions that the signal and noise are orthogonal and that the noisy observation, clean speech and noise, are related by

$$|Y_k(\omega)|^2 \approx |X_k(\omega)|^2 + |N_k(\omega)|^2,$$

we can express the solution for the Wiener filter transfer function $\hat{G}(\omega)$ according to equation (4.16). The transfer function, which is also known as the *gain function* because it indicates the amount of suppression, can then be written as

$$\hat{G}(\omega) = \frac{\Sigma_X(\omega)}{\Sigma_X(\omega) + \Sigma_N(\omega)}. \tag{6.12}$$

Simple substitutions let us alternatively express the gain function as a function of the *a priori* SNR value η according to

$$\hat{G}(\omega) = \frac{\eta(\omega)}{\eta(\omega) + 1}. \tag{6.13}$$

Note that $\hat{G}(\omega)$ must lie in the range $0 \le \hat{G}(\omega) \le 1$. Inaccurate estimates of the noise, however, can cause $|\hat{N}(\omega)|$ to exceed $|Y(\omega)|$, thereby resulting in negative and complex values of $\hat{G}(\omega)$, which must be compensated for.

Parametric Wiener Filters

The *parametric Wiener filter* (Lim and Oppenheim 1979) is a generalization of (6.12) by the introduction of two additional variables a, b and the overestimation factor α similar to spectral subtraction

$$\hat{G}(\omega) = \left(\frac{X^a(\omega)}{X^a(\omega) + \alpha N^a(\omega)} \right)^b. \tag{6.14}$$

As in (6.13), we can write the transfer function of the Wiener filter as a function of *a priori* SNR such that

$$\hat{G}(\omega) = \left(\frac{\eta^a(\omega)}{\eta^a(\omega) + \alpha} \right)^b.$$

The conventional Wiener filter follows by setting $a = b = \alpha = 1$, and the square-root Wiener filter by setting $a = \alpha = 1$ and $b = 1/2$. The free parameters change the attenuation of the signal according to the *a priori* SNR value and may yield improved performance with respect to the conventional Wiener filter.

A detailed discussion about the free parameters can be found in Loizou (2007). There, other extensions such as *constrained* or *codebook-driven* Wiener filtering are also presented.

6.3.2 Gaussian and Super-Gaussian MMSE Estimation

In this section, we briefly present direct estimators which are optimal with respect to the MMSE criterion in the logarithmic spectral domain, assuming that the pdf of speech is Gaussian. The logarithmic spectral domain is not only subjectively more meaningful than the spectral magnitude domain, but direct estimation performed in this domain has also been proven to yield higher noise attenuation, about 3 dB, in contrast to the linear MMSE estimator. Because of space limitations we present only the direct logarithmic MMSE estimation technique (Ephraim and Malah 1985) without derivations. The linear MMSE estimation as well as the logarithmic MMSE estimators, including their derivations, are widely covered in the literature; see Loizou (2007) for example.

The direct logarithmic MMSE estimator calculates the conditional mean

$$\hat{\mathbf{x}}_k = e^{\mathcal{E}\{\log \mathbf{x}_k | \mathbf{y}_k\}}.$$

The solution for the conditional mean, however, is not straightforward and additional constraints must be imposed. As discussed in Section 6.5.4, let η and γ denote the a priori and a posteriori SNR respectively. Assuming a Gaussian distribution for the spectral representation of the speech signal, the solution is given by

$$\hat{\mathbf{x}}_k = \frac{\eta_k}{\eta_k + 1} + \exp\left\{\frac{1}{2}\int_{v_k}^{\infty}\frac{e^{-t}}{t}dt\right\}\mathbf{y}_k \approx \frac{\eta_k}{\eta_k + 1} + \exp\left\{\frac{1}{2}\frac{e^{v_k}}{v_k}\sum_{l}\frac{l!}{v_k^l}\right\}\mathbf{y}_k,$$

where

$$v_k = \frac{\eta_k}{\eta_k + 1}\gamma.$$

From Section 2.2.4 and Figure 2.4 it is clear that the Gaussian assumption does not hold. Thus it has been proposed to use super-Gaussian distributions, such as the Laplacian or Gamma distribution, in the MMSE estimation framework (Lotter and Vary 2005; Martin 2005). Only small improvements have been obtained, however, through the introduction of the non-Gaussian assumption.

6.3.3 RASTA Processing

To increase the independence from constant and slowly varying channel characteristic, *relative spectrum processing* (RASTA) was proposed by Hermansky and Morgan (1994). RASTA applies an infinite impulse response bandpass filter in the logarithmic power domain, whose transfer function is given by

$$H(z) = 0.1z^4\frac{2 + z^{-1} - z^{-3} - 2z^{-4}}{1 - 0.98z^{-1}},$$

to the speech features. This lowpass filter smooths fast frame-to-frame changes, while the highpass filter is intended to remove convolutional noise similar to CMN; see

Section 6.9.2. In fact, it has been shown empirically that RASTA processing provides an effect similar to real-time CMN. Note that these methods are not totally unproblematic in that not only artifacts, but also slowly varying classification-relevant features, may be suppressed.

6.3.4 Stereo-Based Piecewise Linear Compensation for Environments

Stereo-based piecewise linear compensation for environments (SPLICE) is a *nonparametric approach*, which assumes that the distortion environment is known. It makes no assumptions about the distortion, however, but rather learns the mapping of one or more classes of distortions *a priori*. The main idea of SPLICE is to apply a transformation based on a probabilistic model of distortion from clean speech into noisy speech. The latter is learned from a set of stereo training data. SPLICE was first applied in cepstral space, but may also be applied in alternative domains. SPLICE is an extension to the *fixed codeword-dependent cepstral normalization* algorithm (Acero and Stern 1991) which itself is a successor of *codeword-dependent cepstral normalization* (CDCN) (Acero and Stern 1990). The original version of SPLICE as proposed by Deng *et al.* (2000) assumes that the noisy speech vector $\mathbf{y}_k$ lies in one of several partitions of the acoustic space. These partitions are determined from a mixture of M Gaussians. The mean and variances of the correction $\mathbf{r}$ are trained by vectors which have been classified into corresponding codewords. Furthermore, the SPLICE algorithm assumes that the relation between $\mathbf{x}_k$ and $\mathbf{y}_k$ is piecewise linear, according to

$$\mathbf{x}_k = \mathbf{y}_k + \mathbf{r}(\mathbf{y}_k) \approx \mathbf{y}_k + \mathbf{r}_{m(\mathbf{y}_k)},$$

where $m(\mathbf{y}_k)$ determines which part of the local linear approximation is used. Under these assumptions, the enhanced feature can be calculated under the MMSE criterion as follows:

$$\hat{\mathbf{x}}_k = \int_{\mathbf{x}_k} \mathbf{x}_k \, p(\mathbf{x}_k|\mathbf{y}_k) \, d\mathbf{x}_k \approx \int_{\mathbf{x}_k} \left(\mathbf{y}_k + \mathbf{r}_{m(\mathbf{y}_k)}\right) p(\mathbf{x}_k|\mathbf{y}_k) \, d\mathbf{x}_k$$

$$= \mathbf{y}_k + \int_{\mathbf{x}_k} \mathbf{r}_{m(\mathbf{y}_k)} \, p(\mathbf{x}_k|\mathbf{y}_k) \, d\mathbf{x}_k = \mathbf{y}_k + \int_{\mathbf{x}_k} \sum_{m=1}^{M} \mathbf{r}_m \, p(\mathbf{x}_k, m|\mathbf{y}_k) \, d\mathbf{x}_k$$

$$= \mathbf{y}_k + \sum_{m=1}^{M} p(m|\mathbf{y}_k) \, \mathbf{r}_m$$

The posterior probabilities $p(m|\mathbf{y}_k)$ are computed by Bayes' rule using the clustered parameters in the *Gaussian mixture model* (GMM) approximation of $p(\mathbf{y})$.

As mentioned earlier, the major drawback of the earliest versions of SPLICE was their dependence on *stereo data*[3] in order to calculate the *maximum likelihood* (ML) estimate of clean speech features. Two extensions to the original approach have been

---

[3] We refer to stereo data as two time-aligned channels, one providing the original observations, the other containing distorted observations of exactly the same source.

proposed to overcome those limitations, one using a ML criterion (Wu 2004) and one using discriminative training by minimum classification error (Wu and Huo 2002). Deng *et al.* (2005) reports that the latter method is very similar to the fMPE algorithm of Povey *et al.* (2005), which is described in Section 5.8.

6.4   Schematics of Indirect Speech Feature Enhancement

As depicted in Figure 6.2, frame-based indirect feature enhancement methods can be decomposed into four separate processing components. Below we provide a brief description of each component. More detailed descriptions follow in the coming sections.

- *Distortion estimation* – All feature enhancement techniques require an estimate of the distortion which is either represented as a point estimate or as a density. Apart from those techniques based on training with *stereo data* such as SPLICE, the distortion is in general not known *a priori* and thus must be estimated from the distorted observation.
- *Distortion evolution* – If the distortion is assumed to be nonstationary it is not sufficient to have a fixed distortion estimate over all frames, rather the distortion must be tracked and the estimate constantly updated. Tracking the distortion requires the *prediction* $p(\mathbf{d}_k | \mathbf{d}_{k-1})$ of the estimate at time k given the previous distortion estimates $\mathbf{d}_{0:k-1}$.
- *Distortion evaluation* – The second step necessary in tracking is to *update* the prior, which entails the evaluation of the likelihood $p(\mathbf{y}_k | \mathbf{d}_k)$ for each distortion hypothesis based on a model of clean speech.
- *Distortion compensation* – To finally derive a point estimate of clean speech features involves the compensation of the distortion by subtraction or inverse filtering. The original feature estimate can be augmented by the uncertainty of the enhancement process and propagated into the acoustic model of the recognition engine, as explained in Section 6.11.

Some of the components either rely on *a priori* knowledge or estimates derived from the current observation. In the case of stationary distortion or if the distortion estimate is constantly updated, such as with the use of minimum statistics or multi-step linear prediction, the two components *distortion evolution* and *distortion evaluation* can be

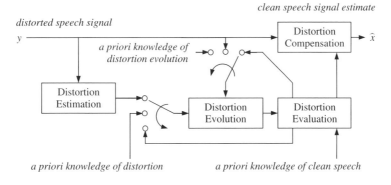

Figure 6.2   Schematics of frame-based speech feature enhancement techniques

eliminated. It is possible that the removal of these components, however, will lead to degradations in system performance.

Note that frame-based speech feature enhancement might be applied in multiple steps in different working domains. A speech recognition front-end with enhancement could, for example, first use multi-step linear prediction to reduce the effect of reverberation in the power spectral domain, then apply a *particle filter* (PF) to track and subtract additive noise in the logarithmic spectral domain. Subsequently, channel effects could be removed through mean and variance normalization of the features in the cepstral domain.

6.5 Estimating Additive Distortion

Neglecting the convolution term in (6.1), the relationship between the input sequence $\mathbf{x}$ and the output sequence $\mathbf{y}$ can be expressed as

$$y[n] = x[n] + n[n].$$

In this case, the only distortion is additive noise which, in general, is assumed to be uncorrelated with the speech signal and unknown.

Feature enhancement techniques require knowledge about the distortion. In the case of additive noise, such knowledge can be represented in the different spectral domains as the *noise floor*. The estimate of the noise floor has a great influence on the overall quality of the enhanced speech signal and thus is the most important component in any additive noise speech enhancement technique. A broad variety of techniques have been proposed for estimating the noise floor, all of which rely on one or more of the following assumptions:

- speech and noise signals are statistically independent;
- the observed noisy speech signals contain regions where the speech is not present;
- the noise signal is more stationary than the speech signal;
- noise statistics are stationary within the analysis window used for the short-time Fourier transform.

In the following sections we describe several popular methods for estimating additive distortions.

6.5.1 Voice Activity Detection-Based Noise Estimation

Under the assumptions that the noise is stationary and that the speech can be reliably separated into regions of speech and non-speech through *voice activity detection* (VAD), a simple and very effective way to obtain a static noise estimate $|\hat{N}(\omega)|$ is based on the average of the power spectrum over K frames,

$$|\hat{N}(\omega)|^2 = \frac{1}{K} \sum_{k=0}^{K-1} |Y_k(\omega)|^2. \tag{6.15}$$

In this technique, the frames used in the noise estimate are assumed to contain only noise. The noise estimate can also be updated sequentially by single-pole recursive averaging, according to

$$\hat{N}_k(\omega) = \begin{cases} \alpha_{\mathrm{a}}\hat{N}_{k-1}(\omega) + (1 - \alpha_{\mathrm{a}})\,X_k(\omega), & X_k(\omega) \geq \hat{N}_{k-1}(\omega) \\ \alpha_{\mathrm{d}}\hat{N}_{k-1}(\omega) + (1 - \alpha_{\mathrm{d}})\,X_k(\omega), & X_k(\omega) < \hat{N}_{k-1}(\omega) \end{cases}$$

where α_{a} and α_{d} are the *attack* and *decay* constants which must be tuned on training data, respectively.

The distinction between speech and noise becomes difficult for low SNR values and thus noise estimation methods based on VAD work well only for medium and high SNRs. A broad variety of VAD methods have been proposed in the literature. Beritelli *et al.* (2001) present and compare different standard VAD algorithms. One class of VAD exploits the bimodal structure of speech pdfs. Whereas the mode with higher energy represents the speech signal, the mode with lower energy reflects the noise signal. Comparing the energy of a given frame with a threshold lying between the two modes is sufficient to determine if speech is present or not.

6.5.2 Minimum Statistics Noise Estimation

Tracking nonstationary noise requires the update of the noise estimate in both the presence and absence of speech. In such cases, it is better not to rely on an explicit threshold between speech activity and speech pauses. One popular method that uses no explicit threshold is known as *minimum statistics*, and was proposed by Martin (1994) and modified by Doblinger (1995). Like other *soft-decision* methods, it can update the estimated noise distribution during speech activity. Thus the minimum statistics algorithm performs well in nonstationary noise, as demonstrated by Meyer *et al.* (1997), among others.

The minimum statistics method rests on three assumptions, namely that

- the speech and the noise signals are statistically independent,
- the power of a noisy speech signal frequently decays to levels which are representative of the noise power level, and
- the noise is stationary within the sliding window.

It is therefore possible to obtain an estimate of the noise floor by tracking the minima of a smoothed version of the spectrum. Since the minimum is smaller than the average value, the minimum tracking method requires a bias compensation.

Recursive estimation of the power spectrum with a time- and frequency-dependent smoothing parameter $\beta_k(\omega)$ leads to

$$\overline{S}_k(\omega) = \beta_k(\omega)\overline{S}_{k-1}(\omega) + (1 - \beta_k(\omega))|Y_k(\omega)|^2. \tag{6.16}$$

Note that realizing the full potential of minimum statistics requires the application of a time- and frequency-dependent smoothing parameter, which was not applied in the earliest versions of the algorithm (Martin 2005). A coarse estimate of the noise floor can then be

Algorithm 6.1   Outline of the minimum statistics algorithm

1. Compute short-term power spectrum $|Y_k(\omega)|^2$
2. Compute time- and frequency-dependent smoothing parameter $\beta_k(\omega)$
3. Compute the magnitude squared spectrum (6.16)
4. Search for the minimum value over M frames (6.17)
5. Compute and compensate for the bias (6.18)

Steps 1 through 5 are repeated with $k \mapsto (k+1)$ until all frames are processed.

derived for each frame by the minimum of the smoothed power spectral estimate over a sliding window of length M:

$$|\hat{N}_k(\omega)|^2 = \min(\overline{S}_{k-M/2}(\omega), \cdots, \overline{S}_{k-1}(\omega), \overline{S}_k(\omega), \overline{S}_{k+1}(\omega), \cdots, \overline{S}_{k+M/2}(\omega)). \quad (6.17)$$

As mentioned before, the noise floor estimated by tracking the smoothed spectral minima is biased toward higher values. This bias can be determined and can be shown to be dependent on $\overline{Y}_k(\omega)$. Compensation for the bias is thus possible by multiplying with the factor

$$Q = 2\frac{\mathcal{E}\left\{|N_k(\omega)|^2\right\}}{\text{variance}\left\{|\hat{N}_k(\omega)|^2\right\}}. \quad (6.18)$$

A summary of minimum statistics noise estimation can be found in Algorithm 6.1.

6.5.3   Histogram- and Quantile-Based Methods

Some of the variations on the previously described minimum statistics approach are the histogram-based methods. These accomplish noise estimation through the analysis of sub-band histograms (Hirsch and Ehrlicher 1995), where it is assumed that the most frequently observed spectral energies in individual frequency bands correspond to noise. In some cases, the histogram can be separated into two modes where the low-energy region represents the energy observed in regions without speech, while the high-energy regions represent the energy observed if speech is present. Alternatively, instead of relying on the maximum, a centroid clustering algorithm, such as K-means, can be used to detect noise, the lower centroid value, and speech regions in the histogram. This algorithm, however, yields larger noise estimation errors (Hirsch and Ehrlicher 1995). Note that the observed histogram patterns are not consistent and typically depend on the type and level of the noise.

For each frame and frequency bin, a histogram of the power spectrum is constructed over a region of several hundred milliseconds. The choice of the observation length involves the well-known tradeoff between a reliable noise estimate and tracking capability. For long observation times, the noise estimates obtained with minimum statistics-based tracking and the histogram-based algorithms become very similar.

To prevent noise overestimation, which is more prominent in low-frequency regions and in particular occurs for short observation windows, it was proposed to exclude frames

Algorithm 6.2 Outline of the histogram-based method

1. Compute short-term power spectrum $|Y_k(\omega)|^2$
2. Smooth the short-term power spectrum by first-order recursion

$$\overline{S}_k(\omega) = \alpha \overline{S}_{k-1}(\omega) + (1 - \alpha) |Y_k(\omega)|^2$$

   with the smoothing constant α, e.g., 0.8
3. Compute the histogram for each frequency band of $\overline{S}(\omega)$ over the last observations $k, k-1, \ldots, k-m$, e.g., several hundred ms
4. Take for each frequency bin the spectral power with the highest number of appearance as noise estimate $\hat{N}_k(\omega)$
5. Smooth the noise estimate by first-order recursion

$$\overline{N}_k(\omega) = \beta \overline{N}_{k-1}(\omega) + (1 - \beta)\hat{N}_k(\omega)$$

   with the smoothing constant β, e.g., 0.9.

Steps 1 through 5 are repeated with $k \mapsto (k + 1)$ until all frames are processed.

Algorithm 6.3 Outlined of the quantile-based method

1. Sort for each frequency band of $Y(\omega)$ the short-term power spectra over the last observations $k, k-1, \ldots, k-m$
2. Take for each frequency bin the median spectral power as the noise estimate $\hat{N}_k(\omega)$

Steps 1 and 2 are repeated with $k \mapsto (k + 1)$ until all frames are processed.

with large power – as they very likely contain speech activity – and perform an update for noise-only frames (Ahmed and Holmes 2004).

A summarization of the *histogram-based* method can be found in Algorithm 6.2. Note that in the original work by Hirsch and Ehrlicher (1995) Step 2 was not performed, which alters the noise estimate of the histogram-based method.

The histogram-based approach has been extended by Stahl *et al.* (2000) to a *quantile-based* approach which assumes that even during active speech regions there are frequency bands not occupied by speech. These unoccupied bands represent the energy level of the noise. Note that the observation windows influence the noise estimate similarly to the influence of the noise estimate in the histogram-based method. A summary of the quantile-based method can be found in Algorithm 6.3.

6.5.4 *Estimation of the a Posteriori and a Priori Signal-to-Noise Ratio*

The SNR estimate is closely related to the estimate of noise inasmuch as it requires the estimate of noise energy. The estimate of the *a posteriori SNR*, at frame k, is defined as the ratio between the noisy signal power and the noise power

$$\gamma_k(\omega) = \frac{|Y_k(\omega)|^2}{\mathcal{E}\left\{|N_k(\omega)|^2\right\}}.$$

The *a priori SNR* is defined as the ratio between the original signal power and the noise power,

$$\eta_k(\omega) = \frac{\mathcal{E}\left\{|X_k(\omega)|^2\right\}}{\mathcal{E}\left\{|N_k(\omega)|^2\right\}}.$$

It has been observed that SNR-based feature enhancement techniques exhibit fewer musical tones, as described in Section 6.9.1, if the SNR estimate has a low variance. The high fluctuation of the SNR estimate is due to the high variance of the spectral estimate of the noise (Vaseghi 2000). To reduce the fluctuations, and thereby the musical tones, a SNR value may be estimated by a weighted combination of the past and present SNR estimates (Ephraim, Y. and Malah, D. 1984):

$$\bar{\eta}_k(\omega) = \alpha(\omega)\,\bar{\eta}_{k-1}(\omega) + (1 - \alpha(\omega))\,\max\left(\gamma_k(\omega), 0\right)$$

where $\alpha(\omega)$ is in the range of 0.9 up to 0.99. To further reduce the variance of the noise estimate, an average over neighboring frequencies may also be useful.

6.6 Estimating Convolutional Distortion

As explained in Section 3.1, the transfer function of a room at a particular position and point in time is completely described by its *impulse response $h[n]$* or the Fourier transform thereof, the so-called *frequency response* or *transfer function* $\mathbf{h}^{(f)}$. Thus the reverberant speech sequence $y[n]$ is the convolution of $h[n]$ with the clean input sequence $x[n]$, such that,

$$y[n] = \sum_l h[l]\,x[n - l]. \tag{6.19}$$

The level of difficulty of estimating and removing channel distortions depends on the length and the rate of change of the impulse response. While the *deconvolution* of slowly varying short-time channel distortions is relatively simple, the deconvolution of rapidly changing long-time channel distortions is one of the fundamental, largely unsolved, problems in DSR. Deconvolution requires finding the inverse filter $\mathbf{h}_{\mathrm{inv}}$ which would enable the estimation of $\mathbf{x}$ given $\mathbf{y}$. An ideal inverse filter would have the following form which can be expressed either in the time domain

$$\sum_l h_{\mathrm{inv}}[l]\,h[n - l] = \delta[n - D]$$

or in the z-domain

$$H(z)\,H_{\mathrm{inv}}(z) = z^{-D}$$

with some delay D.

In a noise-free environment, perfect restoration would be possible if the impulse response were known and its inverse existed. Unfortunately, the inverse filter cannot always be simply realized as $H(z)_{\text{inv}} = H(z)^{-1}z^{-D}$ because some channels are *not* invertible. This happens if (Neely and Allen 1979)

- the transfer function is non-minimum phase, which implies that some zeros are outside the unit circle, and thus would result in an unstable inverse filter as described in Section 3.1.3;
- many inputs are mapped to the same output and therefore a closed-form solution does not exist.

Even if the channel is invertible further problems might arise in frequency regions with very high amplification which is, in particular, apparent for low SNR values.

From the discussion in Section 6.1 it follows that deconvolution can, in principal, be achieved, either by inverse filtering in the time domain, a multiplication in the frequency domain or an addition in the logarithmic spectral or cepstral domain with the appropriate *inverse filter*. The success of inverse filtering depends on the length of the analysis window, on the available knowledge sources and assumptions made about the speech signal and the channel. Unfortunately, the favorable circumstances that would make inverse filtering possible typically cannot be realized in realistic acoustic environments.

Regrettably, little work has been published about dereverberation techniques for large vocabulary speech recognition systems. An exception is the overview paper by Eneman *et al.* (2003) which evaluates and compares different techniques.

6.6.1 Estimating Channel Effects

Assuming that the impulse response does not change during the observation window, the short-time convolution of the signal can be estimated by the cepstral mean vector

$$\hat{\mathbf{h}}^{(c)} \approx \boldsymbol{\mu}^{(c)} = \frac{1}{K} \sum_{k=0}^{K-1} \mathbf{y}_k^{(c)},$$

where only the frames in which the speaker is active are summed. This is because convolution can only be estimated in the presence of speech, which is in contrast to VAD for noise estimation, where the noise must be estimated in the absence of speech. The length of the observation window chosen to calculate the average must be several seconds long to capture sufficient phonetic variability in the observation, but short enough to cope well with the nonstationary effects of the channel impulse response.

Cepstral normalization, as presented in Section 6.9.2, requires an utterance or a whole recording of one speaker to compute the cepstral mean and variance. In a real-time system, those parameters must be estimated on the fly. One convenient way to achieve this is to update the current cepstral mean estimate $\boldsymbol{\mu}_k^{(c)}$ by a new observation according to

$$\boldsymbol{\mu}_k^{(c)} = \alpha \mathbf{y}_k^{(c)} + (1 - \alpha)\,\boldsymbol{\mu}_{k-1}^{(c)}.$$

The free parameter α is set to cover several seconds of speech.

Note that, as already mentioned in Section 6.1, the convolution modeled by $\hat{\mathbf{h}}^{(c)}$ cannot be longer than the observation window. Thus $\hat{\mathbf{h}}^{(c)}$ is *not* sufficient for dereverberation as only convolutional terms no longer than a range between 10 and 32 ms can be covered. Nevertheless, compensating for these short-time distortions with CMN, as described in Section 6.9.2, has proven useful in close and distant ASR systems, as channel effects – such as those introduced by microphone characteristics – can be effectively removed.

6.6.2 Measuring the Impulse Response

The *impulse response* of a system is defined as the signal observed at the system output after its excitation by a Dirac impulse. In a realistic system, however, the Dirac impulse δ undergoes a series of transformations which are not only due to the impulse response of the room h_{room}, but also due to the sensor configuration used for data capture, namely the loudspeaker, the microphone and even the cables and plugs. Thus, the measurement of the impulse response requires high-quality components providing a flat frequency response and introducing only a linear phase shift. Expressing all distortions from the recording equipment used for data capture in h_{record} we can write

$$h(t) = \int_{-\infty}^{\infty} \int_{-\infty}^{\infty} h_{\text{room}}(\tau_2) \, h_{\text{record}}(\tau_1 - \tau_2) \, \delta(t - \tau_1) \, d\tau_1 d\tau_2.$$

To eliminate the effect of background noise, the excitation signal should have as much power as possible. But trying to produce and record a true Dirac impulse is impossible, in that any realizable data capture equipment would be overdriven by a signal of such high power, resulting in clipping. It is, however, possible to spread the energy over time by an arbitrary signal $s(t)$ covering the entire frequency range, such that,

$$\hat{s}(t) = \int_{-\infty}^{\infty} \int_{-\infty}^{\infty} h_{\text{room}}(\tau_2) \, h_{\text{record}}(\tau_1 - \tau_2) \, s(t - \tau_1) \, d\tau_1 d\tau_2.$$

If the transfer function of the recording equipment $H_{\text{record}}(\omega)$ is known, it can easily be removed to obtain the true estimate of the impulse response. Therefore, we can write the transfer function as

$$H_{\text{room}}(\omega) = \frac{\hat{S}(\omega)}{H_{\text{record}}(\omega) \, S(\omega)},$$

where the magnitude of the signal $S(\omega)$ should be nonzero for the frequencies under investigation. Well-suited signals for s are broadband excitation signals with flat power spectra such as white noise or *chirp* signals; i.e., a sine wave whose frequency increases linearly with time. A discussion about suitable signals for s can be found in Griesinger (1996).

Some properties of a transfer function can be found by inspecting the graphical representation of the impulse response, the so-called *reflectogram* (Kuttruff 2000). For example, echoes manifest themselves as large, isolated peaks. In a reflectogram, the influence of reverberation also becomes apparent. Removing insignificant details, as can be accomplished through smoothing techniques or the *Hilbert transform* (Bracewell 1999), can improve the appearance of the reflectogram, and render it easier to analyze.

6.6.3   Harmful Effects of Room Acoustics

It is useful to gain an insight into the harmful effects of reverberation, in order to develop strategies for successfully combating them. Unfortunately, relatively little work has been published in this area focusing on automatic recognition. Pan and Waibel (2000) have investigated the influence of room acoustics by comparing stereo data of close and distant recordings in the mel-scale logarithmic spectral domain derived from truncated mel-frequency cepstral coefficient coefficients. While noise affects mainly the spectral valleys, reverberation may also cause distortions at spectral peaks, i.e., at the fundamental frequency and its harmonics in voiced speech.

Although the definition varies from author to author, we will consider *early reflections* to occur between the arrival of the direct signal and 100 ms thereafter. Similary, we will take *late reflections* as any reflections or reverberations occurring after 100 ms. Tashev and Allred (2005) found that reverberation between 50 ms after the arrival of the direct signal and the time when the sound pressure has dropped 40 dB below its highest level, has the most damaging effect on the word accuracy of a DSR system. Petrick *et al.* (2007) have separately investigated early reflections, late reflections and reflections that are only present in low-or high-frequency regions in the context of ASR. They obtained slightly different results and concluded that late reflections which appear between 100 and 300 ms after the direct signal have the most damaging effect on the classification accuracy. Furthermore, they found that reverberation in the frequencies between 250 Hz and 2.5 kHz leads to poor ASR accuracy, while reverberation frequency components outside that range do not have a significant impact on recognition accuracy.

In our own experiments we found that dereverberation algorithms which begin to estimate the level of the reverberant energy around 60 ms after the direct signal provide the best recognition performance and that the adjustment of this parameter has a slight effect on the enhancement and thus recognition accuracy. The end time of the reverberation estimate should be sufficiently long to contain enough reverberation energy. This parameter, however, has only a limited effect on recognition accuracy and thus is not critical.

Moreover, early reverberation in higher frequencies was found to improve automatic recognition performance. Similar results where found by Nishiura *et al.* (2007), who reported that early reflections within approximately 12.5 ms of the direct signal actually improve recognition accuracy. This is significantly shorter than the 50 ms time frame wherein early reflections were found to improve human recognition accuracy (Kuttruff 2000).

6.6.4   Problem in Speech Dereverberation

As an alternative to addressing the problems inherent with the inversion of an estimated room transfer function, Kinoshita *et al.* (2005) proposed to suppress the influence of reverberation in the power or logarithmic spectral domain. As mentioned in Section 6.6, they proposed to suppress reverberation in a fashion similar to that in which additive noise is typically suppressed. Their approach is only valid under the assumption that the direct signal and reflections are statistically independent, which of course does not hold in general. That is, speech signals are not composed of an i.i.d. sequence as they have inherent features such as periodicity as well as a particular formant structure. Furthermore, the speech production model described in Section 2.2.1 consists of an excitation source $\mathbf{u}$ which is convolved with the glottal, vocal tract, and the lip radiation filters. The joint effect of these can be summarized in the speech production filter $\mathbf{h}_{\text{speech}}$. Thus the desired speech sequence is already a convolved sequence, which can be expressed as

$$x[n] = \sum_{l=0}^{L} h_{\text{speech}}[l]\, u[n - l].$$

Therefore, the observed signal can be described as

$$y[n] = \sum_{m=0}^{M} h_{\text{room}}[m]\, x[n - m] = \sum_{m=0}^{M} \underbrace{\sum_{l=0}^{L} h_{\text{room}}[m - l]\, h_{\text{speech}}[l]}_{h[m]} \, u[n - m], \qquad (6.20)$$

where the impulse response $\mathbf{h}$ is the convolution of the room impulse response $\mathbf{h}_{\text{room}}$ and the speech production filter $\mathbf{h}_{\text{speech}}$. The construction of an inverse filter $\mathbf{h}_{\text{inv}}$ that converts a convolved sequence into a sequence where each component is independent would not only filter out the impulse response of the channel $\mathbf{h}_{\text{room}}$ but also the impulse response of the speech production filter $\mathbf{h}_{\text{speech}}$, and thus would suppress features relevant for classification.

Separating the room impulse response into early and late reflections and assuming that the impulse response of the speech production filter is sufficiently short in comparison to the start time of the late reflection $M_{\text{early}} + 1$, allows (6.20) to be expressed as

$$y[n] \approx \sum_{m=0}^{M_{\text{early}}} \sum_{l=0}^{L} h_{\text{early}}[m - l]\, h_{\text{speech}}[l]\, u[n - m] + \sum_{m=M_{\text{early}}+1}^{\infty} h_{\text{late}}[m]\, u[n - m].$$

With the aid of this equation, it may be possible to develop algorithms intended solely for removing late reflections, which might be sufficient for speech feature enhancement. Gillespie and Atlas (2003) proposed a technique dubbed *correlation shaping* in which it is assumed that the *linear prediction* (LP) residue is only correlated within a short duration due to the speech signal and that the correlation over a long duration is caused by reverberation. Similar assumptions are made for the approach proposed by Kinoshita *et al.* (2006) which will be described in Section 6.6.5.

6.6.5 Estimating Late Reflections

Several algorithms have been proposed to estimate and compensate for harmful late reflections. Probably one of the most promising family of methods assumes that the reverberant power spectrum $\mathbf{r}_k$ is a scaled or weighted summation over previous frames, according to

$$\mathbf{x}_k^{(\text{reverberant})} = \mathbf{x}_k + \mathbf{r}_k = \mathbf{x}_k + \sum_{m=1}^{M} \mathbf{s}_m \, \mathbf{x}_{k-m} \tag{6.21}$$

where k denotes the frame index, the signal is denoted by $\mathbf{x}_k = \mathbf{x}_k^{(p)}$, the reverberation is denoted as $\mathbf{r}_k = \mathbf{r}_k^{(p)}$, and the scale terms $\mathbf{s}_m$ are dependent on the frequency. The scale terms can be determined, for example, by the Rayleigh distribution (Wu and Wang 2006) and adjusted by an estimate of the reverberation time or by more complex methods, such as that proposed by Sehr and Kellermann (2008). In contrast to classical spectral subtraction methods which estimate and subtract additive distortions in the power spectral domain, spectral dereverberation methods estimate and subtract spectral energy caused by reverberation. The advantage of treating the reverberation as additive in the power spectral domain is that the distortions can be easily removed, without the need to estimate and to invert the room impulse response, by simple subtraction. In addition, it has been shown by Lebart *et al.* (2001) that such methods are not sensitive to fluctuations in the impulse response.

Instead of estimating the reverberant power spectrum $\mathbf{r}_k$ by scaled versions of previous frames, as in (6.21), Kinoshita *et al.* (2006) proposed to determine the reflection sequence in the time domain by *multi-step linear prediction* (MSLP) (Gespert and Duhamel 1997) and thereafter convert it into a reverberation estimate $\mathbf{r}_k$ by short-time spectral analysis. In contrast to LP, MSLP aims to predict a signal after a given delay D, the *step-size*. With the prediction error $e[n]$ we can formulate MSLP as

$$y[n] = \sum_{m=1}^{M} a_m \, y[n - m - D] + e[n], \tag{6.22}$$

where $\{a_m\}$ denote the LP coefficients, $y[n]$ the observed signal and M the model order. For $D = 0$ MSLP reduces to LP. The required LP coefficients can be found by minimizing the mean squared error in (6.22). The solution can be expressed in matrix notation as

$$\mathcal{E} \left\{ \mathbf{y}[n - D] \, \mathbf{y}[n - D]^T \right\} \mathbf{a} = \mathcal{E} \left\{ \mathbf{y}[n - D] \, \mathbf{y}[n]^T \right\}$$

with $\mathbf{a} = [a_1, a_2, \cdots, a_M]^T$. Thus, we obtain the MSLP coefficients by

$$\mathbf{a} = \left(\mathcal{E} \left\{ \mathbf{y}[n - D] \, \mathbf{y}[n - D]^T \right\} \right)^{-1} \mathcal{E} \left\{ \mathbf{y}[n - D] \, \mathbf{y}[n]^T \right\}$$

which can be solved efficiently by the Levinson–Durbin recursion given in Algorithm 5.1.

The reflection sequence $r[n]$ can then be obtained by filtering the observation sequence $x[n]$ with the prediction filter using the MSLP coefficients as

$$\mathbf{r} = \mathbf{x} * \mathbf{a}.$$

The resulting reflection sequence $r[n]$ can now be treated like additive noise. Thus, after being converted into the power spectral domain, it gives a noise floor estimate $\hat{\mathbf{r}}_k^{(p)}$ which changes for each frame k. The noise estimate can then be compensated for by various methods, as will be described in Section 6.9.

As stated in Section 6.6.4 the complete impulse response consists of two parts, the room impulse and the vocal tract filter. Thus, the reflection sequence $r[n]$ might not only contain the unwanted distortions due to the impulse response of the room, but also a bias due to the vocal tract. In order to reduce the bias, Kinoshita *et al.* (2006) suggested a pre-whitening step to remove the short-term correlation due to $\mathbf{h}_{\text{speech}}$ prior to the estimation of the MSLP coefficients. The pre-whitening filter can be implemented by a simple LP with a small order covering a time span of 2 ms; e.g., 32 taps for 16 kHz signals.

6.7 Distortion Evolution

If the *distortion* is *non-stationary*, it is not sufficient to have a fixed estimate which is unchanging over time. Rather, the estimate must be constantly updated to track the time evolution of the distortion. In the following, only the evolution of additive noise is presented. The relations developed here, however, hold also for convolutional distortions $\mathbf{h}$. Given the trajectory of the noise $\mathbf{n}_{0:k-1}$ up to time k, the noise transition probability $p(\mathbf{n}_k|\mathbf{n}_{0:k-1})$ can be modeled by a *dynamic system model*. Note that the evolution might be estimated for a number of samples; e.g., using particle filters. For simplicity of notation, however, the sample index will be suppressed, wherever obvious, for the remainder of this section. Table 6.2 summarizes the different dynamic system models which will be presented in the following sections.

6.7.1 Random Walk

The simplest solution for predicting the next state is known as *random walk*. It simply takes the previous state as the estimate of the current state and adds a random variable ϵ_k, which is considered to be i.i.d. zero mean Gaussian, such that

$$\mathbf{n}_k = \mathbf{n}_{k-1} + \epsilon_k.$$

Table 6.2 Summary of the different approaches to model the evolution of the distortion

Noise evolution	Model
Random	$\mathbf{n}_{k-1} + \epsilon_k$
Polyak averaging and feedback	$(1 - \alpha)\,\mathbf{n}_{k-1} + \alpha\boldsymbol{\mu}_{\mathbf{n}_{k-1}} + \beta\left(\mathbf{n}_{k-1}^{\text{average}} - \mathbf{n}_{k-1}\right) + \epsilon_k$
Static autoregressive matrix	$\sum_{m=1}^{M} \mathbf{A}_m \mathbf{n}_{k-m} + \epsilon_k$
Dynamic autoregressive matrix	$\mathbf{A}_k \mathbf{n}_{k-1} + \epsilon_k$
Extended Kalman filter	$\mathbf{n}_{k-1} + \mathbf{G}_{k-1}\left\{\mathbf{y}_k - \mathbf{f}\left(\mathbf{x}_k^{(m)}, \mathbf{n}_{k-1}\right)\right\} + \epsilon_k$

Whenever possible, additional knowledge should be considered to obtain a better prediction. Some approaches, which have been demonstrated to be successful in the prediction of speech or noise spectra, are described in the following sections.

6.7.2 Semi-random Walk by Polyak Averaging and Feedback

To improve over the random walk of the state space model, Polyak averaging and feedback have been proposed by Fujimoto and Nakamura (2005a). The basic idea is to limit the range of the predicted noise hypothesis to within a fixed interval of the preceding frames. The *Polyak average* is calculated over the preceding K frames according to

$$\mathbf{n}_k^{\text{average}} = \frac{1}{K} \sum_{l=k-K+1}^{k} \mathbf{n}_l.$$

Given two real parameters α and β, we can express the state transition equation as

$$\mathbf{n}_k = (1 - \alpha)\, \mathbf{n}_{k-1} + \alpha \boldsymbol{\mu}_{n_{k-1}} + \beta \left(\mathbf{n}_{k-1}^{\text{average}} - \mathbf{n}_{k-1} \right) + \boldsymbol{\epsilon}_k, \tag{6.23}$$

where

$$\boldsymbol{\mu}_{n_k} = \sum_{m=1}^{M} w_k^{(m)} \mathbf{n}_k^{(m)}$$

is the weighted average of noise samples, and w_k denotes the normalized weights. In (6.23), the parameter α determines how much the noise samples are moved to the noise sample average, while β represents the scaling factor of the feedback.

When the noise is varying slowly, the difference between the Polyak average and noise has a small value and thus the parameter range becomes small. For rapidly varying noise, on the other hand, the difference between the Polyak average and noise has a large value and thus the parameter range becomes large. The two different cases are depicted in Figure 6.3.

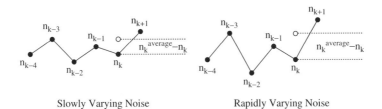

Slowly Varying Noise   Rapidly Varying Noise

Figure 6.3 Polyak averaging and feedback for slowly and rapidly varying noise. Note that the indices of the particles have been dropped and thus the sequence $n_{k-4:n+1}$ represents the evolution of a single noise estimate

To better account for the time variation of the noise, which may well change frame-by-frame, it has also been suggested to vary the parameters of the Polyak averaging on a frame-by-frame basis. This method is known as the *switching dynamic system model* (Fujimoto and Nakamura 2006).

6.7.3 Predicted Walk by Static Autoregressive Processes

Singh and Raj (2003) proposed to model the evolution of noise spectra in the logarithmic spectral domain as an autoregressive process

$$
\mathbf{n}_k = \underbrace{\left[\mathbf{A}^{(1)} \vdots \mathbf{A}^{(2)} \vdots \dots \vdots \mathbf{A}^{(L)}\right]}_{=\mathbf{A}^{(1:L)}} \cdot \underbrace{\begin{bmatrix} \mathbf{n}_{k-1} \\ \mathbf{n}_{k-2} \\ \vdots \\ \mathbf{n}_{k-L} \end{bmatrix}}_{=\mathbf{N}_{k-1:k-L}} + \boldsymbol{\epsilon}_k = \sum_{l=1}^{L} \mathbf{A}^{(l)} \mathbf{n}_{k-l} + \boldsymbol{\epsilon}_k. \tag{6.24}
$$

where $\mathbf{A}^{(1:L)} = \{\mathbf{A}_k\}$ represents a set of regression matrices and L is the model order. Letting B denote the number of spectral bins, the overall size of the prediction matrix $\mathbf{A}^{(1:L)}$ is given by $LB \times B$. Using a model order L larger than 1 circumvents the restrictive assumption of the noise being a Markov chain, which is tantamount to the assumption that the current noise spectrum depends only on the last noise spectrum. This method of merging states to overcome the limitations of the Markov assumption is well known in the statistical literature and, for example, is proposed in Meyn and Tweedie (1993).

Learning the Autoregressive Noise Model

The autoregressive model consists of two components that must be learned:

- The *linear prediction* (LP) *matrix* $\mathbf{A}^{(1:L)}$ which can be calculated by the minimization of the prediction error norm

$$
\mathbf{A}^{(1:L)} = \mathcal{E}\{\mathbf{n}_k \, \mathbf{N}_{k-1:k-L}^T\} \, \mathcal{E}\{\mathbf{N}_{k-1:k-L} \, \mathbf{N}_{k-1:k-L}^T\}^{-1}. \tag{6.25}
$$

Given the noise data $\mathbf{n}_1, \dots, \mathbf{n}_K$, these matrices can be learned according to

$$
\mathcal{E}\{\mathbf{n}_k \, \mathbf{N}_{k-1:k-L}^T\} = \frac{1}{K} \sum_{k=l}^{K} \mathbf{n}_k \, \mathbf{N}_{k-1:k-L}^T
$$

and

$$
\mathcal{E}\{\mathbf{N}_{k-1:k-L} \, \mathbf{N}_{k-1:k-L}^T\} = \frac{1}{K} \sum_{k=l}^{K} \mathbf{N}_{k-1:k-L} \, \mathbf{N}_{k-1:k-L}^T.
$$

Note that the parameters in the matrices can be learned reliably only if the noise data consists of pieces containing sufficient history, which is to say, the number of frames used to form the estimate exceeds the model order L.

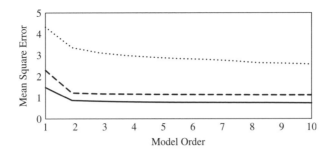

Figure 6.4 Mean squared prediction error for different noise types (destroyer solid line, driving car (Volvo) dashed line and machine gun dotted line – taken from the NOISEX database) with a frame shift of 10 ms vs the model order of the autoregressive process

- The *covariance matrix* can be calculated by

$$\mathbf{\Sigma}_{\Delta\mathbf{n}} = \mathcal{E}\{(\mathbf{n}_k - (\mathbf{A}^{(1:L)}\,\mathbf{N}_{k-1:k-L}))(\mathbf{n}_k - (\mathbf{A}^{(1:L)}\,\mathbf{N}_{k-1:k-L}))^T\}.$$

In the case of particle filtering, however, better results might be obtained by manually increasing the search space by raising the variance.

Learning a LP matrix of order L requires the reliable estimation of $B^2 L$ parameters. This is only possible if a huge amount of training data is available. Figure 6.4 presents the squared prediction error averaged over all frequency bins for static, semi-static and dynamic noise vs different model orders. As is apparent from the figure, an increase in model order results in only a marginal reduction in mean square error. Furthermore, it can happen that errors introduced by the noisy hypotheses could be emphasized, and hence actually degrade the prediction. Therefore, a model order of 1 is typically used in practice, as the LP matrix for $L = 1$ can be robustly estimated with a reasonable amount of training data.

6.7.4 Predicted Walk by Dynamic Autoregressive Processes

In the previous section, we applied a LP matrix estimated prior to its application. This approach has two obvious drawbacks:

- the noise must be known *a priori*, or voice activity detection is required;
- the prediction matrix cannot adjust to different types of distortion in those regions where speech is present.

To overcome the drawbacks apparent in static autoregressive processes, a *dynamic* and thus instantaneous and integrated estimate of the linear prediction matrix

$$\mathbf{A}_k = \mathbf{A}_k^{(1)} = \mathcal{E}\{\mathbf{n}_k\,\mathbf{n}_{k-1}^T\}\,\mathcal{E}\{\mathbf{n}_{k-1}\,\mathbf{n}_{k-1}^T\}^{-1} \tag{6.26}$$

is required for each individual frame k.

In a framework where the likelihood of the noise can be evaluated and where a number of samples can be drawn – such as in the application of PFs – it becomes possible to estimate the two matrices $\mathbf{n}_k\mathbf{n}_{k-1}^T$ and $\mathbf{n}_{k-1}\mathbf{n}_{k-1}^T$ on the current $\mathbf{n}_k^{(m)}$ and previous $\mathbf{n}_{k-1}^{(m)}$ noise estimates for all samples $m = 1, 2, \ldots, M$ (Wölfel 2008a). To ensure that the prediction estimates leading to good noise estimates are emphasized and those predictions leading to poor estimates are suppressed, it is necessary to weight the contribution to the matrices of each noise estimate by their likelihood $p(\mathbf{y}_k|\mathbf{n}_k^{(m)})$, as described in Section 6.8. Thus, the matrices can be evaluated for each frame k by using

$$\mathcal{E}\{\mathbf{n}_k\mathbf{n}_{k-1}^T\} = \frac{1}{M} \sum_{m=1}^{M} w_k^{(m)} \mathbf{n}_k^{(m)} \mathbf{n}_{k-1}^{(m)\,T}$$

and

$$\mathcal{E}\{\mathbf{n}_{k-1}\mathbf{n}_{k-1}^T\} = \frac{1}{M} \sum_{m=1}^{M} w^{(m)} \mathbf{n}_{k-1}^{(m)} \mathbf{n}_{k-1}^{(m)\,T}$$

to solve for (6.26). The weight of the different samples can be determined, for example, by

- the likelihood of the current observation

$$w_k^{(m)} = p(\mathbf{y}_k|\mathbf{n}_k^{(m)}),$$

- or the likelihood of the previous and current observations

$$w_k^{(m)} = p(\mathbf{y}_{k-1}|\mathbf{n}_{k-1}^{(m)})p(\mathbf{y}_k|\mathbf{n}_k^{(m)}),$$

or

$$w_k^{(m)} = \sqrt{p(\mathbf{y}_{k-1}|\mathbf{n}_{k-1}^{(m)})\, p(\mathbf{y}_k|\mathbf{n}_k^{(m)})}.$$

Smoothing over previous frames may help to improve the reliability of the estimate. With the introduction of the forgetting factor α we can write the smoothed matrix $\overline{\mathbf{A}}_k$ as

$$\overline{E}\left\{\mathbf{n}_k\,\mathbf{n}_{k-1}^T\right\} = \alpha\,\mathcal{E}\{\mathbf{n}_k\,\mathbf{n}_{k-1}^T\} + (1-\alpha)\,\overline{E}\left\{\mathbf{n}_{k-1}\,\mathbf{n}_{k-2}^T\right\}$$

and

$$\overline{E}\left\{\mathbf{n}_{k-1}\,\mathbf{n}_{k-1}^T\right\} = \alpha\mathcal{E}\left\{\mathbf{n}_{k-1}\,\mathbf{n}_{k-1}^T\right\} + (1-\alpha)\,\overline{E}\left\{\mathbf{n}_{k-2}\,\mathbf{n}_{k-2}^T\right\}.$$

The *sample variance* can now be calculated according to the normalized weight $w_k^{(m)}$, the likelihood of the mth particle divided by the summation over all likelihoods, as

$$\boldsymbol{\Sigma}_{\Delta\mathbf{n}} = \sum_{m=1}^{M} w_k^{(m)} \left(\mathbf{n}_k^{(m)} - \mathbf{A}_k\,\mathbf{n}_{k-1}^{(m)}\right)\left(\mathbf{n}_k^{(m)} - \mathbf{A}_k\,\mathbf{n}_{k-1}^{(m)}\right)^T \qquad (6.27)$$

or with $\overline{\mathbf{A}}_k$ respectively. Note that the subscript d representing each vector component of the noise $\mathbf{n}_k$ at frame k has been suppressed to improve readability. The noise can now be predicted by

$$\mathbf{n}_k = \mathbf{A}_k \, \mathbf{n}_{k-1} + \boldsymbol{\epsilon}_k.$$

6.7.5 Predicted Walk by Extended Kalman Filters

As an alternative to the prior approach, the evolution of noise spectra can be modeled by an array of extended KFs (Fujimoto and Nakamura 2005a). In this section, we develop this technique in the log-spectral domain. In keeping with the usage in the rest of the chapter, the state of the extended Kalman filter will correspond to the estimate $\mathbf{n}_k$ of the additive noise at time k, the original clean speech will be denoted by $\mathbf{x}_k$, and the noise-corrupted speech by $\mathbf{y}_k$. The enhancement technique proposed by Fujimoto and Nakamura differs from a normal particle filter in that an extended Kalman filter is used to propagate the particles $\left\{ \mathbf{n}_{k-1}^{(m)} \right\}$ forward in time to obtain the original samples $\left\{ \mathbf{n}_{k|k-1}^{(m)} \right\}$. In order to describe this technique, we must specialize the presentation of the Kalman filter in Section 4.3. Let us begin by rewriting the Riccati equation (4.32) for the case $\mathbf{F}_{k|k-1} = \mathbf{I}$ in order to calculate the predicted state estimation error covariance matrix as

$$\mathbf{K}_{k|k-1} = \mathbf{K}_{k-1} + \mathbf{U}_{k-1},$$

where $\mathbf{U}_{k-1}$ is the covariance matrix of the process noise $\boldsymbol{\epsilon}_k$. As indicated by (4.33), the filtered state estimation error covariance matrix $\mathbf{K}_{k-1}$ is obtained from the recursion

$$\mathbf{K}_k^{(m)} = \left[\mathbf{I} - \mathbf{G}_k^{(m)} \overline{\mathbf{H}}_k \left(\mathbf{n}_{k|k-1}^{(m)} \right) \right] \mathbf{K}_{k|k-1}^{(m)}. \tag{6.28}$$

Let $\mathbf{x}_k$ denote a speech sample drawn from a *hidden Markov model* (HMM) in state m, which had previously been trained on clean speech. The linearized observation functional in (6.28) is given by

$$\overline{\mathbf{H}}_k \left(\mathbf{n}_{k|k-1} \right) = \left. \frac{\partial \mathbf{f}\,(\mathbf{x}_k, \mathbf{n})}{\partial \mathbf{n}} \right|_{\mathbf{n} = \mathbf{n}_{k|k-1}},$$

which is a direct specialization of (4.45). The calculation of the Kalman gain $\mathbf{G}_k$ required for the recursion (6.28) is described below. The observation equation can be specialized as

$$\mathbf{y}_k = \mathbf{x}_k^{(m)} + \mathbf{f}\left(\mathbf{x}_k^{(m)}, \mathbf{n}_{k|k-1}^{(m)} \right) + \mathbf{v}_k, \tag{6.29}$$

where

$$\mathbf{f}(\mathbf{x}_k^{(m)}, \mathbf{n}_{k|k-1}^{(m)}) = \mathbf{x}_k^{(m)} + \log \left[\mathbf{I} + \exp \left(\mathbf{n}_k^{(m)} - \mathbf{x}_k^{(m)} \right) \right].$$

As before, all operations on vectors are performed component-by-component. Based on (6.29), the innovation (4.23) can be specialized as

$$\mathbf{s}_k^{(m)} = \left\{ \mathbf{y}_k - \mathbf{f}\left(\mathbf{x}_k^{(m)}, \mathbf{n}_{k|k-1}^{(m)}\right) \right\}.$$

The covariance matrix of $\mathbf{s}_k$ is then given by

$$\mathbf{S}_k = \mathbf{H}_k\left(\mathbf{n}_{k|k-1}^{(m)}\right)\mathbf{K}_{k|k-1}\mathbf{H}_k^T\left(\mathbf{n}_{k|k-1}^{(m)}\right) + \mathbf{V}_k.$$

As indicated by (4.30), the Kalman gain is given by

$$\mathbf{G}_k^{(m)} = \mathbf{K}_{k|k-1}^{(m)}\overline{\mathbf{H}}_k^T\left(\mathbf{n}_{k|k-1}^{(m)}\right)\left(\mathbf{S}_k^{(m)}\right)^{-1},$$

The update formula for the extended KF can, as in (4.28) and (4.29), be expressed as

$$\mathbf{n}_k^{(m)} = \mathbf{n}_{k|k-1}^{(m)} + \mathbf{G}_k^{(m)}\mathbf{s}_k^{(m)}.$$

6.7.6 Correlated Prediction Error Covariance Matrix

In the previous sections, the different dimensions have been assumed to be i.i.d., which contradicts the observation that neighboring spectral bins are correlated. The correlation in the random process, however, can be easily integrated (Wölfel 2008c). As the random process represents only the difference between the true noise and predicted noise $\hat{\mathbf{n}}_k$ we start by writing

$$\Delta\mathbf{n} = \mathbf{n}_k - \hat{\mathbf{n}}_k.$$

The covariance matrix of the random process is then

$$\Sigma_{\Delta\mathbf{n}} = (\Delta\mathbf{n} - \boldsymbol{\mu}_{\Delta\mathbf{n}})(\Delta\mathbf{n} - \boldsymbol{\mu}_{\Delta\mathbf{n}})^T,$$

where the mean values are given by

$$\boldsymbol{\mu}_{\Delta\mathbf{n}} = \frac{1}{M}\sum_{m=1}^{M}\Delta\mathbf{n}^{(m)}.$$

The correlation matrix can now be calculated by normalizing the single bins at position (j, i) of $\Sigma_{\Delta\mathbf{n}}$ by the square root of their variances, $\sigma_{\Delta\mathbf{n}}[j]\sigma_{\Delta\mathbf{n}}[i]$, such that,

$$R_{\Delta\mathbf{n}}[j, i] = \frac{\Sigma_{\Delta\mathbf{n}}[j, i]}{\sigma_{\Delta\mathbf{n}}[j]\sigma_{\Delta\mathbf{n}}[i]} \,\forall\, j, i.$$

We can now calculate the Cholesky decomposition $\mathbf{R}_{\Delta\mathbf{n}}$ by solving for

$$\mathbf{U}_{\Delta\mathbf{n}}^T\mathbf{U}_{\Delta\mathbf{n}} = \mathbf{R}_{\Delta\mathbf{n}}.$$

where $\mathbf{U}_{\Delta\mathbf{n}}$ is upper triangular.

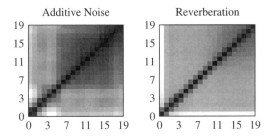

Figure 6.5 Correlation matrices for additive dynamic noise and reverberation. The additive noise shows high correlation over a large number of bins while reverberation is less correlated and mostly limited to neighboring bins

Given the Cholesky decomposition matrix we can draw correlated noise samples from the uncorrelated noise samples, where ϵ_k are identical to the ones used in the uncorrelated case by

$$\epsilon_k^{\text{corr}} = \mathbf{U}_{\Delta \mathbf{n}} \epsilon_k.$$

Note that ϵ_k^{corr} can be obtained from ϵ_k by performing backward substitution on the upper triangular Cholesky factor $\mathbf{U}_{\Delta \mathbf{n}}$, as described in Section B.15.

Figure 6.5 shows two correlation matrices. The first matrix is calculated on dynamic noise while the second matrix is calculated on reverberant data. For the case of dynamic noise, the correlation extends over a large number of bins, while for reverberation the correlation is mainly limited to neighboring regions.

6.8 Distortion Evaluation

Some approaches to speech feature enhancement, such as the KF or the PF, require a quality assessment of the enhanced signal on a frame-by-frame basis. A widely used criterion to judge the enhanced signal is the likelihood $p(\mathbf{y}_k | \mathbf{n}_k)$ which, for example, in the PF framework compares each distortion hypothesis against a *prior* clean speech model $p_x(\cdot)$. To capture the dynamics in the speech signal the prior speech model can be extended by either

- a *switching model* which can change, freely or based on constraints, between different states representing, for example, different phone hypotheses, or
- a *delta feature model* which models the local time difference of static features.

Some combination of the above is also possible.

To evaluate the likelihood, it is common practice to neglect the phase and thus to use the approximate relation between the distortions and clean signal as given in Table 6.1. The phase difference can be introduced as a hidden variable into the ML evaluation, which will be presented in Section 6.8.3.

6.8.1 Likelihood Evaluation

The evaluation of the likelihood of a noisy speech observation $\mathbf{y}$ is a fundamental component in any MMSE estimator. This likelihood can be evaluated with a GMM trained on clean speech according to

$$p(\mathbf{y}|\mathbf{x}, \mathbf{n}, \mathbf{h}) = \sum_{m=1}^{M} w_m \left| \frac{\partial \mathbf{f}(\mathbf{x}, \mathbf{n}, \mathbf{h})}{\partial \mathbf{x}} \right| \mathcal{N}(\mathbf{y} = f(\mathbf{x}, \mathbf{n}, \mathbf{h}); \boldsymbol{\mu}_m, \boldsymbol{\Sigma}_m), \qquad (6.30)$$

where $f(\mathbf{x}, \mathbf{n}, \mathbf{h})$ represents a linear or nonlinear functional. With the fundamental transformation law of probabilities (see Section B.8) we are able to solve for the likelihood function in different domains. A summary of the likelihood function for additive noise in different domains is given in Table 6.3.

Substantial overestimates of the actual noise lead to severe problems with likelihood computations in the frequency, power spectral, logarithmic spectral, as well as in the cepstral domain. That is, the likelihood function cannot be evaluated if the noise $\mathbf{n}_m$ exceeds the observation $\mathbf{y}_m$ in just a single bin b. This is an artifact of treating speech and noise as strictly additive. For those cases wherein the likelihood cannot be evaluated, it must be set to zero,

$$p(\mathbf{y}|\mathbf{n}) = 0 \text{ if } \mathbf{n}_m > \mathbf{y}_m \text{ for at least one bin } b.$$

Häb-Umbach and Schmalenströer (2005) reported that noise overestimation can lead to a severe decimation of the particle population, or even to its complete annihilation.

Figure 6.6 presents the likelihood $p(y|n)$ for one logarithmic spectral bin. The noise distribution is represented by a single Gaussian with a mean of 40 and a variance of 10. The observation of the corrupted speech spectrum is 50. At the point where the noise hypothesis n exceeds the observation y the approximation between x and n drives the likelihood to zero, whereas some nonzero probability mass is maintained with the exact representation (Faubel 2006). Moreover, we observe in Figure 6.6 that the likelihood function is a peaked distribution. This may cause a Bayesian filter to function poorly, or,

Table 6.3 Likelihood function in different domains for additive noise

Domain	Likelihood
time	$p^{(t)}(\mathbf{y}|\mathbf{n}^{(j)}) = p_x\left(\mathbf{y}^{(t)} - \mathbf{n}^{(t)}\right)$
spectra	$p^{(f)}(\mathbf{y}|\mathbf{n}) = p_x\left(\mathbf{y}^{(f)} - \mathbf{n}^{(f)}\right)$
power spectral	$p^{(p)}(\mathbf{y}|\mathbf{n}) = 2\left(\mathbf{y}^{(p)} + \mathbf{n}^{(p)}\right) p_x\left(\left(\mathbf{y}^{(p)} - \mathbf{n}^{(p)}\right)^2\right)$
logarithmic spectral	$p^{(l)}(\mathbf{y}|\mathbf{n}) = \dfrac{p_x\left(\mathbf{y} + \log\left(1 - e^{\mathbf{n}^{(l)} - \mathbf{y}^{(l)}}\right)\right)}{\prod_{b=1}^{B} 1 - e^{n_b^{(l)} - y_b^{(l)}}}$
cepstra	$p^{(c)}(\mathbf{y}|\mathbf{n}) = \dfrac{p_x\left(\mathbf{y} + \mathbf{T}\log\left(1 - e^{\mathbf{T}^{-1}\left(\mathbf{n}^{(c)} - \mathbf{y}^{(c)}\right)}\right)\right)}{\prod_{b=1}^{B} 1 - e^{n_b^{(c)} - y_b^{(c)}}}$

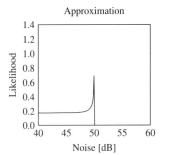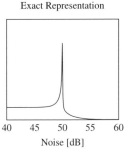

Figure 6.6 Likelihood functions in the logarithmic spectral domain representing the approximation and exact representation

in the case of PFs, that more particles are required (Pitt and Shephard 1999). To cope with peaked likelihood functions, it has been suggested that the likelihood function should be factorized into a number of broader distributions (Maccormick and Blake 2000).

6.8.2 Likelihood Evaluation by a Switching Model

So far a stationary speech model has been used to evaluate the likelihood. This static model, however, systematically ignores the dynamic properties of speech. To account for the dynamics in speech the general model of speech should be replaced by a statistical state sequence, a *switching model*, with phoneme or sub-phoneme classes such as the HMM. This approach has been applied to speech enhancement since the 1980s (Ephraim *et al.* 1989). In contrast to its use in speech recognition, the objective of applying the HMM for speech enhancement is to model the general characteristics of speech independent of the phoneme sequence, to distinguish between speech and noise, and not, between the individual speech units.

As the state sequence is not known *a priori*, it must be estimated and aligned. Note that this step is very critical for the performance of the applied filter. This is because the distorted speech observation is transformed toward the estimated and thus possibly incorrect state sequence which consistently results in an incorrect estimate of the word sequence in the recognition pass.

Various approaches to estimating the state sequence for feature enhancement have been proposed and used either as a switching model in the filter framework directly or to evaluate for the likelihood of the noise estimate. Such approaches include:

- HMMs (Ephraim *et al.* 1989);
- feedback of a previous recognition pass, using the first best hypothesis (Faubel and Wölfel 2006) or the word graph (Yan *et al.* 2007), the advantage over the HMM is the incorporation of additional knowledge sources such as the language model;
- a different modality such as a video stream (Almajai *et al.* 2007).

Switching between different states may cause a sudden change in the state estimate and thereby destabilize the filter. One way to prevent this and to weaken the influence of a wrong phone hypothesis is to interpolate between a specific state, such as a phoneme, and a general state, such as a general speech model, to form a *mixed state* (Faubel and Wölfel 2006)

$$\hat{p}_{\text{mixed}(t)}(x) = \alpha \cdot \hat{p}_{\text{phone}(t)}(x) + (1 - \alpha) \cdot p(x)$$

where α denotes the mixture weight or to calculate a mixed state according to the posterior probabilities.

6.8.3 Incorporating the Phase

It is common practice to consider the relative phase between the speech and noise to be zero, which significantly simplifies the relation between speech, noise and corrupted speech. This simplification, however, may introduce new problems. For example, in the event that the noise hypothesis exceeds the observation, the likelihood cannot be evaluated, as mentioned in Section 6.8.1. In the case of PFs, it can cause severe sample attrition and has been identified by Häb-Umbach and Schmalenströer (2005) as the main problem preventing PFs from functioning effectively.

Assuming that the relative phase θ is independent of $\mathbf{x}$ and $\mathbf{n}$, such that $p(\theta|\mathbf{x}, \mathbf{n}) = p(\theta)$, we can account for the relative phase by its introduction as a hidden variable, according to

$$p(\mathbf{y}|\mathbf{x}, \mathbf{n}) = \int_{-\pi}^{+\pi} p(\mathbf{y}|\mathbf{x}, \mathbf{n}, \theta)\, p(\theta)\, d\theta.$$

Instead of marginalizing over the relative phase, Droppo *et al.* (2002) marginalized over $\alpha \approx \cos\theta$, according to

$$p(\mathbf{y}|\mathbf{x}, \mathbf{n}) = \int_{-\infty}^{+\infty} p(\mathbf{y}|\mathbf{x}, \mathbf{n}, \alpha)\, p_\alpha(\alpha)\, d\alpha.$$

With the conditional probability distribution for logarithmic spectral energies

$$p(\mathbf{y}|\mathbf{x}, \mathbf{n}, \alpha) = \delta\left(\mathbf{y} - \log\left(e^{\mathbf{x}} + e^{\mathbf{x}} + 2\alpha e^{0.5(\mathbf{x}+\mathbf{n})}\right)\right),$$

and the identity

$$\int_{-\infty}^{+\infty} \delta\left(f(\alpha)\right) p_\alpha(\alpha)\, d\alpha = \sum_{\alpha: f(\alpha)=0} \frac{p_\alpha(\alpha)}{\left|\frac{d}{d\alpha} f(\alpha)\right|},$$

it becomes possible to evaluate the likelihood as

$$p(\mathbf{y}|\mathbf{x}, \mathbf{n}) = \frac{1}{2} e^{\mathbf{y}-0.5(\mathbf{x}+\mathbf{n})} + p_\alpha\left(\frac{e^{\mathbf{y}} - e^{\mathbf{x}} - e^{\mathbf{n}}}{2e^{0.5(\mathbf{x}+\mathbf{n})}}\right).$$

Under the Gaussian approximation for $p_\alpha(\alpha) = \mathcal{N}(\alpha; 0, \Sigma_\alpha)$, the likelihood can be calculated as

$$p(\mathbf{y}|\mathbf{x}, \mathbf{n}) = \exp\left\{\mathbf{y} - \frac{\mathbf{x} + \mathbf{n}}{2} - \frac{1}{2}\log 8\pi\,\Sigma_\alpha - \frac{(e^\mathbf{y} - e^\mathbf{x} - e^\mathbf{n})^2}{8\pi\,\Sigma_\alpha e^{\mathbf{x}+\mathbf{n}}}\right\}.$$

6.9   Distortion Compensation

The goal of distortion compensation is to derive either a point estimate of a clean speech feature, or a pdf of the original speech including information about the uncertainty of the enhancement process. The point estimate of the feature, and potentially the uncertainty thereof, is then propagated and evaluated in the acoustic model of the ASR system.

6.9.1   Spectral Subtraction

Boll (1979) proposed *spectral subtraction*, which was one of the earliest and became one of the most widely used approaches to noise suppression and speech enhancement. Subsequently, many enhancement techniques that are variations of spectral subtraction appeared in the literature. For consistency, we describe the conventional implementation as originally proposed by Boll. Assuming that the clean speech signal $\mathbf{x}$ is distorted by uncorrelated additive noise $\mathbf{n}$, the distorted power spectral density $\mathbf{y}$ can be approximated by

$$|Y_k(\omega)|^2 \approx |X_k(\omega)|^2 + |N_k(\omega)|^2. \tag{6.31}$$

Given an estimate of the noise power $|\hat{N}(\omega)|^2$, it is possible to derive a clean spectral estimate by simply removing the additive distortion term through subtraction in the power spectral domain

$$|\hat{X}_k(\omega)|^2 = |Y_k(\omega)|^2 - |\hat{N}_k(\omega)|^2. \tag{6.32}$$

The noise estimate $|\hat{N}(\omega)|^2$ was discussed in Section 6.5 for additive distortions, and in Section 6.6 for estimates of convolutional distortions, which are treated as additive distortions by modeling them as a diffuse noise field using MSLP. A similar approach to spectral subtraction, but yet on the autocorrelation sequences has been proposed by Weiss *et al.* (1974),

$$\mathbf{r}_{\hat{x}\hat{x}} = \mathbf{r}_{yy} - \mathbf{r}_{\hat{n}\hat{n}},$$

where $\mathbf{r}_{\hat{x}\hat{x}}$, $\mathbf{r}_{yy}$ and $\mathbf{r}_{\hat{n}\hat{n}}$ represent the autocorrelation sequences of the estimated clean speech signal, the distorted speech signal and the estimated noise signal respectively. This technique is referred to as the INTEL technique.

Limitations of Spectral Subtraction

Despite its simplicity and effectiveness, spectral subtraction is not without limitations, the most egregious of which is that improvements in ASR performance tend to diminish for

SNR values below zero. Comparing the exact representation of the clean speech

$$X(\omega) = \sqrt{Y(\omega)Y(\omega)^* - N(\omega)N(\omega)^*}e^{j\vartheta(\omega)}$$

$$= \sqrt{|Y(\omega)|^2 - |N(\omega)|^2 - X(\omega)N^*(\omega) - X^*(\omega)N(\omega)}e^{j\vartheta(\omega)}$$

$$= \sqrt{|Y(\omega)|^2 - |N(\omega)|^2 - 2|X(\omega)||N(\omega)|\cos\theta(\omega)}e^{j\vartheta(\omega)}$$

where $\vartheta(\omega)$ represents the phase error and $\theta(\omega)$ represents the phase difference between $X(\omega)$ and $N(\omega)$, let us analytically discuss possible errors. Three sources of error become evident:

- *Phase error* – The phase error is related to the difference between the true and the distorted phase. It has no effect on speech recognition as the phase is neglected in the front-end process.
- *Cross-term error* – The cross-term errors are related to phase difference between clean speech and noise which affect the magnitude of the estimate of clean speech.
- *Magnitude error* – The magnitude error is reflected by the difference between the true noise $N(\omega)$ and noise estimate $\hat{N}(\omega)$.

Evans *et al.* (2006) confirmed that the magnitude error is the greatest source of degradation in ASR performance. For SNRs below 0 dB, however, cross-term errors cause degradation in ASR performance, which are not negligible.

The use of noise subtraction for speech feature enhancement in ASR has proven more successful than it has for speech enhancement (Morii 1988). This is because ASR is most often based on features – i.e., cepstra – for which phase is unimportant. Speech enhancement, on the other hand, requires the resynthesis of the speech signal. The latter can only be accomplished with approximate phase information taken from the distorted signal.

Musical tones

The main drawback of spectral subtraction techniques is that the noise remaining after the processing has a very unnatural quality (Boll 1979; Cappé, O. 1994). This can be explained by the fact that the magnitude of the short-time power spectrum exhibits strong fluctuations in distorted areas. After the spectral attenuation, the frequency bands that originally contained the noise consist of randomly spaced spectral peaks corresponding to the maxima of the short-time power spectrum. Between these peaks, the short-time power spectrum values are close to or below the estimated averaged noise spectrum, which results in strong attenuations. As a result, the residual noise is composed of sinusoidal components with random frequencies that come and go in each short-time frame (Boll 1979). These artifacts are known as *musical*[4] *tones* phenomenon.

---

[4] This term is a reference to the presence of pure tones in the residual noise.

Nonlinear Spectral Subtraction

In order to overcome spectral distortions such as musical tones, caused by *simple* spectral subtraction, many variants of spectral subtraction have been proposed (Boll 1979; Vaseghi, S. and Frayling–Cork, R. 1992). They mainly differ in averaging the noise or in post-processing. None has succeeded, however, in completely eliminating the distortions introduced by the subtraction. A prominent family of approaches applies heuristic methods to solve this problem, and is known as *nonlinear* spectral subtraction.

Three extensions have proven useful in suppressing negative power estimates and in improving the quality of spectral subtraction:

- *Spectral flooring* – To prevent negative signal energy a spectral floor β is introduced which applies a lower bound to the enhanced spectral energy. It has been shown that a value slightly above zero leads to better recognition results, as in clean speech material there is still some noise energy in every frequency band.
- *Noise overestimation* – It might be helpful to adjust the scale of the estimated noise energy. It has been reported in the literature that the overestimation of the noise by a factor α with values between 1 and 2 lead to the best recognition accuracy.

  It has been claimed that overestimation of the noise seems to be particularly helpful for low SNRs. Therefore, to achieve good performance in different environments it has been proposed that the overestimation factor α should depend on the estimated SNR. A SNR-dependent overestimation factor could be (Vaseghi 2000)

$$\alpha(\text{SNR}(\omega)) = 1 + \frac{\sqrt{\mathcal{E}\left\{|\hat{N}(\omega)|^2\right\}}}{|\hat{N}(\omega)|}.$$

- *Nonlinear noise estimate* – The estimate of the noise itself has a nonlinear relationship. Lockwood and Boudy (1992) suggested an estimate for the nonlinear noise power as

$$|\hat{N}(\omega)|^2_{\text{NL}} = \frac{\max_{\text{over } M \text{ frames}} |\hat{N}(\omega)|^2}{1 + \gamma \text{SNR}(\omega)}$$

with γ as a design parameter.

With the previous extensions, we can express nonlinear spectral subtraction as

$$|\hat{X}(\omega)|^2 = \max\left\{|Y(\omega)|^2 - \alpha|\hat{N}(\omega)|^2_{\text{NL}}, \beta \geq 0\right\}. \tag{6.33}$$

6.9.2 Compensating for Channel Effects

In contrast to reverberation, channel effects can be compensated for by some variant of cepstral mean normalization (Atal 1974). Hence, they are much easier to suppress than true reverberation because they occur over a much smaller time window. Similar to spectral subtraction, which subtracts the estimate of the noise **n** in the spectra domain,

CMN subtracts the estimate of the impulse response $\mathbf{h}$, as in Section 6.6.1, in the cepstral domain for each frame k, according to

$$\hat{\mathbf{x}}_k^{(c)} = \mathbf{y}_k^{(c)} - \mathbf{h}_k^{(c)}.$$

CMN is a simple but powerful method for suppressing short-time convolutional distortion. This is apparent upon observing that the sample mean vector $\boldsymbol{\mu}_y^{(c)}$ can be expressed as

$$\boldsymbol{\mu}_y^{(c)} = \frac{1}{K} \sum_{k=0}^{K-1} \mathbf{y}_k^{(c)} = \frac{1}{K} \sum_{k=0}^{K-1} \mathbf{x}_k^{(c)} + \mathbf{h}^{(c)} = \boldsymbol{\mu}_x^{(c)} + \mathbf{h}^{(c)}.$$

Thus the normalized cepstrum subtracts the convolutional channel distortion $\mathbf{h}$ from the signal

$$\hat{\mathbf{y}}_k^{(c)} = \mathbf{y}_k^{(c)} - \boldsymbol{\mu}_y^{(c)} \approx \mathbf{x}_k^{(c)}.$$

In addition to the normalization of the mean, it is also common practice to normalize the variance of the cepstrum by dividing through the variance $\left(\sigma_y^{(c)}\right)^2$ for each cepstral bin b as

$$\hat{x}_k^{\text{norm}}[b] = \frac{y_k^{(c)}[b] - \mu_y^{(c)}[b]}{\left(\sigma_y^{(c)}[b]\right)^2} \; \forall\, b.$$

In ASR, training material is typically assumed to be undistorted. Hence, speech feature enhancement techniques which try to map the noisy observation to clean speech are typically not applied to the training speech. Cepstral normalization, on the other hand, must always be applied during both training and recognition as the mean and variance values of the features are different from those before processing.

6.9.3 Distortion Compensation for Distributions

In contrast to the prior sections, which have described various compensation methods for point estimates, this section introduces compensation methods where the distortion is represented as a pdf; e.g., in the PF or parallel model combination framework.

Vector Taylor Series

The use of the *vector Taylor series* (VTS) was proposed by Moreno *et al.* (1996) to approximate the nonlinearity between the vectors $\mathbf{x}$, $\mathbf{n}$ and $\mathbf{y}$ either in the logarithmic spectral or cepstral domain:

$$\mathbf{y} = \mathbf{x} + \mathbf{h} + f(\mathbf{n} - \mathbf{x} - \mathbf{h}). \tag{6.34}$$

The nonlinear function $f(\mathbf{z})$ is given in the logarithmic spectral domain as

$$\mathbf{f(z)} = \log\left(1 + e^{\mathbf{z}}\right),$$

while in the cepstral domain it is given by

$$\mathbf{f(z)} = \mathbf{T}\log\left(1 + \mathbf{T}^{-1}e^{\mathbf{z}}\right).$$

Under the assumption that $\mathbf{x}$, $\mathbf{n}$ and $\mathbf{h}$ are Gaussian distributed with the vector means $\boldsymbol{\mu}_x, \boldsymbol{\mu}_n$ an $\boldsymbol{\mu}_h$ and covariance matrices $\boldsymbol{\Sigma}_x, \boldsymbol{\Sigma}_n$ and $\boldsymbol{\Sigma}_h$, respectively, the Jacobians of (6.34) can be evaluated at $\boldsymbol{\mu} = \boldsymbol{\mu}_n - \boldsymbol{\mu}_x - \boldsymbol{\mu}_h$ as

$$\left.\frac{\partial\mathbf{y}}{\partial\mathbf{x}}\right|_{(\mu_x,\mu_n,\mu_h)} = \left.\frac{\partial\mathbf{y}}{\partial\mathbf{h}}\right|_{(\mu_x,\mu_n,\mu_h)} = \mathbf{A}; \quad \left.\frac{\partial\mathbf{y}}{\partial\mathbf{n}}\right|_{(\mu_x,\mu_n,\mu_h)} = \mathbf{I} - \mathbf{A}$$

where $\mathbf{I}$ is the unity matrix and $\mathbf{A}$ is given for the logarithmic spectral domain as

$$\mathbf{A} = \mathbf{F}; \quad \text{where } \mathbf{F} \text{ is a diagonal matrix following } \frac{1}{1 + e^{\mu}}$$

or for the cepstral domain as

$$\mathbf{A} = \mathbf{TFT}^{-1}; \quad \text{where } \mathbf{F} \text{ is a diagonal matrix following } \frac{1}{1 + \exp\mathbf{T}^{-1}\boldsymbol{\mu}}.$$

The first-order VTS expansion around $(\boldsymbol{\mu}_x, \boldsymbol{\mu}_n, \boldsymbol{\mu}_h)$ becomes

$$\mathbf{y} \approx \boldsymbol{\mu}_x + \boldsymbol{\mu}_h + f(\boldsymbol{\mu}_n - \boldsymbol{\mu}_x - \boldsymbol{\mu}_h) + \mathbf{A}(\mathbf{x} - \boldsymbol{\mu}_x) + \mathbf{A}(\mathbf{h} - \boldsymbol{\mu}_h) + (\mathbf{I} - \mathbf{A})(\mathbf{n} - \boldsymbol{\mu}_n) \tag{6.35}$$

With (6.35), the mean $\boldsymbol{\mu}_y$ and the variance $\boldsymbol{\Sigma}_y$ can be approximated as

$$\boldsymbol{\mu}_y \approx \boldsymbol{\mu}_x + \boldsymbol{\mu}_h + f(\boldsymbol{\mu}_n - \boldsymbol{\mu}_x - \boldsymbol{\mu}_h)$$

and

$$\boldsymbol{\Sigma}_y \approx \mathbf{A}\boldsymbol{\Sigma}_x\mathbf{A}^T + \mathbf{A}\boldsymbol{\Sigma}_h\mathbf{A}^T + (\mathbf{I} - \mathbf{A})\boldsymbol{\Sigma}_n(\mathbf{I} - \mathbf{A})^T$$

respectively. Note that even though $\boldsymbol{\Sigma}_x$, $\boldsymbol{\Sigma}_n$ and $\boldsymbol{\Sigma}_h$ might be diagonal, $\boldsymbol{\Sigma}_y$ is no longer a diagonal matrix.

To compute the delta and delta–delta parameters, the derivatives of the approximation must be taken

$$\frac{\partial\mathbf{y}}{\partial t} \approx \mathbf{A}.$$

Under the assumption that $\mathbf{h}$ is constant, the delta mean and delta–delta mean are given by

$$\boldsymbol{\mu}_{\Delta y} \approx \boldsymbol{\mu}_{\Delta x},$$

and

$$\boldsymbol{\mu}_{\Delta \Delta y} \approx \boldsymbol{\mu}_{\Delta \Delta x}.$$

Under the same assumption the delta variance and delta–delta variance can be calculated as

$$\boldsymbol{\Sigma}_{\Delta y} \approx \mathbf{A}\boldsymbol{\Sigma}_{\Delta x}\mathbf{A}^T + (\mathbf{I} - \mathbf{A})\boldsymbol{\Sigma}_{\Delta n}(\mathbf{I} - \mathbf{A})^T,$$

and

$$\boldsymbol{\Sigma}_{\Delta \Delta y} \approx \mathbf{A}\boldsymbol{\Sigma}_{\Delta \Delta x}\mathbf{A}^T + (\mathbf{I} - \mathbf{A})\boldsymbol{\Sigma}_{\Delta \Delta n}(\mathbf{I} - \mathbf{A})^T.$$

The VTS can readily be applied to solve for (6.10) by expanding around each Gaussian mean $\boldsymbol{\mu}_m$ of a GMM (Singh and Raj 2003). This holds because the index m of a specific Gaussian in the mixture $p(\mathbf{x}_k)$ can be introduced as a hidden variable m, as $p(\mathbf{x}_k|\mathbf{y}_{1:k}, \mathbf{n}_k, \mathbf{h}_k)$ can be represented as the marginal density

$$p(\mathbf{x}_k|\mathbf{y}_{1:k}, \mathbf{n}_k, \mathbf{h}_k) = \sum_{m=1}^{M} p(\mathbf{x}_k, m|\mathbf{y}_{1:k}, \mathbf{n}_k, \mathbf{h}_k).$$

With the equality

$$p(\mathbf{x}_k, m|\mathbf{y}_{1:k}, \mathbf{n}_k, \mathbf{h}_k) = p(m|\mathbf{y}_{1:k}, \mathbf{n}_k, \mathbf{h}_k)p(\mathbf{x}_k|m, \mathbf{y}_{1:k}, \mathbf{n}_k, \mathbf{h}_k)$$

it is possible to rewrite (6.11) as

$$f_k(\mathbf{y}_{1:k}, \mathbf{n}_k, \mathbf{h}_k) = \sum_{m=1}^{M} p(m|\mathbf{y}_{1:k}, \mathbf{n}_k) \int \mathbf{x}_k p(\mathbf{x}_k|m, \mathbf{y}_{1:k}, \mathbf{n}_k)d\mathbf{x}_k, \qquad (6.36)$$

where the sum over m has been pulled out of the integral. Now, the noise can be considered to shift the means of the clean speech distribution in the spectral domain.

Considering only the additive noise term, we can write the effect of $\mathbf{n}_k$ on the mth Gaussian, for example, in the logarithmic spectral domain as

$$e^{\mu'_m} = e^{\mu_m} + e^{\mathbf{n}_k}.$$

Solving for $\boldsymbol{\mu}'_m$ yields

$$\boldsymbol{\mu}'_m = \boldsymbol{\mu}_m + \underbrace{\log(1 + e^{\mathbf{n}_k - \mu_m})}_{=\Delta_{\mu_m, \mathbf{n}_k}}. \qquad (6.37)$$

Instead of shifting the mean, we can conversely shift the distorted spectrum $\mathbf{y}_k$ in the opposite direction to obtain the clean speech spectrum

$$\mathbf{x}_k = \mathbf{y}_k - \mathbf{\Delta}_{\mu_k, \mathbf{n}_k}. \tag{6.38}$$

This yields

$$p(\mathbf{x}_k | m, \mathbf{y}_{1:k}, \mathbf{n}_k) = \delta(\mathbf{x}_k - (\mathbf{y}_k - \mathbf{\Delta}_{\mu_m, \mathbf{n}_k}))$$

and hence

$$f_k^{\text{vts}}(\mathbf{n}_k) = \sum_{m=1}^{M} p(m | \mathbf{y}_{1:k}, \mathbf{n}_k) \int \mathbf{x}_k \, \delta(\mathbf{x}_k - (\mathbf{y}_k - \mathbf{\Delta}_{\mu_m, \mathbf{n}_k}))) \, d\mathbf{x}_k$$

$$= \sum_{m=1}^{M} p(m | \mathbf{y}_{1:k}, \mathbf{n}_k) \left(\mathbf{y}_k - \mathbf{\Delta}_{\mu_m, \mathbf{n}_k} \right)$$

$$= \mathbf{y}_k - \sum_{m=1}^{M} p(m | \mathbf{y}_{1:k}, \mathbf{n}_k) \mathbf{\Delta}_{\mu_m, \mathbf{n}_k}. \tag{6.39}$$

Comparing (6.39) with equation 19 in Raj *et al.* (2004) shows that the VTS approach approximates $p(m | \mathbf{y}_{1:k}, \mathbf{n}_k)$ by $p(m | \mathbf{y}_k, \mathbf{n}_k)$. This is equivalent to the assumption that m is independent of the preceding distorted speech spectra, if $\mathbf{y}_k$ and $\mathbf{n}_k$ are known. Another assumption implicitly made by Raj *et al.* is that m is independent of the current noise spectrum, i.e., $p(m | \mathbf{n}_k) = p(m) = c_m$, which holds for additive noise, not however for convolutional distortions such as reverberation. Under the previous assumptions, $p(m | \mathbf{y}_{1:k}, \mathbf{n}_k)$ can be calculated as

$$p(m | \mathbf{y}_k, \mathbf{n}_k) = \frac{p(\mathbf{y}_k | \mathbf{n}_k, m) p(m | \mathbf{n}_k)}{p(\mathbf{y}_k | \mathbf{n}_k)} = \frac{c_m p(\mathbf{y}_k | \mathbf{n}_k, m)}{\sum_{m=1}^{M} c_m p(\mathbf{y}_k | \mathbf{n}_k, m)}.$$

Statistical Inference

Using Monte Carlo estimation to approximate the integral in (6.10), it becomes possible to directly use the linear or nonlinear relation given in Table 6.1, as the density is reduced to a finite number of point estimates. Considering once more the previous example, and assuming only additive distortions in the logarithmic spectral domain, we obtain the relation

$$p(\mathbf{x}_k | \mathbf{y}_{1:k}, \mathbf{n}_k) = \delta \left\{ \mathbf{x}_k - \left[\mathbf{y}_k + \log(1 - e^{\mathbf{n}_k - \mathbf{y}_k}) \right] \right\}.$$

Substituting the last equation into (6.11), gives the solution of the *statistical inference approach* (SIA) in the logarithmic spectral domain as

$$f_k^{\text{sia}}(\mathbf{n}_k) = \int \mathbf{x}_k \cdot \delta \left\{ \mathbf{x}_k - \left[\mathbf{y}_k + \log \left(1 - e^{\mathbf{n}_k - \mathbf{y}_k} \right) \right] \right\} \, d\mathbf{x}_k = \mathbf{y}_k + \log(1 - e^{\mathbf{n}_k - \mathbf{y}_k}).$$

When applied within the particle filter framework, the SIA was found to yield better speech recognition performance as compared to the VTS (Faubel and Wölfel 2007).

6.10 Joint Estimation of Additive and Convolutional Distortions

The potential to cope with difficult nonlinear and non-Gaussian problems makes the PF, in contrast to the KF, particularly useful for tracking nonstationary noise signals in the logarithmic or cepstral feature domains. A variety of different PF variants have been proposed and evaluated for the enhancement of speech features: auxiliary and likelihood PFs (Häb-Umbach and Schmalenströer 2005) as well as PFs with an extended KF proposal density (Fujimoto and Nakamura 2005b), or the use of static (Singh and Raj 2003) or dynamic (Wölfel 2008a) autoregressive matrices. All approaches, however, are similar in structure.

This section presents an implementation of a PF that jointly estimates and removes non-stationary noise and reverberation in the logarithmic spectral domain on a frame-by-frame basis (Wölfel 2008b). This, in contrast to the previously mentioned approaches, integrates an additive reverberation estimate derived by MSLP into the PF framework and thus is capable of coping not only with nonstationary noise but also with reverberation. An outline is presented in Algorithm 6.4 and a corresponding sketch is given in Figure 6.7. The proposed approach is evaluated and discussed, as an example system, in Section 14.4.

In the following only the details that have not been presented in other sections are presented. Those are the initialization of the particles, the evolution of particles and distortion estimate, combination and compensation as well as the working domain.

Scaling the Reverberation Estimates

In order to compensate for estimation errors in the estimated reflection energy $\mathbf{r}_k$ which might be due to

- approximation of the reflection energy,
- additive noise in the estimate, as well as
- stationary assumption of the impulse response,

a scaling term is introduced as

$$\mathbf{r}_k^{(\mathrm{PF})} = \log(\mathbf{s}_k)\mathbf{r}_k. \tag{6.40}$$

Note that the scaling term $\mathbf{s}_k$ is different from (6.21), inasmuch as it changes for each frame k while the scale terms in (6.21) are usually constant over long observation windows such as an utterance. Thus $\mathbf{s}_k$ is able to adjust for changes in the room impulse response without updating the parameters of the reverberation models (6.21) or (6.22) which can only be estimated over a much longer time interval as they contain many more free variables; e.g., to cover a reverberation of 200 ms we need to estimate either 129 spectral bins multiplied by 20 frames = 2580 for model (6.21) or 3200 LP coefficients for model (6.22).

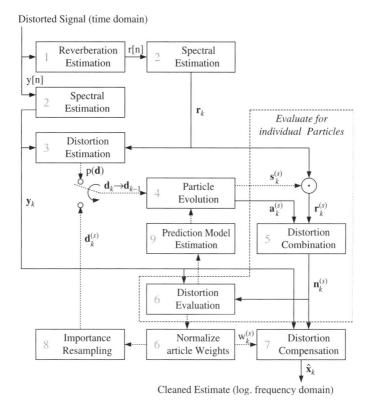

Figure 6.7 Schematics of joint particle filter estimation of additive and reverberant distortions for one frame without initialization. Solid arrows represent the flow of the signal. Dotted arrows represent the flow of particle information such as the particle weight and the particle values which represent estimates for additive distortions for each frequency bin and a scaling factor for the convolutional distortion. Variables are defined as in the text. The individual steps are described in Algorithm 6.4

The reverberation energy estimate in (6.40) can either

- be scaled by a single factor $s[1]$

$$r[b]_k^{(s)} = \log\left(s[1]_k^{(s)}\right) r[b]_k \tag{6.41}$$

adding one dimension to the PF, or
- be scaled by a single factor $s[1]$ and be tilted by $s[2]$ to scale lower and higher frequencies differently

$$r_k^{(s)}[b] = \log\left(s_k^{(s)}[1] + s_k^{(s)}[2](b - \bar{b})\right) r_k[b] \tag{6.42}$$

where $\bar{b} = (B + 1)/2$, adding two dimensions to the PF, or

Algorithm 6.4   Outline of the particle filter for speech feature enhancement to jointly
estimate additive distortions and reverberation

---

1. *Reverberation Estimation* – The reverberation sequence is calculated by MSLP according to (6.22).
2. *Spectra Estimation* – The reverberant and distorted short-time power spectra are estimated for all frames.
3. *Distortion Estimation and Particles Initialization* – The prior additive distortion density $p(\mathbf{a}_0)$ and prior scale density $p(\mathbf{s}_0)$ are set accordingly. Samples $\mathbf{d}_0^{(s)}, s =
   0, \ldots, S - 1$, are drawn from the prior distortion density $p(\mathbf{d}_0)$ as defined in (6.44).
4. *Particle Evolution* – All particles $\mathbf{d}_k^{(s)}, s = 0, \ldots, S - 1$, are propagated by the particle
   transition probability $p(\mathbf{d}_k|\mathbf{d}_{0:k-1})$.
5. *Distortion Combination* – The expected distortion $\mathbf{n} = u(\mathbf{a}, \mathbf{s})$ is calculated as

$$n_k^{(s)}[b] = \log\left(e^{a_k^{(s)}[b]} + e^{r_k^{(s)}[b]}\right) \ \forall\, b \in B$$

   where $a_k^{(s)}[b]$ represents additive distortions and $r_k^{(s)}[b]$ represents the scaled spectral
   distortion due to reverberation as determined by either (6.41) or (6.42).
6. *Distortion Evaluation* – The distortion samples $\mathbf{n}$ are evaluated and normalized.
7. *Distortion Compensation* – The estimated original feature is calculated according to
   either (6.39) or (6.40).
8. *Importance Resampling* – The normalized weights are used to resample among the
   noise particles $\mathbf{d}_k^{(s)}, s = 1, \ldots, S$ to prevent the degeneracy problem.
9. *Prediction Model Estimation* – The dynamic transition probability model matrix $\mathbf{D}_k$
   must be updated according to (6.26).

Steps 4 through 9 are repeated with $k \mapsto (k + 1)$ until either all frames are processed or
the track is lost and must be reinitialized with step 3.

- be scaled for each frequency bin individually $s[b]$

$$r_k^{(s)}[b] = \log\left(s_k^{(s)}[b]\right) r_k[b] \tag{6.43}$$

doubling the dimension of the PF.

Scaling each bin individually significantly increases the search space and thus the
execution time. Moreover, it did not provide performance superior to that achieved with
alternative approaches with lower dimensionality. This technique has been presented here,
however, for the sake of completeness.

Particle Initialization

The first step in any PF framework is the initialization of the particles by drawing samples from the *prior distortion density* $\mathbf{d}_0$. In our framework the *prior distortion density*

$$p(\mathbf{p}_0) = \begin{bmatrix} p(\mathbf{a}_0) \\ \cdots\cdots \\ p(\mathbf{s}_0) \end{bmatrix} \tag{6.44}$$

is a concatenation of the *prior additive distortion density* $p(\mathbf{a}_0)$ and the *prior scale density* $p(\mathbf{s}_0)$ of the estimated late reflection energies. In those cases where the silence region is still dominated by the reverberation, the prior additive distortion density $p(\mathbf{a}_0)$ cannot be estimated directly. It can, however, be decomposed into two densities which can be estimated:

- The *prior overall distortion density* $p(\mathbf{n}_0) = \mathcal{N}(\boldsymbol{\mu}_n, \boldsymbol{\Sigma}_n)$ derived on silence regions of the input signal which contains *additive* and *convolutional* distortions and
- the *prior reverberation density* $p(\mathbf{r}_0) = \mathcal{N}(\boldsymbol{\mu}_r, \boldsymbol{\Sigma}_r)$ which is estimated over all frames derived on the late reflection energy sequence $\mathbf{r}_{0:K}$ estimated by MSLP as described in Section 6.6.5.

With the prior overall distortion density and the prior reverberation density, it is now possible to derive the prior additive distortion density as

$$p(\mathbf{a}_0) = \mathcal{N}(\boldsymbol{\mu}_a, \boldsymbol{\Sigma}_n)$$

by subtracting the mean value of the reverberation energy from the mean value of the noise energy

$$\boldsymbol{\mu}_a = \log\left(e^{\mu_n} - e^{\mu_r}\right).$$

For simplicity, the variance term $\boldsymbol{\Sigma}_a$ has been set to the variance term of the noise term $\boldsymbol{\Sigma}_n$ resulting in an overestimate of the variance. This, however, is not critical here.

The prior scale density $p(\mathbf{s}_0)$ is assumed to be Gaussian $\mathcal{N}(\boldsymbol{\mu}_s, \boldsymbol{\Sigma}_s)$ with $\mu_{s,1} = 1.0$ and $\mu_{s,2} = 0.0$ for the actual scale and tilt terms, respectively, as we assume a correct estimate of the spectral energies which are due to reverberation. The variance term $\boldsymbol{\Sigma}_s$ is set to a small variable or can be learned from the data. In contrast to the correct mean values, however, this is not a critical value.

Particle Evolution

The evolution for each particle $\mathbf{p}_k^{(s)}$, $s = 0, \ldots, S - 1$, is estimated by an autoregressive process

$$\mathbf{p}_k^{(s)} = \mathbf{P}_{k-1}\mathbf{p}_{k-1}^{(s)} + \boldsymbol{\epsilon}_k^{(s)}; \ \boldsymbol{\epsilon}_k^{(s)} \sim \mathcal{N}(0, \sigma_\epsilon).$$

The estimate of the autoregressive matrix $\mathbf{P}_{k-1}$ can be represented by a joint matrix. Better results may be obtained by considering the additive distortion and the scale terms as independent components, such that,

$$
\mathbf{P}_k = \begin{bmatrix} \mathbf{A}_k \vdots \mathbf{0} \\ \cdots\cdots \\ \mathbf{0} \vdots \mathbf{S}_k \end{bmatrix},
$$

where the additive distortion matrix $\mathbf{A}_k$ is recalculated for each frame k by the dynamic autoregressive process and the scale matrix $\mathbf{S}_k$ is modeled by a random walk $\mathbf{S} = \mathbf{S}_k = \mathrm{diag}(1)$.

Distortion Estimation and Combination

In order to allow for a joint evaluation of additive noise and reverberation, the two distortions must be combined into a single distortion estimate. For each particle $\mathbf{p}_k^{(s)}$, $s = 0, \ldots, S - 1$ a corresponding distortion sample $n_k^{(s)}[b]$ must be calculated according to

$$
n_k^{(s)}[b] = \log \left\{ e^{a_k^{(s)}[b]} + s_k^{(s)}[b] \, e^{r_k[b]} \right\},
$$

for every frequency bin $b = 0, \ldots, B - 1$. Here $a[b]$ represents additive distortions, $s[b]$ represents the scale terms and $r[b]$ represents the spectral distortion due to reverberation. The resulting distortion estimate can now be treated just like the noise estimate common to all feature enhancement PF approaches.

Working Domain

PFs for speech feature enhancements have to be applied in a dimension-reduced logarithmic spectral domain. This can be implemented either by a filter bank or by a truncated cepstral sequence. In the reference implementation discussed in Section 14.4, a spectral estimate based on the warped minimum variance distortionless response was used. As the operation of a PF in a space with high-dimensionality such as 129, which was used for spectral estimation, would be infeasible or very slow, the example implementation works in a dimension reduced logarithmic spectral domain. The features in the reduced space are obtained by applying an inverse discrete cosine transform, implemented as a simple 20×20 matrix multiplication, to go back to the cepstral domain then truncating a final length of 20. In the *truncated* logarithmic spectral domain the relation between the noisy observation $\mathbf{y}$, the original feature $\mathbf{x}$, and noise $\mathbf{n}$ can be approximated as

$$
\mathbf{x} = \mathbf{y} + \log(\mathbf{1} - e^{\mathbf{n}-\mathbf{y}}) + \mathbf{e}_\theta + \mathbf{e}_{\text{envelope}} \approx \log(e^{\mathbf{y}} - e^{\mathbf{n}}). \tag{6.45}
$$

The first error term $\mathbf{e}_\theta$ has been presented in Section 6.1.3. The second error term $\mathbf{e}_{\text{envelope}}$ is due to spectral or cepstral envelope techniques and is assumed to have a negligible effect. Thus the approximation in (6.45) is sufficient.

In order to avoid any performance degradation stemming from feature manipulation and the lower dimensionality of the PF working domain with respect to the reverberation

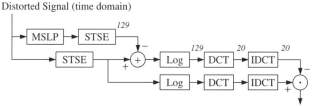

Reverberation Estimate (log. frequency domain)

Figure 6.8 Diagram of the reverberation estimate in the logarithmic spectral domain. STSE stands for short-time spectral analysis, DCT and IDCT for discrete cosine transform and its inverse, respectively, and MSLP for multi-step linear prediction. The small number gives the dimension of the feature stream

estimate, it is important not to simply process the reverberation estimate like the distorted observation. A good approximation in the reduced working domain of the PF is obtained by the process shown in Figure 6.8.

6.11 Observation Uncertainty

To account for uncertainty in the acoustic observation sequence x or enhancement process, the decoding likelihood must be integrated over all possible values of x. Let $p(\mathbf{x}|\mathbf{y})$ represent the distribution of the uncertain acoustic observation sequence and $\mathcal{X}$ all possible values of the acoustic observation. With the assumption that the uncertainty in the acoustic observation is due to additive noise – in particular its estimate – and therefore independent of the word identities W and the model parameters Λ, we can extend the fundamental equation of speech recognition (7.3) to account for the *observation uncertainty* with the noise observation y as

$$W^* = \underset{W \in \mathcal{W}}{\operatorname{argmax}} \left(\int_{\mathbf{x} \in \mathcal{X}} p(\mathbf{x}|\mathbf{y}) p(\mathbf{x}|\Lambda, W) d\mathbf{x} \right) P(W). \tag{6.46}$$

Considering a GMM with Gaussians as the output distribution

$$p(\mathbf{x}|\Lambda) = \sum_m w_m p(\mathbf{x}|\Lambda_m) = \sum_m w_m \mathcal{N}(\mathbf{x}; \boldsymbol{\mu}_m, \boldsymbol{\Sigma}_m)$$

and denoting the estimated noise mean and estimation error, modeled as a single Gaussian distribution where the variance parameter is assumed to provide a complete characterization of the uncertainty, as

$$p(\mathbf{x}|\mathbf{y}) = \mathcal{N}(\mathbf{x}; \hat{\mathbf{x}} = \mathbf{y} - \boldsymbol{\mu}_n, \boldsymbol{\Sigma}_n)$$

we can solve, under the linear assumption – compare Section 9.3.1 – the integral in (6.46) by the well-known equality

$$\int \mathcal{N}(x; \mu_1, \Sigma_1) \cdot \mathcal{N}(x; \mu_2, \Sigma_2) dx = \mathcal{N}(\mu_1; \mu_2, \Sigma_1 + \Sigma_2)$$

as

$$\int_{\mathbf{x} \in \mathcal{X}} p(\mathbf{x}|\mathbf{y}) p(\mathbf{x}|\Lambda) d\mathbf{x} = \sum_m w_m \int_{\mathbf{x}} p(\mathbf{x}|\mathbf{y}) p(\mathbf{x}|\Lambda_m) d\mathbf{x}$$

$$= \sum_m w_m \int_{\mathbf{x}} \mathcal{N}(\mathbf{x}; \hat{\mathbf{x}} = \mathbf{y} - \boldsymbol{\mu}_n, \boldsymbol{\Sigma}_n) \mathcal{N}(\mathbf{x}; \boldsymbol{\mu}_m, \boldsymbol{\Sigma}_m) d\mathbf{x}$$

$$= \sum_m w_m \mathcal{N}(\hat{\mathbf{x}} = \mathbf{y} - \boldsymbol{\mu}_n; \boldsymbol{\mu}_m, \boldsymbol{\Sigma}_m + \boldsymbol{\Sigma}_n). \quad (6.47)$$

From (6.47) it follows that the noisy feature y is enhanced by subtracting the mean noise estimate $\boldsymbol{\mu}_n$. Furthermore, the Gaussian variance in the HMM process is dynamically adjusted by enlarging the acoustic model variance $\boldsymbol{\Sigma}_m$ associated with the original speech by the variance $\boldsymbol{\Sigma}_n$ associated with the uncertainty of the noise estimate.

6.12 Summary and Further Reading

Speech feature enhancement represents a very broad array of techniques, all of which aim at restoring the characteristics of the original speech that has been corrupted by noise or reverberation in one of several domains. A full treatment of these techniques would require an entire book by itself. Thus, due to space constraints we have limited the algorithms presented here to those that work in the feature domain. Furthermore, we have restricted our presentation to those techniques that compensate for distortions that cannot be otherwise compensated for in an ASR system, namely nonstationary additive distortions and reverberation. For example, other distortions are stationary noise which can be compensated for in a state-of-the-art ASR system by various feature or acoustic model adaptation techniques, such as those presented in Chapter 9.

An overview paper which compares several speech feature enhancement techniques for ASR on artificially reverberated and re-recorded speech data in a room is by Eneman *et al.* (2003). An extensive study into additive noise removal is presented in the book by Loizou (2007). Another interesting source for further study about speech enhancement is Benesty *et al.* (2005). Besides a more general treatment of audio signals Hänsler and Schmidt (2008) contains solutions for specific applications including the enhancement of audio signals in automobiles and the automatic evaluation of hands-free systems.

The signal subspace technique to speech enhancement (Ephraim and Van Trees 1995) is a quite novel approach which has, so far, not drawn much attention. This is probably due to the high computational load and the operation in a less intuitive domain as compared to the spectral domain. However, promising results on speech recognition tasks were reported by Huang and Zhao (2000). The reader interested in comparing the performance of KFs and PFs for speech feature enhancement might consider the interacting multiple model as proposed by Kim (1998) wherein KFs in different stages interact with each other. Among others, the Speech Technology Group at Microsoft and the Robust Speech Recognition Group at CMU are good sources to catch on the latest developments.

6.13 Principal Symbols

Symbol	Description
γ	*a posteriori* signal to noise ratio
ϵ	random variation
η	*a priori* signal to noise ratio
$\boldsymbol{\mu}$	mean vector
Λ	model parameter
Ω	discrete angle frequency
Σ	covariance matrix
ω	angle frequency; short for $e^{j\omega}$
$\mathbf{A}$	linear prediction matrix
b	bin
d	dimension
$d(\cdot, \cdot)$	distortion measure
$\mathbf{d}$	distortion
D	delay
e	error term
h	impulse response
H	transfer function
k	frame index
n	discrete-time index
$\mathbf{n}$	noise signal
$p(\mathbf{x})$	prior distribution of speech
$p(\mathbf{x}_k \vert \mathbf{y}_{1:k})$	filtering density
$p(\mathbf{x}_{k+1} \vert \mathbf{x}_k)$	(state) transition probability, evolution
$p(\mathbf{y})$	prior distribution of speech
$p(\mathbf{y}_k \vert \mathbf{x}_k)$	output probability, likelihood function
r	reflection sequence
$\mathbf{R}$	correlation matrix
s	signal
$\mathbf{T}$	discrete cosine transform matrix
$\mathbf{U}$	Cholesky decomposition matrix
w	weight
W	word
x	clean signal, input signal
$\mathbf{x}_{0:k}$	clean signal sequence

Symbol	Description
y	noisy signal, output signal
$\mathbf{y}_{1:k}$	noisy signal sequence
$.^{(c)}$	cepstral domain
$.^{(f)}$	spectral domain
$.^{(l)}$	logarithmic spectral domain
$.^{(p)}$	power spectral domain
$.^{(t)}$	time domain

7

Search: Finding the Best Word Hypothesis

Search is the process by which an *automatic speech recognition* (ASR) system finds the best sequence of words conditioned on a sequence of acoustic observations. In a *distant speech recognition* (DSR) scenario this sequence of acoustic observations would be obtained using the feature extraction and enhancement techniques discussed in Chapters 5 and 6, respectively, perhaps after array processing the output of several microphones as discussed in Chapter 13. For any given sequence $\mathbf{y}_{1:K}$ of *acoustic observations* of length K, an ASR system should hypothesize that *word sequence* $w^*_{1:K_\mathrm{w}}$ which achieves

$$w^*_{1:K_\mathrm{w}} = \operatorname*{argmax}_{w_{1:K_\mathrm{w}}} P(w_{1:K_\mathrm{w}}|\mathbf{y}_{1:K}), \tag{7.1}$$

where K_w is the length of word sequence, which is unknown by assumption. Bayes' rule states

$$P(w_{1:K_\mathrm{w}}|\mathbf{y}_{1:K}) = \frac{p(w_{1:K_\mathrm{w}}, \mathbf{y}_{1:K})}{p(\mathbf{y}_{1:K})} = \frac{p(\mathbf{y}_{1:K}|w_{1:K_\mathrm{w}})\, P(w_{1:K_\mathrm{w}})}{p(\mathbf{y}_{1:K})}, \tag{7.2}$$

where the latter equality follows from the definition of conditional probability. Substituting (7.2) into (7.1) and ignoring the term $p(\mathbf{y}_{1:K})$, which does not depend on the word sequence $w_{1:K_\mathrm{w}}$, we arrive at the *fundamental formula of statistical speech recognition* (Jelinek 1998, sect. 1.2),

$$w^*_{1:K_\mathrm{w}} = \operatorname*{argmax}_{w_{1:K_\mathrm{w}} \in \mathcal{W}} p(\mathbf{y}_{1:K}|w_{1:K_\mathrm{w}})\, P(w_{1:K_\mathrm{w}}), \tag{7.3}$$

where $\mathcal{W}$ is the ensemble of all possible word sequences. Typically, the term $p(\mathbf{y}_{1:K}|w_{1:K_\mathrm{w}})$ in (7.3) is said to be determined by the *acoustic model* (AM) and $P(w_{1:K_\mathrm{w}})$ is determined by the *language model* (LM). As explained in Section 7.1.2, the search for that word sequence maximizing (7.3) is usually achieved by some variant of the *Viterbi algorithm*, which has the advantage of simplicity, but the drawback of only returning the single best hypothesis.

If it is desired to retain not only the single best hypothesis, but also several other hypotheses that were relatively likely, it is necessary to modify the basic Viterbi algorithm. The essential idea is to store not only the single best hypotheses that has reached a given state by a given time, but also multiple other hypotheses that scored well in comparison with the best hypothesis. These methods are discussed in Section 7.1.3. Moreover, it is often undesirable to store the complete state alignment generated during a Viterbi search due to the exorbitant requirements for *random access memory* (RAM). A memory efficient method for retaining only the word hypotheses, without the full state alignment, is discussed in Section 7.1.4. In Section 7.5, techniques are presented for combining word lattices from different systems that have proven effective at reducing the final word error rate.

In Section 7.2, we begin our discussion of *weighted finite-state transducers* (WFSTs). After some initial definitions, we present the all-important operations of weighted composition, weighted determinization, weight pushing, weighted minimization and epsilon removal. These algorithms will prove crucial in building a highly optimized, maximally efficient search engine.

Regardless of the implementation of the Viterbi algorithm, we assume that recognition is based on a search graph. Speech recognition can then be posed as the search for the shortest path through this graph. A search graph is constructed from several knowledge sources, including a grammar or statistical LM, a pronunciation lexicon, and a hidden Markov model, which may include context-dependency information. As described in Section 7.3, all of these knowledge sources can be represented as WFSTs, then combined and optimized with the WFST operations discussed in Section 7.2.

While WFST algorithms minimize execution time, in the most general case, they produce a static search graph that requires a great deal of RAM to store during decoding. Moreover, the WFST algorithms used to construct the search graph require a great deal of memory during their execution. In Section 7.3.6, we discuss how the size of the final search graph can be reduced by eliminating certain indeterminacies that arise due to the modeling of back-off transitions in the LM. Then in Section 7.4 we describe how the necessity of statically expanding the entire search graph can be eliminated – and the RAM required for recognition thereby greatly reduced – by expanding the graph dynamically or on-the-fly during recognition. In Section 7.5, we describe how the hypotheses or word lattices produced by several recognition systems can be combined to reduce the final word error rate.

Anyone with experience in ASR will certainly be familiar with the Viterbi algorithm and, very likely, with several other search techniques. Sections 7.3.6 and 7.4 may still be of interest, however, in that they present research results that have only appeared in the last few years. The use of WFSTs for ASR has become the fashion in recent years; hence, the well-versed reader has a good chance of having had some exposure to that body of work. Nonetheless, the review of WFSTs in Section 7.2 through 7.3.6 will certainly be of use to most all newcomers to ASR, and quite likely even some veterans. Finally, the techniques for word lattice combination presented in Section 7.5, will be familiar to everyone with practical experience in ASR, but may be new to those with less experience.

7.1   Fundamentals of Search

In this section, we discuss the fundamentals of search in ASR, including the Viterbi algorithm, lattice generation, as well as memory efficient word trace decoding. Before proceeding to a discussion of these topics, we must introduce the hidden Markov model.

7.1.1   Hidden Markov Model: Definition

Here we define the model most often used in modern ASR systems, namely, the *hidden Markov model* (HMM). To begin, let us somewhat informally define a *finite-state automaton* (FSA) as consisting of a set of *states* and a set of allowable *transitions* or *arcs* between these states. Each state has an associated *adjacency list* of transitions indicating which other states can be reached from the given state. In turn, each *arc* has an associated *transition probability* indicating how likely it is that the given transition will be taken. It often happens that the adjacency list of a given state will contain a so-called *self-loop*, which is a transition *back* to the same state. To complete this definition, we must designate a single *initial state*, as well as one or more *final states*. Defined as such, the FSA represents a *Markov chain* capable of generating random sequences of states of variable length. The first state in any sequence generated by the Markov chain must be the initial state, and the last state in any such sequence must belong to the set of final states. At each time step, a transition is taken from the current state to one of the states appearing on the current state's adjacency list. The transition taken is chosen randomly according to the probabilities associated with each arc on the adjacency list. The end of the sequence is determined when, after some number of transitions, a final state is reached.

As defined above, a Markov chain is a *random* or *stochastic process* in that the sequence of states is randomly chosen according to a given set of transition probabilities. Nonetheless, the sequence of states, which will denote as

$$\mathbf{x}_{0:K} \triangleq \begin{bmatrix} x_0 & x_1 & \cdots & x_K \end{bmatrix}^T,$$

is *directly* observable. The probability of a given state sequence is determined through the Markov assumption, which can be explicitly stated as

$$p(\mathbf{x}_{0:K}) = \prod_{k=0}^{K-1} p(x_{k+1}|x_k).$$

In order to extend the Markov chain into a HMM, we must introduce a second stochastic process on top of the Markov chain. Conditioned on the sequence $\mathbf{x}_{0:K}$ of states, this second stochastic process generates a sequence of *observations*

$$\mathbf{y}_{1:K} \triangleq \begin{bmatrix} \mathbf{y}_1 & \mathbf{y}_2 & \cdots & \mathbf{y}_K \end{bmatrix}^T.$$

The individual observations $\mathbf{y}_k$ may be drawn from a discrete set, or they may be continuous-valued, but for the purposes of this chapter and the next, we will universally assume the latter. The *hidden* Markov model is referred to as such because the state sequence $\mathbf{x}_{0:K}$ can no longer be observed directly, rather it must be inferred, based on the sequence $\mathbf{y}_{1:K}$ of observations, which *can* be directly observed. Hence, a HMM is a doubly stochastic process, whereby the first process determines the state sequence $\mathbf{x}_{0:K}$ according to a discrete probability distribution $p(\mathbf{x}_{0:K})$, and the second process determines the observation sequence $\mathbf{y}_{1:K}$ according to a continuous, conditional pdf $p(\mathbf{y}_{1:K}|\mathbf{x}_{0:K})$. The reader will note that, so-defined, the HMM already begins to display its suitability for calculating the component probabilities appearing in (7.3).

There are two standard formulations of the HMM, which are distinguished from one another by the way in which $p(\mathbf{y}_{1:K}|\mathbf{x}_{0:K})$ is defined. In the *Moore machine*, shown in Figure 7.1, the observation pdfs are associated with the states (Hopcroft and Ullman 1979, sect. 2.7); hence, an observation is generated upon transitioning into a new state. For the Moore form of the HMM, the *conditional independence* assumption can be expressed as

$$p(\mathbf{y}_{1:K}|\mathbf{x}_{0:K}) = \prod_{k=0}^{K-1} p(\mathbf{y}_{k+1}|\mathbf{x}_{k+1}). \qquad (7.4)$$

As we will learn in the next chapter, the pdf $p(\mathbf{y}_{k+1}|\mathbf{x}_{k+1})$ will typically be defined through a *Gaussian mixture model* (GMM). In the *Mealy machine*, on the other hand, the observation pdfs are associated with the transitions themselves, as shown in Figure 7.2; hence, an observation is generated when a given transition is taken. The conditional independence assumption for the Mealy form of the HMM can be written as

$$p(\mathbf{y}_{1:K}|\mathbf{x}_{0:K}) = \prod_{k=0}^{K-1} p(\mathbf{y}_{k+1}|\mathbf{x}_{k+1}, \mathbf{x}_k). \qquad (7.5)$$

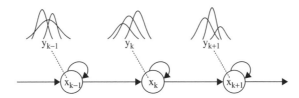

Figure 7.1 Moore formulation of the hidden Markov model whereby observation probabilities are associated with *states*

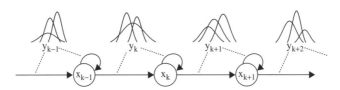

Figure 7.2 Mealy formulation of the hidden Markov model whereby observation probabilities are associated with *transitions* between states

   The two forms of the HMM are equivalent (Hopcroft and Ullman 1979, Thms 2.6–2.7), so which one is chosen is a matter of convenience. As we will discuss in Section 7.2, the traditional finite-state automata theory is formulated in terms of input and output symbols that appear on the arcs of an automaton. Hence, in this chapter, we will make exclusive use of the Mealy formulation. In Chapter 8, we will consider the problem of training the parameters of a HMM, whereby we will find it more convenient to associate the observation pdfs with states, as this will enable us to specify an observation pdf with a *one*-state index instead of two. Hence, in Chapter 8 we will make use of the Moore formulation.

   Depending on the author, there are either two or three problems to be solved in applying HMMs to speech recognition; see (Deller Jr *et al.* 1993, sect. 12) and (Rabiner 1989). We have chosen the two-problem formulation:

1. Given a sequence $\mathbf{y}_{1:K}$ of acoustic observations, how can the state sequence

$$\mathbf{x}_{0:K}^* = \underset{\mathbf{x}_{0:K}}{\operatorname{argmax}} \; p(\mathbf{y}_{1:K}|\mathbf{x}_{0:K}) \, P(\mathbf{x}_{0:K})$$

   be determined that was most likely to have generated $\mathbf{y}_{1:K}$? We will refer to this as the HMM *recognition* problem.
2. Given a sequence $w_{1:K_w}$ of words and a sequence $\mathbf{y}_{1:K}$ of acoustic observations, how can the parameters of a HMM be chosen so as to optimally represent $P(w_{0:K_w})$ and $p(\mathbf{y}_{1:K}|\mathbf{x}_{0:K})$? We will refer to this as the HMM *training* problem.

So-formulated, the first problem, which comprises the subject of the current chapter, differs from that considered in Chapter 4 only inasmuch as $\mathbf{x}_{0:K}$ must be drawn from a finite set instead of an infinite continuum of possibilities. The second problem will comprise the exclusive topic of Chapter 8.

7.1.2   Viterbi Algorithm

As explained in subsequent sections, a *search graph* maps a sequence of names or indices of GMMs to a sequence of words along with a weight corresponding to the negative log-probability of the word sequence. Ideally, we seek that word sequence which maximizes the criterion (7.1). As multiple paths through the search graph and multiple time alignments could in fact correspond to the *same* word sequence, finding the optimal word sequence according to (7.1) would entail summing over all paths and all time alignments corresponding to the same word sequence. A more tractable approach is to find the optimal *alignment* of features to HMM states by applying the *Viterbi algorithm*. Thereafter, the word string associated with this alignment of states to observations would be considered as optimal. Such an optimal alignment or *Viterbi path* is illustrated in Figure 7.3. Hence, the search problem, in some sense, reduces to one of efficiently implementing the Viterbi algorithm.

   The Viterbi algorithm is an instance of *dynamic programming* specialized for the HMM. The Viterbi algorithm and its variants will comprise our principal tools for solving the

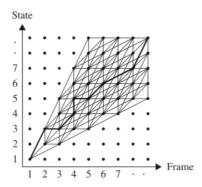

Figure 7.3 Paths investigated during a Viterbi search. The single best path is indicated by the thick line

HMM recognition problem. The maximization inherent in the Viterbi algorithm begins by initializing the *forward probability* $\alpha(\mathbf{y}_0, i) = 0$, where i is the single initial state of the HMM. Thereafter, it proceeds by iterating according to

$$\alpha(\mathbf{y}_{1:k+1}, n) = \min_q \left[\alpha(\mathbf{y}_{1:k}, q) - \log P_{n|q} - \log p(\mathbf{y}_{k+1}; \Lambda_{n|q}) \right], \tag{7.6}$$

where n is the next state to be reached in the search, $\Lambda_{n|q}$ are the parameters of the GMM associated with a transition from q to n, and $P_{n|q}$ is the transition probability from q to n. In (7.6), $\mathbf{y}_{k+1}$ is the next acoustic feature, and $\mathbf{y}_{1:k}$ denotes all acoustic features observed up to time k, as mentioned previously.

In this section, we introduce the basic concepts and notation of finding the shortest path through a search graph with the Viterbi algorithm. The simplified presentation of the search graph used here will be made more rigorous in Section 7.2. Let E and Q, respectively, denote the set of edges and states in a search graph. Similarly, let i denote the single valid initial state, and let F denote the set of valid end states.

The Viterbi algorithm is perhaps most often implemented as a *token-passing decoder* (Young *et al.* 1989). Each *token* v_k active at a particular time instant k is a data structure consisting of the following members:

- an accumulated AM score v_{AM};
- an accumulated LM score v_{LM};
- a back pointer to the prior token b;
- a pointer to the edge e in the search graph over which the current state was reached.

Each token is associated with a single state in the search graph. In order to advance the search to the next time step, a copy of a token v_k is passed from its associated state p, to all states that can be reached through transitions of the form

$$e = (p[e], l_i[e], l_o[e], w[e], n[e]) \in E$$

where

- $p[e]$ is the *previous state* in the decoding graph;
- $l_i[e]$ is the *input symbol*, in this case, the GMM index;
- $l_o[e]$ is the *ouptut symbol*, in this case, either the null symbol, denoted as ϵ, or a word index;
- $w[e]$ is the *weight* on the edge e;
- $n[e]$ is the *next state* in the decoding graph.

Propagating a token v_k forward in time implies iterating over the adjacency list of the associated state p in the search graph. For each edge $e = (p, l_i, l_o, w, n)$ in the adjacency list of p, the input symbol is read, and, provided the input symbol is not ϵ, the acoustic likelihood is evaluated with the corresponding GMM. Then a new token v_{k+1} is created. The AM score $v_{AM}(v_{k+1})$ of the new token is given by the AM score of the current token plus the new acoustic likelihood. The LM score $v_{LM}(v_{k+1})$ of the new token is given by the LM score of the current token plus the weight $w[e]$ on the edge e scaled by a LM weight. The new token also contains a back pointer $b(v_{k+1})$ to the current token as well as a pointer $e(v_{k+1})$ to the current edge. This is to say,

$$v_{AM}(v_{k+1}) = v_{AM}(v_k) - \log p(\mathbf{y}_{k+1}; \Lambda_{l_i}), \tag{7.7}$$

$$v_{LM}(v_{k+1}) = v_{LM}(v_k) + \beta \cdot w[e], \tag{7.8}$$

$$b(v_{k+1}) = v_k, \tag{7.9}$$

$$e(v_{k+1}) = e, \tag{7.10}$$

where $\beta > 1$ is the *LM weight*. The latter is applied to make the LM play a stronger role in the search. The combined score $v(v_{k+1}) = v_{AM}(v_{k+1}) + v_{LM}(v_{k+1})$ of the new token is compared to that of the best-scoring token to have reached the new state n so far, and if $v(v_{k+1})$ is *lower*, then v_{k+1} is retained, otherwise it is discarded.

Note that it is often beneficial to allow ϵ-symbols on the input side of the edges in a search graph. This is in part due to the introduction of auxiliary symbols during the construction of the search graph, as discussed in Section 7.3.2, which must subsequently be replaced with ϵ. In those cases wherein an edge $e = (p, \epsilon, l_o, w, n)$ is encountered, no acoustic likelihood is evaluated, rather the acoustic score from the prior token is retained such that, $v(v_{k+1}) = v(v_k)$, and the LM weight is updated according to (7.8). As the ϵ-transition consumes no input, the adjacency list of n is immediately expanded. This process continues recursively until a non-ϵ-transition is taken.

The iteration described above is repeated until all frames of speech have been processed. When the end of the speech is reached, the token with the lowest score is chosen among all tokens that have reached valid end states in the search graph. Then a *trace back* is performed beginning from the back pointer of this single best token, following the edges through the search graph back to the initial state. The output symbols on these edges are read, and every time a symbol other than ϵ is encountered on the output side of an edge, the corresponding word is added to the best hypothesis.

At each time step k, the token v_k^* with the current best combined AM and LM score $v(v_k^*)$ is determined, and thereafter all tokens with a combined score within a predefined *beam* of the best score are propagated forward in time. Clearly, setting a smaller beam reduces the number of tokens propagated forward at any given time step, and thereby reduces the computational expense of the search. Hence, the search runs faster. But it also increases the chance of commiting a *search error*, whereby a hypothesis with fewer word errors that would have eventually proven to have a lower score than the hypothesis chosen as the best is *discarded* because it fell outside the beam at some time step.

7.1.3 Word Lattice Generation

A *word lattice* is a memory efficient way of representing multiple word hypotheses, which may have many lengthy substrings in common. Formally, a lattice is a *directed acyclic graph* (Cormen *et al.* 2001, sect. 22.4), wherein each node is associated with a time instant and a node in the original search graph, and each edge is associated with an input and an output label in the original search graph. Depending on the application, the lattice may contain all information necessary to recreate the state alignment of the Viterbi search, or it may contain only the word identities. Such lattices are useful for *lattice rescoring*. Because a lattice represents a greatly constrained search space, knowledge sources, such as large LMs and long-span AMs, that are intractable to apply during an initial recognition pass, can be used during lattice rescoring to enhance the accuracy of a recognition engine. As discussed in Section 8.2, such lattices are also ideally suited for efficiently representing a set of competing hypotheses for *discriminative training*, each of which might plausibly be mistaken for the correct hypothesis during decoding. Finally, lattices are useful for accumulating the statistics required for unsupervised speaker adaptation, as described in Chapter 9, especially when the initial *word error rate* (WER) is very high, implying that the single best hypothesis contains many errors.

A lattice can be generated as follows. During the forward Viterbi search, the tokens reaching each state that have a *worse* total score than the best token are not discarded. Rather, they are stored in a linked list on the best token. At the end of the utterance a trace back is conducted not only for the single best path, but also for every path represented by a token in the linked list on the best token. As mentioned previously, a node in the lattice is uniquely defined by a pair consisting of a time instant and a node in the original search graph. As the search graph contains many self loops, it often happens that observations from several successive time instants are associated with the same state in the search graph. It is not necessary to form lattice nodes for all time instants associated with the same node in the search graph. Rather, unique lattice nodes are created only when the trace back encounters a *change* of states in the original search graph. Each edge in the lattice then contains the following information:

- The starting time
- The ending time
- The AM score
- The LM score
- The GMM index
- The word index.

In the vast majority of cases, the word index will be that corresponding to the ϵ-symbol.

As mentioned previously, it is often the case that all of the information contained in the complete lattice is not needed. For example, in many instances only the word identities are required. In such cases, a simpler lattice representation can be obtained through a procedure proposed by Ljolje *et al.* (1999). Firstly, the lattice is *projected* onto the output side by discarding all information save for the word identities. Secondly, as most of the edges in the projected lattice are labeled with ϵ, the epsilon removal algorithm described in Section 7.2.6 is performed to reduce the size of the lattice. An example of a word lattice after the epsilon removal operation is shown in Figure 7.4. Finally, the size of the lattice can be further reduced through determinization then minimization, which, as discussed in Section 7.2, are both WFST equivalence transformations. The final lattice after determinization and minimization is illustrated in Figure 7.5. The lattices in Figures 7.4 and 7.5 are *equivalent* inasmuch as they encode or accept exactly the same set of word hypotheses. The second lattice, however, clearly has significantly fewer nodes and edges.

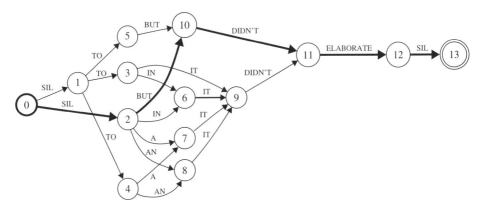

Figure 7.4   Initial word lattice after epsilon removal. The correct transcription of the utterance corresponds to the path shown in bold

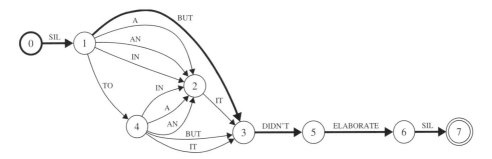

Figure 7.5   Final word lattice after weighted determinization and minimization. The correct transcription of the utterance corresponds to the path shown in bold

7.1.4 Word Trace Decoding

While maintaining the complete alignment of time frames to states is useful for HMM training, it can become prohibitive in terms of the RAM required to store all the tokens during decoding. An alternative approach was proposed by Saon *et al.* (2005), wherein backpointers for all active states are *not* stored. Rather, the backpointers to every preceding state are replaced with *word trace* structures, which store the end time of the current word along with a backpointer to the previous word represented by its own word trace structure. Each token in the search then contains only a pointer to the corresponding word trace structure, and a new word trace structure is created only when the search encounters a non-ϵ-symbol on the output side of the search graph.

To generate lattices with such a word trace structure, not only the single best token is stored at each state, but the N-best tokens. Typically N can be set to a relatively small value such as 5 or 10. As the recognition graph can have merges in the middle of words due to the minimization procedure described in Section 7.2.5, it is necessary to perform a *mergesort-unique operation* at states where several tokens meet in order to go from $2N$ potential tokens back to N. During this operation, two sorted lists of tokens are merged into a single sorted list of *unique* tokens. Thereafter, only tokens associated with the top N word sequences are retained. Unique word sequences are retained by performing a *hashing operation* on each new word index. A word sequence is then represented by its hash value, which is stored in the word trace structure. As mentioned, a new word trace structure is created every time a word label is encountered on the output side, and only the top-scoring token is propagated beyond the word boundary.

The lattice generation procedure described above is depicted in Figure 7.6 for the case of $N = 2$. At time $k - 1$, the top $2N$ best hypotheses together with their scores are ('A HORSE', score $= 3$) and ("A MOUSE", score $= 5$) on one N-best list, and ("THE HORSE", score $= 1$) and ("ONE HORSE", score $= 4$) on the other N-best list. When the two N-best lists are merged at time k with the mergesort-unique operation, only the top-N entries from both lists are retained, and the remaining entries are ("THE HORSE", score $= 3$) and ("A HORSE", score $= 4$). At time $k + 1$, a boundary for the word "ATE" is encountered, a new word trace structure is created, and only the best hypothesis is propagated forward, resulting in the hypothesis/score pair ("THE HORSE ATE", score $= 6$). The search then continues in this fashion with the receipt of each new frame of speech.

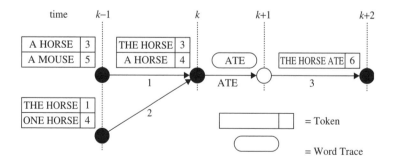

Figure 7.6 Word lattice generation with the word trace decoder for $N = 2$, after Saon *et al.* (2005)

7.2 Weighted Finite-State Transducers

In Section 7.3 we will discuss the knowledge sources necessary to construct a complete ASR system. In this section, on the other hand, we will discuss how each of these knowledge sources can be conveniently represented as a WFST; i.e., a directed graph with an input symbol, an output symbol and a weight associated with each edge. We will see that two or more WFSTs can be combined with the weighted composition algorithm presented in Section 7.2.2, then optimized through a set of equivalence transformations presented in Sections 7.2.3–7.2.6. As described in Section 7.1.2, the shortest path through the resulting WFST conditioned on a sequence of acoustic observations can be found with the Viterbi search. Once the shortest path has been found, the most likely word sequence can be determined by simply tracing back from the final state to the initial state and reading off the output symbols on the arcs along the way. We begin by introducing the basic notation and definitions of FSA. Thereafter we present the algorithms themselves.

This section can be read at several different levels. The first and simplest is that of "speaking the language" of WFSTs, and having an intuitive feel for what each of the basic algorithms does. To cater to this level of understanding, we have provided simple examples[1] illustrating the application of each of the algorithms presented here on small transducers. Such examples should make clear the effect that a given algorithm produces. The second level of understanding is that of knowing how each algorithm is implemented. Readers desiring this information will find pseudocode for most of the algorithms along with a discussion thereof. The third level of understanding is that of being able to prove that each of the algorithms is *correct*. Those desiring such deep understanding will, unfortunately, find at most sketches of the required correctness proofs. We have, however, been at pains to provide references to the original work in order to lay the basis for further reading.

7.2.1 Definitions

As we treat *weighted* finite-state automata in this section, a formalism for combining and manipulating weights will be necessary. Thus we begin with a definition.

Definition 7.2.1 **(semiring)** *A semiring* $K = (\mathbb{K}, \oplus, \otimes, \overline{0}, \overline{1})$ *consists of a set* $\mathbb{K}$, *an associative and commutative operation* $\oplus$, *an associative operation* $\otimes$, *the identity* $\overline{0}$ *under* $\oplus$, *and the identity* $\overline{1}$ *under* $\otimes$. *By definition,* $\otimes$ *distributes over* $\oplus$ *and*

$$\overline{0} \otimes a = a \otimes \overline{0} = \overline{0}.$$

A semiring is a ring that may lack negation. While this definition may seem excessively formal, it will prove useful in that operations on FSAs can be defined in terms of the operations on an abstract semiring. Thereafter, the definitions of the several algorithms need not be modified when the semiring is changed.

---

[1] Most of the examples of transducer operations found in this section are based on those in the excellent tutorial on WFSTs presented by Mehryar Mohri and Michael Riley at Interspeech in Aalborg, Denmark in 2002.

A simple example is the semiring of natural numbers $(\mathcal{N}, +, \cdot, 0, 1)$. In ASR we typically use one of two semirings, depending on the operation. The *tropical semiring* $(\mathbb{R}^+, \min, +, 0, 1)$, where $\mathbb{R}^+$ denotes the set of non-negative real numbers, is useful in finding the shortest path through a search graph based on the Viterbi algorithm presented in Section 7.1.2. The set $\mathbb{R}^+$ is used in the tropical semiring because the hypothesis scores represent negative log-likelihoods. The two operations on weights correspond to the multiplication of two probabilities, which is equivalent to addition in the negative log-likelihood domain, and discarding all but the lowest weight, such as is done by the Viterbi algorithm. The *log-probability semiring* $(\mathbb{R}^+, \oplus_{\log}, +, 0, 1)$ differs from the tropical semiring only inasmuch as the min operation has been replaced with the *log-add operation* $\oplus_{\log}$, which is defined as

$$a \oplus_{\log} b \triangleq -\log(e^{-a} + e^{-b}).$$

The log-probability semiring is typically used for the weight pushing equivalence transformation discussed in Section 7.2.4. In addition to the tropical and log-probability semiring which clearly operate on real numbers, it is also possible to define the *string semiring* wherein the weights are in fact strings (Mohri 1997), and the operation $\oplus = \wedge$ corresponds to taking the longest common substring, while $\odot = \cdot$ corresponds to concatenation of two strings. Hence, the string semiring can be expressed as $K_{\text{string}} = (\Sigma^* \cup \infty, \wedge, \cdot, \infty, \epsilon)$. This semiring will prove useful in Section 7.2.3 during our discussion of weighted determinization.

We now define our first automaton.

Definition 7.2.2 (weighted finite-state acceptor) *A weighted finite-state acceptor (WFSA)* $A = (\Sigma, Q, E, i, F, \lambda, \rho)$ *on the semiring* $K = (\mathbb{K}, \oplus, \otimes, \bar{0}, \bar{1})$ *consists of*

- *an alphabet* Σ,
- *a finite set of states* Q,
- *a finite set of transitions* $E \subseteq Q \times (\Sigma \cup \{\epsilon\}) \times \mathbb{K} \times Q$,
- *a initial state* $i \in Q$ *with weight* λ,
- *a set of end states* $F \subseteq Q$,
- *and a function* ρ *mapping from* F *to* $\mathbb{R}^+$.

A transition or edge $e = (p[e], l[e], w[e], n[e]) \in E$ *consists of*

- *a previous state* $p[e]$,
- *a next state* $n[e]$,
- *a label* $l[e] \in \Sigma$, *and*
- *a weight* $w[e] \in \mathbb{K}$.

A final state $n \in F$ *may have an associated weight* $\rho(n)$.

A simple WFSA is shown in Figure 7.7. This acceptor would assign the input string "red white blue" a weight of $0.5 + 0.3 + 0.2 + 0.8 = 1.8$.

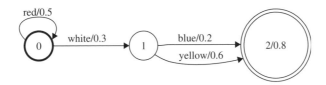

Figure 7.7   A simple weighted finite-state acceptor

As already explained, speech recognition will be posed as the problem of finding the shortest path through a WFSA, where the length of a path will be determined by a combined AM and LM score. Hence, we will require a formal definition of a path.

Definition 7.2.3 (**successful path**) *A path π through an acceptor A is a sequence of transitions $e_1 \cdots e_K$, such that*

$$n[e_k] = p[e_{k+1}] \forall k = 1, \ldots, K - 1.$$

A successful path $\pi = e_1 \cdots e_K$ is a path from the initial state i to an end state $f \in F$.

A weighted finite-state acceptor is so-named because it *accepts* strings from Σ^*, the *Kleene closure* (Aho *et al.* 1974) of the alphabet Σ, and assigns a weight to each accepted string. A string s is accepted by A iff there is a successful path π labeled with s through A. The label $l[\pi]$ for an entire path $\pi = e_1 \cdots e_K$ can be formed through the concatenation of all labels on the individual transitions:

$$l[\pi] \triangleq l[e_1] \cdots l[e_K].$$

The weight $w[\pi]$ of a path π can be represented as

$$w[\pi] \triangleq \lambda \otimes w[e_1] \otimes \cdots \otimes w[e_K] \otimes \rho(n[e_K]),$$

where $\rho(n[e_K])$ is the final weight. Typically, Σ contains ϵ, which, as stated before, denotes the null symbol. Any transition in A with the label ϵ consumes no symbol from s when taken.

We now generalize our notion of a WFSA in order to consider machines that translate one string of symbols into a second string of symbols from a different alphabet along with a weight.

Definition 7.2.4 (**weighted finite-state transducer**) *A WFST $T = (\Sigma, \Omega, Q, E, i, F, \lambda, \rho)$ on the semiring $\mathbb{K}$ consists of*

- *an input alphabet Σ,*
- *an output alphabet Ω,*
- *a set of states Q,*
- *a set of transitions $E \subseteq Q \times (\Sigma \cup \{\epsilon\}) \times (\Omega \cup \{\epsilon\}) \times \mathbb{K} \times Q$*
- *an initial state $i \in Q$ with weight λ,*
- *a set of final states $F \subseteq Q$,*

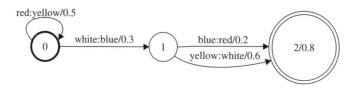

Figure 7.8 A simple weighted finite-state transducer

- *and a function ρ mapping from F to $\mathbb{R}^+$.*

A transition $e = (p[e], l_i[e], l_o[e], w[e], n[e]) \in E$ consists of

- *a previous state $p[e]$,*
- *a next state $n[e]$,*
- *an input symbol $l_i[e]$,*
- *an output symbol $l_o[e]$, and*
- *a weight $w[e]$.*

A WFST, such as that shown in Figure 7.8 maps an input string to an output string and a weight. For example, such a transducer would map the input string "red white blue" to the output string "yellow blue red" with a weight of $0.5 + 0.3 + 0.2 + 0.8 = 1.8$. It differs from the WFSA only in that the edges of the WFST have *two* labels, an input and an output, rather than one. As with the WFSA, a string s is accepted by a WFST T iff there is a successful path π labeled with $l_i[\pi] = s$. The weight of this path is $w[\pi]$, and its output string is

$$l_o[\pi] \triangleq l_o[e_1] \cdots l_o[e_K].$$

Any ϵ-symbols appearing in $l_o[\pi]$ can be ignored. Note that we will define other refinements of the general WFST in subsequent sections. Each redefinition of the most general WFST will be intended to illustrate the function of a particular equivalence transformation.

7.2.2 Weighted Composition

We now define the most fundamental operation on WFSTs.

Definition 7.2.5 (**weighted composition**) *Consider a transducer S which maps an input string u to an output string v with a weight of w_1. Consider also a transducer T which maps input string v to output string y with weight w_2. The composition*

$$R = S \circ T$$

of S and T maps string u directly to y with weight

$$w = w_1 \otimes w_2.$$

In what follows, we will adopt the convention that the components of particular transducer are denoted by subscripts; e.g., Q_R denotes the set of states of the transducer R. In the

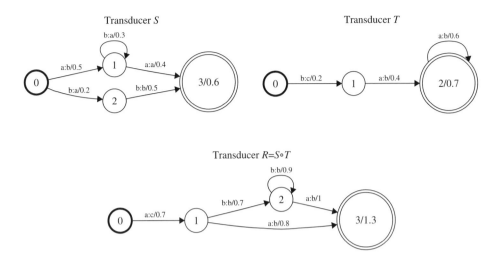

Figure 7.9   Weighted composition of two simple transducers

absence of ϵ-transitions, the construction of such a transducer R is straightforward. It entails simply pairing the output symbols on the transitions of a node $n_S \in Q_S$ with the input symbols on the transitions of a node $n_T \in Q_T$, beginning with the initial nodes i_S and i_T. Each $n_R \in Q_R$ is uniquely determined by the pair (n_S, n_T). The composition of two simple transducers is shown in Figure 7.9; for simplicity, all examples shown in this chapter are based on the tropical semiring. From the figure, it is clear that the transition from State 0 labeled with a:b/0.5 in S has been paired with the transition from State 0 labeled with b:c/0.2 in T, resulting in the transition labeled a:c/0.7 in R. After each successful pairing, the new node $n_R = (n_S, n_T)$ is placed on a queue to eventually have its adjacency list expanded.

The pairing of the transitions of n_S with those of n_T is *local*, inasmuch as it only entails the consideration of the adjacency lists of two nodes at a time. This fact provides for the so-called lazy implementation of weighted composition (Mohri *et al.* 2002). As R is constructed it can so happen that nodes are created that do not lie on a successful path; i.e., from such a node, there is no path to an end state. Such nodes are typically removed or *purged* from the graph as a final step. It is worth noting, however, that this purge step is *not* a local operation as it is necessary to consider the entire transducer R in order to determine if any given node is on a successful path. We will return to this point in Section 7.4.

When ϵ-symbols are introduced, composition becomes more complicated, as it is necessary to specify when an ϵ-symbol on the output of a transition in n_S can be combined with an ϵ-symbol on the input of n_T. As observed by Pereira and Riley (1997), in order to avoid the creation of redundant paths through R, it is necessary to replace the composition $S \circ T$ with $S \circ V \circ T$, where V is a filter. One possibility for V is shown in Figure 7.10. Practical implementations of the general weighted composition algorithm do not actually use such a filter, but instead simply keep track of the filter's *state*. Hence, a node $n_R \in Q_R$ is specified by a triple (n_S, n_T, f), where $f \in \{0, 1, 2\}$ is an index indicating the state of V. In effect, the filter specifies that after a lone ϵ-transition on either the input or output

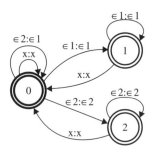

Figure 7.10 Filter used during composition with ϵ-symbols, after Pereira and Riley (1997)

side is taken, placing the filter in State 1 or State 2 respectively, an ϵ-transition on the *other* side may not be taken until a non-ϵ match between input and output occurs, thereby returning the filter to State 0 over one of the edges labeled with x:x.

7.2.3 Weighted Determinization

Having presented weighted composition, by which two transducers can be combined, we now present the first of a series of equivalence transformations. We begin with a pair of definitions.

Definition 7.2.6 **(equivalent)** *Two WFSAs are equivalent if for any accepted input sequence, they produce the same weight. Two WFSTs are equivalent if for any accepted input sequence they produce the same output sequence and the same weight.*

Definition 7.2.7 **(deterministic)** *A WFST is deterministic[2] if at most one transition from any node is labeled with any given input symbol.*

As we will see when discussing the construction of a recognition graph, it is typically advantageous to work with deterministic WFSTs, because there is *at most* one path through the transducer labeled with a given input string. This implies that, in general, the effort required to learn if a given string is accepted by a transducer, and to calculate the associated weight and output string, is *linear* with the length of the string, and *does not* depend on the size of the transducer. More to the point, it implies that the acoustic likelihood that must be calculated when taking a transition during decoding need only be calculated *once*. This has a decisive impact on the efficiency of the search process inherent in ASR. Thus we are led to consider our first equivalence operation, *determinization*, which produces a deterministic transducer τ_2 that is equivalent to some given transducer τ_1. The following discussion of the determinization algorithm is based on Mohri (1997). We begin with a definition.

---

[2] Strictly speaking, deterministic and *sequential* as given in Definition 7.3.1, are equivalent. The distinction between them was effectively introduced by Mohri and Riley *et al.* in writing the enormously popular AT&T *finite-state machine* (FSM) library, inasmuch as `fsmdeterminize` treats the 0-index, which is reserved for ϵ, the same as any other input index.

Definition 7.2.8 (**string-to-weight subsequential transducer**) *A string-to-weight subsequential transducer on the semiring* $\mathbb{K}$ *is an 8–tuple* $\tau = (\Sigma, Q, i, F, \delta, \sigma, \lambda, \rho)$ *consisting of*

- *an input alphabet* Σ,
- *a set of states* Q,
- *an initial state* $i \in Q$ *with weight* $\lambda \in \mathbb{R}^*$,
- *a set of final states* $F \subseteq Q$,
- *a transition function* δ *mapping* $Q \times \Sigma$ *to* Q,
- *a output function* σ *mapping* $Q \times \Sigma$ *to* $\mathbb{R}^+$,
- *and a final weight function* ρ *mapping from* F *to* $\mathbb{R}^+$.

The determinization algorithm for weighted automata proposed by Mohri (1997) is similar to the classical powerset construction for the determinization of conventional automata (Hopcroft and Ullman 1979, sect. 2). The states in the determinized transducer correspond to *subsets* of states in the original transducer, together with a residual weight. The initial state i_2 in τ_2 corresponds only to the initial state i_1 of τ_1. The subset of states together with their residual weights that can be reached from i_1 through a transition with the input label a then form a state in τ_2. As there may be several transitions with input label a having different weights, the output of the transition from i_2 labeled with a can only have the minimum weight of all transitions from i_1 labeled with a. The *residual weight* above this minimum must then be carried along in the definition of the subset to be applied later. Each time a new state in τ_2, consisting of a subset of the states of τ_1 together with their residual weights, is defined, it is added to a queue **Q**, so that it will eventually have its adjacency list expanded. When the adjacency lists of all states in τ_2 have been expanded and **Q** has been depleted, the algorithm terminates.

In order to clearly describe such an algorithm, let us define the following sets:

- $\Gamma(q_2, a) = \{(q, x) \in q_2 : \exists e = (q, a, \sigma[e], n_1[e]) \in E_1\}$ denotes the set of pairs (q, x) which are elements of q_2 where q has at least one edge labeled with a;
- $\gamma(q_2, a) = \{(q, x, e) \in q_2 \times E_1 : e = (q, a, \sigma_1[e], n_1[e]) \in E_1\}$ denotes the set of triples (q, x, e) where (q, x) is a pair in q_2 such that q admits a transition with input label a;
- $\nu(q_2, a) = \{q' \in Q_1 : \exists(q, x) \in q_2, \exists e = (q, a, \sigma_1[e], q') \in E_1\}$ is the set of states q' in Q_1 that can be reached by transitions labeled with a from the states of subset q_2.

Pseudocode for the complete algorithm is provided in Listing 7.1.

The weighted determinization algorithm is perhaps most easily understood by specializing all operations for the tropical semiring. This implies $\oplus$ is replaced by min and $\odot$ is replaced by $+$. The algorithm begins by initializing the set F_2 of final states of τ_2 to $\emptyset$ in Line 01, and equating the initial state and weight i_2 and λ_2 respectively to their counterparts in τ_1 in Lines 02–03. The initial state i_2 is then pushed onto the queue **Q** in Line 04. In Line 05, the next subset q_2 to have its adjacency list expanded is popped from **Q**. If q_2 contains one or more pairs (q, x) comprising a state $q \in Q_1$ and residual weight x whereby $q \in F_1$, then q_2 is added to the set of final states F_2 in Line 08 and assigned a final weight $\rho_2(q_2)$ equivalent to the minimum of all $x \odot \rho_1(q)$, where $(q, x) \in q_2$ and $q \in F_1$ in Line 09.

Listing 7.1 Pseudocode for weighted determinization

```
00  def determinize (τ₁, τ₂):
01      F₂ ← ∅
02      i₂ ← i₁
03      λ₂ ← λ₁
04      Q ← { i₂ }
05      while |Q| > 0:
06          pop q₂ from Q
07          if ∃ (q, x) ∈ q₂ such that q ∈ F₁ :
08              F₂ ← F₂ ∪ {q₂}
09              ρ₂(q₂) ←        ⊕        x ⊙ ρ₁(q)
                          q∈F₁,(q,x)∈ q₂
10          for a such that Γ(q₂, a) ≠ ∅:
11              σ₂(q₂, a) ← ⊕_{(q,x)∈Γ(q₂,a)} [ x ⊙        ⊕            σ₁[e]]
                                                     e=(q,a,σ₁[e],n₁[e])∈ E₁
12              δ₂(q₂, a) ←    ∪    { ( q̂, ⊕              [σ₂(q₂,a)]⁻¹ ⊙ x ⊙σ₁[e] ) }
                            q̂∈ ν(q₂,a)    (q,x,t) ∈ γ(q₂,a),n₁[e]=q̂
13              if δ₂(q₂, a) ∉ Q₂:
14                  Q₂ ← Q₂ ∪ { δ₂(q₂, a) }
15                  push δ₂(q₂, a) on Q
```

The next step is to begin expanding the adjacency list of q_2 in Line 10, which specifies that the input symbols on the edges of the adjacency list of q_2 are obtained from the union of the input symbols on the adjacency lists of all q such that there exists $(q, x) \in q_2$. In Line 11, the weight assigned the edge labeled with a on the adjacency list of q_2 is obtained by considering each $(q, x) \in \Gamma(q_2, a)$ and finding the edge with the minimum weight on the adjacency list of q that is labeled with a and multiplying this minimum weight with the residual weight x. Thereafter, the minimum of all the weights x is taken for all pairs (q, x) in $\Gamma(q_2, a)$. In Line 12, the identity of the new subset of $(q, x) \in Q_2$ is determined and assigned to $\delta_2(q_2, a)$. If this new subset is previously unseen, it is added to the set Q_2 of states of τ_2 in Line 14 and pushed onto the queue $\mathbf{Q}$ in Line 15 to have its adjacency list expanded in due course. Mohri (1997) proved the following theorem.

Theorem 7.2.9 (weighted determinization) *If the weighted determinization algorithm terminates, then the resulting transducer τ_2 is deterministic and equivalent to the original transducer τ_1.*

Not all transducers can be determinized. For instance, transducers admitting more than one successful path labeled with the same input sequence but producing different output sequences cannot be determinized. We will discuss a simple remedy for this problem, which arises when representing a pronunciation lexicon as a WFST, in Section 7.3.2.

As mentioned in Section 7.2.1, the weights in a semiring may also be strings. Hence, the algorithm described in Listing 7.1 is also valid for string to string transducers. Moreover, the algorithm is also valid when the semiring is the cross product of the string $(\Sigma^* \cup \infty, \wedge, \cdot, \infty, \epsilon)$ and tropical $(\mathbb{R}^+ \cup \{\infty\}, \min, +, \infty, 0)$ semirings. Hence, it is possible to determinize the WFSTs defined in Section 7.2.

A simple example of weighted determinization is shown in Figure 7.11. The two WFSTs in the figure are equivalent over the tropical semiring in that they both accept the same

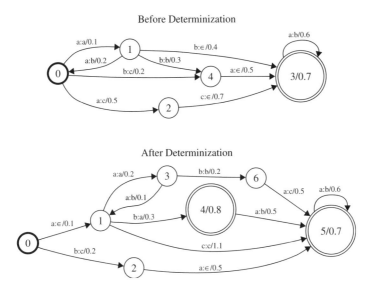

Figure 7.11 Weighted determinization of a simple transducer

input strings, and for any given input string, produce the same output string and the same weight. For example, the original transducer will accept the input string *aba* along either of two successful paths, namely, using the state sequence $0 \rightarrow 1 \rightarrow 3 \rightarrow 3$ or the state sequence $0 \rightarrow 1 \rightarrow 4 \rightarrow 3$. Both sequences produce the string *ab* as output, but the former yields a weight of $0.1 + 0.4 + 0.6 = 1.1$, while the latter assigns a weight of $0.1 + 0.3 + 0.5 = 0.9$. Hence, given that these WFSTs are defined over the tropical semiring, the final weight assigned to the input *aba* is 0.9, the minimum of the weights along the two successful paths. The second transducer also accepts the input string *aba*. There is, however, a single sequence labeled with this input, namely, that with the state sequence $0 \rightarrow 1 \rightarrow 4 \rightarrow 5$, which produces a weight of $0.1 + 0.3 + 0.5 = 0.9$. Hence, for the input string *aba*, both transducers produce the same output string *ab* and weight 0.9. For such small transducers, it is not difficult to verify that the same output and same weights are produced for all other accepted strings as well.

Provided that every node in the initial transducer τ_1 is on a successful path, every node on the determinized transducer τ_2 will likewise lie on a successful path. This implies that the weighted determinization algorithm described above is completely *local*. This point will have important implications in Section 7.3.4, where we consider the construction and immediate determinization of a transducer HC mapping directly from GMM to phone indices.

7.2.4 Weight Pushing

For any given transducer, there are many equivalent transducers that differ only in the distribution of weights along their edges. As discussed in Sections 7.1.2 through 7.1.4, an ASR system typically uses a beam search to find the most likely word sequence. The efficiency of the beam search depends very strongly on eliminating unlikely hypotheses

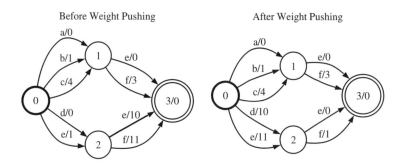

Figure 7.12 Weight pushing over the tropical semiring for a simple transducer

as early as possible from the beam. This implies that the weights should be *pushed* as far toward the initial node as possible to achieve the most efficient search. In this section, we discuss an algorithm proposed by Mohri and Riley (2001) for achieving this optimal distribution of weights. An example of *weight pushing* over the tropical semiring for a simple transducer is provided in Figure 7.12.

The weight pushing algorithm proposed by Mohri and Riley (2001) begins with the definition of a *potential function* $V : Q \rightarrow \mathbb{K} - \{\bar{0}\}$. The weights of the transducer are then reassigned according to

$$\lambda \leftarrow \lambda \otimes V(i),$$

$$\forall e \in E, w[e] \leftarrow [V(p[e])]^{-1} \otimes (w[e] \otimes V(n[e])),$$

$$\forall f \in F, \rho(f) \leftarrow [V(f)]^{-1} \otimes \rho[f].$$

A moment's thought will reveal that this reassignment has no effect on the weight assigned to any accepted string, as each weight from V is added and subtracted once. For optimal weight pushing, we assign a potential to a state q to be equal to the weight of the shortest path from q to the set of final states F, such that

$$V(q) = \bigoplus_{\pi \in P(q)} w[\pi],$$

where $P(q)$ denotes the set of all paths from q to F. Mohri and Riley (2001) noted that the general all pairs shortest path algorithm (Cormen *et al.* 2001, sect. 26.2) is too inefficient to enable weight pushing on very large transducers, but proposed instead the *approximate* shortest path algorithm in Listing 7.2. The algorithm functions by first assigning all states q a potential of $\bar{0}$ in Lines 01–02, and placing the initial state i on a queue $\mathbf{S}$ of states that are to be *relaxed* in Line 03. For each node q, the current potential $d[q]$ as well as the amount of weight $r[q]$ that has been added since the last relaxation step are maintained. When q is popped from $\mathbf{S}$, all nodes $n[e]$ that can be reached from the adjacency list $E[q]$ are tested in Line 09 to determine whether they should be relaxed. The relaxation itself occurs in Lines 10 and 11. Thereafter the relaxed node $n[e]$ is placed on $\mathbf{S}$ if not already there in Lines 12 and 13. The algorithm terminates when $\mathbf{S}$ is depleted. The approximation

Listing 7.2 Approximate shortest path algorithm

```
00   def shortestDistance():
01     for j in 1 to |Q|:
02       d[j] ← r[j] ← 0̄
03     S ← { i }
04     while |S| > 0:
05       pop q from S
06       R ← r[q]
07       r[q] ← 0̄
08       for e ∈ E[q]:
09         if d[n[e]] ≠ d[n[e]] ⊕ (R ⊗ w[e]):
10           d[n[e]] ← d[n[e]] ⊕ (R ⊗ w[e])
11           r[n[e]] ← r[n[e]] ⊕ (R ⊗ w[e])
12           if n[e] ∉ S:
13             push n[e] on S
14     d[i] ← 1̄
```

in this algorithm involves the test in Line 09, which, strictly speaking, must always be true implying, that the algorithm will never terminate. In practice, however, a small threshold on the deviation from equality can be set so that the algorithm terminates after a finite number of relaxations.

In the experience of the present authors, it is necessary to first *reverse* the graph before calculating the potential of each node, which implies that for every edge $e = (p, l_i, l_o, w, n)$ in the original graph R there will be an edge $e_{reverse} = (n, l_i, l_o, w, p)$ in $R_{reverse}$. This point is not mentioned by Mohri and Riley (2001). They, however, report the importance of pushing weights over the log-probability semiring. Moreover, they provide empirical results indicating that pushing weights over the tropical semiring can actually lead to *reduced* search efficiency.

7.2.5 Weighted Minimization

Minimization entails constructing the transducer equivalent to a given transducer with the minimal number of arcs and states. The importance of minimization is two-fold. Firstly, minimal transducers require less RAM to store and manipulate, in some cases, an order of magnitude less RAM. Secondly, using a minimal search graph during recognition can substantially reduce run-time, because fewer hypotheses must be maintained and propagated at each time step. Minimization is readily accomplished through a straightforward modification of the classical *set partitioning algorithm* (Aho *et al.* 1974, sect. 4.13). Before describing this algorithm, we must define a conventional automaton *without* weights.

***Definition 7.2.10* (finite-state machine)** *A FSM is a 5-tuple $A = (\Sigma, Q, E, i, F)$ consisting of*

- *an alphabet Σ,*
- *a finite set of states Q,*
- *a finite set of transitions $E \subseteq Q \times (\Sigma \cup \{\epsilon\}) \times Q$,*
- *a initial state $i \in Q$,*

- *and a set of end states $F \subseteq Q$.*

A transition $e = (p[e], l[e], n[e]) \in E$ consists of

- *a previous state $p[e] \in Q$,*
- *a next state $n[e] \in Q$,*
- *a label $l[e] \in \Sigma$,*

A final state $q \in F$ may have an associated label $a \in \Sigma$.

We now pose the problem as follows. Consider a FSM with the set of states Q. We wish to partition Q into subsets $M = \{Q_i\}$ such that $\forall a : \exists e_1 = (p_1, a, n_1), e_2 = (p_2, a, n_2) \in E$, it holds that

$$p_1, p_2 \in Q_j \Rightarrow n_1, n_2 \in Q_i \tag{7.11}$$

for some i. We seek the *coarsest partition* $\{Q_i\}$ of Q, which is by definition the partion with fewest elements, that satisfies (7.11). Let v be a partition of Q and let f be a function mapping $Q \times \Sigma$ to Q. In the present case, f is defined implicitly through the transitions $E \subseteq Q \times (\Sigma \cup \{\epsilon\}) \times Q$. For each $Q_i \in v$ define the sets

$$\mathrm{symbol}(Q_i) = \{a \in \Sigma : \exists e = (p, a, n) \in E, n \in Q_i, p \in Q\}, \tag{7.12}$$

$$f^{-1}(Q_i, a) = \{p \in Q : \exists e = (p, a, n) \in E, n \in Q_i\}. \tag{7.13}$$

So defined $\mathrm{symbol}(Q_i)$ is subset of symbols used as input labels on at least one edge into a node in Q_i. Similarly, $f^{-1}(Q_i, a)$ is the set of nodes having at least one transition labeled with a into a node in Q_i.

Pseudocode for the partitioning algorithm is given in Listing 7.3. To apply the partition algorithm to WFSTs, it is first necessary to *encode* both input and output symbols, along with the weight on each edge, as a *single symbol*. It is a subset of these encoded symbols that is returned by $\mathrm{symbol}(Q_i)$. After the partition algorithm has completed, the composite symbols are *decoded* to obtain the final WFST.

We will say the set $T \subseteq Q$ is *safe* for v if for every $B \in v$, either $B \subseteq f^{-1}(T, a)$ or $B \cap f^{-1}(T, a) = \emptyset \forall a \in \Sigma$. The key of the algorithm is the partitioning of Q_j in Lines 12–13, which ensures that there are no transitions of the form $e_1 = (p_1, a, n_1)$ and $e_2 = (p_2, a, n_2)$, where either $p_1, p_2 \in Q_j$ or $p_1, p_2 \in Q_n$, for which (7.11) does not hold. Hence, Lines 12–13 ensure that P is safe for the resulting partition, inasmuch as if $Q_j \cap f^{-1}(P, a) \neq \emptyset$ for some Q_j, then either $Q_j \subseteq f^{-1}(P, a)$, or else Q_j is split into two blocks in Lines 12–13, the first of which is a subset of $f^{-1}(P, a)$, and the second of which is disjoint from that subset. For reasons of efficiency, the smaller of Q_j and Q_n is placed on **S** in Lines 17–20, unless Q_j is already on **S**, in which case Q_n is placed on **S** in Lines 14–15 regardless of whether or not $|Q_n| < |Q_j|$. Aho *et al.* (1974, sect. 4.13) proved the following lemma.

Lemma 7.2.11 (**set partitioning**) *After the algorithm in Listing 7.3 terminates, every block Q_i in the resulting partition v' is safe for the partition v'.*

Listing 7.3 Partition algorithm

```
00   def partition():
01       Q_0 ← Q − F
02       Q_1 ← F
03       push Q_0 on S
04       push Q_1 on S
05       n ← 1
06       while |S| > 0:
07          pop P from S
08          for a in  symbol (P):
09             for Q_j such that Q_j ∩ f^{-1}(P,a) ≠ ∅
10             and Q_j ⊄ f^{-1}(P,a):
11                n + = 1
12                Q_n ← Q_j ∩ f^{-1}(P, a)
13                Q_j ← Q_j − Q_n
14                if Q_j ∈ S:
15                   push Q_n on S
16                else:
17                   if |Q_n| < |Q_j|:
18                      push Q_n on S
19                   else:
20                      push Q_j on S
```

Mohri (1997) proved that *weighted minimization* can be accomplished through a sequence of three steps, namely,

- weighted determinization,
- weight pushing,
- classical minimization as described above.

As mentioned previously, the graph must be first encoded, then minimized, and finally decoded. The effects of this procedure on a simple transducer are illustrated in Figure 7.13. As the initial transducer in the figure is deterministic, it is not necessary to perform the first step. Hence, only the results of weight pushing and minimization are shown in the second and third transducers, respectively.

It is worth noting that applying the classical minimization procedure to the original transducer *without* first pushing the weights yields a transducer that is identical to the original unpushed version. Pushing weights prior to classical minimization, however, yields an equivalent transducer with significantly fewer nodes and edges.

7.2.6   Epsilon Removal

Epsilon removal is by definition the construction of a transducer τ_2 that is equivalent to τ_1, but which contains no ϵ-symbols as inputs. As discussed in Section 7.1.3, it is often useful to perform such an operation on word lattices. We will encounter another use of epsilon removal in Sections 7.3.6 and 7.4 when discussing the construction of compact transducers, especially for fast on-the-fly composition. This procedure is demonstrated on a simple transducer in Figure 7.14. In this section, we discuss an algorithm for epsilon removal proposed by Mohri (2002).

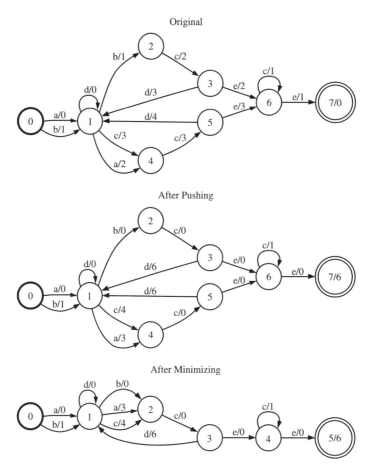

Figure 7.13 Weighted minimization of a simple transducer

Let us begin by defining the ϵ-distance from state p to state q in τ_1 as

$$d[p,q] \triangleq \bigoplus_{\pi \in P(p,q), i[\pi]=\epsilon} w[\pi],$$

where $P(p,q)$ is the set of all paths from p to q and $i[\pi] = \epsilon$ indicates that the path π is labeled solely with ϵ. The algorithm then works in two steps:

1. The ϵ-closure of a state p is discovered, which by definition is

$$C[p] \triangleq \{(q,w) : q \in \epsilon[p], d[p,q] = w \in \mathbb{K} - \{\bar{0}\}\},$$

where $\epsilon[p]$ denotes the set of states reachable from p by paths labeled solely with ϵ. The ϵ-closure of p is the set of states q that can be reached from p entirely over transitions labeled with ϵ-symbols, together with the combined weight $d[p,q]$.

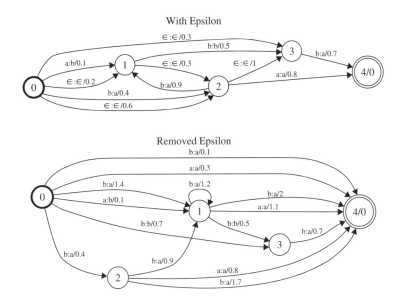

Figure 7.14 Epsilon removal for a simple transducer

Listing 7.4 Epsilon removal algorithm

```
00   def epsilonRemoval(τ):
01     for p ∈ Q:
02       E[p] ← { e ∈ E[p] : i[e] ≠ ε }
03       for (q, w) ∈ C[p]:
04         E[p] ← E[p] ∪ {(p, a, w ⊗ w₁, r) : (q, a, w₁, r) ∈ E[q], a ≠ ε}
05         if q ∈ F and p ∉ F:
06           F ← F ∪ { p }
07           ρ[p] ← ρ[p] ⊕ (w ⊗ ρ[q])
```

2. The transitions from p labeled with ϵ-symbols are replaced by non-ϵ-symbols with their weights $\otimes$-multiplied by $d[p, q]$. State p then becomes a final state if some state $q \in \epsilon[p]$ is final and its final weight is

$$\rho[p] \triangleq \bigoplus_{q \in \epsilon[p] \cap F} d[p, q] \otimes \rho[q].$$

Mohri (2002) demonstrates that this algorithm produces an automaton that is equivalent to the original τ. The required distances $d[p, q]$ can be found with the same approximate shortest path algorithm given in Listing 7.2. The pseudocode of the algorithm is given in Listing 7.4.

7.3 Knowledge Sources

At this point, we have posed the speech recognition problem as that of finding the shortest path through a search graph based on the likelihoods of the acoustic features of an utterance together with the score returned by a LM for a given word sequence. The acoustic likelihoods are calculated with the GMMs whose names appear as input labels on the edges of the search graph, as indicated by (7.7). The LM scores appear as weights on these same edges. We have also introduced the theory of WFSTs and the related equivalence transformations that are useful for constructing and optimizing the search graph for maximum efficiency during decoding. What is still necessary to complete this exposition is to describe the *knowledge sources* that are needed to create the search graph. Briefly stated, the required knowledge sources include:

1. The *grammar G*, which accepts sequences of words in a language;
2. The *pronunciation lexicon L*, which specifies how sequences of words are expanded into sequences of phonemes;
3. The *hidden Markov model H*, which determines how the phonemes are expanded into state sequences;
4. The *context-dependency transducer HC*, which specifies how sequences of phones are expanded into sequences of context-dependent GMMs;
5. The *acoustic model*, which assigns a likelihood to sequences of acoustic observations conditioned on sequences of HMM states.

The training of the AM is the subject of Chapter 8. The other knowledge sources along with their representations as WFSTs are described in Sections 7.3.1 through 7.3.4. Their static combination is then described in Section 7.3.5, along with techniques for limiting the size of the final search graph in Section 7.3.6. In Section 7.4, we will describe how the single large search graph can be factored into two much smaller graphs that are then composed on-the-fly during decoding. Depending on the dimensions of the AM and LM, this dynamic expansion algorithm can be more efficient in terms of both RAM requirements and execution speed than its static expansion counterpart.

7.3.1 Grammar

The first knowledge source is the *grammar*, of which there are two primary sorts. A *finite-state grammar* (FSG) is crafted from rules or other expert knowledge, and typically only accepts a very restricted set of word sequences. Such a grammar is illustrated in Figure 7.15 for a hypothetical travel assistance application. The FSG in the figure accepts strings such as "SHOW ME THE QUICKEST WAY FROM ALEWIFE TO BOSTON." FSGs are useful in that their constrained languages help to prevent recognition errors. Unfortunately, these constrained languages also cause all formulations of queries, responses, or other verbal interactions falling outside of the language accepted by the grammar to be rejected or misrecognized.

The second type of grammar is a *statistical language model* or *N-gram*, which assigns negative log-probabilities to sequences of words. The primary difference between the FSG and the *N*-gram is that the *N*-gram typically includes a *backoff* node, which enables it to accept *any* sequence of words. The structure of the simplest *N*-gram, namely the *bigram*, in which the probability of the current word w_k is conditioned solely on the prior word w_{k-1}, is shown in Figure 7.16. From the figure, it is clear that the bigram contains two kinds of nodes. The first type is the actual bigram node, for which all incoming transitions must be labeled with the same prior word w_{k-1} in order to uniquely specify the context. In a real bigram, the transitions leaving the bigram

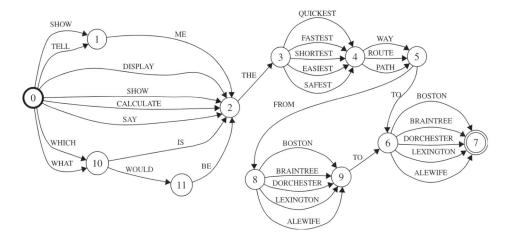

Figure 7.15   Finite-state grammar for a hypothetical travel assistance application

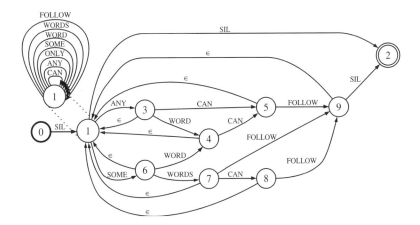

Figure 7.16   Statistical bigram. Node 1, the backoff node, is broken out for clarity

nodes would, in the simplest case, carry a weight determined from the bigram frequency statistics of a large training corpus of the form,

$$P(w_i|w_j) \approx \frac{N_{i|j}}{N_j},$$

where $N_{i|j}$ is the number of times w_i was observed to follow w_j, and N_j is the total number of times w_j was observed in the context of any following word.

As mentioned previously, the backoff node allows transitions labeled with any word in the vocabulary. In a bigram, the weights on the transition from the backoff node labeled with word w_i are estimated from unigram word frequency counts of the form,

$$P(w_i) \approx \frac{N_i}{N},$$

where N_i is the number of times w_i was observed in a training text in *any* context, and N is the total number of tokens in the training text. Transitions from bigram nodes to the backoff node are typically labeled with the null symbol ϵ, so that they can be taken without consuming any input. In practice, the transitions labeled with ϵ-symbols also carry weights determined by one of a number of backoff schemes. We will subsequently consider two such schemes for estimating backoff weights, namely, the methods of Good–Turing and Kneser–Ney.

As both FSGs and statistical N-grams can be represented as WFSAs, hybrid structures with the features of both can also be constructed. In this way, it is possible to combine a well-structured task grammar with a statistical LM to provide flexibility when users stray outside the task grammar.

We will now consider more closely how the weights in a statistical LM can be robustly estimated. In general, the probability of a word w_i is conditioned on a *history*, which we denote as $h(w_1, \ldots, w_{K-1})$, where $w_1, \ldots, w_{K-1}$ are the preceding words (Jelinek 1998, sect. 4.2). The probability of the word sequence $w_{1:K}$ can be represented as

$$P(w_{1:K}) = \prod_{k=1}^{K} P(w_k | h(w_1, \ldots, w_{k-1})). \tag{7.14}$$

If the grammar is in state h_{k-1} at time $k-1$, then the next word w_k causes a change to state h_k, such that (7.14) can be expressed as

$$P(w_{1:K}) = \prod_{k=1}^{K} P(w_k | h_{k-1}).$$

The simplest method of estimating the probabilities $P(w_k | h_{k-1})$ is then to count the number of times $N(w, h)$ that word w followed history h, as well as the total number of times

$$N(h) \triangleq \sum_{w} N(w, h)$$

that h occurred with any following word, and define the *relative frequency estimate*

$$f(w_k | h_{k-1}) \triangleq \frac{N(w, h_{k-1})}{N(h_{k-1})}. \tag{7.15}$$

Typically, the context h_{k-1} of a word w_k is chosen to be the two or three preceding words, such that $h_{k-1} = (w_{k-1}, w_{k-2})$ or $h_{k-1} = (w_{k-1}, w_{k-2}), w_{k-3})$, thereby yielding a *trigram* or *fourgram*, although in theory any number of preceding words might be used. The simple relative frequency estimate (7.15) is, however, usually insufficient, inasmuch as many trigrams or fourgrams that occur in the test set will *never* be observed in the training text regardless of the size of the corpus or corpora used for training. Jelinek (1998, sect. 4.3) reports the results of an experiment conducted at IBM in the 1970s, whereby researchers divided a text corpus of patent descriptions based on a 1000-word vocabulary into 300,000-word and 1,500,000-word test and training sets, respectively. They found that 23% of the trigrams that occurred in the test set never appeared in the training set, and this is for a vocabulary of extremely modest size. In other words, the problem of building a LM reduces to one of solving the *data sparsity* problem, whereby nonzero probabilities must be assigned to events that have been observed few times or not at all. As mentioned previously, a very common solution to the data sparsity problem is to define the *backoff probability*, as

$$\hat{P}(w_k|w_{k-1}, w_{k-2}) \triangleq \begin{cases} \alpha(w_k|w_{k-1}, w_{k-2}), & \text{if } N(w_k, w_{k-1}, w_{k-2}) > M, \\ \gamma(w_{k-1}, w_{k-2})\,\hat{P}(w_k|w_{k-1}), & \text{otherwise,} \end{cases}$$

$$(7.16)$$

where α and γ are chosen such that $\hat{P}(w_k|w_{k-1}, w_{k-2})$ is properly normalized, as we will shortly discuss. The probability $\hat{P}(w_k|w_{k-1})$ in (7.16) has the same structure as $\hat{P}(w_k|w_{k-1}, w_{k-2})$, namely,

$$\hat{P}(w_k|w_{k-1}) \triangleq \begin{cases} \alpha(w_k|w_{k-1}), & \forall\, N(w_{k-1}, w_{k-2}) > L, \\ \gamma(w_{k-1})\,\hat{P}(w_k), & \text{otherwise,} \end{cases} \qquad (7.17)$$

where M and L are, respectively, the thresholds for "frequently-seen" events on the trigram and bigram probabilities. Backoff schemes differ in how they assign the terms α and γ in (7.16) and (7.17). What all such schemes have in common, however, is that they draw probability away from frequently seen events and reassign it to seldom or unseen events.

Good–Turing Estimation

We now describe the *Good–Turing estimation*, which is a common technique for dealing with the data sparsity problem. Let $\mathcal{D}$ denote a set of *development* data of size $N_\mathcal{D} = |\mathcal{D}|$, and let $\mathcal{H}$ denote a set of *held-out* data. Let us denote as $x_i = (w_k, w_{k-1}, w_{k-2})$ each trigram in the respective data sets. Moreover, let $N_\mathcal{D}(x_i)$ denote the number of times x occurs in $\mathcal{D}$. Given the threshold $M > 0$, the probability estimates will have the form

$$\hat{P}(x_i) = \begin{cases} \alpha f_\mathcal{D}(x_i), & \forall\, \{x|N_\mathcal{D}(x_i) > M\}, \\ q_j, & \text{for } N_\mathcal{D}(x_i) = j, \end{cases} \qquad (7.18)$$

where we have defined the relative frequency estimate,

$$f_\mathcal{D}(x_i) \triangleq \frac{N_\mathcal{D}(x_i)}{N_\mathcal{D}}.$$

These relative frequency estimates will be determined on the basis of the word counts in $\mathcal{D}$, while the additional parameters $\{q_j\}$ and α will be estimated from the statistics in $\mathcal{H}$.

Let us define the Kronecker delta function

$$\delta(a, b) = \begin{cases} 1, & \text{if } a = b, \\ 0, & \text{otherwise.} \end{cases}$$

Moreover, let $N_{\mathcal{H}}(x_i)$ denote the number of times x_i has occurred in $\mathcal{H}$, and let r_j denote the number of times a sequence of symbols x_i has occurred in $\mathcal{H}$, such that $N_D(x_i) = j$; i.e., such that

$$r_j = \sum_{x_i \in \mathcal{H}} N_{\mathcal{H}}(x_i) \, \delta(N_D(x_i), j).$$

Let r^* denote the number of instances x_i such that $N_D(x_i) > M$,

$$r^* \triangleq \sum_{j>M} r_j.$$

The total size $|\mathcal{H}|$ of the held-out data set is then

$$|\mathcal{H}| = \sum_{j=0}^{M} r_j + r^*.$$

Finally, let us define

$$P_M \triangleq \sum_{x_i : N_D(x_i) > M} f_D(x_i) = \frac{1}{N_D} \sum_{j>M} j \, n_j,$$

and require that

$$\sum_{j=1}^{M} n_j \, q_j + \alpha \, P_M = 1, \tag{7.19}$$

where n_j denotes the number of unique elements x such that $N(x) = j$. We require that the condition (7.20) is satisfied in order to ensure that $\hat{P}(x_i)$ in (7.19) is a valid probability distribution. With the foregoing definitions, it is straightforward to demonstrate (Jelinek 1998, sect. 15.2) that the maximum likelihood estimates of $\{q_j\}$ and α, respectively, are given by

$$q_j = \frac{1}{n_j} \frac{r_j}{|\mathcal{H}|} \, \forall \, j = 0, \ldots, M, \tag{7.20}$$

$$\alpha = \frac{1}{P_M} \frac{r^*}{|\mathcal{H}|}. \tag{7.21}$$

Let us now define a training set denoted as $\mathcal{T}$. The Good–Turing estimate is obtained by first forming $N = |\mathcal{T}|$ distinct held-out sets by simply omitting one element $x_i \in \mathcal{T}$, such that the ith development and held-out sets are, respectively,

$$\mathcal{D}_i \triangleq \mathcal{T} - x_i, \qquad \mathcal{H}_i \triangleq \{x_i\} \, \forall \, i = 1, \ldots, N. \tag{7.22}$$

Let $N(x)$ denote the number of instances of the sequence x in $\mathcal{T}$, and, as mentioned previously, let n_j denote the number of unique elements x such that $N(x) = j$. Let us denote with $N_j(x)$ the number of times x occurs in $\mathcal{D}_j$, and let r_m denote the number of instances in the held-out sets $\mathcal{H}_j \, \forall \, j = 1, 2, \ldots, N$ that sequences of symbols x have occurred such that $N_j(x) = m$. Indeed, if $\mathcal{D}_j = \{x\}$, then $N_j(x) = N(x) - 1$, and if $N_j(x) = m$, then $N(x) = m + 1$. Moreover, there are n_{m+1} such elements x in $\mathcal{T}$, and hence $(m + 1) n_{m+1}$ values of $j \in \{1, \ldots, N\}$ for which $\mathcal{H}_j = \{x\}$ such that $N_j(x) = m$. It then follows that

$$r_m = (m + 1) n_{m+1}. \tag{7.23}$$

Substituting (7.24) into (7.21) provides

$$q_j = \frac{n_{j+1}}{n_j} \frac{j+1}{N} \forall j = 0, 1, \ldots, M.$$

The value of the normalization constant α is obtained by setting

$$\sum_{j=0}^{M} q_j n_j + \alpha \sum_{j>M} \frac{j}{N} n_j = 1.$$

The optimal value is then given by

$$\alpha = \frac{\sum_{j>M+1} j n_j}{\sum_{j>M} j n_j}.$$

Kneser–Ney Backoff

In the Kneser and Ney (1995) marginal constraint backoff procedure the threshold in (7.16) is set to $M = 0$, and absolute discounting (Ney *et al.* 1994) is used to determine the probability of the frequently seen events, according to

$$\alpha(w|h) = \frac{N(h, w) - d}{N(h)},$$

where $0 < d < 1$. Under the Kneser–Ney scheme, the backoff probability can be expressed as

$$\hat{P}(w|h) \triangleq \begin{cases} \alpha(w_k|h), & \text{if } N(w, h) > M, \\ \gamma(h) \beta(w|\hat{h}), & \text{otherwise,} \end{cases} \tag{7.24}$$

where $\hat{h}$ is the less detailed equivalence corresponding to the specific class h. Under a trigram model, for example, $h = (w_k, w_{k-1}, w_{k-2})$ would imply that $\hat{h} = (w_k, w_{k-1})$. Moreover, the term $\gamma(w_{k-1}, w_{k-2})$ in (7.25) is uniquely determined from

$$\gamma(h) = \frac{1 - \sum_{w:N(w,h) > 0} \alpha(w|h)}{\sum_{w:N(w,h)=0} \beta(w|\hat{h})}.$$

The Kneser–Ney scheme differs from other backoff schemes in that the parameters of the β distribution are not fixed, but instead are optimized along with the other parameters.

This optimization is performed by determining that $\beta(w|\hat{h})$ achieves

$$p(w|\hat{h}) = \sum_g p(g, w|\hat{h}).\qquad(7.25)$$

The marginal constraint (7.26) is shown to result in the solution (Kneser and Ney 1995)

$$\beta(w|\hat{h}) = \frac{N_+(\cdot, \hat{h}, w)}{N_+(\cdot, \hat{h}, \cdot)},\qquad(7.26)$$

where

$$N_+(\cdot, \hat{h}, w) \triangleq \sum_{g:\hat{g}=\hat{h}, N(g,w)>0} 1,$$

$$N_+(\cdot, \hat{h}, \cdot) \triangleq \sum_w N_+(\cdot, \hat{h}, w).$$

Hence, the probability distribution $\beta(w|\hat{h})$ is significantly different from $p(w|\hat{h})$. That is, in (7.27) only the fact that a word w has been observed in some coarse context $\hat{h}$ is taken into account. The frequency of such an event is entirely ignored.

Language Model Perplexity and Out of Vocabulary Rate

For present purposes, we are interested exclusively in the end-to-end performance of a complete DSR system, and hence will uniformly hold that LM to be the best which provides the best end-to-end performance. As mentioned earlier, the latter will typically be measured in terms of WER. Nonetheless, it is useful to have other, simpler metrics for judging the quality of a LM that can be calculated without running the entire system.

Consider that the log-probability of a set $\mathcal{W}$ of test text, as determined by the LM of a DSR system, can be expressed as

$$\mathcal{L}(\mathcal{W}; \Lambda) \triangleq -\frac{1}{K_\mathcal{W}} \sum_{k=1}^{K} \log P(w_k|w_{k-1}, \ldots, w_1),$$

where $K_\mathcal{W}$ is the total number of tokens in the test text. The most frequently-quoted metric for the quality of a LM is known as *perplexity* (Jelinek 1998, sect. 8.3), which is defined as

$$\mathcal{P}(\mathcal{W}; \Lambda) \triangleq \exp \mathcal{L}(\mathcal{W}; \Lambda).\qquad(7.27)$$

Perplexity can be equated to the average length of an imaginary list of equally-probable words from which the recognizer must choose the next word of the best hypothesis. Hence, a higher perplexity is indicative of either a worse LM, or a more difficult word prediction task.

The *out of vocabulary* (OOV) *rate* defines the number of words which are not present in the dictionary, but appear in the test word sequence. Usually an OOV word causes more than one error in the word sequence of the recognition output due to the correlation of the current word to the following words by the LM.

7.3.2 Pronunciation Lexicon

As described in Section 2.2.2, the words of natural languages are composed of subword units in a particular sequence. In modern ASR systems, each word is typically represented as a concatenation of subword units based on its phonetic transcription. For example, the word "man" would be phonetically transcribed as "M AE N", which is the representation in the recognition dictionary instead of the IPA symbols. The words of nearly any language can be covered by approximately 40 to 45 distinct phones. For example, the well–known Carnegie Mellon University dictionary of American English, which has been a standard since the early 1990s, contains over 100,000 words and their phonetic transcriptions specified in terms of 39 phones. The *British English Example Pronunciation Dictionary* was developed to support large vocabulary speech recognition on the WSJCAM0 set. The latter data set was collected from British English speakers, who read sentences from the standard *Wall Street Journal* (WSJ) data base. In order to accommodate the British English pronunciations, the standard *Carnegie Mellon University* (CMU) phone set was supplemented with four additional phones: "OH" for the vowel in "pot", and "IA", "EA", and "ua" for the diphthongs in "peer", "pair" and "poor" respectively.

As shown in Figure 7.17 a pronunciation lexicon can be readily represented as a finite-state transducer L, wherein each word is encoded along a different branch. In order to allow for word sequences instead of only individual words, the ϵ-transitions from the last state of each word transcription back to the initial node are included. Typically L is *not* determinized, as this would delay the pairing of word symbols when L and G are composed as described in Section 7.3.5.

One more aspect of constructing a pronunciation lexicon is noteworthy. It stems from the fact that languages such as English contain many *homonyms*, like "read" and "red".

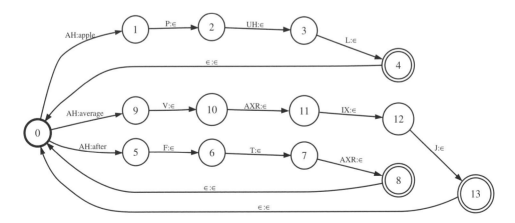

Figure 7.17 A pronunciation lexicon represented as a finite-state transducer

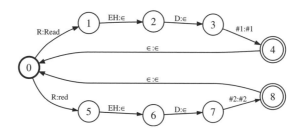

Figure 7.18   A pronunciation transducer with the auxiliary symbols #1 and #2

Such words have the *same* phonetic transcription, namely "R EH D", but *different* spellings. Were the same phonetic transcription for both words simply added to a pronunciation transducer, and composed with a grammar G, the result would not be determinizable, for the reason explained in Section 7.2.3. As a simple remedy, auxiliary symbols such as #1 and #2 are typically introduced into the pronunciation transducer in order to disambiguate the two phonetic transcriptions (Mohri *et al.* 2002), as shown in Figure 7.18.

7.3.3   Hidden Markov Model

As explained in the last section, the acoustic representation of a word is constructed from a set of subword units known as phones. Each phone in turn is represented as a HMM, most often consisting of three states. The transducer H that expands each context independent phoneme into a three-state HMM is shown in Figure 7.19. In the figure, the input symbols such as "AH-b", "AH-m", and "AH-e" are the names of GMMs. It is these GMMs that are used to evaluate the likelihoods of the acoustic features during the search process. The acoustic likelihoods are then combined with the LM weights appearing on the edges of the search graph in order to determine the shortest successful path through the graph for a given utterance, and therewith the most likely word sequence.

7.3.4   Context Dependency Decision Tree

As coarticulation effects are prevalent in all human speech, a phone must be modeled in its left and right context to achieve optimal recognition performance. A *triphone model* uses one phone to the left and one to the right as the context of a given phone. Similarly, a *pentaphone model* considers two phones to the left and two to the right; a *septaphone model* considers three phones to the left and three to the right. Using even a triphone model, however, requires the contexts to be clustered. This follows from the fact that if 45 phones are needed to phonetically transcribe all the words of a language, and if the HMM representing each context has three states, then there will be a total of $3 \cdot 45^3 = 273{,}375$ GMMs in the complete AM, all of which need to be trained. Such training could not be robustly accomplished with any reasonable amount of training data. Moreover, many of these contexts will never occur in any given training set for two reasons:

1. It is common to use different pronunciation lexicons during training and test, primarily because the vocabularies required to cover the training and test sets are often different.

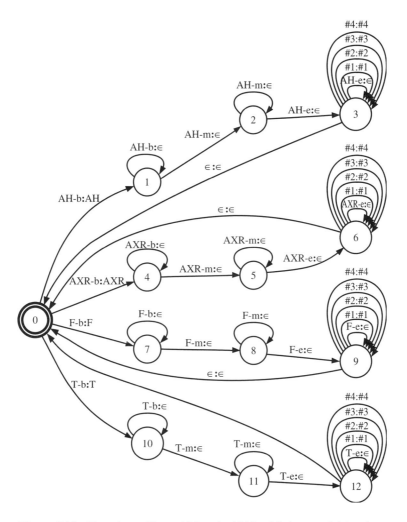

Figure 7.19 Transducer *H* specifying the hidden Markov model topology

2. State-of-the-art ASR systems typically use *crossword* contexts to model coarticulation effects between words.

From the latter point it is clear that even if the training and test vocabularies are exactly the same, new contexts can be introduced during the test if the same words appear in a *different order*.

A popular solution to these problems is to use triphone, pentaphone, or even septaphone contexts, but to use such context together with *context* or *state clustering*. With this technique, sets of contexts are grouped or clustered together, and all contexts in a given cluster share the same GMM parameters. The relevant context clusters are most often chosen with a *decision tree* (Young *et al.* 1994) such as that depicted in Figure 7.20. As shown in the figure, each node in the decision tree is associated with a question about the

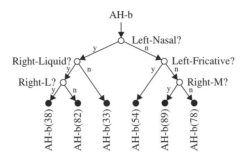

Figure 7.20 A decision tree for modeling context dependency

phonetic context. The question "Left-Nasal" at the root node of the tree is to be interpreted as, "Is the left phone a nasal?" Those phonetic contexts for which this is true are sent to the left, and those for which it is false to the right. This process of posing questions and partitioning contexts based on the answer continues until a *leaf* node is reached, whereupon all contexts clustered to a given leaf are assigned the same set of GMM parameters. Usually the clusters are estimated with the same training set as that used for HMM parameters. In addition to ensuring that each state cluster has sufficient training data for reliable parameter estimation, this decision-tree technique provides a convenient way of assigning contexts not seen in the training data to an appropriate cluster. In order to model coarticulation effects during training and test, the context-independent transducer H depicted in Figure 7.19 is replaced with the context-dependent transducer HC shown in Figure 7.21. The edges of HC are labeled on the input side with the GMM names (e.g., "AH-b(82)", "AH-m(32)", and "AH-e(43)") associated with the leaf nodes of a decision tree, such as that depicted in Figure 7.20.

Typically, one decision tree is estimated for each context-independent state; e.g., there would be one tree for "AH-b", one for "AH-m", and another for "AH-e". As mentioned previously, these trees are estimated on the same training set used for HMM parameter estimation based on a likelihood measure (Young *et al.* 1994). Let C denote the set of all contexts seen in the training set for a given context-independent state, and let $\mathcal{L}(C)$ denote the likelihood of all acoustic features assigned to the contexts in C based on a Viterbi alignment of the training set $\mathcal{Y} = \{y_1, y_2, \ldots, y_K\}$. Similarly, let $\mu(C)$ and $\Sigma(C)$ respectively denote the pooled mean and variance of these acoustic features. Then $\mathcal{L}(C)$ can be expressed as

$$\mathcal{L}(C) = -\frac{1}{2}\left\{\log[(2\pi)^n \,|\Sigma(C)| + n]\right\} \sum_{c \in C}\sum_{y \in \mathcal{Y}} \gamma_c(y),$$

where $|\cdot|$ denotes the determinant operation, $\gamma_c(y)$ is the posterior probability that feature y was assigned to cluster c, and n is length of y. For a given question q, the improvement in the training set likelihood can be expressed as

$$\Delta L_q = L(C_y) + L(C_n) - L(C),$$

where C_y and C_n, respectively, denote the contexts seen for the YES and NO clauses of q. Beginning from the root node, wherein all contexts corresponding to a given

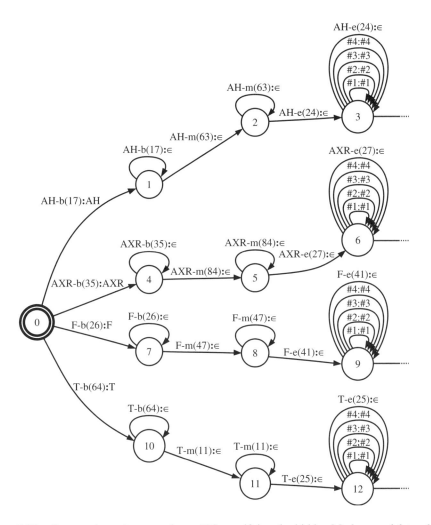

Figure 7.21 Context-dependent transducer HC specifying the hidden Markov model topology

context-independent state are clustered together, the complete decision tree is grown by successively posing all questions at each node and choosing the question that achieves

$$q^* = \operatorname*{argmax}_q \Delta L_q,$$

where the maximum is taken over the entire decision tree. That node with the highest likelihood of improvement is then split according to its best question. If the feature occupancy count for a given node falls below some predefined threshold, it is no longer considered for splitting. Similarly, when the best likelihood improvement over all nodes falls below a threshold, no more splitting is performed, and the existing leaf nodes are used to define the final state clusters.

In the balance of this section, we develop an algorithm for constructing the transducer HC that maps directly from sequences of GMMs to phone sequences based on the decision tree described above. This algorithm, which was proposed by Stoimenov and McDonough (2006), was a modification of a prior algorithm proposed by Schuster and Hori (2005). As HC is the most difficult of the knowledge sources considered here to construct, first-time readers are perhaps best advised to skip directly to Section 7.3.5 at this point, and return to these details after having gained an understanding of how a search graph in its entirety is constructed.

It is of historical interest that C and H were at one time constructed separately, where C mapped from polyphones to phones, and H mapped from (names of) GMMs to polyphones (Mohri and Riley 1998). Such a construction is limited, however, in that it is only practical for triphones; using a larger phonetic context such as pentaphones or septaphones results in a C transducer that is intractably large. Hence, we present here the more modern construction technique whereby HC is constructed jointly. This resulting transducer HC remains of manageable size regardless of the length of the phonetic context due to the fact that many polyphones will cluster to the same sequence of GMMs.

The algorithm begins by calculating a *bit matrix* **B** for each leaf node in a decision tree that specifies which phones are allowed in which positions. Each row of **B** corresponds to a phone and each column corresponds to a position in the polyphone context. As shown in Table 7.1, position (m, n) of **B** is 1 iff the mth phone is allowed in the nth position. The bit matrices are easily calculated by walking down the decision tree(s) from the root node to the leaves, and unsetting the bits corresponding to disallowed phones at each juncture.

Let A and B represent two possible questions in a decision tree, and consider a node in a decision tree in which the compound question A AND B is posed. Interpreting the YES clause in the tree is straightforward; it is only necessary to reset all bits for which either A or B is false. The NO clause for this question, on the other hand, is not so straightforward, as $!(A \text{ and } B) = !A \text{ or } !B$. This implies that for the NO clause we must reset the bits where either A or B is true, which means that separate bit matrices must be retained to represent the cases $!A$ and $!B$. Hence, to handle decision trees with compound contexts, it is necessary to extend the notion of a bit matrix to include a list **L** of bit matrices.

We say two bit matrices $\mathbf{B}_i$ and $\mathbf{B}_j$ are *equivalent* if all bits in all locations have equal values, which we denote as $\mathbf{B}_i == \mathbf{B}_j$. We say **B** is *valid* if at least one bit is set in each

Table 7.1 Typical bit matrix corresponding to center phone "AH"

Phone	Polyphone position				
	0	1	2	3	4
AH	0	0	1	1	0
B	0	0	0	1	0
⋮			⋮		
Z	1	0	0	1	0

column. Let $\mathbf{L}_n = \{\mathbf{B}_i\}$ be a *bit matrix list*. We say two bit matrix lists $\mathbf{L}_n$ and $\mathbf{L}_m$ are equivalent if $|\mathbf{L}_n| = |\mathbf{L}_m|$ and each $\mathbf{B}_i \in \mathbf{L}_n$ is equivalent to exactly one $\mathbf{B}_j \in \mathbf{L}_m$. We can assume without loss of generality that $\mathbf{B}_i \neq \mathbf{B}_j$ for any $\mathbf{B}_i, \mathbf{B}_j \in \mathbf{L}_m$, where $i \neq j$.

Metastate Enumeration

Let p denote the center phone for any given polyphone context. Let s_i denote the leaf node in a decision tree associated with the ith state in a HMM. Assuming for simplicity that all HMMs have three states, define a *metastate* $\mathbf{s}$ as a quintuple $\mathbf{s} = (p, s_1, s_2, s_3, \mathbf{L})$ where $\mathbf{L}$ is the list of *valid* bit matrices corresponding to the state sequence s_1, s_2, s_3. Let $\mathbf{L}' = \mathbf{L} \gg$ be the list of bit matrices obtained by right shifting each $\mathbf{B} \in \mathbf{L}$ and let $\mathbf{L}'' = \mathbf{L}_n \,\&\, \mathbf{L}_m$ denote the list of valid bit matrices obtained by performing the bitwise operation on each $\mathbf{B}_i \in \mathbf{L}_n$ with every $\mathbf{B}_j \in \mathbf{L}_m$. We can enumerate a set $\mathbf{S}$ of valid metastates as follows. Begin with a bit matrix list $\mathbf{L}_{s_1}$ corresponding to the leaf node associated with the first state of a three-state sequence for a polyphone with center phone p. Similarly, let $\mathbf{L}_{s_2}$ and $\mathbf{L}_{s_3}$ be the bit matrix lists for the second and third states for such a three-state sequence for a polyphone with center phone p. If

$$\mathbf{L} = \mathbf{L}_{s_1} \,\&\, \mathbf{L}_{s_2} \,\&\, \mathbf{L}_{s_3}$$

is nonempty, then s_1, s_2, s_3 is a valid three-state sequence and the metastate $(p, s_1, s_2, s_3, \mathbf{L})$ can be added to $\mathbf{S}$. As discussed in Schuster and Hori (2005), all such valid metastates can be enumerated by first enumerating the valid two-state sequences, then building three-state sequences. We say that two metastates are equivalent if they have the same phone p, the same three-state sequence s_1, s_2, s_3 and equivalent bit matrix lists $\mathbf{L}$.

Metastate Connection

Let $\mathbf{S} = \{\mathbf{s}_i\}$ denote the set of valid metastates obtained from the metastate enumeration algorithm described above, and let $\mathbf{T}$ be a second, initially empty, set of metastates. Let $\mathbf{Q}$ be a queue using any discipline and let SIL denote the initial silence metastate. The initial and end nodes of HC are denoted as INITIAL and FINAL respectively. Additionally, let $\mathbf{E}$ denote the set of edges in HC. Denoting an input dictionary of names of GMMs and an output dictionary of phones as $\mathbf{D}$ and $\mathbf{P}$, respectively, we can express each edge $\mathbf{e} \in \mathbf{E}$ as a four-tuple,

$$\mathbf{e} = (s_{\text{from}}, s_{\text{to}}, d, p)$$

where s_{from} is the previous state, s_{to} is the following state, $d \in \mathbf{D}$ is the input symbol and $p \in \mathbf{P}$ is either an output symbol or ϵ.

Consider now the algorithm for metastate connection in Listing 7.5. In this listing, $\mathbf{Q}$ is a queue of metastates whose connections to other metastates have yet to be determined. The algorithm begins by initializing the final set $\mathbf{E}$ of edges to $\emptyset$ and the set of metastates $\mathbf{T}$ to SIL in Lines 02 and 03. In Line 06, the next metastate $\mathbf{q}$ is popped from $\mathbf{Q}$ and connected in Lines 07–08 to FINAL if $\mathbf{q}$ corresponds to SIL. In the loop that begins at Line 09, each $\mathbf{s} \in \mathbf{S}$ is tested to find if a new metastate $\mathbf{t}$ can be *derived* from $\mathbf{s}$, as in Line

Listing 7.5 Metastate connection

```
00   def connectMetastates(SIL, S):
01      push SIL on Q
02      E ← ∅
03      T ← {SIL}
04      connect INITIAL to SIL
05      while |Q| > 0:
06         pop q from Q
07         if q.p == SIL:
08            connect q to FINAL
09         for s ∈ S:
10            L ← (q.L ≫) & s.L
11            if |L| > 0:
12               t ← (s.p, s.s₁, s.s₂, s.s₃, L)
13               if t ∉ T:
14                  T ← T ∪ { t }
15                  push t on Q
16               e ← (q.s₃, t.s₁, t.s₁.g, t.p)
17               E ← E ∪ { e }
18      return (E, T)
```

12, to which **q** should be connected. This test consists of forming the new list **L** of valid bit matrices in Line 10, and checking if **L** is nonempty in Line 11. Note that the right shift $\gg$ in Line 10 is to be understood as shifting in a column of *ones*. If $|\mathbf{L}| > 0$, then the *name* of the new metastate **t** is formed in Line 12, and **T** is searched to determine if this **t** already exists. If **t** does *not* exist, then it is added to **T** and placed on the queue **Q** in Lines 13–15 in order to eventually have its adjacency list expanded. This ensures that the connections for each $\mathbf{t} \in \mathbf{T}$ are created exactly *once*. The new edge **e** from the last state of **q** to the first state of **t** is created in Lines 16–17, where $\mathbf{t}.s_1.g$ is the name of the GMM associated with the latter.

When a metastate **t** is defined as in Line 12 of Listing 7.5, we will say that **t** is *derived* from **s**, and will denote this relation with the functional notation $\mathbf{s} \leftarrow \texttt{from}(\mathbf{t}, \mathbf{S})$.

Bit Masks

The algorithm for metastate connection described above is correct but impractical, inasmuch as for any reasonably sized decision tree, the number of metastates will quickly become intractably large and deplete all available memory; the algorithm does not finish. Here we consider two modifications to the algorithm: the first is a pure speedup, the second limits the inordinate growth of the number of metastates.

Consider a metastate $\mathbf{s} \in \mathbf{S}$ where **S** is once more the set of metastates obtained from the metastate enumeration algorithm. The set

$$\mathbf{N}(\mathbf{s}, \mathbf{S}) = \{\mathbf{n} \in \mathbf{S} : |(\mathbf{s}.\mathbf{L} \gg) \,\&\, \mathbf{n}.\mathbf{L}| > 0\}$$

is readily seen to be the list of *possible* following metastates for any $\mathbf{t} \in \mathbf{T}$ derived from $\mathbf{s}_F \leftarrow \texttt{from}(\mathbf{t}, \mathbf{S})$. Hence, searching only over $\mathbf{N}(\mathbf{s}_F, \mathbf{S})$ in Line 09 of Listing 7.5 instead

Listing 7.6   Bit mask function

```
00   def bitMask(s, S):
01      set all bits in M to zero
02      for n in N(s, S):
03        for M₁ in n.L:
04          M ← (M | M₁)
05      return (M ≪)
```

Listing 7.7   Efficient metastate connection

```
00   def connectMetastates(SIL, S):
01      push SIL on Q
02      E ← ∅
03      T ← {SIL}
04      connect INITIAL to SIL
05      while |Q| > 0:
06        pop q from Q
07        if q.p == SIL:
08          connect q to FINAL
09        s_F ← from(q, S)
10        for s ∈ N(s_F, S):
11          M ← bitMask (s, S)
12          L ← (q.L ≫) & s.L & M
13          if |L| > 0 :
14            t ← (s.p, s.s₁, s.s₂, s.s₃, L)
15            if t ∉ T:
16              T ← T ∪ { t }
17              push t on Q
18            e ← (q.s₃, t.s₁, t.s₁.g, t.p)
19            E ← E ∪ { e }
20      return (T, E)
```

of over all $\mathbf{S}$ results in a significant speedup. Moreover, in so doing, we run no risk of omitting any possible connections from $\mathbf{t}$: If $\mathbf{t}$ is derived from $\mathbf{s}$ as in Line 12 of Listing 7.5, then the bit matrices appearing in $\mathbf{t.L}$ can only have 1's in a subset of the positions where $\mathbf{s.L}$ has 1's. This implies that $\mathbf{t}$ will connect only to a subset of those metastates derived from the elements of $\mathbf{N(s, S)}$.

Consider the definition of the function `bitMask` in Listing 7.6. The left shift operation in Line 05 of this listing is to be understood as shifting in a column of *zeros*. It is not difficult to see that Line 10 of Listing 7.5 can be replaced with

$$\mathbf{L} \leftarrow (\mathbf{q.L} \gg) \ \& \ \mathbf{s.L} \ \& \ \mathbf{M},$$

where $\mathbf{M} \leftarrow$ `bitMask`$(\mathbf{s, S})$: Applying $\mathbf{M}$ to the prior definition of $\mathbf{L}$ unsets those bits that will be unset in any event as soon as $\mathbf{L}$ is right shifted and multiplied with any $\mathbf{n} \in \mathbf{N(s, S)}$. Leaving these bits set only causes an unneeded increase in $|\mathbf{T}|$, inasmuch as metastates in $\mathbf{T}$ which are essentially equivalent will be treated as different; these metastates will be combined in any event when HC is determinized and minimized.

Table 7.2 Sizes of HC for a
pentaphone context dependency tree

Graph	States	Arcs
HC	975,838	63,178,405
det(HC)	406,173	8,199,840
min(det(HC))	81,499	968,078

With the two changes described here, the metastate connection algorithm can now be reformulated as in Listing 7.7. The most important differences between Listings 7.5 and 7.7 lie in the loop over $N(s_F, S)$ in Lines 09–10 of the latter, which provides a speedup, and the application of the bit mask in Lines 11–12, which inhibits the growth of $|T|$. For efficiency, $N(s, S)$ and bitMask(s, S) are precalculated for all $s \in S$ and stored, so that Lines 10 and 11 only involve table lookups.

Table 7.2 shows the sizes of HC for a pentaphone context dependency tree containing 3500 leaves after each stage in its construction. In this case, the HC transducer was compiled statically from a pentaphone distribution tree.

Dynamic Expansion of *HC*

It is clear from the sizes tabulated in Table 7.2 that the combined transducer HC is much smaller after determinization than when it is initially expanded, and smaller still after minimization. As explained in Section 7.2.3, weighted determinization can be performed incrementally inasmuch as it is *not* necessary to see the entire graph in order to determine the adjacency list for a given node in the determinized graph. Rather, only the nodes in the original graph comprising the *subset* corresponding to a node in the determinized graph and their adjacency lists are required.

The capability of performing incremental determinization prompted Stoimenov and McDonough (2007) to consider the possibility of incrementally expanding HC and simultaneously incrementally determinizing it. Such an incremental expansion of HC can be achieved by expanding the set **S** in the constructor of HC. The queue **Q** in Listing 7.7 would become unnecessary, as the adjacency lists of the nodes in HC would be expanded in the order required by the incremental determinization. To expand the adjacency list in the original HC, the steps in Lines 07-19 in Listing 7.7 would be executed. A caching scheme can then be implemented whereby the connections between metastates in **T** that have not been accessed recently are periodically deleted and their memory recovered. Should the connections be needed in future, they can always be regenerated from the corresponding bit matrix list.

Consider the algorithm for expanding the adjacency list of a node in Listing 7.8. Lines 01-02 test if the adjacency list of **q** has already been expanded, and returns **q**.E in the event that it has. Lines 03-14 are equivalent to Lines 06-16 of Listing 7.7, with the exception of Line 17 in Listing 7.7, which is no longer needed.

As an initial test of the incremental construction algorithm, Stoimenov and McDonough (2007) expanded HC for the same pentaphone decision tree used to generate the statistics

Listing 7.8 Efficient adjacency list expansion

```
00   def edges(q):
01      if q.E ≠ ∅:
02         return q.E
03      if q.p == SIL:
04         connect q to FINAL
05      s_F ← from(q, S)
06      for s ∈ N(s_F, S):
07         M ← bitMask(s, S)
08         L ← (q.L ≫) & s.L & M
09         if |L| > 0:
10            t ← (s.p, s.s₁, s.s₂, s.s₃, L)
11            if t ∉ T:
12               T ← T ∪ { t }
13               e ← (q.s₃, t.s₁, t.s₁.g, t.p)
14               q.E ← q.E ∪ { e }
15      return q.E
```

Table 7.3 Memory usage and run-time requirements for static and dynamic construction of det(HC)

Algorithm	Memory usage (Gb)	Run time (minutes)
Static expansion	7.70	50
Dynamic expansion	1.42	56

in Table 7.2. The run-time and memory usage statistics for the various build scenarios are given in Table 7.3.

As is clear from the results in Table 7.3, the dynamic expansion of the graph reduces memory usage from 7.70 to 1.42 Gb, which represents a factor of 5.42 reduction. This large decline in the size of the task image is accompanied by a modest increase in run time from 50 to 56 minutes. Hence, dynamic expansion of HC is very worth while.

7.3.5 Combination of Knowledge Sources

In order to construct the final search graph, the knowledge sources described above must be combined as follows. The grammar G and pronunciation lexicon L are first composed and determinized to form $\det(L \circ G)$. After HC has been constructed as described in Section 7.3.4, it must be determinized and minimized to form $\min(\det(HC))$. As explained previously, the former can be done incrementally during the construction of HC in order to reduce the memory footprint. In the final sequence of steps $\min(\det(HC))$ and $\det(L \circ G)$ are composed, determinized, pushed and minimized. The complete construction sequence is then

$$R = \min \text{push} \det(\min(\det(HC)) \circ \det(L \circ G)). \qquad (7.28)$$

The final operation is to replace the word boundary symbols #1, #2, etc., with ϵ-symbols. As desired, the final search graph maps directly from names of GMMs to words, which is exactly what is required for the search process. The GMM names enable the evaluation of the likelihood of acoustic features. These likelihoods are combined with the LM weights stored on the edges of the search graph, such that the shortest path through the search can be found, and thereby the most likely word sequence.

Note that if the search graph is intended for recognition with the fast on-the-fly composition algorithm described in Section 7.4, the most efficient search is obtained if $\det(L \circ G)$ is minimized to form $B = \min \text{push} \det(L \circ G)$. If the search graph is only intended for static recognition, it is sufficient to push and minimize only at the end of the complete construction sequence. Performing these operations on the intermediate product $\det(L \circ G)$ and then again after the final determinization and minimization will result in exactly the same graph as if they are only performed as final operations.

7.3.6 Reducing Search Graph Size

In Section 7.2.3, we defined a deterministic WFST as having at most one edge with a given input symbol, *including* the ϵ-symbol, in the adjacency list of any node. Consider now the following definition and related theorem.

Definition 7.3.1 **(sequential transducer)** *A sequential transducer is deterministic and has no edges with ϵ as input symbol.*

Theorem 7.3.2 **(Mohri)** *The composition of two sequential transducers is sequential.*

While seemingly simple, Mohri's theorem (Mohri 1997) has deep practical implications. First of all, consider the graph construction sequence specified in (7.28). By far, the most resource intensive operation in terms of both computation and main memory is the determinization after the composition of $\min(\det(HC))$ and $\det(L \circ G)$. According to Mohri's theorem, if both $\min(\det(HC))$ and $\det(L \circ G)$ were sequential, this determinization could be eliminated entirely. This poses no problem for the context dependency transducer $\min(\det(HC))$, as its construction ensures that it is sequential. More problematic is $L \circ G$ because, as explained in Section 7.3.1, ϵ-transitions in G are typically used to enable transitions to the backoff node. While the ϵ-transitions can be removed with the epsilon removal algorithm, this causes a massive increase in the number of edges in the graph.

The foregoing considerations led McDonough and Stoimenov (2007) to consider the following modifications to the search graph construction procedure. Firstly, the ϵ-symbols in G shown in Figure 7.16 were replaced with a *backoff symbol* %. Then, at the end of each word sequence in L shown in Figure 7.18, a self-loop with % as input and output was added. With these modifications, the $L \circ G$ component was constructed according to

$$B = \det(\epsilon\text{- removal}(L \circ G)). \tag{7.29}$$

As the ϵ-transitions were replaced with explicit back symbols in G, the epsilon removal operation did not cause a massive increase in the number of transitions in the model. Rather, the only remaining ϵ-transitions were those stemming from the ϵ-symbols on the transitions back to the branch node in L. As before, the HC component was constructed according to

$$A = \min(\det(HC)). \tag{7.30}$$

Now, however, to the end of each three-state sequence in HC described in Section 7.3.4, a self-loop with % as input and output was added. Then the complete search graph was constructed according to

$$R = \min(\text{push}(A \circ B)). \tag{7.31}$$

As a final operation, the backoff % and word boundary, #1, #2, ..., symbols are removed from R prior to its use in recognition.

Dimensions of the search graphs constructed as described in this chapter beginning from bigram or trigram LMs of various sizes are given in Table 7.4 for a *Wall Street Journal* ASR system with a 5000 word vocabulary. The tabulated dimensions indicate that the size of the initial LM has a large impact on the size of the final search graph. The search graph for the shrunken bigram, which was constructed without the devices for size reduction described in this section, was actually larger than that for the full bigram, which was constructed using these size reduction techniques. The results of a series of DSR experiments on data from the Speech Separation Challenge, Part II using these LMs are presented in Section 14.9. Those results indicate that the quality of the LM has a significant impact on final system performance.

While the algorithm described above is undoubtedly beneficial in terms of reducing the size of the final search graph, the graph obtained using the full *WSJ* LM still had nearly 50 million states and over 100 million edges. Moreover, this enormous graph was for an ASR system with a very modest vocabulary. Recognition with this graph could only be performed on a 64-bit workstation, and the size of the graph in RAM was nearly 7 Gb, which can be prohibitive even for research purposes, and impossible on platforms having only more modest memory. Hence, in the next section, we will consider how the static expansion of such an enormous graph can be avoided entirely through fast on-the-fly composition.

7.4 Fast On-the-Fly Composition

Consider once more the composition of A and B as defined in (7.31). As B has no ϵ-symbols on the input side, each node $n_R \in Q_{A \circ B}$ is uniquely defined by the pair

Table 7.4 Sizes of shrunken and full trigram language models and search graphs. After McDonough and Stoimenov (2007)

Language model	G		$HC \circ L \circ G$	
	Bigrams	Trigrams	Nodes	Arcs
Shrunken bigram	323,703	0	4,974,987	16,672,798
Full bigram	835,688	0	4,366,485	10,639,728
Shrunken trigram	431,131	435,420	14,187,005	32,533,593
Full trigram	1,639,687	2,684,151	49,082,515	114,304,406

(n_A, n_B), where $n_A \in Q_A$ and $n_B \in Q_B$. This implies that the complexity of the general composition algorithm introduced by the ϵ-symbols has been eliminated; B has no ϵ-symbols on the input side, and an ϵ-symbol on the output side of A can always be taken. As R is typically many times larger than A and B prior to their composition, and remains so even after determinization, pushing, and minimization, several authors, including Dolfing and Hetherington (2001), Willett and Katagiri (2002), Hori and Nakamura (2005), Caseiro and Trancoso (2006), and Cheng *et al.* (2007), have proposed algorithms for *fast on-the-fly composition*. The common element among these algorithms is that the single enormous search graph is never statically expanded. Rather a search is conducted simultaneously through two or more smaller search graphs. In this section, we consider a method proposed by McDonough and Stoimenov (2007) for performing such on-the-fly expansion.

First of all, let us redefine B in (7.29) as

$$B = \min \operatorname{push} \det(\epsilon\text{-} \operatorname{removal}(L \circ G)).$$

Now, instead of *statically* composing A and B, we perform recognition with the on-the-fly-composition. For the sequential A and B transducers considered here, the latter is a straightforward modification of the token-passing algorithm. The composition R is not actually constructed, rather each token simply maintains a pointer to an edge in both E_A and E_B, and each active hypothesis is associated with a state $n_R = (n_A, n_B) \in Q_R$ where $n_A \in Q_A$ and $n_B \in Q_B$. As the ϵ-transitions from the input side of B have been removed, the filter shown in Figure 7.10 is not required; an ϵ-transition on the output of A can *always* be taken. Moreover, as the filter state is no longer needed, n_R is *uniquely* specified by the pair (n_A, n_B). As the search progresses, every time a non-ϵ phone symbol is encountered on the output side of A, the adjacency list of the corresponding node of B must be searched for a matching symbol. If a match is found, a new token with a pointer to the edge in B with the matching symbol is created.

Such dynamic composition has the potential to be still faster than static expansion of the entire search graph, inasmuch as it enables the application of the full N-gram LM during the initial recognition passes. Use of the full LM greatly improves the efficiency of the beam search during recognition, because it allows for the identification of unlikely search hypotheses, which can then be pruned away at an early stage. There remains one further problem to be solved, however, in order to efficiently implement such an on-the-fly algorithm. As mentioned in Section 7.2.2, nodes can be formed during the composition of A and B that *do not* lie on a successful path. After static composition, such nodes are typically purged. Expanding the set of active hypotheses across transitions from nodes that are not on successful paths during on-the-fly composition would clearly result in wasted computation and hence prove detrimental to the efficiency of the search. Hence, this is to be avoided at all costs.

A naive solution to this problem would be to simply enumerate the set of nodes that are not on successful paths, and to search no further when such a node is reached. A more memory efficient solution was proposed by McDonough and Stoimenov (2007). In order to present this solution, we begin with a set of definitions.

Definition 7.4.1 (**white, gray, black nodes**) *A node is white iff all edges on its adjacency list are on successful paths. A node is black iff none of the edges on its adjacency list* *is on a successful path. A node is gray iff at least one edge on its adjacency list is on a successful path, and at least one edge on its adjacency list is not on a successful path.*

We now state a simple theorem with interesting implications.

Theorem 7.4.2 *All paths from the initial node $i \in R$ to a black node must go through a gray node.*

Proof: Without loss of generality, the initial node can be assumed to be white. Assuming that a path from the initial node to a black node would never cross a gray node leads immediately to a contradiction with the definition of a white node. $\square$

Now we state a definition and another theorem.

Definition 7.4.3 (fence) *The fence F is that subset of black nodes that can be reached by a single transition from a gray node.*

Theorem 7.4.4 *All paths from i to a black node must cross a node $n_F \in F$.*

Proof: Follows as a corollary of Theorem 7.4.2. $\square$

Theorem 7.4.4 clearly implies that in order to avoid expanding any black node during on-the-fly composition, we need not store the indices of *all* black nodes, but rather only the indices of the fence nodes, which can be found quite simply. Starting from the initial node $i_R = (i_A, i_B)$, perform a breadth-first search (Cormen *et al.* 2001, sect. 22.2) to discover all nodes in the set $A \subset Q_R$ that are *accessible* from i_R as well as the end nodes $F_R \subset A$. Now reverse both A and B and searching backwards from each $n_F \in F_R$, discover all nodes in the set C that are both accessible from i_R and *coaccessible* from F_R. Clearly the set of black nodes is then $B = A - C$. Now a third breadth-first search can be conducted to discover the gray nodes, and therewith the fence F.

Shown in Table 7.5 are the task image sizes for the static and dynamic decoders reported by McDonough and Stoimenov (2007) on a 5K vocabulary task.

Table 7.5 Task image sizes in Mb for the static and on-the-fly recognition engines at various beam settings. After McDonough and Stoimenov (2007)

Beam	Static	Dynamic	
	Small trigram	Small trigram	Full trigram
120.0	1380	179	470
130.0	1381	182	473
140.0	1384	187	476
155.0	1389	200	485

As is clear from the tabulated sizes, when the same small trigram is used for both the static and on-the-fly recognizers, the on-the-fly recognizer requires a factor of approximately seven less RAM.

7.5 Word and Lattice Combination

In this section we describe two widely used methods to combine the hypotheses – on a word level – of different recognition systems. These methods have proven to be useful for combining systems varying to different degrees; e.g., acoustic channel, front-end, randomized decision trees or hypothesis from independent systems.

ROVER

Probably the most widely-used word hypothesis combination method, due to its simplicity, is *ROVER* (Fiscus 1997) the *recognizer output voting error reduction* method. To form a word transition network the single word sequences from different recognition outputs must be aligned in basic units dubbed *correspondence sets*. Each correspondence set for each individual feature stream can either represent a word or a silence region, but "filler" phones such as breath or mumble can also be represented. In each correspondence set the word or silence region is picked providing the highest score

$$\text{score}[W, i] = \alpha M[W, i] + (1 - \alpha)C[W, i],$$

where $M[W, i]$ and $C[W, i]$ represent the occurrence of appearance and the confidence score of word W in correspondence set i. The weight parameter α must be determined *a priori* on development data.

ROVER is the method of choice when a large number of word hypotheses or lattices coming from very different system structures must be combined. As it utilizes information on the word identities at the time of combination, however, other relevant information might potentially be discarded and the benefit of this lattice combination technique may fail to reduce the word error rate.

Confusion Network Combination

Confusion networks reduce the complexity of lattice representations, which are orders of magnitude higher than a N-best list, to a simpler form that maintains all possible paths in the lattice. In so doing, the search space is transformed into a series of slots, where each slot consists of word hypotheses – and possibly null arcs – and associated posterior probabilities. Therefore, by combining the hypotheses or lattices of the same time segment from different tiers into a single word confusion network, the networks can be used to optimize the WER by selecting the word with the highest posterior probability in each particular slot. This hypothesis has been dubbed by Mangu *et al.* (2000) the *consensus hypothesis*.

A brief description of the *confusion network combination* algorithm is given in Algorithm 7.1. A detailed description can be found in the original work by Mangu *et al.* (2000).

Algorithm 7.1 Confusion network combination

1. *Lattice alignment* – The alignment consists of an equivalence relation over the word hypotheses together with an ordering of equivalence classes which must be consistent with the order of the original lattice.
2. *Intra-word clustering* – This step groups all the links corresponding to the same word instance. The similarity measure between two sets of links is given by

$$\text{SIM}(E_1, E_2) = \max_{e_1 \in E_1, e_2 \in E_2} \text{overlap}(e_1, e_2) p(e_1) p(e_2),$$

where the normalized overlap, $\text{overlap}(e_1, e_2)$, between the two links is weighted by the link posteriors $p(e_1)$ and $p(e_2)$ to be less sensitive to unlikely word hypotheses. The overlap, $\text{overlap}(e_1, e_2)$, is normalized by the sum of the lengths of the two hypotheses.
3. *Inter-word clustering* – This step clusters equivalence classes corresponding to similar words based on phonetic similarity which can be defined with $p_F(W) = p(e \in F : \text{words}(e) = W)$ as

$$\text{SIM}(F_1, F_2) = \underset{w_1 \in \text{Words}(F_1), W_2 \in \text{Words}(F_2)}{\text{avg}} \text{sim}(W_1, W_2) p_{F_1}(W_1) p_{F_1}(W_2).$$

4. *Pruning* – Word lattices might contain links with very low posterior probabilities which are negligible for the total posterior probabilities. For the correct alignment, however, they can have a detrimental effect as it conserves consistency with the lattice order independent of the probabilities. It has been experimentally demonstrated that pruning prior to the class initialization and merging reduces the word error rate.
5. *The consensus hypothesis* – The best word hypothesis is read from the path through the confusion graph with the highest combined link weights.

7.6 Summary and Further Reading

In this chapter, we have learned that search is the process by which an ASR system finds the best sequence of words conditioned on a sequence of acoustic observations. For any given sequence $y_{1:K}$ of acoustic observations, an ASR system should hypothesize that word sequence $w^*_{1:K_w}$ which achieves $w^*_{1:K_w} = \text{argmax}_{w_{1:K_w}} P(w_{1:K_w} | y_{1:K})$, where $\mathcal{W}$ is the ensemble of all possible word sequences. Typically, the term $p(\text{by}_{1:K} | w^*_{1:K_w})$ is said to be determined by the acoustic model and $P(w_{1:K_w})$ is determined by the language model.

One of the earliest publications addressing the sparse data problem in language modeling was by Katz (1987). Another important work on the use of N-grams for statistical language modeling was by Brown *et al.* (1992). Good surveys of smoothing techniques and the language modeling field in general are provided by Chen and Goodman (1999) and Rosenfeld (2000), respectively. Another good survey of the field is given by Goodman (2001). The use of context-free grammars for language modeling is described in Charniak (2001).

Much of the problem of language modeling lies in finding suitable training text. With the advent of powerful and publicly available engines for searching the Internet, it has become very popular to use such searches to amass text for LM training. Representative work in this area was proposed by Bulyko *et al.* (2003, 2007). In cases where insufficient LM training text is available for a given domain, it is common to adapt LM trained on a much larger corpus of general text. Such adaptation techniques are described in Wang and Stolcke (2007).

Several authors have investigated the application of neural networks to the language modeling problem. Good descriptions of these techniques are given by Bengio *et al.* (2003) and Schwenk and Gauvain (2004).

While the techniques for estimating LM probabilities described in Section 7.3.1 were based on a maximum likelihood criterion, it is also possible to use other criteria for estimating such weights. In particular, Jelinek (1998, sect. 13) describes a technique for estimating probabilities of a LM whereby the marginal probabilities are constrained to match relative frequency estimates from a set of training text, and subject to these constraints the probability distribution of the LM must diverge minimally from a known distribution. Jelinek (1998, sect. 13) also explains that under the assumption of a uniform prior probability distribution, minimizing the divergence is equivalent to maximizing the entropy of the resulting LM.

Comprehensive overviews of the language modeling field, as well as the use of statistical methods for information retrieval, are provided by Manning and Schütze (1999) and Manning *et al.* (2008).

When a static search graph is constructed using the knowledge sources and techniques described in Section 7.3, it is often necessary to reduce the size of the N-gram LM in order to obtain a final graph of tractable size. While this has become less of an issue for the large workstations typically found in research labs, it remains a problem for reduced footprint devices such as PDAs and cell phones. Seymore and Rosenfeld (1996) investigated the effect on recognition performance of shrinking a statistical N-gram through various methods.

The application of the theory of WFSTs to ASR was largely due to the seminal contributions of several very talented computer scientists at the AT&T Bell Labs. Early work on weighted composition, as described in Section 7.2.2, was published by Pereira and Riley (1997), which was followed shortly thereafter by Mohri (1997), who proposed the weighted determinization algorithm described in Section 7.2.3. The latter work also demonstrated that weighted minimization could be performed as a sequence of three steps, as discussed in Section 7.2.5. It was not until the appearance of Mohri and Riley (2001), however, that an efficient algorithm for the all-important weight-pushing step appeared. Other classic works describing the details of building search graphs are Mohri and Riley (1998) and Mohri *et al.* (2002). The design of the well-known AT&T finite-state machine library is described in Mohri *et al.* (1998, 2000).

Like a word lattice, a N-best list is a way of representing multiple alternative word hypothesis. Unlike a word lattice, the common substrings are not shared between hypotheses. Rather, each unique hypothesis is written out separately in its entirety. Mohri and Riley (2002) describe a way to efficiently generate such N-best lists using WFST techniques.

The search algorithms described in this chapter all assume that the underlying LM can be represented as a weighted finite-state acceptor, which is certainly not the most general representation of a natural language. As described in Pereira and Wright (1991), however, it is possible to approximate more general context-free grammars as FSAs.

Early work on the compilation of context dependency decision trees into WFSTs, as described in Section 7.3.4, was presented by Sproat and Riley (1996), and later followed up by Chen (2003). A simpler technique based on bit masks was proposed by Schuster and Hori (2005), and later corrected by Stoimenov and McDonough (2006, 2007).

Prior to the turn of the century, most ASR research sites used dynamic decoders that were based on fairly ad hoc techniques for expanding the search space. A summary of such techniques is given in Aubert (2000). The use of WFST technology for ASR was confined largely to the group at AT&T who had originally developed such techniques. That began to change with the appearance of Kanthak *et al.* (2002), which was the first systematic attempt to compare the efficiency of conventional search algorithms based on dynamic expansion with those based on WFSTs. What made the results reported in Kanthak *et al.* (2002) so compelling was that they were obtained with the active cooperation of leading proponents of both technologies. After the appearance of this work, several large ASR research sites began to jump on the WFST band wagon, including IBM; see Saon *et al.* (2005).

Since the turn of the century, several authors have addressed the problem of using WFSTs to perform on-the-fly-composition in an attempt to retain the efficiency of this approach while reducing the great amount of RAM required to fully expand the search graph. Contributors to the state-of-the-art in this field include Dolfing and Hetherington (2001), Willett and Katagiri (2002), Hori and Nakamura (2005), Caseiro and Trancoso (2006), and Cheng *et al.* (2007). The on-the-fly composition technique presented in Section 7.4 is due to McDonough and Stoimenov (2007).

As formulated in this chapter, WFST and search algorithms operate on graphs. Hence, it is worth-while for those readers interested in these topics to have a general reference on graph algorithms such as Cormen *et al.* (2001, Part VI).

7.7 Principal Symbols

Symbol	Description
ρ	final weight function in finite-state transducer (acceptor)
β	language model weight
ϵ	null symbol
Ω	output alphabet in finite-state transducer
Σ	(input) alphabet in finite-state transducer (acceptor)
A, B	factored search graphs
$b(v_{k+1})$	backpointer for token v_{k+1}
e	edge in a search graph

Symbol	Description
$e(v_{k+1})$	edge pointer for token v_{k+1}
E	set of transitions in finite-state transducer
F	set of final states finite-state transducer (acceptor)
i	initial state in finite-state transducer (acceptor)
k	time index
$\mathbb{K}$	set in definition of semiring
$l_\mathrm{i}[e]$	input symbol of e
$l_\mathrm{o}[e]$	output symbol of e
n	node in a search graph
$n[e]$	next state of e
$p[e]$	previous state of e
Q	set of states in finite-state transducer
R	complete search graph
v_k	token at time k
$v_\mathrm{AM}(v_k)$	acoustic model score for token v_k
$v_\mathrm{LM}(v_k)$	language model score for token v_k
$w[e]$	weight of e
$\overline{0}$	identity under $\oplus$ in semiring
$\overline{1}$	identity under $\otimes$ in semiring
$\oplus$	an associative and commutative operation in semiring
$\otimes$	an associative operation in semiring

8

Hidden Markov Model Parameter Estimation

As discussed in the last chapter, after all speaker tracking, beamforming, speech enhancement, and cepstral feature extraction has taken place, an *automatic speech recognition* (ASR) engine is needed to transform the final sequence of acoustic features associated with each utterance into a word hypothesis or a word lattice. Most modern recognition engines are based on the *hidden Markov model* (HMM), which can contain millions of free parameters that must be estimated from dozens or hundreds of hours of training data. How this parameter estimation may be efficiently performed is the subject of this chapter. We consider two primary parameter estimation criteria: *maximum likelihood* (ML) estimation, and discriminative estimation. The latter includes techniques based on such criteria as maximum mutual information, as well as minimum word and minimum phone error.

In ML parameter estimation, only the correct transcription of each training utterance is considered. As we will learn in this chapter, ML estimation attempts to associate as much probability mass with this correct transcription as possible. In contrast to ML estimation, discriminative estimation considers not only the correct transcription of each utterance, but also incorrect hypotheses that could be mistaken for the correct one. A set of likely but incorrect hypotheses is typically determined by running the recognition engine on all training utterances, then generating word lattices as described in Section 7.1.3. The HMM *acoustic model* (AM) can then be discriminately trained to assign a higher likelihood to the correct transcription than to any of its incorrect competitors. Discriminative training typically produces an AM that makes significantly fewer recognition errors than its ML counterpart. This performance improvement, however, comes at the price of a more computationally expensive parameter estimation procedure, primarily due to the necessity of running the recognizer over the entire training set. But with availability of ever more and ever cheaper computational power, this additional computation is no longer considered prohibitive.

We now summarize the balance of this chapter. In Section 8.1, we provide a discussion of ML parameter estimation methods. This begins with a proof of convergence of the *expectation maximization* (EM) algorithm. Due to its simplicity, the EM algorithm is almost invariably used for ML parameter estimation in the ASR field. We then

---

demonstrate how the EM algorithm can be applied to estimate the parameters of a Gaussian mixture model in Section 8.1.1. Forward–backward estimation, which is the primary technique required to extend the EM algorithm to the HMM, is described in Section 8.1.2. Section 8.1.3 presents a discussion of *speaker-adapted training* (SAT), which enables the incorporation of speaker adaptation parameters into HMM training. This has proven to provide lower word error rates than when speaker adaptation is conducted only during testing. The extension of SAT to include the optimal assignment of regression classes is discussed in Section 8.1.4. The last section on ML parameter estimation, Section 8.1.5, discusses how the computational expense of large-scale training can be reduced through Viterbi and label training.

Section 8.2 presents techniques for discriminative training. In Section 8.2.1, the conventional maximum mutual information re-estimation formulae, as first proposed by Normandin (1991), are presented. Discriminative training on word lattices, as originally proposed by Valtchev *et al.* (1997), is described in Section 8.2.2. The use of word lattices significantly reduces the computational expense of discriminative training as the training set need only be decoded once with the recognizer. Minimum word and phone error training are presented in Section 8.2.3. The combination of maximum mutual information training with SAT is then presented in Section 8.2.4.

The final section of this chapter summarizes the presentation here and provides suggestions for further reading.

8.1   Maximum Likelihood Parameter Estimation

As implied by its name, ML parameter estimation attempts to adjust the parameters of a parametric probability model so as to maximize the likelihood of a set of training data. To begin our discussion of this technique, let us consider a simple example of ML parameter estimation. Assume we have a multidimensional Gaussian pdf of the form

$$\mathcal{N}(\mathbf{y}; \boldsymbol{\mu}, \boldsymbol{\Sigma}) \triangleq \frac{1}{\sqrt{|2\pi\boldsymbol{\Sigma}|}} \exp\left[-\frac{1}{2}(\mathbf{y} - \boldsymbol{\mu})^T \boldsymbol{\Sigma}^{-1} (\mathbf{y} - \boldsymbol{\mu})\right], \tag{8.1}$$

where $\boldsymbol{\mu}$ and $\boldsymbol{\Sigma}$ are the mean vector and covariance matrix, respectively, and $\mathbf{y}$ is an acoustic observation.

For some *training set* $\mathcal{Y} = \{\mathbf{y}_k\}_{k=0}^{K-1}$ of i.i.d. observations, we wish to chose $\boldsymbol{\mu}$ and $\boldsymbol{\Sigma}$ so as to maximize the log-likelihood

$$\log p(\mathcal{Y}; \boldsymbol{\mu}, \boldsymbol{\Sigma}) = \sum_{k=0}^{K-1} \log p(\mathbf{y}_k; \boldsymbol{\mu}, \boldsymbol{\Sigma}),$$

where $p(\mathbf{y}_k; \boldsymbol{\mu}, \boldsymbol{\Sigma})$ is the Gaussian pdf (8.1). As shown in Anderson (1984, sect. 3.2), the ML parameter estimates are given by the *sample mean* and *sample covariance matrix*

$$\hat{\boldsymbol{\mu}} = \frac{1}{K} \sum_{k=0}^{K-1} \mathbf{y}_k, \tag{8.2}$$

$$\hat{\boldsymbol{\Sigma}} = \frac{1}{K} \sum_{k=0}^{K-1} (\mathbf{y}_k - \hat{\boldsymbol{\mu}}) (\mathbf{y}_k - \hat{\boldsymbol{\mu}})^T , \tag{8.3}$$

respectively. Hence, if there is only a single Gaussian, the ML estimates for the mean vector and covariance matrix can be written in closed form.

As we will see in the sequel, ML estimation in the ASR field is almost invariably accomplished with the EM algorithm. In this chapter, we will use the EM algorithm to estimate the parameters of a HMM. In Chapter 9, the EM algorithm will be applied to the estimation of speaker-dependent adaptation parameters. Hence, let us begin by proving that the EM algorithm produces ML parameter estimates. We begin our exposition with the following important result.

Lemma 8.1.1 (**Jensen's Inequality**) *Let X be a discrete-valued random variable with probability distribution $P(x)$, and let $F(x)$ be any other discrete probability distribution. Then*

$$\sum_{x} P(x) \log P(x) \geq \sum_{x} P(x) \log F(x),$$

with equality iff $P(x) = F(x) \,\forall\, x$.

Let Λ^0 denote the *current* set of parameter values, and consider a continuous r.v. Y and let X denote a discrete-valued r.v. of so-called *hidden variables*. Moreover, assume that there is a training set containing many realizations of Y. Our objective is to chose the parameters Λ of a pdf so as to maximize the average log-likelihood $\log p(y; \Lambda)$ over the entire training set. In the sequel, we will refer to Y as the *incomplete observation*, and the pair $Z = (Y, X)$ as the *complete observation*. Moreover, let $\mathcal{Y} = \{y_k\}$ and $\mathcal{X} = \{x_k\}$ denote the *incomplete* and *hidden training sets*. As we will soon see, ML parameter estimation is most conveniently accomplished with the EM algorithm through a suitable choice of hidden variables X to complement the incomplete observation Y.

The proof that the EM algorithm converges to a set of ML parameter estimates begins then with a consideration of the following chain of equalities:

$$\log p(y; \Lambda) - \log p(y; \Lambda^0)$$

$$= \sum_{x} P(x|y; \Lambda^0) \log p(y; \Lambda) \frac{p(x, y; \Lambda)}{p(x, y; \Lambda)} - \sum_{x} P(x|y; \Lambda^0) \log p(y; \Lambda^0) \frac{p(x, y; \Lambda^0)}{p(x, y; \Lambda^0)}$$

$$= \sum_{x} P(x|y; \Lambda^0) \log \frac{p(x, y; \Lambda)}{P(x|y; \Lambda)} - \sum_{x} P(x|y; \Lambda^0) \log \frac{p(x, y; \Lambda^0)}{P(x|y; \Lambda^0)}$$

$$= \sum_{x} P(x|y; \Lambda^0) \log p(x, y; \Lambda) - \sum_{x} P(x|y; \Lambda^0) \log p(x, y; \Lambda^0)$$

$$+ \sum_{x} P(x|y; \Lambda^0) \log P(x|y; \Lambda^0) - \sum_{x} P(x|y; \Lambda^0) \log P(x|y; \Lambda). \tag{8.4}$$

Associating $P(x)$ with $P(x|y; \Lambda^0)$ and $F(x)$ with $P(x|y; \Lambda)$, it follows from Jensen's inequality that

$$\sum_x P(x|y; \Lambda^0) \log P(x|y; \Lambda^0) - \sum_x P(x|y; \Lambda^0) \log P(x|y; \Lambda) \geq 0.$$

Hence, based on the final equality in (8.4), we can write

$$\log p(y; \Lambda) - \log p(y; \Lambda^0)$$

$$\geq \sum_x P(x|y; \Lambda^0) \log p(x, y; \Lambda) - \sum_x P(x|y; \Lambda^0) \log p(x, y; \Lambda^0). \qquad (8.5)$$

Now let us define the *auxiliary function* (Dempster *et al.* 1977)

$$Q(\Lambda|\Lambda^0) \triangleq \mathcal{E}\{\log p(\{\mathcal{Y}, \mathcal{X}\}; \Lambda)|\mathcal{Y}; \Lambda^0\} = \sum_x P(x|y; \Lambda^0) \log p(x, y; \Lambda). \qquad (8.6)$$

We have then proven the following theorem, which forms the basis for all ML estimation techniques to be discussed subsequently.

Theorem 8.1.2 **(convergence of the EM algorithm)** *For $Q(\Lambda|\Lambda^0)$ defined as in (8.6),*

$$Q(\Lambda|\Lambda^0) > Q(\Lambda^0|\Lambda^0) \Rightarrow \log p(y; \Lambda) > \log p(y; \Lambda^0).$$

The utility of the EM lies in the fact that it is often much easier to maximize $Q(\Lambda|\Lambda^0)$ with respect to Λ than it is to maximize $\log p(y; \Lambda)$ directly. As is clear from (8.6), the auxiliary function is equivalent to the expected value of the log-likelihood of the complete training set $\mathcal{Z} = (\mathcal{Y}, \mathcal{X})$ conditioned on the incomplete observation y given the current estimate Λ^0 of the model's parameters. The importance of the auxiliary function stems from the fact that through successive iterations involving its calculation, the *E-step*, and subsequent maximization, the *M-step*, with respect to the set Λ of model parameters, the EM algorithm achieves the desired (local) maximum of the training set likelihood. The generality of the EM algorithm was first discussed by Dempster *et al.* (1977).

In the following sections, we will see how a suitable auxiliary function can be used to perform ML parameter estimation on HMMs. We will begin, however, with a simpler case, namely, the Gaussian mixture model.

8.1.1 Gaussian Mixture Model Parameter Estimation

In the Moore formulation of the HMM, there is an observation pdf associated with each state. We will uniformly assume that the observation pdf can be represented as a *Gaussian mixture model* (GMM), which we will now define. Let $\{w_m\}$, $\boldsymbol{\mu}_m$, and $\boldsymbol{\Sigma}_m$, denote respectively the *a priori* probability, mean vector, and covariance matrix of the mth component of the GMM, and let L denote the length of the observation vector $\mathbf{y}$. The *a priori* probability w_m is also known as the *mixture weight*. In the ASR field, Σ_m is most often assumed to be diagonal such that

$$\Sigma_m = \text{diag} \left\{ \sigma_{m,0}^2 \, \sigma_{m,1}^2 \, \cdots \, \sigma_{m,L-1}^2 \right\}. \tag{8.7}$$

The total likelihood returned by the GMM for observation $\mathbf{y}$ can then be expressed as

$$p(\mathbf{y}; \Lambda) = \sum_{m=1}^{M} w_m \, \mathcal{N}(\mathbf{y}; \Lambda_m), \tag{8.8}$$

where M is the total number of mixture components and $\Lambda_m = (\boldsymbol{\mu}_m, \Sigma_m)$.

Equation (8.8) can be interpreted as follows: Based on the set of *a priori* probabilities $\{w_m\}$, choose a mixture component. Based on the chosen component, generate a normally distributed observation $\mathbf{y}_k$ with mean $\boldsymbol{\mu}_m$ and covariance Σ_m. That is, choose $\mathbf{y}$ according to the normal distribution $\mathcal{N}(\mathbf{y}; \boldsymbol{\mu}_m, \Sigma_m)$. The following development will address the task of obtaining ML estimates of the speaker-independent parameters $\Lambda = \{(w_m, \boldsymbol{\mu}_m, \Sigma_m)\}$.

Let us define as before a set $\mathcal{Y} = \{\mathbf{y}_k\}$ of training data and observe that the log-likelihood of $\mathcal{Y}$ under the pdf (8.8) is

$$\log p(\mathcal{Y}; \Lambda) = \sum_k \log p(\mathbf{y}_k; \Lambda) = \sum_k \log \left[\sum_m w_m \, p(\mathbf{y}_k; \Lambda_m) \right], \tag{8.9}$$

where $p(\mathbf{y}_k; \Lambda_m)$ is defined in (8.1). This task of estimating Λ would be greatly simplified if we were so fortunate as to know which Gaussian component g_m had generated a given observation $\mathbf{y}_k$. In this case, ML parameter estimation could be performed by simply applying (8.2–8.3) separately to the parameters Λ_m of each Gaussian component. As we do not possess this knowledge, let us postulate a set $\mathcal{X} = \{x_{m,k}\}$ of hidden variables which contain this missing information. We will define these hidden variables as

$$x_{m,k} \triangleq \begin{cases} 1, & \text{for } g_k = g_m, \\ 0, & \text{otherwise}, \end{cases}$$

where k is an index over acoustic observations and m is an index over Gaussian components. The notation $g_k = g_m$ indicates that $\mathbf{y}_k$ was drawn from the *mth* Gaussian component. With this definition, the joint log-likelihood of the complete training set $\{\mathcal{Y}, \mathcal{X}\}$ is given by

$$\log p(\{\mathcal{Y}, \mathcal{X}\}; \Lambda) = \sum_k \log \left[\sum_m x_{m,k} \, w_m \, p(\mathbf{y}_k; \Lambda_m) \right] \tag{8.10}$$

$$= \sum_{m,k} x_{m,k} \left[\log w_k + \log p(\mathbf{y}_k; \Lambda_m) \right], \tag{8.11}$$

where, in writing (8.11) we have exploited the fact that exactly one of the terms in the bracketed summation in (8.10) is nonzero. Let us then define the expectation

$$c_{m,k} = \mathcal{E}\{x_{m,k}|\mathbf{y}_k; \Lambda^0\}$$

as the *a posteriori* probability that $\mathbf{y}_k$ was drawn from the mth Gaussian component, which is given by

$$c_{m,k} \triangleq p(g_k = g_m|\mathbf{y}_k; \Lambda^0) = \frac{w_m p(\mathbf{y}_k; \Lambda_m^0)}{\sum_{m'} w_{m'} p(\mathbf{y}_k; \Lambda_{m'}^0)}.$$

Similarly, let us now define the *occupancy counts* as the sum of the *a posteriori* probabilities:

$$c_m \triangleq \sum_k c_{m,k}.$$

As indicated by (8.6), the desired auxiliary function is obtained by taking the conditional expectation of the complete log-likelihood according to

$$Q(\Lambda|\Lambda^0) \triangleq \mathcal{E}\{\log p(\{\mathcal{Y}, \mathcal{X}\}; \Lambda)|\mathcal{Y}; \Lambda^0\} = K_1(\mathcal{Y}; \Lambda, \Lambda^0) + K_2(\mathcal{Y}; \Lambda, \Lambda^0), \quad (8.12)$$

where

$$K_1(\mathcal{Y}; \Lambda, \Lambda^0) \triangleq \sum_{m,k} c_{m,k} \log w_m = \sum_m c_m \log w_m, \quad (8.13)$$

$$K_2(\mathcal{Y}; \Lambda, \Lambda^0) \triangleq \sum_{m,k} c_{m,k} \log p(\mathbf{y}_k; \Lambda_m). \quad (8.14)$$

Use of EM algorithm in GMM parameter estimation entails the calculation of the auxiliary function (8.12) and its subsequent maximization with respect to the parameters $\Lambda = \{(w_m, \Lambda_m)\}$. Observe that the terms $K_1(\mathcal{Y}; \Lambda, \Lambda^0)$ and $K_2(\mathcal{Y}; \Lambda, \Lambda^0)$ can be maximized independently to obtain the updated parameters $\{w_m\}$ and $\{\Lambda_m\}$. Maximization of $K_1(\mathcal{Y}; \Lambda, \Lambda^0)$ under the constraint

$$\sum_m w_m = 1 \quad (8.15)$$

is readily accomplished via the method of *undetermined Lagrangian multipliers* (Amazigo and Rubenfeld 1980, sect. 4.3). To do so, we form the modified objective function

$$K_1'(\mathcal{Y}; \Lambda, \Lambda^0, \lambda) = \sum_m c_m \log w_m + \lambda \left(1 - \sum_m w_m\right),$$

where λ is the *Lagrangian multiplier*. Maximizing K_1' can be achieved by calculating the partial derivative $\partial K_1'/\partial w_m$ and equating it to zero, such that,

$$\frac{\partial K_1'(\mathcal{Y}; \Lambda, \Lambda^0, \lambda)}{\partial w_m} = \frac{c_m}{w_m} - \lambda = 0,$$

which implies

$$\hat{w}_m = \frac{c_m}{\lambda}. \tag{8.16}$$

To solve for λ, we substitute (8.16) into the constraint (8.15) to obtain

$$\sum_m \hat{w}_m = \frac{1}{\lambda} \sum_m c_m = 1,$$

or, equivalently,

$$\lambda = \sum_m c_m.$$

Substituting for λ in (8.16) provides the parameter estimate

$$\hat{w}_m = \frac{c_m}{c}, \tag{8.17}$$

where

$$c = \sum_m c_m. \tag{8.18}$$

From (8.19), it is apparent that

$$\log p(\mathbf{y}; \Lambda_m) = -\frac{1}{2} \left[\log |2\pi \boldsymbol{\Sigma}_m| + (\mathbf{y} - \boldsymbol{\mu}_m)^T \boldsymbol{\Sigma}_m^{-1} (\mathbf{y} - \boldsymbol{\mu}_m) \right]. \tag{8.19}$$

To solve for the optimal mean and covariance matrix, we substitute (8.19) into (8.14) to obtain

$$K_2 = -\frac{1}{2} \sum_{m,k} c_{m,k} \left[\log |2\pi \boldsymbol{\Sigma}_m| + (\mathbf{y}_k - \boldsymbol{\mu}_m)^T \boldsymbol{\Sigma}_m^{-1} (\mathbf{y}_k - \boldsymbol{\mu}_m) \right]. \tag{8.20}$$

From (8.20), it is straightforward to demonstrate that the optimal means can be calculated as

$$\hat{\boldsymbol{\mu}}_m = \frac{1}{c_m} \sum_k c_{m,k} \mathbf{y}_k \tag{8.21}$$

Representing $\hat{\boldsymbol{\Sigma}}_m$ as in (8.7), it follows

$$\hat{\sigma}_{m,n}^2 = \frac{1}{c_m} \sum_k c_{m,k} (y_{k,n} - \hat{\mu}_{m,n})^2$$

$$= \left(\frac{1}{c_m} \sum_k c_{m,k} \, y_{k,n}^2 \right) - \hat{\mu}_{m,n}^2 \; \forall n = 0, \ldots, L - 1, \qquad (8.22)$$

where $y_{k,n}$ and $\hat{\mu}_{m,n}$ are the nth components of $\mathbf{y}_k$ and $\hat{\boldsymbol{\mu}}_m$, respectively. Equations (8.21)–(8.22) are the desired parameter estimates, which illustrate the advantage inherent in using the EM algorithm for ML parameter estimation. That is, the joint optimization of all parameters in the GMM has been decoupled into the separate optimization of the parameters of each Gaussian component. Only the mixture weights must still be estimated jointly, which is readily accomplished.

8.1.2 Forward–Backward Estimation

As mentioned in Section 7.1.1, we will here adopt the Moore form of the HMM, which consists of a Markov chain whereby a GMM is associated with each *state*. For the purpose of ML parameter estimation, the HMM is structured so that only the sequence of GMMs associated with the correct word sequence for a give training utterance are admitted. Constructing such a HMM requires that the grammar G described in Section 7.3.1 be replaced with a simpler transducer consisting merely of transitions labeled with the correct words strung together between an initial and final node; one such transducer is created for each unique training utterance. Discriminative parameter estimation, which we will discuss in Section 8.2, is more akin to actual recognition in that the recognizer must be run over all training utterances to generate word lattices. Given that the word lattices contain many hypotheses that are likely to be mistaken by the recognizer for the correct hypothesis, they are useful for estimating the parameters of the AM such that the recognizer is steered away from the false but likely candidates, and toward the true hypothesis.

In this section, we derive the relations necessary to form ML estimates of the parameters of a HMM using the EM algorithm. We begin by introducing some notation. Consider a directed graph representing a HMM comprising a set $\{n_i\}$ of *nodes* and a set $\{e_{j|i}\}$ of *edges*, where n_i is the ith node and $e_{j|i}$ is the directed edge from n_i to n_j. Let us define the *transition probability*

$$p_{j|i} = p(n_j|n_i) = p(e_{j|i}|n_i). \qquad (8.23)$$

In other words, $p_{j|i}$ is the probability of going from state n_i to state n_j, which is independent of the frame index k by assumption. The final equality in (8.23) follows upon assuming that at most one edge joins any two nodes. Let $g_{i,m}$ denote the mth Gaussian component associated with node n_i and – as in the last section – define the mixture weight

$$w_{m|i} = p(g_{i,m}|n_i).$$

In keeping with (8.8), we will express the likelihood assigned an observation $\mathbf{y}$ by the GMM associated with node n_i as

$$p(\mathbf{y}; \Lambda_i) = \sum_m w_{m|i}\, p(\mathbf{y}; \Lambda_{i,m}),$$

where $\Lambda_i = \{(w_{m|i}, \Lambda_{i,m})\}$ are the corresponding GMM parameters. Defining the mean vector $\boldsymbol{\mu}_{i,m}$ and covariance matrix $\boldsymbol{\Sigma}_{i,m}$, such that $\Lambda_{i,m} = (\boldsymbol{\mu}_{i,m}, \boldsymbol{\Sigma}_{i,m})$, the acoustic likelihood of a single Gaussian component can be expressed much as in the case of the simple GMM. Moreover, let $\mathbf{y}_{1:K}$ denote the entire sequence of observations for a given utterance, and let $\mathbf{y}_{k_1:k_2}$ denote an observation subsequence from time k_1 to time k_2.

Let us define the *forward probability* $\alpha(\mathbf{y}_{1:k}, i)$ as the joint likelihood of generating the observation subsequence $\mathbf{y}_{1:k}$ and arriving at state n_i at time k given the current model parameters Λ:

$$\alpha(\mathbf{y}_{1:k}, i) \triangleq p(n_k = n_i, \mathbf{y}_{1:k}; \Lambda), \tag{8.24}$$

where – in a slight abuse of notation – we have used n_k to denote the HMM state associated with observation $\mathbf{y}_k$, and n_i to denote the ith state in the HMM. Similarly, the *backward probability* is the likelihood of generating the observation subsequence $\mathbf{y}_{k+1:K}$ conditioned on beginning from state n_j at time k:

$$\beta(\mathbf{y}_{k+1:K}|j) \triangleq p(\mathbf{y}_{k+1:K}|n_k = n_j; \Lambda) \tag{8.25}$$

These probabilities can be calculated via the well-known forward and backward recursions (Baum *et al.* 1970),

$$\alpha(\mathbf{y}_{1:k}, i) = \sum_j \alpha(\mathbf{y}_{1:k-1}, j)\, p(n_k = n_i | n_{k-1} = n_j)\, p(\mathbf{y}_k | n_k = n_i)$$

$$= p(\mathbf{y}_k; \Lambda_i) \sum_j \alpha(\mathbf{y}_{1:k-1}, j)\, p_{i|j}, \tag{8.26}$$

$$\beta(\mathbf{y}_{k+1:K}|i), = \sum_j \beta(\mathbf{y}_{k+2}^T|j)\, p(n_{k+1} = n_j | n_k = n_i)\, p(\mathbf{y}_{k+1} | n_k = n_j)$$

$$= \sum_j \beta(\mathbf{y}_{k+2:K}|j)\, p_{j|i}\, p(\mathbf{y}_{k+1}; \Lambda_j). \tag{8.27}$$

The forward and backward probabilities are initialized according to

$$\alpha(\mathbf{y}_0, i) = \begin{cases} 1, & \text{for } n_i \text{ the valid initial state,} \\ 0, & \text{otherwise.} \end{cases} \tag{8.28}$$

$$\beta(\mathbf{y}_{K+1}|j) = \begin{cases} 1, & \text{for } n_j \text{ a valid end state,} \\ 0, & \text{otherwise.} \end{cases} \tag{8.29}$$

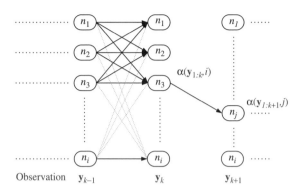

Figure 8.1 Trellis calculation of forward probabilities

We will assume as in Chapter 7 that there is only a single valid initial state but there are several valid end states. This "trellis" calculation is illustrated in Figure 8.1 for $\alpha(\mathbf{y}_{1:k+1}, j)$. The calculation for $\beta(\mathbf{y}_{k+1:K}|j)$ proceeds similarly.

In their direct form (8.26–8.27), the forward–backward probabilities are prone to numerical underflow, in that they require the multiplication of many small probabilities. For this reason, the negative logarithms of such quantities are typically manipulated during parameter estimation. Clearly, in the log domain, multiplication is replaced by addition. Addition, on the other hand, is replaced by the so-called *log-add function*. Let p_A and p_B denote two nonzero probabilities, and set

$$m_A = \log p_A,$$
$$m_B = \log p_B,$$

where, by assumption, $p_A > p_B$. Then it is readily verified that for $p_C = p_A + p_B$,

$$m_C = \log p_C = m_A + \log(1 + e^{(m_B - m_A)}).$$

For reasons that will soon become apparent, we are interested in calculating the probability

$$p(n_k = n_i, \mathbf{y}_{1:k}; \Lambda) = \alpha(\mathbf{y}_{1:k}, i)\,\beta(\mathbf{y}_{k+1:K}|i). \tag{8.30}$$

Toward this end, consider that the likelihood of the entire utterance $\mathbf{y}_{1:K}$ is given by

$$p(\mathbf{y}_{1:K}; \Lambda) = \sum_i \alpha(\mathbf{y}_{1:k}, i)\,\beta(\mathbf{y}_{k+1:K}|i)$$

for any time k. In particular, if $k = K$ the previous equation reduces to

$$p(\mathbf{y}_{1:K}; \Lambda) = \sum_i \alpha(\mathbf{y}_{1:K}, i).$$

Let g_k denote the Gaussian component associated with observation $\mathbf{y}_k$. The joint likelihood of a complete observation sequence $\{\mathbf{y}_{1:K}, g_{1:K}\}$ is then given by

$$\log p(\mathbf{y}_{1:K}; \Lambda) = \sum_{k=1}^{K} \{\log p(g_k|n_k) + \log p(\mathbf{y}_k|g_k) + \log p(n_k|n_{k-1})\}.$$

As in our treatment of parameter estimation for the simple GMM, we will define a set $\mathcal{X}$ of hidden variables for each utterance. In this case, however, the hidden variables will specify both the state and the Gaussian mixture component associated with each feature in $\mathbf{y}_{1:K}$:

$$x_{jm,k} \triangleq \begin{cases} 1, & \text{for } g_k = g_{jm}, \\ 0, & \text{otherwise}, \end{cases},$$

$$x_{j|i,k} \triangleq \begin{cases} 1, & \text{for } e_k = e_{j|i}, \\ 0, & \text{otherwise}. \end{cases}.$$

The hidden variable $x_{jm,k}$ specifies the Gaussian component, just as in the case of the simple GMM. Note that specifying the component g_{jm} also uniquely specifies the state n_j. The new hidden variable $x_{j|i,k}$ is required to indicate which sequence of transitions through the Markov chain were taken. With these definitions, the likelihood of the sequence of complete observations is

$$\log p(\{\mathcal{X}, \mathcal{Y}\}; \Lambda) = \sum_{k} \left\{ \sum_{j,m} x_{jm,k} \left[\log p(g_k = g_{jm}|n_k = n_j) + \log p(\mathbf{y}_k|g_k = g_{jm}) \right] \right.$$

$$\left. + \sum_{i,j} x_{j|i,k} \log p(n_k = n_j|n_{k-1} = n_i) \right\}$$

$$= \sum_{k} \left\{ \sum_{j,m} x_{jm,k} \left[\log w_{m|j} + \log p(\mathbf{y}_k; \Lambda_{jm}) \right] + \sum_{i,j} x_{j|i,k} \log p_{j|i} \right\}.$$

Taking the expectation of the last expression provides

$$\mathcal{E}\{\log p(\{\mathcal{Y}, \mathcal{X}\}; \Lambda)|\mathcal{Y}; \Lambda^0\} =$$

$$\sum_{k} \left\{ \sum_{j,m} c_{jm,k} \left[\log w_{m|j} + \log p(\mathbf{y}_k; \Lambda_{jm}) \right] + \sum_{i,j} c_{j|i,k} \log p_{j|i} \right\} \quad (8.31)$$

where

$$c_{jm,k} = \mathcal{E}\{x_{jm,k}|\mathcal{Y}; \Lambda^0\},$$

and

$$c_{j|i,k} = \mathcal{E}\{x_{j|i,k}|\mathcal{Y}; \Lambda^0\}$$

are the relevant *a posteriori* probabilities. The first of these *a posteriori* probabilities can be calculated according to

$$c_{jm,k} = p(g_k = g_{jm} | \mathbf{y}_{1:K}; \Lambda)$$
$$= p(n_k = n_j, g_k = g_{jm} | \mathbf{y}_{1:K}; \Lambda)$$
$$= p(n_k = n_j | \mathbf{y}_{1:K}; \Lambda) \, p(g_k = g_{jm} | n_k = n_j, \mathbf{y}_{1:K}; \Lambda)$$
$$= \frac{p(n_k = n_j, \mathbf{y}_{1:K}; \Lambda)}{p(\mathbf{y}_{1:K}; \Lambda)} \, \frac{p(g_k = g_{jm}, \mathbf{y}_k | n_k = n_j; \Lambda_j)}{p(\mathbf{y}_k | n_k = n_j; \Lambda_j)}.$$

In light of (8.30), the last line reduces to

$$c_{jm,k} = \frac{\alpha(\mathbf{y}_{1:k}, i) \, \beta(\mathbf{y}_{k+1:K} | i)}{p(\mathbf{y}_{1:K}; \Lambda)} \, \frac{w_{m|j} \, p(\mathbf{y}_k; \Lambda_{jm})}{\sum_{m'} w_{m'|j} \, p(\mathbf{y}_k; \Lambda_{jm'})}. \tag{8.32}$$

Similarly,

$$c_{j|i,k} = p(n_k = n_i, n_{k+1} = n_j | \mathbf{y}_{1:K}; \Lambda)$$
$$= \frac{p(n_k = n_i, n_{k+1} = n_j, \mathbf{y}_{1:K}; \Lambda)}{p(\mathbf{y}_{1:K}; \Lambda)}$$
$$= \frac{p(n_k = n_i, \mathbf{y}_{1:k}) \, p(n_{k+1} = n_j, \mathbf{y}_{k+1} | n_k = n_i) \, p(\mathbf{y}_{k+2:K} | n_{k+1} = n_j)}{p(\mathbf{y}_{1:K}; \Lambda)}, \tag{8.33}$$

where the dependence on Λ in the last equation has been suppressed for convenience. Note, however, that

$$p(n_{k+1} = n_j, \mathbf{y}_{k+1} | n_k = n_i) = p(n_{k+1} = n_j | n_k = n_i) \, p(\mathbf{y}_{k+1} | n_{k+1} = n_j)$$
$$= p_{j|i} \, p(\mathbf{y}_{k+1}; \Lambda_j).$$

Substituting this expression and (8.24–8.25) into (8.33) we find

$$c_{j|i,k} = \frac{\alpha(\mathbf{y}_{1:k}, i) \, p_{j|i} \, p(\mathbf{y}_{k+1}; \Lambda_j) \, \beta(\mathbf{y}_{k+2:K} | j)}{p(\mathbf{y}_{1:K}; \Lambda)}. \tag{8.34}$$

Hence, once more define the occupancy counts

$$c_{jm} \triangleq \sum_k c_{jm,k}, \tag{8.35}$$

$$c_{j|i} \triangleq \sum_k c_{j|i,k}. \tag{8.36}$$

It is then possible to rewrite (8.31) as

$$\mathcal{E}\{\log p(\{\mathcal{X}, \mathcal{Z}\}; \Lambda) | \mathcal{Y}, \Lambda^0\} = K_1(\mathcal{Y}; \Lambda, \Lambda^0) + K_2(\mathcal{Y}; \Lambda, \Lambda^0) + K_3(\mathcal{Y}; \Lambda, \Lambda^0),$$

where

$$K_1(\mathcal{Y}) = \sum_{j,m} c_{jm} \log w_{m|j}, \tag{8.37}$$

$$K_2(\mathcal{Y}) = \sum_{j,m,k} c_{jm,k} \log p(\mathbf{y}_k; \Lambda_{jm})$$

$$= -\frac{1}{2} \sum_{j,m,k} c_{jm,k} \left[\log |2\pi \boldsymbol{\Sigma}_{jm}| + (\mathbf{y}_k - \boldsymbol{\mu}_{jm})^T \boldsymbol{\Sigma}_{jm}^{-1} (\mathbf{y}_k - \boldsymbol{\mu}_{jm}) \right], \tag{8.38}$$

$$K_3(\mathcal{Y}) = \sum_{i,j} c_{j|i} \log p_{j|i}. \tag{8.39}$$

The dependence on Λ and Λ^0 in (8.37–8.39) has been suppressed out of convenience. Optimizing K_1 with respect to $\{w_{m|i}\}$ as in (8.17–8.18) provides

$$\hat{w}_{m|i} = \frac{c_{im}}{c_i}, \tag{8.40}$$

where

$$c_i = \sum_m c_{im}. \tag{8.41}$$

The similarity of (8.40–8.41) to (8.17–8.18) is unmistakable. Solving K_2 for the optimal Gaussian parameters yields

$$\hat{\boldsymbol{\mu}}_{jm} = \frac{1}{c_{jm}} \sum_k c_{jm,k} \mathbf{y}_k, \tag{8.42}$$

$$\hat{\sigma}^2_{jm,n} = \frac{1}{c_{jm}} \sum_k c_{jm,k} (y_{kn} - \hat{\mu}_{jm,n})^2$$

$$= \left(\frac{1}{c_{jm}} \sum_k c_{jm,k} y^2_{kn} \right) - \hat{\mu}^2_{jm,n}. \tag{8.43}$$

These formulae are nearly identical with the GMM update equations (8.21)–(8.22). The only new term is $K_3(\mathcal{Y}; \Lambda, \Lambda^0)$, which can readily be optimized with respect to the transition probabilities $\{p_{j|i}\}$ such that

$$p_{j|i} = \frac{c_{j|i}}{c_{.|i}}, \tag{8.44}$$

where

$$c_{.|i} = \sum_j c_{j|i}. \tag{8.45}$$

Note that the sums in (8.35–8.36) and (8.42–8.43) can be calculated by partitioning the training utterances into subsets, accumulating partial sums for each on different processors, then combining these statistics during the maximization step. This is the basis of so-called *parallel training*.

8.1.3 Speaker-Adapted Training

As we will learn in Chapter 9, model-based *speaker adaptation* is a technique for adapting the parameters of a HMM to better match the characteristics of a particular speaker's voice. Both *maximum likelihood linear regression* (MLLR) and *all-pass transform* (APT) adaptation, as described in Sections 9.2.1 and 9.2.2, respectively, are based on a linear transformation of the speaker-independent means according to

$$\hat{\boldsymbol{\mu}}_m^{(s)} \triangleq \mathbf{A}^{(s)} \boldsymbol{\mu}_m^{(s)} + \mathbf{b}^{(s)}, \tag{8.46}$$

where the transformation matrix and linear shift $\mathbf{A}^{(s)}$ and $\mathbf{b}^{(s)}$, respectively, are estimated according to a ML criterion. Often, MLLR is performed on the speaker-independent model trained with the conventional forward–backward algorithm presented in the last section. In this case, however, the conditions for training differ from those for test inasmuch as speaker adaptation is used only in the latter. Better results are obtained when conditions for both training and test are matched. *Speaker-adapted training* (SAT) (Anastasakos *et al.* 1996) is a method for the ML estimation of the parameters of a HMM that achieves the desired match. As shown in Figure 8.2, SAT, which is also based on the EM algorithm, proceeds along much the same lines as conventional HMM training, with a forward–backward step followed by a parameter update designed to maximize an appropriate auxiliary function. Before training, all utterances in the training set are partitioned by speaker, and before the forward–backward pass over each such partition the adaptation parameters for the given speaker are used to transform the means of the *speaker-independent* (SI) model as in (8.46). With the completion of the forward–backward step, the *speaker-dependent* transformation parameters for the relevant speaker are re-estimated, just as in normal speaker adaptation. The maximization step in SAT is an iterative parameter update wherein the SI means and variances are each updated in turn while holding all other HMM parameters fixed at their current values. The advantage of the iterative approach lies in the fact that a closed-form solution exists for the optimal values for each set of parameters when the other sets are held constant.

To derive the mean update formulae for SAT, let us define a set $\mathcal{Y} = \{\mathcal{Y}^{(s)}\}$ of training data contributed by several speakers, where $\mathcal{Y}^{(s)} = \{\mathbf{y}_k^{(s)}\}$ are the observations from speaker s. Consider the auxiliary function (8.38). Suppressing the index over states, replacing the speaker-independent mean $\boldsymbol{\mu}_{jm}$ with the adapted mean $\hat{\boldsymbol{\mu}}_m^{(s)}$ from (8.46) and summing over all speakers, we obtain,

$$\mathcal{G}(\mathcal{Y}; \{(\mathbf{A}^{(s)}, \mathbf{b}^{(s)})\}, \Lambda) = \sum_{m,k,s} c_{m,k}^{(s)} (\mathbf{y}_k^{(s)} - \mathbf{A}^{(s)} \boldsymbol{\mu}_m - \mathbf{b}^{(s)})^T \boldsymbol{\Sigma}_m^{-1} (\mathbf{y}_k^{(s)} - \mathbf{A}^{(s)} \boldsymbol{\mu}_m - \mathbf{b}^{(s)}),$$

where we have suppressed the Gaussian normalization constants and the negative sign, as they do not affect the optimization of the mean vectors. Through straightforward algebraic

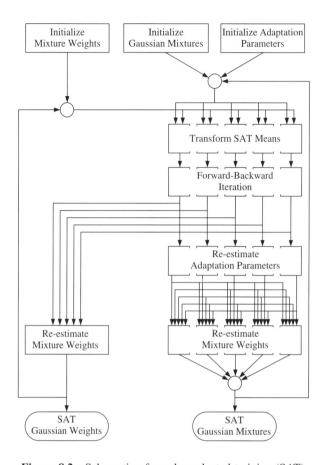

Figure 8.2   Schematic of speaker-adapted training (SAT)

manipulations, the last equation can be rewritten as

$$\mathcal{G}(\mathcal{Y}; \{(\mathbf{A}^{(s)}, \mathbf{b}^{(s)})\}, \Lambda) = \sum_{s,m} c_m^{(s)} [(\tilde{\boldsymbol{\mu}}_m^{(s)} - \mathbf{A}^{(s)} \boldsymbol{\mu}_m - \mathbf{b}^{(s)})^T \boldsymbol{\Sigma}_m^{-1} (\tilde{\boldsymbol{\mu}}_m^{(s)} - \mathbf{A}^{(s)} \boldsymbol{\mu}_m - \mathbf{b}^{(s)})],$$

(8.47)

where

$$\tilde{\boldsymbol{\mu}}_m^{(s)} = \frac{1}{c_m^{(s)}} \sum_k c_{m,k}^{(s)} \mathbf{y}_k^{(s)},$$

and

$$c_m^{(s)} = \sum_k c_{m,k}^{(s)}.$$

Equation (8.47) can be further manipulated to yield

$$\mathcal{G}(\mathcal{Y}; \{(\mathbf{A}^{(s)}, \mathbf{b}^{(s)})\}, \Lambda) = \sum_m \mathcal{G}_m(\mathcal{Y}; \{(\mathbf{A}^{(s)}, \mathbf{b}^{(s)})\}, \Lambda_m),$$

where

$$\mathcal{G}_m(\mathcal{Y}; \{(\mathbf{A}^{(s)}, \mathbf{b}^{(s)})\}, \Lambda_m)$$

$$= \sum_s c_m^{(s)}[(\tilde{\boldsymbol{\mu}}_m^{(s)} - \mathbf{A}^{(s)}\boldsymbol{\mu}_m - \mathbf{b}^{(s)})^T \boldsymbol{\Sigma}_m^{-1}(\tilde{\boldsymbol{\mu}}_m^{(s)} - \mathbf{A}^{(s)}\boldsymbol{\mu}_m - \mathbf{b}^{(s)})]. \quad (8.48)$$

From (8.48) it is clear that the parameters $\Lambda_m = (\boldsymbol{\mu}_m, \boldsymbol{\Sigma}_m)$ of each Gaussian component can be optimized independently. Differentiating both sides of this equation with respect to components of the SI means and setting the result to zero, we obtain the mean re-estimation formula

$$\boldsymbol{\mu}_m = \left(\underline{\mathbf{A}}^T \underline{\boldsymbol{\Sigma}}_m \underline{\mathbf{A}}\right)^{-1} \left[\underline{\mathbf{A}}^T \underline{\boldsymbol{\Sigma}}_m \left(\underline{\tilde{\boldsymbol{\mu}}}_m - \underline{\mathbf{b}}\right)\right], \quad (8.49)$$

where

$$\underline{\mathbf{A}} = \begin{bmatrix} \mathbf{A}^{(1)} \\ \mathbf{A}^{(2)} \\ \vdots \\ \mathbf{A}^{(S)} \end{bmatrix}, \underline{\boldsymbol{\Sigma}}_m = \begin{bmatrix} c_m^{(1)}\boldsymbol{\Sigma}_m^{-1} & 0 & \cdots & 0 \\ 0 & c_m^{(2)}\boldsymbol{\Sigma}_m^{-1} & \cdots & 0 \\ \vdots & \vdots & \ddots & \vdots \\ 0 & 0 & \cdots & c_m^{(S)}\boldsymbol{\Sigma}_m^{-1} \end{bmatrix}, \underline{\tilde{\boldsymbol{\mu}}}_m = \begin{bmatrix} \tilde{\boldsymbol{\mu}}_m^{(1)} \\ \tilde{\boldsymbol{\mu}}_m^{(2)} \\ \vdots \\ \tilde{\boldsymbol{\mu}}_m^{(S)} \end{bmatrix}, \underline{\mathbf{b}} = \begin{bmatrix} \mathbf{b}^{(1)} \\ \mathbf{b}^{(2)} \\ \vdots \\ \mathbf{b}^{(S)} \end{bmatrix}.$$

$$(8.50)$$

The solution for $\boldsymbol{\mu}_m$ is written in this form to emphasize its equivalence to the weighted least squares solution of an overdetermined system of linear equations. This equivalence comes as little surprise if we reason as follows: As is apparent from prior definitions, both $\underline{\tilde{\boldsymbol{\mu}}}_m$ and the matrix-vector product $\underline{\mathbf{A}}\boldsymbol{\mu}_m + \underline{\mathbf{b}}$ are elements of $\mathbb{R}^{(L \times S)}$, where L is the dimensionality of the feature vector and S the total number of speakers. The SI mean $\boldsymbol{\mu}_m$, however, is an element of $\mathbb{R}^L$. Hence, it comes to light that we seek that $\boldsymbol{\mu}_m$ that achieves a minimum of the weighted Euclidean norm $|(\underline{\tilde{\boldsymbol{\mu}}}_m - \underline{\mathbf{b}}) - \underline{\mathbf{A}}\boldsymbol{\mu}_m|$ where the weighting is determined by the "concatenated" covariance matrix $\underline{\boldsymbol{\Sigma}}_m$. The optimal solution for $\boldsymbol{\mu}_m$ is given by the perpendicular projection of $\underline{\tilde{\boldsymbol{\mu}}}_m - \underline{\mathbf{b}}$ onto the subspace spanned by the columns of $\underline{\mathbf{A}}$ (see Strang 1980, sect. 3.4). In the statistics literature, (8.49–8.50) would be interpreted as a specialization of the *Gauss–Markov theorem* (Stark and Woods 1994, sect. 6.6).

Casting aside all pedagogic considerations, it is possible to express the intermediate sums appearing in (8.49) more simply as

$$\mathbf{M}_m \triangleq \underline{\mathbf{A}}^T \underline{\boldsymbol{\Sigma}}_m \underline{\mathbf{A}} = \sum_s c_m^{(s)} \mathbf{A}^{(s)T} \boldsymbol{\Sigma}_m \mathbf{A}^{(s)}, \quad (8.51)$$

$$\mathbf{v}_m \triangleq \underline{\mathbf{A}}^T \underline{\boldsymbol{\Sigma}}_m (\underline{\tilde{\boldsymbol{\mu}}}_m - \underline{\mathbf{b}}) = \sum_s c_m^{(s)} \mathbf{A}^{(s)T} \boldsymbol{\Sigma}_m \left(\tilde{\boldsymbol{\mu}}_m^{(s)} - \mathbf{b}^{(s)}\right), \quad (8.52)$$

whereupon the SI mean estimate can be expressed as

$$\boldsymbol{\mu}_m = \mathbf{M}_m^{-1} \mathbf{v}_m. \quad (8.53)$$

Equations (8.51–8.53) provide a solution for the optimal SI mean. A similarly straightforward expression can be obtained for the SI variances (McDonough 1998b). The variance $\tilde{\sigma}_{mn}^{(s)2}$ of the observations from speaker s is given by

$$\tilde{\sigma}_{mn}^{(s)2} = \frac{1}{c_m^{(s)}} \sum_k c_{m,k}^{(s)} \left(y_{kn}^{(s)} - \tilde{\mu}_{mn}^{(s)} \right)^2 , \tag{8.54}$$

where $y_{kn}^{(s)}$ and $\tilde{\mu}_{mn}^{(s)}$ are the nth components of $\mathbf{y}_k$ and $\tilde{\boldsymbol{\mu}}_m$, respectively. The SI variances are then available from

$$\sigma_{mn}^2 = \frac{1}{c_m} \sum_s c_m^{(s)} \left[\tilde{\sigma}_{mn}^{(s)2} + \left(\tilde{\mu}_{mn}^{(s)} - \hat{\mu}_{mn}^{(s)} \right)^2 \right], \tag{8.55}$$

where

$$c_m = \sum_s c_m^{(s)},$$

and $\hat{\mu}_{mn}^{(s)}$ is the nth component of $\hat{\boldsymbol{\mu}}^{(s)}$. Implicit in (8.55) is that the transformed mean $\hat{\boldsymbol{\mu}}_m^{(s)} = \mathbf{A}^{(s)} \boldsymbol{\mu}_m + \mathbf{b}^{(s)}$ is calculated from the optimal speaker-independent mean $\boldsymbol{\mu}_m$ given by (8.53).

From the foregoing, it is clear that SAT requires accumulating the statistics

$$c_m^{(s)} = \sum_k c_{m,k}^{(s)}, \tag{8.56}$$

$$\mathbf{o}_m^{(s)} = \sum_k c_{m,k}^{(s)} \mathbf{y}_k^{(s)}, \tag{8.57}$$

$$\mathbf{s}_m^{(s)} = \sum_k c_{m,k}^{(s)} \mathbf{y}_k^{(s)2}, \tag{8.58}$$

where $\mathbf{y}_k^{(s)2}$ denotes the vector obtained by squaring $\mathbf{y}_k^{(s)}$ component-wise. A typical implementation of SAT requires writing out the quantities $c_m^{(s)}$, $\mathbf{o}_m^{(s)}$, and $\mathbf{s}_m^{(s)}$, in addition to $(\mathbf{A}^{(s)}, \mathbf{b}^{(s)})$, for every speaker in the training set. A moment's thought will reveal, however, that $\mathbf{M}_m$ can be accumulated from only $\{(\mathbf{A}^{(s)}, \mathbf{b}^{(s)})\}$ and $\{c_m^{(s)}\}$. Moreover, the sum $\mathbf{v}_m$ in (8.52) can be accumulated for all speakers and written to disk just once per training iteration, or once per iteration for each processor used in parallel forward–backward training, after which these partial sums can be added together. For a training corpus with several thousand speakers such as *Broadcast News*, this results in a tremendous savings in the disk space required for SAT, as first noted by McDonough and Waibel (2004). Although the sum in (8.55) requires that the *newly-updated* mean $\boldsymbol{\mu}_m$ be used in calculating $\hat{\mu}_{mn}^{(s)}$, experience has proven that the re-estimation works just as well if the *prior* value of $\boldsymbol{\mu}_m$ is used instead, in which case the partial sums in (8.55) can also be written out just once for each parallel processor. This novel and useful approximation has been dubbed *fast* SAT (McDonough *et al.* 2007).

As discussed in Section 9.2.2, the APT can be used to formulate a very effective means of speaker adaptation based on very few free parameters as compared to the more conventional MLLR. Its specification through very few free parameters implies that the APT is highly constrained. This in turn implies that if the APT is to be applied during test, HMM training *must* be conducted with SAT, as the transformations will otherwise be effective identity, and no improvement in speech recognition performance will be realized through the application of the transform (McDonough 2000).

8.1.4 Optimal Regression Class Estimation

In the foregoing development, we have assumed that a single, global transformation matrix $\mathbf{A}^{(s)}$ is estimated for each speaker s. This assumption, however, can be relaxed as follows: Partition the Gaussian components of the HMM into R mutually-exclusive subsets or *regression classes*, and for each subset estimate a unique transformation matrix $\mathbf{A}_r^{(s)}$.

Once an initial assignment of Gaussian components to regression classes has been made, it is possible to update these assignments using a ML criterion (McDonough 1998a). The *optimal regression class* (ORC) estimation procedure described in McDonough (1998a) and summarized here is a slight departure from that presented in Gales (1996) inasmuch as the mean and class assignment of a Gaussian are updated jointly rather than sequentially. Let $\rho(m)$ be a function returning the index of the regression class to which the mth Gaussian component is assigned, and let $\{\mathbf{A}_t^{(s)}\}$ denote the set of transformations available for speaker s. Let us then condition the auxiliary function (8.48) on the assignment $\rho(m)$ of the regression class, such that

$$\mathcal{G}_m(\mathcal{Y}; \{\mathbf{A}_t^{(s)}\}, \Lambda_m, r = \rho(m)) = \mathcal{G}_m(\mathcal{Y}; \{\mathbf{A}_t^{(s)}\}, (\boldsymbol{\mu}_m, \boldsymbol{\Sigma}_m), r), \qquad (8.59)$$

where

$$\begin{aligned}
&\mathcal{G}_m(\mathcal{Y}; \{\mathbf{A}_t^{(s)}\}, (\boldsymbol{\mu}_m, \boldsymbol{\Sigma}_m), r) \\
&= \sum_s c_m^{(s)} [(\tilde{\boldsymbol{\mu}}_m^{(s)} - \mathbf{A}_r^{(s)} \boldsymbol{\mu}_m - \mathbf{b}_r^{(s)})^T \boldsymbol{\Sigma}_m^{-1} (\tilde{\boldsymbol{\mu}}_m^{(s)} - \mathbf{A}_r^{(s)} \boldsymbol{\mu}_m - \mathbf{b}_r^{(s)})].
\end{aligned} \qquad (8.60)$$

In order to determine the optimal mean $\boldsymbol{\mu}_{m;r}^*$ conditioned on the assignment of the mth Gaussian component to the rth regression class, we differentiate both sides of (8.60) with respect to $\boldsymbol{\mu}_m$ and equate to zero, which provides

$$\boldsymbol{\mu}_{m;r}^* = \left(\mathbf{A}_r^T \boldsymbol{\Sigma}_m \mathbf{A}_r\right)^{-1} \mathbf{A}_r^T \boldsymbol{\Sigma}_m \left(\tilde{\boldsymbol{\mu}}_m - \mathbf{b}_r^{(s)}\right),$$

where

$$\mathbf{A}_r^T \boldsymbol{\Sigma}_m \mathbf{A}_r = \sum_s c_m^{(s)} \mathbf{A}_r^{(s)\,T} \boldsymbol{\Sigma}_m \mathbf{A}_r^{(s)},$$

$$\mathbf{A}_r^T \boldsymbol{\Sigma}_m \left(\tilde{\boldsymbol{\mu}}_m - \mathbf{b}_r^{(s)}\right) = \sum_s c_m^{(s)} \mathbf{A}_r^{(s)\,T} \boldsymbol{\Sigma}_m \left(\tilde{\boldsymbol{\mu}}_m^{(s)} - \mathbf{b}_r^{(s)}\right).$$

Substituting $\boldsymbol{\mu}^*_{m;r}$ back into (8.59–8.60) provides

$$
\mathcal{G}_m(\mathcal{Y}; \{\mathbf{A}^{(s)}_r\}, (\boldsymbol{\mu}^*_{m;r}, \boldsymbol{\Sigma}_m), r)
$$

$$
= \frac{1}{2} \sum_s c^{(s)}_m [(\tilde{\boldsymbol{\mu}}^{(s)}_m - \mathbf{A}^{(s)}_r \boldsymbol{\mu}^*_{m;r} - \mathbf{b}^{(s)}_r)^T \boldsymbol{\Sigma}^{-1}_m (\tilde{\boldsymbol{\mu}}^{(s)}_m - \mathbf{A}^{(s)}_r \boldsymbol{\mu}^*_{m;r} - \mathbf{b}^{(s)}_r)];
$$

Finally, define the optimal regression class r^* as that achieving

$$
r^*(m) = \underset{r}{\operatorname{argmin}} \ \mathcal{G}_m(\mathcal{Y}; \{\mathbf{A}^{(s)}_t\}, (\boldsymbol{\mu}^*_{m;r}, \boldsymbol{\Sigma}_m), r). \tag{8.61}
$$

This definition corresponds to the joint optimization of the mean and regression class assignment while holding the covariance fixed. Subsequently the covariances can be assigned in accordance with (8.55–8.54). It is worth noting that ORC estimation can be easily incorporated into the SAT paradigm discussed in Secton 8.1.3.

That the procedure given above provides the ML assignment of Gaussian components to regression classes is clear from the following. Suppose several iterations of SAT have been used to train an adapted HMM with fixed regression classes, so that the parameters of the model have converged to their optimal values. If another pass of SAT is conducted wherein the regression class assignments are re-estimated using the procedure above, the likelihood of the training set must either increase or remain the same, as guaranteed by Theorem 8.1.2. This will also hold true for subsequent SAT iterations using the ORC procedure. Moreover, if the training set likelihood does not increase, then a (local) maximum has been reached. Hence, the ORC procedure is guaranteed to find a local maximum of the training set likelihood.

8.1.5 Viterbi and Label Training

While forward–backward training is optimal, it is also computationally expensive. Two common simplifications are possible, namely, *Viterbi* and *label training*. As discussed in Section 7.1.2, the Viterbi algorithm is the application of dynamic programming to find the most likely state sequence through a HMM. It can be simply formulated as follows. Consider once more the forward probability $\alpha(\mathbf{y}_{1:k}, i)$ defined in (8.26). Instead of actually summing the probabilities of all transitions reaching a given state, we can approximate the sum with the *largest* probability, such that

$$
\alpha(\mathbf{y}_{1:k}, i) \approx p(\mathbf{y}_k; \Lambda_i) \max_j \alpha(\mathbf{y}_{1:k-1}, j) \, p_{i|j}.
$$

As these approximate forward probabilities are calculated for each time instant k and each state n_i, a back pointer is maintained to that predecessor state that had the highest probability. When the final frame has been processed, the probability

$$
p(\mathbf{y}_{1:K}; \Lambda) \approx \max_i \alpha(\mathbf{y}_{1:K}, i)
$$

is evaluated for all valid final states, and a trace back is performed to find the ML state sequence. Thereafter, the estimates (8.40–8.45) can be performed under the assumption that the posterior probabilities in (8.32) of the states on the Viterbi path are identically equal to 1, and the posterior probabilities of all other states are identically equal to 0. While this is clearly a simplification, experience has shown that it works well in practice, and provides performance nearly identical to the full forward–backward algorithm. Moreover, it uses significantly less computation than full forward–backward training.

Label training differs from Viterbi training in that the Viterbi path is calculated just once for each utterance, and this state sequence is fixed for all iterations of HMM parameter estimation. As such, it is even less costly in terms of computation than Viterbi training. Nonetheless, it provides a very reasonable performance.

8.2 Discriminative Parameter Estimation

As shown in the prior development, ML estimation considers only the correct transcription during training. It can be shown that ML estimation is in fact optimal, provided two conditions hold (Gopalakrishnan *et al.* 1991):

1. The statistical model is *correct*;
2. Unlimited training material is available.

Unfortunately, for ASR based on the HMM, neither of these conditions holds. Firstly, the HMM is clearly *not* the correct model of human speech production. Secondly, the data available for HMM training is almost invariably woefully limited.

In *discriminative* parameter estimation, not only is the correct transcription considered, but also those hypotheses that might be confused by the recognizer with the correct transcription. This is known to provide superior generalization and recognition performance when the conditions mentioned above are not fulfilled. In order to efficiently encode those hypotheses which may be mistaken for the correct transcription, word lattices, as described in Section 7.1.3, are typically used. One of the earliest optimization criterion used for such discriminative parameter estimation was *maximum mutual information* (MaxMI). More recently other criteria have been investigated by several researchers, including *minimum word error* (MWE) and *minimum phone error* (MPE).

The balance of this portion of the chapter is organized as follows. In Section 8.2.1, we review the conventional MaxMI re-estimation formulae. How the statistics for MaxMI training are accumulated using word lattices is the topic of Section 8.2.2. In Section 8.2.3 we will consider the recently proposed minimum word and phone error optimization criteria. Finally, the combination of MaxMI parameter estimation with SAT is described in Section 8.2.4. This combination of two training techniques has been shown to be more effective than either technique individually.

8.2.1 Conventional Maximum Mutual Information Estimation Formulae

Mutual information (MI) is a well-known statistic from information theory (Gallager 1968). Roughly speaking, it indicates how much two random variables have "in common", or how much knowledge one random variable conveys about another. In Section 13.5.4, we will seek to *minimize* the MI between the outputs of two beamformers as a means of separating the voices of different speakers. Here we seek to *maximize* the MI between a sequence of words and a sequence of acoustic features in order to improve the accuracy of an ASR system.

To begin our development, let $\mathbf{y}_{1:K}^{(s)}$, $g_{1:K}^{(s)}$, and $w_{1:K_w}^{(s)}$ respectively denote observation, Gaussian component, and word *sequences* associated with an utterance of speaker s, where K denotes the length of $\mathbf{y}_{1:K}^{(s)}$ and $g_{1:K}^{(s)}$, and K_w denotes the length of $w_{1:K_w}^{(s)}$. Let us define the *empirical mutual information* between the ensemble of words W and that of observations Y as

$$I(W, Y; \Lambda) \triangleq \sum_s \log \frac{p(w_{1:K_w}^{(s)}, \mathbf{y}_{1:K}^{(s)}; \Lambda)}{p(w_{1:K_w}^{(s)}) \, p(\mathbf{y}_{1:K}^{(s)}; \Lambda)}, \qquad (8.62)$$

where Λ is the set of HMM parameters as before. Let Λ^0 denote the current set of parameter values, and define the *difference* in posterior probabilities of $g_{1:K}^{(s)}$ that comes from knowledge of the correct word transcription as

$$c_{1:K}^{(s)} \triangleq p(g_{1:K}^{(s)} | w_{1:K_w}^{(s)}, \mathbf{y}_{1:K}^{(s)}; \Lambda^0) - p(g_{1:K}^{(s)} | \mathbf{y}_{1:K}^{(s)}; \Lambda^0). \qquad (8.63)$$

As with ML parameter estimation, we also define an *auxiliary function*, which in this case can be expressed as

$$Q(\Lambda | \Lambda^0) \triangleq S^{(1)}(\Lambda | \Lambda^0) + S^{(2)}(\Lambda | \Lambda^0), \qquad (8.64)$$

where,

$$S^{(1)}(\Lambda | \Lambda^0) \triangleq \sum_{s, g_{1:K}^{(s)}} c_{1:K}^{(s)} \log p(\mathbf{y}_{1:K}^{(s)} | g_{1:K}^{(s)}; \Lambda),$$

$$S^{(2)}(\Lambda | \Lambda^0) \triangleq \sum_{s, g_{1:K}^{(s)}} d'(g_{1:K}^{(s)}) \int_{\mathbf{y}_{1:K}^{(s)}} p(\mathbf{y}_{1:K}^{(s)} | g_{1:K}^{(s)}; \Lambda^0) \log p(\mathbf{y}_{1:K}^{(s)} | g_{1:K}^{(s)}; \Lambda) \, d\mathbf{y}_{1:K}^{(s)},$$

and $d'(g_{1:K}^{(s)})$ is a normalization constant that is typically chosen heuristically to achieve a balance between speed of convergence and robustness. Gunawardana (2001) sought to demonstrate that

$$Q(\Lambda | \Lambda^0) > Q(\Lambda^0 | \Lambda^0) \Rightarrow I(W, Y; \Lambda) > I(W, Y; \Lambda^0). \qquad (8.65)$$

While a fallacy in Gunawardana's proof was revealed, Axelrod *et al.* (2007) have devised a correct proof of (8.65) that is valid for a very general class of AMs.

Once more it is necessary to collect the statistics (8.56–8.58), but now $c_{m,k}^{(s)}$ must in each case be redefined as

$$c_{m,k}^{(s)} = p(g_k^{(s)} = g_m | w_{1:K_w}^{(s)}, \mathbf{y}_{1:K}^{(s)}; \Lambda^0) - p(g_k^{(s)} = g_m | \mathbf{y}_{1:K}^{(s)}; \Lambda^0). \qquad (8.66)$$

Let us additionally define c_m, $\mathbf{o}_m$, and $\mathbf{s}_m$, which are obtained by summing the relevant quantities in (8.56–8.58) over all speakers in the training set, according to

$$c_m \triangleq \sum_s c_m^{(s)}, \qquad \mathbf{o}_m \triangleq \sum_s \mathbf{o}_m^{(s)} \qquad \text{and} \qquad \mathbf{s}_m \triangleq \sum_s \mathbf{s}_m^{(s)}. \tag{8.67}$$

As shown in the seminal work of Gopalakrishnan *et al.* (1991), the re-estimation formulae for the Gaussian mixture weights is given by

$$w_m = \frac{w_m^0 \{f_m + C\}}{\sum_{m'} w_{m'}^0 \{f_{m'} + C\}}, \tag{8.68}$$

where w_m^0 is the current value of the mixture weight, and

$$f_m = \frac{c_m}{w_m^0}. \tag{8.69}$$

The constant C must be chosen to ensure that all mixture weights are positive.

We will now rigorously derive the MaxMI re-estimation formulae for the Gaussian mean μ_m. To simplify what follows, let us decompose (8.64) as a sum over individual Gaussian components. From the conditional independence assumption (7.4) inherent in the HMM, we know that

$$\log p(\mathbf{y}_{1:K}^{(s)}|g_{1:K}^{(s)}; \Lambda) = \sum_{k=1}^{K} \log p(\mathbf{y}_k^{(s)}|g_k^{(s)}; \Lambda) = \sum_{k=1}^{K} \log p(\mathbf{y}_k^{(s)}; \Lambda_m)\Big|_{g_m=g_k^{(s)}},$$

where Λ_m once more denotes the parameters of the mth Gaussian component. Hence,

$$S^{(1)}(\Lambda|\Lambda^0) = \sum_{k,s,g_{1:K}^{(s)}} c_{1:K}^{(s)} \log p(\mathbf{y}_k^{(s)}; \Lambda_m)\Big|_{g_m=g_k^{(s)}}$$

$$= \sum_m \sum_{k,s,g_{1:K}^{(s)}:g_k^{(s)}=g_m} c_{1:K}^{(s)} \log p(\mathbf{y}_k^{(s)}; \Lambda_m)$$

$$= \sum_m \sum_{k,s} c_{m,k}^{(s)} \log p(\mathbf{y}_k^{(s)}; \Lambda_m). \tag{8.70}$$

Also invoking conditional independence, we have

$$p(\mathbf{y}_{1:K}^{(s)}|g_{1:K}^{(s)}; \Lambda^0) \log p(\mathbf{y}_{1:K}^{(s)}|g_{1:K}^{(s)}; \Lambda)$$

$$= \prod_k p(\mathbf{y}_k^{(s)}|g_k^{(s)}; \Lambda_m^0)\Big|_{g_m=g_k^{(s)}} \times \sum_k \log p(\mathbf{y}_k^{(s)}|g_k^{(s)}; \Lambda_m)\Big|_{g_m=g_k^{(s)}}, \tag{8.71}$$

which implies

$$S^{(2)}(\Lambda|\Lambda^0) = \sum_m \sum_{k,s,\mathbf{g}_{1:K}^{(s)}:g_k^{(s)}=g_m} d'(g_{1:K}^{(s)}) \int_{\mathbf{y}} p(\mathbf{y}; \Lambda_m^0) \log p(\mathbf{y}; \Lambda_m) d\mathbf{y}$$

$$= \sum_m \sum_s d_m^{(s)} \int_{\mathbf{y}} p(\mathbf{y}; \Lambda_m^0) \log p(\mathbf{y}; \Lambda_m) \, d\mathbf{y}, \tag{8.72}$$

where

$$d_m^{(s)} = \sum_{k,\mathbf{g}_{1:K}^{(s)}:g_k^{(s)}=g_m} d'(g_{1:K}^{(s)}). \tag{8.73}$$

Finally, we have shown,

$$Q(\Lambda|\Lambda^0) = \sum_m Q_m(\Lambda_m|\Lambda_m^0) = \sum_m \left[S_m^{(1)}(\Lambda|\Lambda^0) + S_m^{(2)}(\Lambda|\Lambda^0) \right], \tag{8.74}$$

where

$$S_m^{(1)}(\Lambda|\Lambda^0) = \sum_{k,s} c_{m,k}^{(s)} \log p(\mathbf{y}_k^{(s)}; \Lambda_m), \tag{8.75}$$

$$S_m^{(2)}(\Lambda|\Lambda^0) = \sum_s d_m^{(s)} \int_{\mathbf{y}} p(\mathbf{y}; \Lambda_m^0) \log p(\mathbf{y}; \Lambda_m) \, d\mathbf{y}. \tag{8.76}$$

Next let us set $\Lambda_m = (\boldsymbol{\mu}_m, \boldsymbol{\Sigma}_m)$ as before and observe that the acoustic log-likelihood for the mth Gaussian can be expressed as in (8.19), from which it follows that

$$\nabla_{\boldsymbol{\mu}_m} \log p(\mathbf{y}; \Lambda_m) = \boldsymbol{\Sigma}_m^{-1}(\mathbf{y} - \boldsymbol{\mu}_m), \tag{8.77}$$

$$\nabla_{\sigma_{m,n}^2} \log p(\mathbf{y}; \Lambda_m), = \frac{1}{2} \left[\frac{\left(y_n - \mu_{m,n}\right)^2}{(\sigma_{m,n}^2)^2} - \frac{1}{\sigma_{m,n}^2} \right], \tag{8.78}$$

where y_n and $\mu_{m,n}$ are the nth components of $\mathbf{y}$ and $\boldsymbol{\mu}_m$, respectively. From (8.74–8.76), it follows that

$$\nabla_{\boldsymbol{\mu}_m} Q_m(\Lambda_m|\Lambda_m^0) = \sum_{k,s} c_{m,k}^{(s)} \nabla_{\boldsymbol{\mu}_m} \log p(\mathbf{y}_k^{(s)}; \Lambda_m)$$

$$+ \sum_s d_m^{(s)} \int_{\mathbf{y}} p(\mathbf{y}; \Lambda_m^0) \nabla_{\boldsymbol{\mu}_m} \log p(\mathbf{y}; \Lambda_m) d\mathbf{y}.$$

Substituting (8.77) into the last equation gives

$$\nabla_{\boldsymbol{\mu}_m} Q_m(\Lambda_m|\Lambda_m^0) = \sum_{k,s} c_{m,k}^{(s)} \boldsymbol{\Sigma}_m^{-1}(\mathbf{y}_m^{(s)} - \boldsymbol{\mu}_m) + \sum_s d_m^{(s)} \boldsymbol{\Sigma}_m^{-1} \left[\int_{\mathbf{y}} p(\mathbf{y}; \Lambda_m^0) \mathbf{y} \, d\mathbf{y} - \boldsymbol{\mu}_m \right]$$

$$= \boldsymbol{\Sigma}_m^{-1} \left[\mathbf{o}_m + d_m \boldsymbol{\mu}_m^0 - (c_m + d_m) \boldsymbol{\mu}_m \right],$$

where

$$d_m = \sum_s d_m^{(s)},$$

and $\boldsymbol{\mu}_m^0$ is the current value of the mth mean. Upon equating the right-hand side of the last equation to zero, it becomes clear that the conventional re-estimation formulae for the mean $\boldsymbol{\mu}_m$ under the MaxMI criterion can be expressed as

$$\boldsymbol{\mu}_m = \frac{\mathbf{o}_m + d_m \boldsymbol{\mu}_m^0}{c_m + d_m}. \tag{8.79}$$

It remains only to choose a value for $d_m^{(s)}$. Woodland and Povey (2000) recommended setting $d_m^{(s)}$ as a multiple of the occupancy count of the denominator lattice, according to

$$d_m^{(s)} = E \sum_k p(g_k^{(s)} = g_m | \mathbf{y}_k^{(s)}; \Lambda^0),$$

for $E = 1.0$ or 2.0. Typically the factor E is increased for each iteration of MaxMI training. Manipulations similar to those above readily yield the re-estimation formula for the diagonal covariance components as

$$\sigma_{m,n}^2 = \frac{s_{m,n} + d_m \left(\sigma_{m,n}^{02} + \mu_{m,n}^2 \right)}{c_m + d_m} - \mu_{m,n}^{02} \tag{8.80}$$

where $\sigma_{m,n}^{02}$ is the diagonal covariance component from the prior iteration, and $s_{m,n}$ and $\mu_{m,n}^2$ and the nth components of $\mathbf{s}_m$ and $\boldsymbol{\mu}_m$, respectively.

It is interesting to note that the original derivation of (8.79) and (8.80) in Normandin (1991) was based on an approximation, but a simpler and more satisfying derivation is provided by Axelrod et al. (2007). While (8.79–8.80) were originally derived for estimating the parameters of an unadapted HMM, these updated formulae also lend themselves to use with constrained maximum likelihood linear regression, as described in Section 9.1.2, during training.

Woodland and Povey (2000) also noted that discriminative parameter estimation is more effective if the acoustic log-likelihoods (8.19) are scaled by some factor $\kappa \approx 0.1$ during training. This simple heuristic prevents overtraining to a large extent and helps to increase generalization.

8.2.2 Maximum Mutual Information Training on Word Lattices

We will now describe how the statistics defined in (8.67) can be accumulated using word lattices, which were discussed in Section 7.1.3. In general, the accumulation of the relevant statistics on any single utterance requires two lattices, hereafter referred to as the *numerator* and *denominator lattices*, which are respectively associated with the numerator and denominator of (8.62). Alternatively, the respective lattices are associated with the two terms in (8.63). The numerator lattice is obtained solely from the correct transcription and hence is trivial to generate. The denominator lattice, on the other hand, must efficiently encode all alternative hypotheses that have a non-negligible chance of being mistaken for the correct hypothesis. Hence, these denominator lattices must be obtained by actually decoding the training data.

In writing and manipulating word lattices, it is beneficial to apply the *weighted finite-state transducer* (WFST) techniques discussed in Section 7.2 in order to maximize computational efficiency. Our description of the steps necessary for MaxMI training based on word lattices follows roughly that given in Valtchev *et al.* (1997). We indicate, however, where the WFST algorithms described can be applied to obtain more efficient training procedures. Although currently widespread, the use of WFSTs in ASR was still in its infancy when Valtchev *et al.* (1997) appeared. Hence, that earlier work was based on ad hoc techniques for manipulating word lattices.

1. For each utterance in the training set, generate a pair of *numerator* and *denominator* word lattices. As mentioned previously, the numerator word lattice is constructed solely from the correct transcription. The denominator lattice is obtained by performing an unconstrained decoding on the utterance with some initial AM. In order to obtain the maximum possible acoustic diversity, the unconstrained decoding is often performed with a bigram or even unigram *language model* (LM). To generate the initial reduced word lattice, the AM and LM scores are discarded, as is all information pertaining to the actual state sequence. Only the word identities themselves are retained. Thereafter, unneeded lattice edges are deleted through the epsilon removal procedure described in Section 7.2.6.

2. If the correct transcription is not found in one of the initial reduced denominator lattices, it must be added. All of these denominator lattices are then compacted through the determinization and minimization algorithms described in Sections 7.2.3 and 7.2.5 respectively. Thereafter, both the numerator and denominator lattices are used to construct constrained search graphs as described in Section 7.3. For the numerator, the word lattice directly replaces the grammar G. For the denominator, the word lattice must be composed with G, before constructing the constrained search graph.

3. The constrained search graph for each utterance is used to generate a new lattice with detailed timing information, as well as the full AM and LM scores. This ensues by conducting a full Viterbi search over the constrained search graph with the acoustic model from the prior iteration. Subsequently, the lattice forward and backward probabilities, denoted respectively as $\overline{\alpha}$ and $\overline{\beta}$, can be computed. The forward probability is calculated recursively beginning from the starting node according to

$$\overline{\alpha}_i = \sum_j \overline{\alpha}_j v_{\mathrm{AM}}(e_{i|j})\, v_{\mathrm{LM}}(e_{i|j}),$$

where $v_{\mathrm{AM}}(e_{i|j})$ and $v_{\mathrm{LM}}(e_{i|j})$, respectively, are the AM and LM scores associated with the edge from the jth to the ith node in the lattice. The backward probabilities $\overline{\beta}_j$ can be recursively calculated in similar fashion beginning at the end of the lattice. The forward and backward probabilities are initialized as indicated in (8.28–8.29).

4. The *a posteriori* probability for each edge $e_{i|j}$ in the lattice is given by

$$\overline{c}_{i|j} = \frac{\overline{\alpha}_j \nu_{\text{AM}}(e_{i|j}) \, \nu_{\text{LM}}(e_{i|j}) \, \overline{\beta}_i}{\mathcal{L}} \tag{8.81}$$

where $\mathcal{L}$ is the total lattice likelihood computed by summing the forward probabilities over all hypothesized end words. Such an *a posteriori* probability must be calculated separately for the numerator and denominator lattices, and the statistics must be accumulated for each as indicated in (8.56–8.58).

5. The Gaussian component occupancy statistics (8.63) are then accumulated based on the posterior probabilities $\overline{c}_{i|j}$ calculated for both numerator and denominator lattices from the prior step.

6. After all training utterances have been processed, a parameter update occurs based on (8.68–8.69) and (8.79–8.80). Thereafter, the next iteration begins from Step 3 above.

By composing the word lattice with G as described in Step 2, a LM is obtained with the same weights as G, but that only accepts the set of word sequences contained in the orginal word lattice. In order to increase the acoustic diversity during discriminative training, it is common to use a unigram for G. Note that new lattices and constrained search graphs are typically *not* generated between iterations. This greatly reduces the computational expense of MaxMI training. It is, however, possible to conduct a full Viterbi search over the constrained search graphs in Step 3, inasmuch as this operation is very efficient once the search graphs have been optimized with WFST operations.

8.2.3 Minimum Word and Phone Error Training

Povey and Woodland (2002) proposed two discriminative optimization criteria, namely, *minimum word error* (MWE) and *minimum phone error* (MPE), and observed that these criteria provide an acoustic model that achieves recognition performance superior to that obtained with the MaxMI criterion. In this section, we will describe the MPW and MPE criteria, both of which are frequently augmented with a technique known as I-smoothing in order to improve the robustness of the final acoustic model.

To begin our exposition, let us express the MWE optimization criterion as

$$\mathcal{F}_{\text{MWE}}(\mathcal{Y}; \Lambda) = \frac{\sum_{w_{1:K_w}} p^\kappa(\mathbf{y}_{1:K} | w_{1:K_w}) P(w_{1:K_w}) \text{RawAccuracy}(w_{1:K_w})}{\sum_{w_{1:K_w}} p^\kappa(\mathbf{y}_{1:K} | w_{1:K_w}) P(w_{1:K_w})} \tag{8.82}$$

where κ is the acoustic likelihood scale factor that is typically used to prevent overtraining, and $\text{RawAccuracy}(w_{1:K_w})$ is a measure of the number of words accurately transcribed in *hypothesis* $w_{1:K_w}$. From (8.82) it is clear that the MWE criterion amounts to a weighted average over all possible hypotheses $w_{1:K_w}$ of $\text{RawAccuracy}(w_{1:K_w})$. For $\kappa \to \infty$, maximizing (8.82) is equivalent to minimizing the word error rate.

Having defined MWE, it is straightforward to extend this definition to that of MPE. That is, MPE is identical to MWE, but the errors are calculated at the phone level instead of word level. Povey and Woodland investigated both context-independent and context-dependent definitions of MPE.

Let q denote a phone arc corresponding to a word w_k. In order to optimize the MWE objective function it is necessary to calculate

$$c_q^{\text{MWE}} = \frac{1}{\kappa} \frac{\partial \mathcal{F}_{\text{MWE}}(\mathcal{Y}; \Lambda)}{\partial \log(q)}, \tag{8.83}$$

which is known as the *MWE arc occupancy*. Once the arc occupancy, whether positive or negative, has been calculated, it is used to accumulate the relevant statistics in (8.56–8.58) and (8.67). Moreover, if $c_q^{\text{MWE}} < 0$, the denominator statistic d_q is updated according to $d_q \leftarrow d_q - c_q^{\text{MWE}}$. When all training utterances have been processed, these statistics are then used in the parameter update equations (8.79)–(8.80).

MWE arc occupancies can be easily computed if the RawAccuracy(s) can be expressed as a sum of terms, each of which corresponds to a word w regardless of the context. Ideally, this would imply

$$\text{RawAccuracy}(s) = \sum_{w \in s} \text{WordAccuracy}(s),$$

where

$$\text{WordAccuracy}(s) = \begin{cases} 1, & \text{if correct word,} \\ 0, & \text{if substitution,} \\ -1, & \text{if insertion.} \end{cases} \tag{8.84}$$

The correct calculation of (8.84), however, actually requires dynamic programming, which would be computationally prohibitive. In order to avoid this computational expense, Povey and Woodland proposed an approximation, namely a word z is found in the reference transcript which overlaps in time with hypothesis word w, then letting e denote the proportion of the length of z which overlaps with w, the word accuracy is approximated as

$$\text{WordAccuracy}(s) = \begin{cases} -1 + 2e, & \text{if same word,} \\ -1 + e, & \text{if different word.} \end{cases} \tag{8.85}$$

The word z is then chosen to make the approximation (8.85) for the word accuracy as large as possible. Equation (8.85) represents a tradeoff between an insertion and a correct word or substitution, respectively. This tradeoff is necessary in that a single reference word may be used more than by a hypothesis sentence. As originally proposed, the reference word z was chosen from a lattice encoding alternate alignments of the correct sentence.

Let $\bar{c}_q$ denote the arc occupancy derived from a forward–backward pass on the lattice as in (8.81), let $c(q)$ denote the average value of RawAccuracy($w_{1:K_w}$) for sentences $w_{1:K_w}$ containing arc q weighted by log-likelihood scaled by κ of those sentences, and let c_{avg} denote the weighted average RawAccuracy($w_{1:K_w}$) for all hypotheses in the lattice, which is equivalent to the MWE criterion for the utterance $w_{1:K_w}$. It follows that $\mathcal{F}_{\text{MWE}}(\mathcal{Y}; \Lambda)$ can then be calculated from

$$\mathcal{F}_{\text{MWE}}(\mathcal{Y}; \Lambda) = \bar{c}_q (c(q) - c_{\text{avg}}).$$

The value of $c(w)$ can be efficiently calculated with another lattice forward–backward pass. As (8.85) is defined for words and the forward–backward algorithm functions at the phone level, PhoneAccuracy(q) is defined to be WordAccuracy(q) if q is the first phone of w and zero otherwise. In the case of MPE, on the other hand, PhoneAccuracy(q) can be calculated directly from (8.85). That is, if α_q and β_q are the forward and backward likelihoods used to calculate normal arc posterior probabilities, let

$$\alpha'_q = \frac{\sum_{r \text{ preceding } q} \alpha'_r \alpha_r t^K_{q|r}}{\sum_{r \text{ preceding } q} \alpha_r t^K_{q|r}} + \text{PhoneAccuracy}(q), \tag{8.86}$$

$$\beta'_q = \frac{\sum_{r \text{ following } q} t^K_{q|r} p^K(r) \beta_r (\beta'_r + \text{PhoneAccuracy}(q))}{\sum_{r \text{ following } q} t^K_{q|r} p^K(r) \beta_r}, \tag{8.87}$$

$$c(q) = \alpha'_q + \beta'_q, \tag{8.88}$$

where $t_{r|q}$ are lattice transition probabilities derived from the LM.

While discriminative training techniques have proven effective at reducing the word error rate of large vocabulary continuous speech recognition systems, they are also notoriously prone to *overtraining*, whereby performance improvements on the training set do not generalize to unseen data. To counteract this effect, it is common to use a mixed optimization criterion consisting of both ML and MaxMI components. I-smoothing is one technique for implementing such a mixed criterion. In the context of MWE and MPE training, I-smoothing is equivalent to increasing the mass of the statistics (8.56–8.58) with some number of expected counts τ based on the alignment of the correct transcription; see Povey and Woodland (2002) and Woodland and Povey (2002).

8.2.4 *Maximum Mutual Information Speaker-Adapted Training*

The re-estimation formulae for *maximum mutual information speaker-adapted training* (MaxMI-SAT) are quite similar to their maximum likelihood counterparts. In this section, we provide the full derivation for re-estimating the HMM mean components under an MaxMI criterion. The formulae for re-estimation of the diagonal covariances, regression classes and semi-tied covariance transformation matrices can be obtained through similar development. To avoid obscuring our main arguments under a mass of detail, we only quote the re-estimation formulae for these latter components here; a full derivation is provided in McDonough *et al.* (2007).

To derive the re-estimation formulae under MaxMI-SAT, we must begin by adding the conditioning on the speaker adaptation parameters in (8.46) to (8.66) and (8.75–8.76), such that,

$$c^{(s)}_{m,k} = p(g^{(s)}_k = g_m | w^{(s)}_{1:K_w}, \mathbf{y}^{(s)}_{1:K}; (\mathbf{A}^{(s)}, \mathbf{b}^{(s)}), \Lambda^0) - p(g^{(s)}_k = g_m | \mathbf{y}^{(s)}_{1:K}; (\mathbf{A}^{(s)}, \mathbf{b}^{(s)}), \Lambda^0). \tag{8.89}$$

and

$$S^{(1)}_m(\Lambda|\Lambda^0) = \sum_{k,s} c^{(s)}_{m,k} \log p(\mathbf{y}^{(s)}_k; (\mathbf{A}^{(s)}, \mathbf{b}^{(s)}), \Lambda_m), \tag{8.90}$$

$$S_m^{(2)}(\Lambda|\Lambda^0) = \sum_s d_m^{(s)} \int_\mathbf{y} p(\mathbf{y}; (\mathbf{A}^{(s)}, \mathbf{b}^{(s)}), \Lambda_m^0) \log p(\mathbf{y}; (\mathbf{A}^{(s)}, \mathbf{b}^{(s)}), \Lambda_m) \, d\mathbf{y}. \quad (8.91)$$

The acoustic log-likelihood for the mth Gaussian can be expressed as

$$\log p(\mathbf{y}; (\mathbf{A}^{(s)}, \mathbf{b}^{(s)}), \Lambda_m)$$

$$= -\frac{1}{2} \left[\log |2\pi \boldsymbol{\Sigma}_m| + (\mathbf{y} - \mathbf{A}^{(s)} \boldsymbol{\mu}_m - \mathbf{b}^{(s)})^T \boldsymbol{\Sigma}_m^{-1} (\mathbf{y} - \mathbf{A}^{(s)} \boldsymbol{\mu}_m - \mathbf{b}^{(s)}) \right],$$

from which it follows that

$$\nabla_{\boldsymbol{\mu}_m} \log p(\mathbf{y}; (\mathbf{A}^{(s)}, \mathbf{b}^{(s)}), \Lambda_m) = \mathbf{A}^{(s)\,T} \boldsymbol{\Sigma}_m^{-1} \mathbf{y} - \mathbf{A}^{(s)\,T} \boldsymbol{\Sigma}_m^{-1} (\mathbf{A}^{(s)} \boldsymbol{\mu}_m + \mathbf{b}^{(s)}), \quad (8.92)$$

$$\nabla_{\sigma_{m,n}^2} \log p(\mathbf{y}; (\mathbf{A}^{(s)}, \mathbf{b}^{(s)}), \Lambda_m), = \frac{1}{2} \left[\frac{\left(y_n - \hat{\mu}_{mn}^{(s)} \right)^2}{(\sigma_{m,n}^2)^2} - \frac{1}{\sigma_{m,n}^2} \right], \quad (8.93)$$

where $\hat{\mu}_{mn}^{(s)}$ is the nth component of the adapted mean $\hat{\boldsymbol{\mu}}_m$ defined in (8.46).

Mean Estimation

From (8.74–8.91) it follows that

$$\nabla_{\boldsymbol{\mu}_m} Q_m(\Lambda_m|\Lambda_m^0) = \sum_{k,s} c_{m,k}^{(s)} \nabla_{\boldsymbol{\mu}_m} \log p(\mathbf{y}_k^{(s)}; (\mathbf{A}^{(s)}, \mathbf{b}^{(s)}), \Lambda_m)$$

$$+ \sum_s d_m^{(s)} \int_\mathbf{y} p(\mathbf{y}; (\mathbf{A}^{(s)}, \mathbf{b}^{(s)}), \Lambda_m^0) \nabla_{\boldsymbol{\mu}_m} \log p(\mathbf{y}; (\mathbf{A}^{(s)}, \mathbf{b}^{(s)}), \Lambda_m) d\mathbf{y}.$$

Substituting (8.92) into the last equation gives

$$\nabla_{\boldsymbol{\mu}_m} Q_m(\Lambda_m|\Lambda_m^0) = \sum_{k,s} c_{m,k}^{(s)} \left[\mathbf{A}^{(s)\,T} \boldsymbol{\Sigma}_m^{-1} \mathbf{y}_k^{(s)} - \mathbf{A}^{(s)\,T} \boldsymbol{\Sigma}_m^{-1} (\mathbf{A}^{(s)} \boldsymbol{\mu}_m + \mathbf{b}^{(s)}) \right]$$

$$+ \sum_s d_m^{(s)} \left[\mathbf{A}^{(s)\,T} \boldsymbol{\Sigma}_m^{-1} \int_\mathbf{y} p(\mathbf{y}; (\mathbf{A}^{(s)}, \mathbf{b}^{(s)}), \Lambda_m^0) \, \mathbf{y} \, d\mathbf{y} \right.$$

$$\left. - \mathbf{A}^{(s)\,T} \boldsymbol{\Sigma}_m^{-1} (\mathbf{A}^{(s)} \boldsymbol{\mu}_m + \mathbf{b}^{(s)}) \right]$$

$$= \sum_s \left[\mathbf{A}^{(s)\,T} \boldsymbol{\Sigma}_m^{-1} \mathbf{o}_m^{(s)} - c_m^{(s)} \mathbf{A}^{(s)\,T} \boldsymbol{\Sigma}_m^{-1} (\mathbf{A}^{(s)} \boldsymbol{\mu}_m + \mathbf{b}^{(s)}) \right]$$

$$+ \sum_s d_m^{(s)} \left[\mathbf{A}^{(s)\,T} \boldsymbol{\Sigma}_m^{-1} (\mathbf{A}^{(s)} \boldsymbol{\mu}_m^0 + \mathbf{b}^{(s)}) - \mathbf{A}^{(s)\,T} \boldsymbol{\Sigma}_m^{-1} (\mathbf{A}^{(s)} \boldsymbol{\mu}_m + \mathbf{b}^{(s)}) \right]$$

where μ_m^0 is the current value of the mth mean. Grouping terms and equating the result to zero, we find that the new value of μ_m can be calculated as in (8.53), provided that we define

$$\mathbf{M}_m \triangleq \sum_s \left(c_m^{(s)} + d_m^{(s)} \right) \mathbf{A}^{(s)\,T} \mathbf{\Sigma}_m^{-1} \mathbf{A}^{(s)}, \tag{8.94}$$

$$\mathbf{v}_m \triangleq \sum_s \mathbf{A}^{(s)\,T} \mathbf{\Sigma}_m^{-1} \left[\left(\mathbf{o}_m^{(s)} - c_m^{(s)} \, \mathbf{b}^{(s)} \right) + d_m^{(s)} \mathbf{A}^{(s)} \mu_m^0 \right]. \tag{8.95}$$

It remains only to choose a value for $d_m^{(s)}$. Good results have been obtained by once more setting $d_m^{(s)}$ as a multiple of the occupancy count of the denominator lattice, according to

$$d_m^{(s)} = E \sum_k p(g_k^{(s)} = g_m | \mathbf{y}_k^{(s)}; (\mathbf{A}^{(s)}, \mathbf{b}^{(s)}), \Lambda^0),$$

for $E = 1.0$ or 2.0 as recommended in Woodland and Povey (2000). Setting $E \geq 1.0$ also ensures that $\mathbf{M}_m$ is positive definite, which is necessary if $\mu_m = \mathbf{M}_m^{-1} \mathbf{v}_m$ is to be an optimal solution.

Diagonal Covariance Estimation

Using development similar to that in the prior section, it can be shown that the diagonal covariance matrices can be estimated under an MaxMI criterion according to the formula:

$$\sigma_{mn}^2 = \frac{\sum_s \left\{ \left(s_{mn}^{(s)} - 2 o_{mn}^{(s)} \hat{\mu}_{mn}^{(s)} + c_m^{(s)} \hat{\mu}_{mn}^{(s)2} \right) + d_m^{(s)} \left[\sigma_{kn}^{0\,2} + \left(\hat{\mu}_{mn}^{0(s)} - \hat{\mu}_{mn}^{(s)} \right)^2 \right] \right\}}{\sum_s (c_m^{(s)} + d_m^{(s)})}, \tag{8.96}$$

where $\hat{\mu}_{mn}^{0(s)}$ is the nth component of $\hat{\mu}_m^0 = \mathbf{A}^{(s)} \mu_m^0 + \mathbf{b}^{(s)}$ and $\sigma_{mn}^{0\,2}$ is the current value of the variance. As before, $s_m^{(s)}$ is defined with $c_{m,k}^{(s)}$ as given in (8.89). In accumulating the term

$$\sum_s \left(s_m^{(s)} - 2 \mathbf{o}_m^{(s)} \hat{\mu}_{mn}^{(s)} + c_m^{(s)} \hat{\mu}_{mn}^{(s)2} \right)$$

the same SAT approximation can be made as before in order to economize on disk space; that is, $\hat{\mu}_m^{(s)}$ can be calculated with μ_m^0 instead of μ_m.

Semi-Tied Covariance Estimation

Gales (1999) defines a *semi-tied covariance matrix* as

$$\mathbf{\Sigma}_m = \mathbf{P} \, \mathbf{\Sigma}_m \mathbf{P}^T$$

where $\mathbf{\Sigma}_m$ is, as before, the diagonal covariance matrix for the mth Gaussian component, and $\mathbf{P}$ is a transformation matrix shared by many Gaussian components. Both $\mathbf{\Sigma}_m$ and $\mathbf{P}$ can be updated using a ML criterion. Assuming $\mathbf{P}$ is nonsingular, however, it is possible to define

$$\mathbf{M} = \mathbf{P}^{-1}$$

The transformation $\mathbf{M}$ can then be applied to each *feature* $\mathbf{y}_k$, which is more computationally efficient than transforming the covariances. Let $\mathbf{m}_i^T$ denote the ith row of $\mathbf{M}$, and let $\mathbf{f}_i^T$ denote the ith row in the *cofactor matrix* of $\mathbf{M}$. Moreover, define

$$\mathbf{W}_j = \sum_m \frac{1}{\sigma_{m,j}^2} \left[\sum_{k,s} c_{m,k}^{(s)} \mathbf{o}_{m,k}^{(s)} \mathbf{o}_{m,k}^{(s)T} + d_m \sum_i \sigma_{m,i}^{0\,2} \, p_i^0 \, p_i^{0\,T} \right] \qquad (8.97)$$

where p_i^0 is the ith column of $\mathbf{P}$,

$$c = \sum_m c_m, \qquad \text{and} \qquad d = \sum_m d_m.$$

A procedure for estimating an optimal transformation $\mathbf{M}$ can be developed directly along the lines proposed by Gales (1999). As shown in McDonough *et al.* (2007), such a procedure proceeds by iteratively updating each row $\mathbf{m}_j^T$ according to

$$\mathbf{m}_j^T = \mathbf{f}_j^T \mathbf{W}_j^{-1} \sqrt{\frac{c+d}{\mathbf{f}_j^T \mathbf{W}_j^{-1} \mathbf{f}_j}}.$$

As explained in McDonough *et al.* (2007), in order to assure convergence, d_m in (8.97) must be chosen to ensure that $\mathbf{W}_j$ is positive definite.

8.3 Summary and Further Reading

In the previous chapter, we learned that most modern speech recognition engines are based on the HMM, which can contain millions of free parameters that must be estimated from tens or hundreds of hours of training data. How this parameter estimation may be efficiently performed was the subject of this chapter. We considered two primary parameter estimation criteria: ML estimation, and MaxMI estimation.

In ML parameter estimation, only the correct transcription of each training utterance is considered. As described in this chapter, ML estimation attempts to associate as much probability mass with this correct transcription as possible. We also discussed how HMM parameter estimation based on the ML criterion can be supplemented with speaker adaptation parameters, the estimation of which will be described in Chapter 9.

In contrast to ML estimation, MaxMI estimation considers not only the correct transcription of each utterance, but also incorrect hypotheses that could be mistaken for the correct one. As we have discussed, a set of possible incorrect hypotheses is typically determined by running the recognition engine on all training utterances, then generating word lattices as described in Section 7.1.3. The HMM AM can then be discriminatively trained to assign a higher likelihood to the correct transcription than to any of its incorrect competitors. MaxMI training, which can also be supplemented with speaker adaptation parameters, typically produces an AM that makes significantly fewer recognition errors than its ML counterpart. This performance improvement, however, comes at the price of a more computationally expensive parameter estimation procedure, primarily due to the necessity of running the recognizer over the entire training set. But with availability of ever more and ever cheaper computational power, this additional computation is no longer considered prohibitive.

The use of speaker adaptation during discriminative training has a large effect on final system performance, as shown in McDonough *et al.* (2007). In this chapter, we have provided a full exposition of the steps necessary to perform speaker-adapted training under a MaxMI criterion. As has been shown, this can be achieved by performing *unsupervised* parameter estimation on the test data, a distinct advantage for many recognition tasks involving conversational speech. We have also derived re-estimation formulae for the SI means and variances of a continuous density HMM when SAT is conducted on the latter under an MaxMI criterion. We also described an approximation to the basic SAT re-estimation formulae that greatly reduces the amount of disk space required to conduct training. We also presented re-estimation formulae for STC transformation matrices based on an MaxMI criterion. Moreover, we presented a positive definiteness criterion, with which the regularization constant present in all MaxMI re-estimation formulae can be reliably set to provide both consistent improvements in the total MI of the training set, as well as fast convergence. We also combined the STC re-estimation formulae with their like for the SI means and variances, and update *all* parameters during MaxMI speaker-adapted training.

In the recent past, a great deal of research effort has been devoted to the development of optimization criteria and paradigms for the discriminative training of HMMs. While many of the optimization criteria and training algorithms have proven better than the MaxMI criterion investigated by Normandin (1991), most authors have ignored speaker adaptation during discriminative HMM training.

Gopalakrishnan *et al.* (1991) first proposed a practical technique for performing MaxMI training of HMMs, and commented on the fact that MaxMI is superior to ML-based parameter estimation given that the amount of available training data is always limited, and that the HMM is not the actual model of speech production. Gopalakrishnan's development was subsequently extended by Normandin (1991) to the case of continuous density HMMs. While these initial works were of theoretical interest, for several years it was believed that the marginal performance gains that could be obtained with MaxMI did not justify the increase in computational effort it entailed with respect to ML training. This changed when Woodland and Povey (2000) discovered that these gains could be greatly increased by scaling all acoustic log-likelihoods during training. Since the publication of Woodland and Povey (2000), MaxMI training has enjoyed a spate of renewed interest and a concomitant flurry of publications, including Uebel and Woodland (2001) in which an MaxMI criterion is used for estimating linear regression parameters, and Zheng *et al.* (2001) in which different update formulae are proposed for the standard MaxMI-based mean and covariance re-estimation. Schlüter (2000) expounds a unifying framework for a variety of popular discriminative training techniques. Gunawardana (2001) sets forth a much simplified derivation of Normandin's original continuous density re-estimation formulae which does not require the discrete density approximations Normandin used. While Gunawardana's original proof contained a fallacy, a correct proof of Gunawardana's main result was provided by Axelrod *et al.* (2007).

Other works that have appeared in the recent past include Povey and Woodland (2002), who explored the notion of using *minimum word* or *phone error* as an optimization criterion during parameter estimation, and found that this provided a significant reduction in word error rate with respect to MaxMI training. Povey *et al.* (2005) then extended this notion to the estimation of feature transformations. More recently, several authors have applied the concept of *large margin estimation*, on which the field of *support vector machines* is based, to discriminative training of HMMs. Representative works in this direction are those of Liu *et al.* (2005) and Sha and Saul (2006). Another approach to discriminative training is based on the concept of *conditional random fields*, which produce the conditional probability of an entire state sequence given a sequence of observations, has been explored by Gunawardana *et al.* (2005) and Mahajan *et al.* (2006). A recent survey of work on discriminative HMM training was provided by He *et al.* (2008).

8.4 Principal Symbols

Symbol	Description
Λ	updated HMM parameter set
Λ^0	initial HMM parameter set
$\mathbf{A}^{(s)}$	transformation matrix for speaker s
$\mathbf{b}^{(s)}$	additive bias for speaker s
g_k	Gaussian component chosen for kth observation
g_m	mth Gaussian component
i, j	indices over HMM states
$I(W, Y; \Lambda)$	mutual information between W and Y with HMM parameters Λ
k	time index
m	index over Gaussian components
$\mathbf{n}^{(s)}_{1:K}$	state sequence for speaker s
$Q(\Lambda \mid \Lambda^0)$	auxiliary function
$\mathbf{y}_k$	kth observation
$\mathbf{y}_{1:K}$	sequence of observations from 1 to K
s	index over speakers
W	ensemble of words
$\mathbf{w}^{(s)}_{1:K_w}$	word sequence for speaker s
Y	ensemble of observations

9

Feature and Model Transformation

The unique characteristics of the voice of a particular speaker are what allow us humans to identify the voice of a person, for example calling on a telephone, as soon as a few syllables have been spoken. These characteristics include fundamental frequency, rate of speaking, accent, and word usage, among others. While lending each voice its own individuality and charm, such characteristics are a hindrance to automatic recognition, inasmuch as they introduce variability in the speech that is of no use in distinguishing between different phonemes and words. To enhance the performance of an automatic recognition system that must function well for many speakers, various transformations are typically applied either to the features used for recognition, the means and covariances of the *hidden Markov model* (HMM) used to evaluate the conditional likelihood of a sequence of features, or to both. The body of techniques used to estimate and apply such transformations to compensate for speaker dependent characteristics fall under the rubrik *speaker adaptation*. In addition, some of the feature and model adaptation techniques are not only capable of compensating for a mismatch in speaker characteristics, but also for a stationary or slowly evolving mismatch in the acoustic environment.

In Section 9.1, we begin our discussion of *feature transformation* techniques. Section 9.1.1 describes the simplest form of feature adaptation, namely, *vocal tract length normalization*, which compensates for the lengths of the vocal tracts of different speakers by applying a linear warping of the frequency axis prior to the extraction of cepstral features. A more powerful, but complementary, feature space transformation, *constrained maximum likelihood linear regression* is discussed in Section 9.1.2, whereby a linear transformation is applied to a sequence of features to maximize their likelihood with respect to the HMM used for recognition. Section 9.2 takes up the discussion of model transformation techniques. Supervised enrollment data implies that the *correct* transcription of the enrollment utterances is available. As explained in Section 9.2.1, *maximum likelihood linear regression* (MLLR) applies a linear transformation to the means of an HMM to maximize a *maximum likelihood* (ML) criterion with respect to some *unsupervised* enrollment data, whereby the errorful transcript or word lattice from a prior recognition pass must be used to collect adaptation statistics. Speaker adaptation based on the *all-pass transform* (APT) is described in Section 9.2.2. Such adaptation can be viewed as a constrained form of MLLR, where the full linear transformation applied to the HMM means is determined by only a handful of free parameters. It is interesting to note that APT adaptation has proven to be just as powerful and more robust than MLLR in many cases of interest. In contrast to model transformation techniques Section 9.3 introduces a technique which separately models speech and noise.

9.1   Feature Transformation Techniques

This section discusses the most common feature space adaptation techniques, whereby a transformation is applied to intermediate or the final features used for recognition. Speaker and/or environment-dependent transformations with several different forms have been proposed in the literature. For example the simplest speaker-dependent transformation involves a simple shift or interpolation of the frequency components in the power spectrum, and is known as *vocal tract length normalization*. Another complementary transformation involves applying a linear scaling to the features in the cepstral domain and is known as *constrained maximum likelihood linear regression*. In addition alternative transformations, which will not be discussed here, have been proposed in the literature to compensate, for example, for the variance in fundamental frequency (Wölfel 2004).

9.1.1   Vocal Tract Length Normalization

Perhaps the simplest and most widely used form of compensation for the characteristics of an individual speaker is *vocal tract length normalization* (VTLN) (Andreou *et al.* 1994; Lee and Rose 1996). As its name implies, VTLN is based on the observation that there are wide variations in the heights of different speakers, and such height differences are also reflected in the length of the speakers' vocal tract. Much like a longer pipe in an organ produces a lower tone than a short pipe, the resonances or *formants* produced by the vocal tract of a taller speaker will generally be lower than those of a shorter speaker, simply because the former will, on average, have a longer vocal tract.

Interspeaker differences are irrelevant for recognition either human or automatic, and tend to degrade the performance of automatic speech recognition systems. VTLN attempts to compensate for such differences by remapping the spectral energy to produce features that appear to have been generated by some "average speaker". As described by Pye and Woodland (1997), this is most often accomplished by adjusting the center frequencies of the mel filter banks discussed in Section 5.2.2. Usually, this entails generating cepstral features with several different *warping parameters*, which control whether the formants are mapped up or down and how much, then choosing that speaker-dependent parameter maximizing the likelihood of the resulting features with respect to a *Gaussian mixture model* (GMM) or HMM. Acero (1990a) and McDonough *et al.* (1998) have also proposed applying the *bilinear transform* (BLT) (5.2.3) as a means of achieving a frequency warping effect. Figure 9.1 compares a linear mapping with a nonlinear mapping provided by the BLT. The BLT has the useful property that such warping can be achieved through a linear transformation of the cepstral coefficients. Depending on the bandwidth of the sampled speech, this may or may not be as effective as applying the frequency warping directly to the power spectrum. Figure 9.1 displays the effect of warping on the frequency under a linear transformation on the left side and under the BLT on the right.

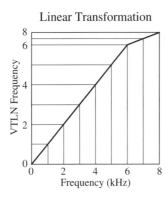
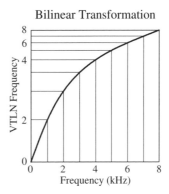

Figure 9.1   Mapping of the vocal tract length with a linear and a bilinear transform

Most investigators have reported that the reductions in word error rate achieved with VTLN are additive with those obtained using the other forms of normalization and speaker adaptation discussed in the balance of this chapter.

9.1.2 Constrained Maximum Likelihood Linear Regression

In this section, we consider another feature transformation based on a simple matrix multiplication. This technique, known as *constrained maximum likelihood linear regression* (CMLLR) was proposed by Gales (1998), whose article actually appeared after the initial work on maximum likelihood linear regression by Leggetter and Woodland (1995a). We discuss the latter technique in Section 9.2.1.

Let us begin by postulating a linear transformation $\overline{\mathbf{A}}$ and additive bias $\overline{\mathbf{b}}$ that must operate simultaneously on both the mean vector and covariance matrices of an HMM. The transformed mean vector and covariance matrix will then be given by

$$\hat{\boldsymbol{\mu}}_m = \overline{\mathbf{A}}\,\boldsymbol{\mu}_m - \overline{\mathbf{b}}, \tag{9.1}$$

$$\hat{\boldsymbol{\Sigma}}_m = \overline{\mathbf{A}}\,\boldsymbol{\Sigma}_m\,\overline{\mathbf{A}}^T, \tag{9.2}$$

respectively, where $\boldsymbol{\mu}_m$ and $\boldsymbol{\Sigma}_m$ are the speaker-independent mean vector and covariance matrix. Upon substituting (9.1–9.2) into (8.38) and suppressing the state index j, we obtain

$$K_2(\mathcal{Y}; \Lambda) = -\frac{1}{2}\sum_{m,k} c_{m,k}\left[\log|2\pi\,\boldsymbol{\Sigma}_m| - \log|\mathbf{A}|^2 + (\hat{\mathbf{y}}_k - \boldsymbol{\mu}_m)^T\boldsymbol{\Sigma}_m^{-1}(\hat{\mathbf{y}}_k - \boldsymbol{\mu}_m)\right], \tag{9.3}$$

where

$$\hat{\mathbf{y}}_k = \overline{\mathbf{A}}^{-1}(\mathbf{y}_k + \overline{\mathbf{b}}).$$

Let us define $\mathbf{A} \triangleq \overline{\mathbf{A}}^{-1}$ and $\mathbf{b} \triangleq \overline{\mathbf{A}}^{-1}\overline{\mathbf{b}}$ and rewrite the last equation as

$$\hat{\mathbf{y}}_k = \mathbf{A}\mathbf{y}_k + \mathbf{b} = \mathbf{W}\boldsymbol{\psi}_k, \tag{9.4}$$

where we have additionally defined the extended transformation matrix and extended feature vector as

$$\mathbf{W} \triangleq \begin{bmatrix} \mathbf{b}^T & \mathbf{A}^T \end{bmatrix}^T, \tag{9.5}$$

$$\boldsymbol{\psi}_k \triangleq \begin{bmatrix} 1 & \mathbf{y}_k \end{bmatrix}^T. \tag{9.6}$$

Gales (1998) demonstrated that the components of $\mathbf{A}$ can be found in an iterative manner, which can be described as follows. Let us define the vector

$$\mathbf{p}^{(i)} \triangleq \begin{bmatrix} 0 & f_{i1} & f_{i2} & \cdots & f_{in} \end{bmatrix}$$

for the cofactors $f_{ij} = \text{cof}(\mathbf{A}_{ij})$. Then the ith row of $\mathbf{W}$ is given by

$$\mathbf{w}^{(i)} = \left(\alpha \mathbf{p}^{(i)} + \mathbf{k}^{(i)} \right) \mathbf{G}^{(i)-1}, \tag{9.7}$$

where

$$\mathbf{G}^{(i)} = \sum_k \frac{1}{\sigma_k^2} \sum_k c_{k,t} \boldsymbol{\psi}_t \boldsymbol{\psi}_k^T \tag{9.8}$$

and

$$\mathbf{k}^{(i)} = \sum_k \frac{1}{\sigma_k^2} \mu_k \sum_k c_{k,t} \boldsymbol{\psi}_k^T. \tag{9.9}$$

Equation (9.3) illustrates one possible advantage of using a constrained model space adaptation, inasmuch as the transformation can actually be implemented as a feature space transformation, which has the potential to greatly reduce computation. During recognition, the log-likelihoods can be evaluated according to

$$\log p(\mathbf{y}_k; \boldsymbol{\mu}, \boldsymbol{\Sigma}, \mathbf{A}, \mathbf{b}) = \log \mathcal{N}(\mathbf{A}\mathbf{y}_k + \mathbf{b}; \boldsymbol{\mu}, \boldsymbol{\Sigma}) + \log |\mathbf{A}|.$$

Moreover, it is not necessary to adapt the model parameters, which could potentially be computationally expensive.

9.2 Model Transformation Techniques

A second class of transformations are applied directly to the final features, e.g., cepstral means and covariances of the HMM used for speech recognition. We will refer to this class of techniques as *model transformation techniques*. We will present two such techniques here, both based on a linear transformation of the cepstral means. The first is the well-known *maximum likelihood linear regression*. The second is based on the *all-pass transform*.

9.2.1 Maximum Likelihood Linear Regression

Now assume the Gaussian means are to be adapted for a particular speaker, denoted by the index s, prior to speech recognition. The adaptation of a single mean is achieved by forming the product $\hat{\mu}_m = \mathbf{A}\mu_m$ for some speaker-dependent transformation matrix $\mathbf{A}$. Very often the transformation is assumed to include an additive shift $\mathbf{b}$ such that

$$\hat{\mu}_m = \mathbf{A}\mu_m + \mathbf{b}. \tag{9.10}$$

We shall account for this case, however, by assuming the shift is represented by the last column of $\mathbf{A}$, and by appending a final component of unity to μ_k. The technique of estimating the components of $\mathbf{A}$ and $\mathbf{b}$ directly based on a ML criterion was originally proposed by Leggetter and Woodland (1995a), who called this technique maximum likelihood linear regression. Representing the additive bias by the final component of μ_k and suppressing the summation over s enables (8.47) to be rewritten as

$$\mathcal{G}(\mathcal{Y}; \{(\mathbf{A}^{(s)}, \mathbf{b}^{(s)})\}, \Lambda) = \frac{1}{2} \sum_{s,m} c_m [(\tilde{\mu}_m^{(s)} - \mathbf{A}^{(s)}\mu_m)^T \Sigma_m^{-1}(\tilde{\mu}_m^{(s)} - \mathbf{A}^{(s)}\mu_m)], \tag{9.11}$$

where

$$\tilde{\mu}_m = \frac{1}{c_m} \sum_k c_{m,k}\, \mathbf{y}_k, \tag{9.12}$$

and

$$c_m = \sum_{m,k} c_{m,k}, \tag{9.13}$$

as before. If the covariance matrix Σ_m is again assumed to be diagonal, then (9.11) can once more be decomposed row by row, and each row can be optimized independently. Let $\mathbf{a}_n^T$ denote the nth row of $\mathbf{A}$, such that

$$\mathbf{A} = \begin{bmatrix} \mathbf{a}_0^T \\ \mathbf{a}_1^T \\ \vdots \\ \mathbf{a}_{L-1}^T \end{bmatrix},$$

where L is the length of the acoustic feature. Then (9.11) can be rewritten as

$$\mathcal{G}(\mathcal{Y}; \{(\mathbf{A}^{(s)}, \mathbf{b}^{(s)})\}, \Lambda) = \sum_{n=0}^{L-1} \mathcal{G}_n(\mathcal{Y}; \mathbf{a}_n, \Lambda),$$

where

$$\mathcal{G}_n(\mathcal{Y}; \mathbf{a}_n, \Lambda) = \frac{1}{2} \sum_m \frac{\hat{c}_m}{\sigma_{m,n}^2} (\tilde{\mu}_m - \mathbf{a}_n^T \mu_m). \tag{9.14}$$

Taking the derivative of (9.14) with respect to $\mathbf{a}_n^T$ yields

$$\frac{\partial \mathcal{G}_n(\mathcal{Y}; \mathbf{a}_n, \Lambda)}{\partial \mathbf{a}_n} = \frac{1}{2} \sum_m \frac{\hat{c}_m}{\sigma_{m,n}^2} \left[-\tilde{\mu}_{m,n} \, \boldsymbol{\mu}_m + \left(\boldsymbol{\mu}_m \boldsymbol{\mu}_m^T \right) \mathbf{a}_n^T \right]. \tag{9.15}$$

Upon equating the right hand side of (9.15) to zero and solving, we find

$$\mathbf{a}_n = \tilde{\mathbf{H}}_n^{-1} \tilde{\mathbf{v}}_n,$$

where

$$\tilde{\mathbf{H}}_n^{-1} = \sum_m \frac{\hat{c}_m}{\sigma_{m,n}^2} \, \boldsymbol{\mu}_m \boldsymbol{\mu}_m^T,$$

$$\tilde{\mathbf{v}}_n = \sum_m \frac{\hat{c}_m}{\sigma_{m,n}^2} \, \tilde{\mu}_{m,n} \, \boldsymbol{\mu}_m,$$

where $\tilde{\mu}_{m,n}$ is the nth component of $\tilde{\boldsymbol{\mu}}_m$.

9.2.2 All-Pass Transform Adaptation

Here we set forth the characteristics of a class of mappings which are designated *all-pass transforms* for reasons which will emerge presently. We also describe how these mappings can be used in transforming cepstral sequences, and in adapting the means of a HMM. This approach was originally proposed by McDonough (2000). As we will learn, the resulting transformation is linear as in the case of MLLR. The transformation matrix specified by the all-pass transform, however, is specified by far fewer parameters than that of MLLR. This sparsity leads to more robust adaptation and, in many cases, better recognition performance.

Sequence Transformation

Consider an arbitrary double-sided, real-valued time sequence $c[n]$ and its z-transform $C(z)$, which are related by the equations

$$C(z) \triangleq \sum_{n=-\infty}^{\infty} c[n] z^n, \tag{9.16}$$

$$c[n] \triangleq \frac{1}{2\pi j} \oint C(z) \, z^{-(n+1)} dz, \tag{9.17}$$

where the contour of integration in (9.17) is assumed to be the unit circle. Note that the definition of the z–transform and its inverse given in (9.16–9.17) differ from the standard definitions presented in Section 3.1.3. The reasons for introducing these non-standard definitions originally arose from the necessity of making certain arguments from the field of complex analysis required to establish several desirable properties of the all-pass transform. As space does not permit us to delve into such mathematical intricacies here, the interested reader is referred to McDonough (2000). Suffice it to say that electrical engineers are accustomed to formulating transform functions that are analytic on the *exterior* of a circle. Mathematicians, on the other hand, would rather manipulate functions of a complex argument that are analytic on the *interior* of a circle. These contradictory tendencies give rise to the nonstandard definitions in (9.16) and (9.17).

For some mapping function Q, assume we wish to form the composition $\hat{C}(z) = C(Q(z))$, which we will also denote as $\hat{C} = C \circ Q$. If Q satisfies suitable analyticity conditions, then $\hat{C} = C \circ Q$ also admits a Laurent series representation,

$$\hat{C}(z) = \sum_{n=-\infty}^{\infty} \hat{c}[n] \, z^n.$$

McDonough (2000), showed that the series coefficients $\hat{c}[n]$ appearing above can be calculated from

$$\hat{c}[n] = \sum_{m=-\infty}^{\infty} c[m] \, q^{(m)}[n], \tag{9.18}$$

where the intermediate sequences $\{q^{(m)}[n]\}$ are specified by

$$q^{(m)}[n] \triangleq \frac{1}{2\pi j} \oint Q^m(z) \, z^{-(n+1)} \, dz. \tag{9.19}$$

Furthermore, the several sequences $\{q^{(m)}[n]\}$ satisfy

$$q^{(m)}[n] = \sum_{k=-\infty}^{\infty} q^{(m-1)}[n] \, q^{(1)}[n - k], \tag{9.20}$$

and $q^{(0)}[n]$ is equivalent to the unit sample sequence

$$q^{(0)}[n] \triangleq \begin{cases} 1, & \text{for } n = 0, \\ 0, & \text{otherwise.} \end{cases} \tag{9.21}$$

From (9.20) and (9.21) it is clear that the sequences $\{q^{(m)}[n]\} \, \forall \, m = 2, 3, \ldots$ can be recursively calculated once $q^{(1)}[n] = q[n]$ is known. Moreover, as (9.20) is clearly equivalent to the convolution (3.8), which we will denote with the shorthand notation

$$q^{(m)} = q^{(m-1)} * q. \tag{9.22}$$

In the complex analysis literature (9.22) is known as a *Cauchy product*.

In the following sections, we show how $q[n]$ can be obtained for both rational and sine-log all-pass transforms.

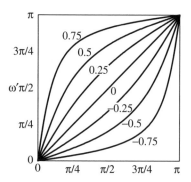

Figure 9.2   Warping of the frequency axis under the bilinear transform

Rational All-Pass Transforms

From Section 5.2.3 it is already known that the *bilinear transform* applies a nonlinear
frequency mapping as plotted for different values of α in Figure 9.2. From the figure it is
apparent that the frequency axis can be warped up or down through suitable settings of
α, much as in traditional VTLN described in Section 9.1.1. From preceding development,
it is clear that $\hat{c}[n]$ can be readily calculated as soon as the coefficients q in the series
expansion of Q are known. Let us rewrite (5.8) as

$$Q(z) = \frac{z - \alpha}{1 - \alpha z}. \tag{9.23}$$

In order to calculate q, we begin with the well-known *geometric series*,

$$\frac{1}{1 - z} = \sum_{n=0}^{\infty} z^n \ \forall \ |z| < 1.$$

Using this series, it is possible to rewrite $Q(z)$ in (9.23) as

$$Q(z) = (z - \alpha) \sum_{n=0}^{\infty} \alpha^n z^n \ \forall \ |z| < \alpha^{-1}.$$

From this equation, the individual coefficients of the series expansion of Q can be determined by inspection as

$$q[n] = \begin{cases} -\alpha, & \text{for } n = 0, \\ \alpha^{n-1}(1 - \alpha^2), & \forall n > 0. \end{cases} \tag{9.24}$$

It is possible to formulate a more general class of mappings that share many of the
desirable characteristics of the bilinear transform. The mappings are dubbed *rational*

all-pass transforms (RAPT)s, and have the functional form

$$Q(z) = \underbrace{\frac{z-\alpha}{1-\alpha z}}_{= A(z;\,\alpha)} \times \underbrace{\frac{z-\beta}{1-\beta^*z}\frac{z-\beta^*}{1-\beta z}}_{B(z;\,\beta)} \times \underbrace{\frac{1-\gamma^*z}{z-\gamma}\frac{1-\gamma z}{z-\gamma^*}}_{G(z;\,\gamma)},$$ (9.25)

where $\alpha, \beta, \gamma \in \mathbb{C}$ satisfy $|\alpha|, |\beta|, |\gamma| < 1$. From (9.23) and (9.25) it is apparent that the latter mapping subsumes the former, and the two are equivalent whenever $\beta = \gamma$. In the sequel, the dependence of $A(z;\alpha)$, $B(z;\beta)$, and $G(z;\gamma)$ on α, β, and γ shall be suppressed whenever it is possible to do so without ambiguity.

The general RAPT in (9.25) has two important characteristics:

1. Q is an all-pass function such that

$$\left|Q(e^{j\omega})\right| = 1 \,\forall\, \omega \in \mathbb{R}.$$ (9.26)

2. The inverse of Q is available from

$$\frac{1}{Q(z)} = Q(z^{-1}).$$ (9.27)

Discrete-time systems having transfer functions that can be represented as a product of terms of the type seen in (9.25) are frequently used for phase compensation of digital filters (Oppenheim and Schafer 1989, sect. 5.5).[1] As implied by (9.26), cascading a phase compensator of this type with an arbitrary linear time-invariant filter does not alter the spectral *magnitude* of the latter. For this reason, such a system is described as *all-pass*; that is, it passes all frequencies without attenuation. In the sequel, we will use the term *all-pass transform* to refer to any conformal map satisfying conditions (9.26–9.27).

The placement of poles and zeros in (9.25) is dictated by the *argument principle* (Churchill and Brown 1990). In particular, we require that the number of zeros within the unit circle exceeds the number of poles by exactly *one*. Moreover, as a consequence of condition (9.26), the effect of any APT can be equated to a nonlinear warping of the frequency axis, just as was previously done with the BLT. Details are provided in (McDonough 2000, sect. 3.2).

Suppose that Q is an RAPT as in (9.25) and that $|\alpha|, |\beta|$, and $|\gamma| < 1$. Then Q admits a Laurent series representation

$$Q(z) = \sum_{n=-\infty}^{\infty} q[n]\, z^n,$$ (9.28)

whose coefficients are given by

$$q = a * b * g,$$ (9.29)

---

[1] In the mathematical literature, such functions are known as Blaschke factors (Greene and Krantz 1997, sect. 9.1)

where $a \leftrightarrow A$, $b \leftrightarrow B$, and $g \leftrightarrow G$. The components of a were given in (9.24). Comparable expressions for the components of b were derived by McDonough (2000). We will only quote the results here:

$$B(z) = \frac{z^2 - 2\rho z \cos\theta + \rho^2}{\sin\theta} \sum_{n=0}^{\infty} \rho^n \sin((n+1)\theta) z^n,$$

where we have set $\beta = \rho e^{j\theta}$. Solving for the individual coefficients results in the expression

$$B(z) = \begin{cases} 0, & \forall n < 0, \\ \rho^2, & \text{for } n = 0, \\ 2\rho(\rho^2 - 1)\cos\theta, & \text{for } n = 1, \\ \frac{1}{\sin\theta}[\rho^{n+2}\sin(\theta(n+1)) - 2\rho^n \cos\theta \sin(\theta n) + \rho^{n-2}\sin(\theta(n-1))], & \forall n \geq 2. \end{cases}$$

The components of g are readily obtained from the relation $G(z) = B(z^{-1})$, as is clear from (9.25) and (9.27).

As we are transforming a cepstral sequence $c[n]$ which is inherently double-sided, it is necessary to calculate $q^{(m)}[n]$ for both positive and *negative* integers m. We can, however, exploit the special structure of the APT in order to relate the components of $q^{(m)}$ to those of $q^{(-m)}$ for $m \geq 1$. Note that

$$Q^{-m}(z) = \left[\frac{1}{Q(z)}\right]^m = [Q(z^{-1})]^m,$$

where the final equality follows from (9.27). Hence,

$$Q^{-m}(z) = Q^m(z^{-1}),$$

which implies

$$q^{(-m)}[n] = q^{(m)}[-n]. \tag{9.30}$$

The importance of (9.30) is that only the set of sequences $\{q^{(m)}[n]\} \forall m \geq 0$ need be calculated directly, and (9.22) provides the means to accomplish this once $q^{(1)}[n] = q[n]$ is known; the latter is available from (9.29).

Cepstral Sequence Transformation

We will now specialize the development above for the unique characteristics of cepstral coefficients. Hence, let us define $\hat{X}(z) = \log \hat{H}(z)$ and $X(z) = \log H(z)$ for some linear time-invariant system function $H(z)$, so that $\hat{X} = X \circ Q$. If $c[n]$ is the real cepstrum corresponding to some windowed segment of speech, then $c[n]$ must be even. As in

(5.45), define $x[n]$ as the *minimum phase* equivalent of $c[n]$, such that

$$x[n] = \begin{cases} 0, & \text{for } n < 0, \\ c[0], & \text{for } n = 0, \\ 2c[n], & \text{for } n > 0, \end{cases}$$

and

$$c[n] = \begin{cases} \frac{1}{2}x[-n], & \text{for } n < 0, \\ x[0], & \text{for } n = 0, \\ \frac{1}{2}x[n], & \text{for } n > 0. \end{cases} \tag{9.31}$$

Exploiting the fact that $c[n]$ is even, we can rewrite (9.18) as

$$\hat{c}[n] = q^{(0)}[n]\, c[0] + \sum_{m=1}^{\infty} \left(q^{(m)}[n] + q^{(-m)}[n] \right) c[m]$$

$$= q^{(0)}[n]\, c[0] + \sum_{m=1}^{\infty} \left(q^{(m)}[n] + q^{(m)}[-n] \right) c[m], \tag{9.32}$$

where the latter equality follows from (9.30). Substituting (9.31) into (9.32) then provides

$$\hat{c}[n] = q^{(0)}[n]\, x[0] + \frac{1}{2} \sum_{m=1}^{\infty} \left(q^{(m)}[n] + q^{(-m)}[n] \right) x[m]. \tag{9.33}$$

Now let us define $\hat{x}[n]$ as the causal portion of $\hat{c}[n]$, so that

$$\hat{x}[n] = \begin{cases} 0, & \text{for } n < 0, \\ \hat{c}[0], & \text{for } n = 0, \\ 2\hat{c}[n], & \text{for } n > 0. \end{cases} \tag{9.34}$$

Substituting (9.33) into (9.34) provides

$$\hat{x}[n] = \begin{cases} \sum_{m=0}^{\infty} q^{(m)}[0]\, x[m], & \text{for } n = 0, \\ \sum_{m=1}^{\infty} \left(q^{(m)}[n] + q^{(m)}[-n] \right) x[m], & \forall n = 0. \end{cases} \tag{9.35}$$

These relations can be stated more succinctly by defining the *transformation matrix* $\mathbf{A} = \{a_{nm}\}$ where

$$a_{nm} = \begin{cases} q^{(m)}[0], & \text{for } n = 0, m \geq 0, \\ 0, & \text{for } n > 0, m = 0, \\ \left(q^{(m)}[n] + q^{(m)}[-n] \right), & \text{for } n, m > 0. \end{cases} \tag{9.36}$$

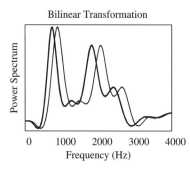
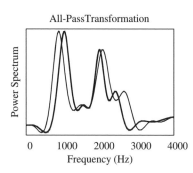

Figure 9.3 Original (thin line) and transformed (thick line) short-term spectra for a male test speaker regenerated from cepstral coefficients 0–14. The transformed spectrum was produced with either the BLT or the all-pass transformation by setting $\alpha = 0.10$

Hence, it is possible to obtain $\hat{x}[n]$ from

$$\hat{x}[n] = \sum_{m=0}^{\infty} a_{nm} \, x[m].$$ (9.37)

From (9.37) it is clear that the composition $\hat{X} = X \circ Q$ reduces to a linear transformation in cepstral space. It is worth noting that this is a consequence of the fact that Q

(1) is analytic on an annular region that includes the unit circle, and
(2) preserves the unit circle,

where the latter characteristic implies that the unit circle is mapped back onto the unit circle by Q.

The left side of Figure 9.3 shows the original and transformed spectra for a windowed segment of male speech sampled at 8 kHz. Both spectra were generated from the first 15 components of the original cepstral sequence. The operations employed in calculating the transformed cepstra $\hat{x}[n]$ were those set forth above. The conformal map used in this case was a BLT with $\alpha = 0.10$. As implied by (9.37), some of the information contained in $x[n]$ for $n = 0, 1, \ldots, N - 1$ is "encoded" in $\hat{x}[n]$ for all $n \geq N$; thus, 25 rather than 15 transformed cepstral coefficients were retained in generating the composite spectrum plotted in the figure. It is clear from a comparison of the respective spectra that all formants have been shifted downward by the transformation, and that the extent of the shift is frequency dependent. Qualitatively, this is just what we should expect based on the curves plotted on the right side of Figure 9.1.

Shown on the right side of Figure 9.3 are the original and all-pass transformed spectra for the same segment of male speech. As in the prior case, these plots were generated from the first 15 components of the original cepstral sequence, but 25 components were retained in the transformed sequence. The latter was obtained in the manner suggested by the development above. The conformal map used in this instance was an RAPT with the general form in (9.25). From the figure it is apparent that whereas the higher formants have been shifted *down*, the lower formants have been shifted *up*. This stands in sharp contrast to the effect produced by the BLT, for which the shift depends on frequency but is always in the same direction, and serves to illustrate the greater generality of the APT.

Sine-Log All-Pass Transforms

In the final portion of this section, we consider a different type of all-pass transform that shares many characteristics of the RAPT. Its chief advantage over the RAPT is its simplicity of form and amenability to numerical computation. Regrettably, this simplicity is not immediately apparent from the abbreviated presentation given here. The interested reader is referred to McDonough (2000) for further details.

Let us begin by defining the *sine-log all-pass transform* as

$$Q(z) \triangleq z \exp F(z), \qquad (9.38)$$

where

$$F(z) \triangleq \sum_{k=1}^{K} \alpha_k \, F_k(z) \text{ for } \alpha_1, \ldots, \alpha_K \in \mathbb{R}, \qquad (9.39)$$

$$F_k(z) \triangleq j \, \pi \, \sin\left(\frac{k}{j} \log z\right), \qquad (9.40)$$

and K is the number of free parameters in the transform. The designation "sine-log" is due to the functional form of $F_k(z)$. It is worth noting that $F_k(z)$ is single-valued even though $\log z$ is multiple-valued (Churchill and Brown 1990, sect. 26). Moreover, applying the well-known relation,

$$\sin z = \frac{1}{2j} \left(e^{jz} - e^{-jz}\right),$$

to (9.40) provides

$$F_k(z) = \frac{\pi}{2} \left(z^k - z^{-k}\right), \qquad (9.41)$$

which is a more tractable form for computation. It can be readily verified that Q as defined (9.38) satisfies (9.26–9.27), just as the RAPTs considered earlier. Moreover, as z traverses the unit circle, $Q(z)$ also winds exactly *once* about the origin, which is necessary to ensure that spectral content is not doubled or tripled (McDonough 2000, sect. 3.5).

In order to calculate the coefficients of a transformed cepstral sequence in the manner described above, it is first necessary to calculate the coefficients q in the Laurent series expansion of Q; this can be done as follows: for F as in (9.39) set

$$G(z) = \exp F(z). \qquad (9.42)$$

Let g denote the coefficients of the Laurent series expansion of G valid in an annular region including the unit circle. Then,

$$g[n] = \frac{1}{2\pi j} \oint G(z) \, z^{-(n+1)} \, dz. \tag{9.43}$$

Moreover, the natural exponential admits the series expansion

$$e^z = \sum_{m=0}^{\infty} \frac{z^m}{m!},$$

so that

$$G(z) = \sum_{m=0}^{\infty} \frac{F^m(z)}{m!}. \tag{9.44}$$

Substituting (9.44) into (9.43) provides

$$g[n] = \frac{1}{2\pi j} \oint \sum_{m=0}^{\infty} \frac{F^m(z)}{m!} \, z^{-(n+1)} \, dz$$

$$= \sum_{m=0}^{\infty} \frac{1}{m!} \frac{1}{2\pi j} \oint F^m(z) \, z^{-(n+1)} \, dz \tag{9.45}$$

The sequence f of coefficients in the series expansion of F are available by inspection from (9.39) and (9.41). Letting $f^{(m)}$ denote the coefficients in the series expansion of F^m, we have

$$f^{(m)}[n] = \frac{1}{2\pi j} \oint F^m(z) \, z^{-(n+1)} \, dz,$$

and upon substituting this into (9.45) we find

$$g[n] = \sum_{m=0}^{\infty} \frac{1}{m!} f^{(m)}[n].$$

Moreover, from the Cauchy product it follows

$$f^{(m)} = f * f^{(m-1)} \, \forall \, m = 1, 2, 3, \ldots .$$

Equations (9.38) and (9.42) imply that $Q(z) = z \, G(z)$, so the desired coefficients are given by

$$q[n] = g[n-1] \, \forall \, n = 0, \pm 1, \pm 2, \ldots .$$

Parameter Estimation

Let us assume that the parameters specifying a conformal map Q are to be chosen in order to maximize the likelihood of a set of training data. The likelihood will be calculated with respect to one or more GMMs. We will assume that a GMM is associated with each state of a HMM. We will also assume that the covariance matrix $\mathbf{\Sigma}_m$ is diagonal, as in Chapter 8, such that

$$\mathbf{\Sigma}_m = \mathrm{diag}\left\{\sigma_{m0}^2, \sigma_{m1}^2, \ldots, \sigma_{m,L-1}^2\right\},$$

where L is the (original) feature length.

The adaptation of a single mean is achieved by forming the product $\hat{\boldsymbol{\mu}}_m = \mathbf{A}^{(s)}\boldsymbol{\mu}_m$ for some speaker-dependent transformation matrix $\mathbf{A}^{(s)} = \mathbf{A}(\alpha)$. More precisely,

$$\mu_{kn} = \sum_{m=0}^{L-1} a_{nm}\,\mu_{km} \ \forall\, n = 0, 1, \ldots, L'-1, \tag{9.46}$$

where the components $\{a_{nm}\}$ of the transformation matrix are given by (9.36). Hence, the likelihood $p(\mathbf{y}; \alpha, \Lambda)$ of a cepstral feature $\mathbf{y}$ is given by

$$p(\mathbf{y}; \alpha, \Lambda) = \sum_{m=1}^{M} w_m \, \mathcal{N}(\mathbf{y}; \mathbf{A}^{(s)}\boldsymbol{\mu}_m, \mathbf{\Sigma}_m).$$

Parameter optimization is most easily accomplished through recourse to the *expectation-maximization* (EM) algorithm. As discussed in Section 8.1, the EM algorithm requires the formulation of an auxiliary function (Dempster *et al.* 1977), which is equivalent to the expected value of the log-likelihood of some set of training data given the current estimate of the model's parameters. Hence, define a set $\mathcal{Y} = \{\mathbf{y}_k\}$ of training data contributed by a single speaker. Ignoring the dependence on the HMM states, the log-likelihood of this set can be expressed as

$$\log p(\mathcal{Y}^{(s)}; \alpha, \Lambda) = \sum_k \log p(\mathbf{y}_k; \alpha, \Lambda)$$

$$= \sum_k \log\left[\sum_m w_m\, p(\mathbf{y}_k; \alpha, \Lambda_m)\right]$$

In McDonough (2000), the relevant auxiliary function is shown to be

$$\mathcal{G}(\mathcal{Y}^{(s)}; \alpha, \Lambda) = \frac{1}{2}\sum_{m,n} \frac{c_m^{(s)}}{\sigma_{m,n}^2}\,(\tilde{\mu}_{mn} - \hat{\mu}_{mn})^2, \tag{9.47}$$

where $\tilde{\mu}_{mn}$ is the nth component of $\tilde{\boldsymbol{\mu}}_m$ in (9.12). It is this objective function that is to be minimized in the second step of the EM algorithm. As given above, $\mathcal{G}(\alpha) = \mathcal{G}(\mathcal{X}^{(s)}; \alpha, \Lambda)$

represents a continuous and continuously differentiable function, and thus is amenable to optimization by any of a number of numerical methods (Gill *et al.* 1981; Luenberger 1984). In order to apply such a method, valid expressions for the gradient and (possibly) Hessian of $\mathcal{G}(\alpha)$ must be available. For reasons of brevity, the derivation of such expressions is not included here. The interested reader should see McDonough (2000, sect. 5.2).

Inclusion of an Additive Bias

As mentioned in Section 9.2.1, very often a cepstral mean transformation of the form $\hat{\mu}_k = \mathbf{A}\mu_k$ is augmented with an additive bias to model the effect of a channel or any other filtering to which the original speech signal may be subject. This bias is easily incorporated into our prior analysis. Let us define $\check{\mu}_m$ as

$$\check{\mu}_m = \hat{\mu}_m + \mathbf{b}, \tag{9.48}$$

where $\mathbf{b}$ is a bias vector whose components are to be estimated along with the other transformation parameters α. Replacing $\hat{\mu}$ with $\check{\mu}$ in (9.47) provides

$$\mathcal{G}(\mathcal{X}^{(s)}; \mathbf{A}^{(s)}, \Lambda) = \frac{1}{2} \sum_{m,n} \frac{c_m^{(s)}}{\sigma_{mn}^2} (\tilde{\mu}_{mn} - \check{\mu}_{mn})^2, \tag{9.49}$$

For any given α, it is straightforward to solve for the optimal $\mathbf{b}$ by taking partial derivatives with respect to the components b_n on both sides of (9.49) and equating to zero:

$$\frac{\partial \mathcal{G}}{\partial b_n} = -\sum_m \frac{c_m}{\sigma_{m,n}^2} \left[\tilde{\mu}_{mn} - (\hat{\mu}_{mn} + b_n) \right] = 0.$$

A trivial rearrangement of the last equation is sufficient to demonstrate that the optimal bias components for a specified α are given by

$$b_n(\alpha) = \frac{\displaystyle\sum_m \frac{c_m}{\sigma_{mn}^2} (\tilde{\mu}_{mn} - \hat{\mu}_{mn})}{\displaystyle\sum_m \frac{c_m}{\sigma_{mn}^2}}. \tag{9.50}$$

9.3 Acoustic Model Combination

In contrast to the adaptation techniques described previously, which modify the means and variances of the acoustic model, the HMM decomposition approach proposed by Varga (1990) uses separate models for speech and noise and searches the combined state space with an extended Viterbi algorithm during decoding. Further work in this direction was undertaken by Gales and Young (1992, 1993) under the name *parallel model combination*, which approximates the distribution of noisy speech by a combination of the clean speech and noise distribution. This latter approach is illustrated pictorially in Figure 9.4. As x

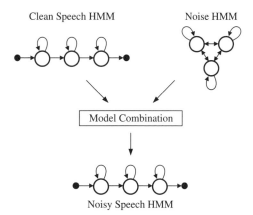

Figure 9.4 Schematics of parallel model combination. In combining the models, the domain of each may need to be altered as in the original approach by Gales and Young (1992)

and n are assumed to be independent we can obtain the noisy speech mean and covariance matrix in the linear domain as

$$\boldsymbol{\mu}_y = \boldsymbol{\mu}_x + \boldsymbol{\mu}_n$$

and

$$\boldsymbol{\Sigma}_y = \boldsymbol{\Sigma}_x + \boldsymbol{\Sigma}_n.$$

If the distribution of speech consists of a GMM with M Gaussians and the noise GMM consists of N Gaussians, the parallel model, representing the noisy speech, contains $M \cdot N$ Gaussians, which potentially represents a significant increase. It is easily observed that the drawback of this method is its computationally expense for a complex noise HMM.

9.3.1 Combination of Gaussians in the Logarithmic Domain

Figure 9.5 plots the effect of adding noise to speech in the logarithmic spectral domain. In this case, both noise and speech are modeled by Gaussian distributions. The speech Gaussian has a mean of 10 and a variance of 6. The noise mean value is varied as indicated in the figure, while the variance is maintained at 1. Monte Carlo simulation of the combined distributions reveals that the combined distribution does not follow a Gaussian curve and that the distribution becomes increasingly bimodal as the noise level increases (Moreno *et al.* 1995). When the noise comes to dominate the speech, however, the combined distribution becomes unimodal once more.

Upon approximating the real distribution by a Gaussian in the logarithmic domain, each component d of the mean vector and each diagonal component d, d of the covariance matrix can be calculated in closed form according to Gales (1995):

$$\mu^{(l)}[d] = \log(\mu^{(f)}[d]) + \frac{1}{2} \log \left(\frac{\Sigma^{(f)}[d, d]}{(\mu^{(f)}[d])^2} + 1 \right),$$

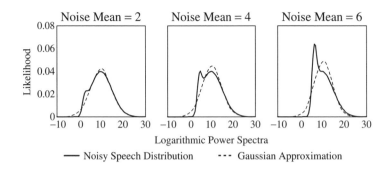

Figure 9.5 Effect of noise modeled by a Gaussian distribution on speech also modeled by a Gaussian distribution in the logarithmic spectral domain

and

$$\Sigma^{(l)}[i, j] = \log \left(\frac{\Sigma^{(f)}[i, j]}{\mu^{(f)}[i]\mu^{(f)}[j]} + 1 \right).$$

9.4   Summary and Further Reading

In this chapter, we have discussed two forms of feature adaptation, namely, VTLN and CMLLR. We have also presented two forms of speaker adaptation based on the transformation of the cepstral means of a HMM, namely, MLLR and APT adaptation. With the exception of VTLN, all forms of adaptation were based on linear transformations. The effectiveness of VTLN, CMLLR, and MLLR is demonstrated through the series of distant speech recognition experiments to be described in Chapter 14. The last section has presented a model transformation technique which keeps separate models for speech and noise.

Masry *et al.* (1968) considered the possibility of representing a continuous-time signal as a discrete-time sequence. Their approach to this problem was posed in terms of defining a basis of orthonormal functions that is complete for signals with particular smoothness properties. Oppenheim and Johnson (1972) took Masry *et al.* (1968) as their starting point in deriving a class of transformations that preserve convolution. They found that one of the principal requirements for such a class is that it have the form of the composition of two functions. Oppenheim and Johnson (1972) also developed the mathematical basis for using the BLT to transform discrete-time sequences, and showed this transformation could be accomplished via a cascade of first-order difference equations.

Zue (1971) used the technique of Oppenheim and Johnson (1972) to restore the speech of divers breathing helium-rich gas mixtures. Shikano (1986, sect. 7) noticed the similarity of the BLT to the mel scale and used it to apply a speaker-independent warp to the short-time spectrum of speech prior to recognition. Acero (1990, sect. 7) first proposed using a speaker-dependent BLT to correct for inter-speaker differences in formant frequency locations; in that work, the optimal BLT parameter for each speaker was estimated by minimizing a vector quantization distortion measure.

Maximum *a posteriori* adaptation, as implied by its name, applies a transformation – most often a simple shift – to the means of an HMM to maximize the *a posteriori* probability of the transformation, rather than the likelihood, based on some *supervised* enrollment data (Lee and Gauvain 1993).

After lying dormant for several years, the use of VTLN to enhance the performance of large vocabulary conversational speech recognition systems was reintroduced by Andreou *et al.* (1994). Their technique had no recourse to the BLT. Instead, a speech waveform was sampled at various rates to induce a linear scaling on the frequency axis of the short-time Fourier transform; the final sampling rate for a particular speaker was chosen to minimize the number of errors made by an HMM-based large vocabulary conversational speech recognition system. The publication of Andreou *et al.* (1994) sparked a flurry of activity: Eide and Gish (1996) proposed a nonlinear warping of the short-time frequency axis implemented in the spectral domain; the choice of warp factor was based on explicit estimates of speaker-dependent formant frequencies. Wegmann *et al.* (1996) and Lee and Rose (1996) independently proposed the use of a GMM to obtain ML estimates of the optimal warping parameters. Pye and Woodland (1997) investigated the use of VTLN together with MLLR adaptation; their findings indicated that the reductions in word error rate achieved by VTLN and MLLR when used in isolation were largely additive when these techniques were combined.

Digalakis *et al.* (1995) introduced transformed-based adaptation of Gaussian mixtures. In this scheme, the mth Gaussian mean vector and covariance matrix were transformed as in (9.1–9.2), where the transform matrix $\mathbf{A}^{(s)}$ was assumed to be diagonal. Hence, the transformation applied to the covariance matrix was completely determined by that applied to the mean; for this reason, the approach of Digalakis *et al.* (1995) came to be known as a *constrained adaptation* of Gaussian mixtures.

Leggetter and Woodland (1995b) proposed the highly successful MLLR adaptation. Their technique was similar to that of Digalakis *et al.* (1995) in that the means of a speaker-independent HMM were transformed as in (9.10), but differed in that $\mathbf{A}^{(s)}$ was taken as a full, instead of diagonal, matrix. In this initial work, only the Gaussian mean was transformed; a covariance transform was subsequently added by Gales and Woodland (1996). In the latter work, the transform applied to the covariance matrix was not explicitly tied to that applied to the mean; hence, this was the first instance of what came to be known as an *unconstrained adaptation*.

The adaptation techniques mentioned above all transform a conventionally-trained speaker-independent model as described in Section 8.1.2. Anastasakos *et al.* (1996) first considered the possibility of training a speaker-independent HMM specifically for use with speaker adaptation. In their technique, transform parameters are first estimated for all speakers in a training set. Then the Gaussian means and variances of a speaker-independent HMM are iteratively re-estimated using the transform parameters of the training set speakers along with the usual forward–backward statistics, as described in Section 8.1.3.

An excellent review of the aforementioned transformation-based approaches to speaker adaptation, along with the requirements of each in terms of computation and memory, is given by Gales (1998). Another valuable reference is Sankar and Lee (1996), who formulate a unified basis for ML speaker normalization and adaptation.

More recently there has been a growing interest in performing speaker adaptation with very limited amounts of enrollment data; e.g., 30 s or less. The results of some preliminary investigations in this area have been reported by Digalakis *et al.* (1996), Kannan and

Khudanpur (1999), and by Bocchieri *et al.* (1999). A distinctly different approach to the problem of rapid adaptation is formulated by Byrne *et al.* (2000); it involves the use of a *discounted likelihood* criterion to achieve robust parameter estimation. Another popular and effective approach to very rapid adaptation, dubbed *eigenvoices*, was developed by Kuhn *et al.* (2000). The theory of speaker adaptation with APTs was extended beyond the simple BLT by McDonough (2000), who noted that it is essential to combine APT adaptation with speaker-adapted training, as discussed in Section 8.1.3, due to the highly constrained nature of the transform. Comparisons between the performance of MLLR and APT adaptation can be found in McDonough and Waibel (2004).

9.5 Principal Symbols

Symbol	Description
μ	mean vector
μ_m	mth mean vector
Σ	covariance matrix
Σ_m	covariance matrix of mth component
A	transformation matrix for MLLR and APT adaptation
b	additive shift vector for MLLR and APT adaptation
b	additive shift vector for CMLLR
$c[n]$	sequence of cepstral coefficients
$\hat{c}[n]$	transformed sequence of cepstral coefficients
$C(z)$	z-transform of cepstral coefficients
$\hat{C}(z)$	z-transform of transformed cepstral coefficients
$q^{(n)}[n]$	components of the APT transformation matrix
$Q(z)$	conformal mapping
W	transformation matrix for CMLLR
y	cepstral feature
$A(z; \alpha)$, $B(z; \beta)$, $G(z; \gamma)$	components specifying a RAPT conformal map

10

Speaker Localization and Tracking

While a recognition engine is needed to convert waveforms into word hypotheses, the speech recognizer by itself is not the only component of a *distant speech recognition* (DSR) system. In this chapter, we introduce the first supporting technology required for a complete DSR system, namely, algorithms for determining the physical positions of one or more speakers in a room, and tracking changes in these positions with time if the speakers are moving. Speaker localization and tracking – whether based on acoustic features, video features, or both – are important technologies, because the beamforming algorithms discussed in Chapter 13 all assume that the position of the desired speaker is *known*. Moreover, the accuracy of a speaker tracking system has a very significant influence on the recognition accuracy of the entire system. This can be easily observed from the experiments conducted in Section 14.7

For present purposes, we will distinguish between speaker localization and speaker tracking as follows. We will say speaker localization is based on an "instantaneous" estimate of the speaker's positions, which implies that the data used to form the estimate lies within a time window of 15 to 25 ms. Speaking tracking systems, on the other hand, may use single observation windows on the order of 15 to 25 ms, but may combine multiple observations in order to track a speaker's trajectory through time.

The balance of this chapter is organized as follows. In Section 10.1, we review the process of source localization based on the conventional techniques, namely, spherical intersection, spherical interpolation, and linear intersection. All those techniques use time delays of arrival between microphone pairs as features for speaker localization. After this presentation of the classical techniques, we begin our exposition of techniques based on the family of Bayesian filters described in Chapter 4. In Section 10.2, we formulate source localization as a problem in nonlinear least squares estimation, then develop an appropriate linearized model. Thereafter we present a simple model for speaker motion, and discuss how such a model can be incorporated into the variant of the *Kalman filter* (KF) described in Section 4.3.3 to create an acoustic localization system capable of tracking a moving speaker. Section 10.2.1 presents a numerically stable implementation of the KF algorithm based on the Cholesky decomposition, which can be used for speaker tracking. In Section 10.3, we consider how the tracking system can be extended to track multiple simultaneously active speakers. This extension is based on the joint probabilistic

Distant Speech Recognition   Matthias Wölfel and John McDonough
© 2009 John Wiley & Sons, Ltd

data association filter described in Section 4.3.6. Section 10.4 describes how the audio data streams used for tracking in prior sections can be augmented with video information from calibrated cameras. The inclusion of such video information conduces to robuster tracking, particularly during intervals when the speaker is silent. Section 10.5 describes how the second class of Bayesian filters, namely particle filters, can also be used for speaker tracking. Section 10.6 summarizes the chapter provides some recommendations for further reading.

The three conventional techniques discussed in Section 10.1 should be well known to anyone well familiar with the source localization literature. Although it has appeared much more recently in the literature, the use of *particle filters* (PFs) for acoustic source tracking, as discussed in Section 10.5, has attracted a great deal of attention within the research community, and hence is likely to be known to anyone familiar with the field. The material presented in Section 10.2, on the other hand, all of which is based on the use of variants of the KF for various tracking scenarios, has appeared in the literature only with the last few years, and is hence of potential interest even to readers well-versed with other algorithms for speaker tracking.

10.1 Conventional Techniques

We begin here our discussion of speaker tracking with a presentation of the conventional techniques that have been in the literature for 10 years or more. The material in Sections 10.1.1 and 10.1.2 is based partially on the treatment of these subjects by Huang *et al.* (2004).

Consider then a sensor array consisting of $N + 1$ microphones located at positions $\mathbf{m}_i \, \forall i = 0, \ldots, N$, and let $\mathbf{x} \in \mathbb{R}^3$ denote the position of the speaker in a three-dimensional space, such that,

$$\mathbf{x} \triangleq \begin{bmatrix} x \\ y \\ z \end{bmatrix}, \quad \text{and} \quad \mathbf{m}_n \triangleq \begin{bmatrix} m_{n,x} \\ m_{n,y} \\ m_{n,z} \end{bmatrix}.$$

Then the *time delay of arrival* (TDOA) between the microphones at positions $\mathbf{m}_1$ and $\mathbf{m}_2$ can be expressed as

$$T(\mathbf{m}_1, \mathbf{m}_2, \mathbf{x}) = \frac{|\mathbf{x} - \mathbf{m}_1| - |\mathbf{x} - \mathbf{m}_2|}{c} \tag{10.1}$$

where c is the speed of sound, which, as mentioned in Section 2.1.2, is approximately 344 m/s at sea level. Equation (10.1) can be rewritten as

$$T_{mn}(\mathbf{x}) = T(\mathbf{m}_m, \mathbf{m}_n, \mathbf{x}) = \frac{1}{c}(D_m - D_n), \tag{10.2}$$

where

$$D_n = \sqrt{(x - m_{n,x})^2 + (y - m_{n,y})^2 + (z - m_{n,z})^2} = |\mathbf{x} - \mathbf{m}_n| \, \forall n = 0, \ldots, N \tag{10.3}$$

is the distance from the speaker to the microphone located at $\mathbf{m}_n$.

Let $\hat{\tau}_{mn}$ denote the observed TDOA for the mth and nth microphones. The TDOAs can be observed or estimated with a variety of well-known techniques. Perhaps the most popular method involves the *phase transform* (PHAT), a variant of the *generalized cross-correlation* (GCC), which can be expressed as (Carter 1981)

$$\rho_{mn}(\tau) \triangleq \frac{1}{2\pi} \int_{-\pi}^{\pi} \frac{Y_m(e^{j\omega\tau})Y_n^*(e^{j\omega\tau})}{\left|Y_m(e^{j\omega\tau})Y_n^*(e^{j\omega\tau})\right|} e^{j\omega\tau} \, d\omega. \tag{10.4}$$

where $Y_n(e^{j\omega\tau})$ denotes the short-time Fourier transform of the signal arriving at the nth sensor in the array (Omologo and Svaizer 1994). The definition of the GCC in (10.4) follows directly from the frequency domain calculation (3.29) of the cross-correlation (3.27) of two sequences. The normalization term $\left|Y_m(e^{j\omega\tau})Y_n^*(e^{j\omega\tau})\right|$ in the denominator of the integrand in (10.4) is intended to weight all frequencies equally. It has been shown that such a weighting leads to more robust TDOA estimates in noisy and reverberant environments (DiBiase *et al.* 2001). Once $\rho_{mn}(\tau)$ has been calculated, the TDOA estimate is obtained from

$$\hat{\tau}_{mn} = \max_{\tau} \; \rho_{mn}(\tau). \tag{10.5}$$

In other words, the "true" TDOA is taken as that which maximizes the PHAT $\rho_{mn}(\tau)$. Thereafter, an interpolation is performed to overcome the granularity in the estimate corresponding to the sampling interval (Omologo and Svaizer 1994). Usually, $Y_n(e^{j\omega_k})$ appearing in (10.4) are calculated with a Hamming analysis window of 15 to 25 ms in duration (DiBiase *et al.* 2001).

There are several other popular methods for calculating TDOAs, including the *adaptive eigenvalue decomposition* algorithm proposed by Benesty (2000), as well as the TDOA estimator based on mutual information proposed by Talantzis *et al.* (2005). Chen *et al.* (2003b) have also proposed a technique for TDOA estimation that exploits the redundancies in the signals collected by several microphones. A good review of recent advances in the field of TDOA estimation is given by Chen *et al.* (2004b).

10.1.1 Spherical Intersection Estimator

The *spherical intersection* (SX) method proposed by Schau and Robinson (1987) provides a closed-form estimate of the speaker's position obtained at each time instant. To describe the technique, we begin by assuming that the zeroth microphone is located at the origin of the coordinate system, such that $\mathbf{m}_0 = \begin{bmatrix} 0 & 0 & 0 \end{bmatrix}^T$. Let us denote the distances from the origin to the nth microphone and from origin to source as

$$R_n \triangleq |\mathbf{m}_n| = \sqrt{m_{n,x}^2 + m_{n,y}^2 + m_{n,z}^2}, \tag{10.6}$$

$$R_s \triangleq |\mathbf{x}| = \sqrt{x^2 + y^2 + z^2}, \tag{10.7}$$

respectively. The *range difference* is defined as

$$d_{mn} \triangleq D_m - D_n \; \forall \, m, n = 0, \ldots, N \tag{10.8}$$

for D_n as in (10.3).

It follows that, in the absence of estimation errors, the range difference can be determined from

$$d_{mn} = c \cdot T_{mn}(\mathbf{x}),$$

where $T_{mn}(\mathbf{x})$ is defined in (10.2). From (10.8) and the fact that $R_s = D_0$, it follows

$$\hat{D}_n = R_s + d_{n0}, \tag{10.9}$$

where $\hat{D}_n$ denotes an observation based on the *measured* range difference d_{n0}. That D_n *predicted* by the spherical signal model in the absence of observation and tracking errors is

$$D_n^2 = |\mathbf{m}_n - \mathbf{x}|^2 = R_n^2 - 2\mathbf{m}_n^T \mathbf{x} + R_s^2. \tag{10.10}$$

The spherical error function for the nth microphone is then the difference between hypothesized and measured values

$$e_{sp,n}(\mathbf{x}) \triangleq \frac{1}{2}\left(\hat{D}_n^2 - D_n^2\right)$$

$$= \mathbf{m}_n^T \mathbf{x} + d_{n0}R_s - \frac{1}{2}(R_n^2 - d_{n0}^2) \,\forall\, n = 1, \ldots, N. \tag{10.11}$$

The total error from all microphone pairs can be expressed in vector form as

$$\mathbf{e}_{sp}(\mathbf{x}) = \mathbf{A}\boldsymbol{\theta} - \mathbf{b} \tag{10.12}$$

where

$$\mathbf{A} \triangleq [\mathbf{S}|\mathbf{d}], \mathbf{S} \triangleq \begin{bmatrix} m_{1,x} & m_{1,x} & m_{1,z} \\ m_{2,x} & m_{2,y} & m_{2,z} \\ \vdots & \vdots & \vdots \\ m_{N,x} & m_{N,y} & m_{N,z} \end{bmatrix}, \mathbf{d} \triangleq \begin{bmatrix} d_{10} \\ d_{20} \\ \vdots \\ d_{N0} \end{bmatrix}, \boldsymbol{\theta} \triangleq \begin{bmatrix} x \\ y \\ z \\ R_s \end{bmatrix}, \mathbf{b} \triangleq \frac{1}{2} \begin{bmatrix} R_1^2 - d_{10}^2 \\ R_2^2 - d_{20}^2 \\ \vdots \\ R_N^2 - d_{N0}^2 \end{bmatrix}.$$

The least squares criterion is then

$$J_{sp} = \mathbf{e}_{sp}^T \mathbf{e}_{sp} = (\mathbf{A}\boldsymbol{\theta} - \mathbf{b})^T (\mathbf{A}\boldsymbol{\theta} - \mathbf{b}). \tag{10.13}$$

The SX algorithm functions in two steps (Schau and Robinson 1987). Firstly, the least-squares solution for $\mathbf{x}$ in terms of R_s is found according to

$$\mathbf{x} = \check{\mathbf{S}}(\mathbf{b} - R_s\mathbf{d}), \tag{10.14}$$

where

$$\check{\mathbf{S}} \triangleq (\mathbf{S}^T\mathbf{S})^{-1}\mathbf{S}^T$$

is the pseudo-inverse of $\mathbf{S}$. Thereafter, substituting (10.14) into the constraint $R_s^2 = \mathbf{x}_s^T \mathbf{x}_s$ provides a quadratic equation

$$R_s^2 = \left[\check{\mathbf{S}} \left(\mathbf{b} - R_s \mathbf{d} \right) \right]^T \left[\check{\mathbf{S}} \left(\mathbf{b} - R_s \mathbf{d} \right) \right],$$

The last equation can be expressed as

$$a\, R_s^2 + b R_s + c = 0, \tag{10.15}$$

where

$$a = 1 - |\check{\mathbf{S}}|^2, \qquad b = 2\mathbf{b}^T \check{\mathbf{S}}^T \check{\mathbf{S}} \mathbf{d}, \qquad c = -|\check{\mathbf{S}} \mathbf{b}|^2.$$

The unique, real, positive root of (10.14) is then taken as the SX estimate of the source range. Note that this implies that the SX algorithm will fail to provide an estimate of the speaker's position if either

- there is *no* real, positive root, or,
- if there are *two* real, positive roots.

This is one of the principal drawbacks of this technique; it can simply fail to provide an estimate of the speaker's position.

10.1.2 Spherical Interpolation Estimator

The primary limitation of the SX estimator is the restriction that $R_s = |\mathbf{x}|$. In an attempt to overcome the problems associated therewith, Abel and Smith (1987) proposed the *spherical interpolation* (SI) estimator. To develop this method, we begin by substituting the least-squares solution (10.14) into the spherical model $\mathbf{A}\theta = \mathbf{b}$ to obtain

$$R_s\, \mathbf{P}_{\mathbf{S}}^{\perp} \mathbf{d} = \mathbf{P}_{\mathbf{S}}^{\perp} \mathbf{b}, \tag{10.16}$$

where

$$\mathbf{P}_{\mathbf{S}}^{\perp} = \mathbf{I} - \mathbf{S}\check{\mathbf{S}} = \mathbf{I} - \mathbf{S}(\mathbf{S}^T \mathbf{S})^{-1} \mathbf{S}^T$$

is the perpendicular projection operator onto the space orthogonal to the column space of $\mathbf{S}$; see Section B.17. Then the least squares solution to (10.14) can be expressed as

$$\hat{R}_{s,SI} = \frac{\mathbf{d}^T \mathbf{P}_{\mathbf{S}}^{\perp} \mathbf{b}}{\mathbf{d}^T \mathbf{P}_{\mathbf{S}}^{\perp} \mathbf{d}}. \tag{10.17}$$

Substituting (10.17) into (10.14) yields the SI estimate

$$\hat{\mathbf{x}}_{s,SI} = \check{\mathbf{S}} \left[\mathbf{I} - \left(\frac{\mathbf{d}\mathbf{d}^T \mathbf{P}_{\mathbf{S}}^{\perp}}{\mathbf{d}^T \mathbf{P}_{\mathbf{S}}^{\perp} \mathbf{d}} \right) \right] \mathbf{b}. \tag{10.18}$$

Huang *et al.* (2004) comment that the SI estimator outperforms the SX estimator, but is computationally more demanding.

10.1.3 Linear Intersection Estimator

The final conventional method for source localization we will consider is the *linear intersection* (LI) method proposed by Brandstein (1995), which also returns a closed-form estimate of the speaker's position. This technique is based on the fact that if a sensor pair has locations $\mathbf{m}_1, \mathbf{m}_2$, and a TDOA τ is observed for this pair, then the locus of potential source locations forms one-half of a hyperboloid of two sheets, which is centered about the mid-point of $\mathbf{m}_1$ and $\mathbf{m}_2$ and has the line segment from $\mathbf{m}_1$ to $\mathbf{m}_2$ as its axis of symmetry. The bearing line to the source, measured as a deviation from the line segment between $\mathbf{m}_1$ and $\mathbf{m}_2$, can be approximated as

$$\hat{\theta}_{12} = \cos^{-1}\left(\frac{c \cdot \hat{\tau}_{12}}{|\mathbf{m}_1 - \mathbf{m}_2|}\right). \tag{10.19}$$

As originally proposed by Brandstein, the LI localization algorithm is based on the particular microphone configuration shown in Figure 10.1. There are two sensor pairs $(\mathbf{m}_{n1}, \mathbf{m}_{n2})$ and $(\mathbf{m}_{n3}, \mathbf{m}_{n4})$, where n is an index over sensor quadruples. The centroid of the nth sensor quadruple is denoted as $\mathbf{c}_n$. The first microphone pair in each quadruple determines, in an approximate sense, a cone with constant direction angle α_n relative to the $\mathbf{x}_n$ axis. The second pair determines a cone with direction angle β_n relative to the $\mathbf{y}_n$ axis. Assuming the location of the source is restricted to the positive z-space, the locus of potential source points is a bearing line $\mathbf{l}'_n$, with remaining direction angle γ_n that can be calculated from the identity,

$$\cos^2 \alpha_n + \cos^2 \beta_n + \cos^2 \gamma_n = 1, \tag{10.20}$$

where $0 \le \gamma_n \le \pi/2$. The line segment may be expressed in terms of the local coordinate system as $\mathbf{l}'_n = \begin{bmatrix} x_n & y_n & z_n \end{bmatrix}^T = r_n \mathbf{a}'_n$, where r_n is the range of the point on the line from

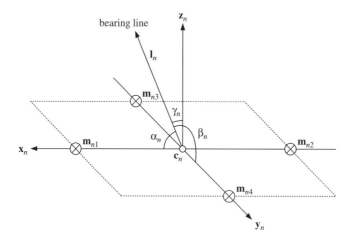

Figure 10.1 Small array with four microphones for speaker tracking

the local origin and $\mathbf{a}'_n$ is the vector of direction cosines

$$\mathbf{a}'_n \triangleq [\cos \alpha_n \quad \cos \beta_n \quad \cos \gamma_n].$$

To express the same bearing line in terms of global coordinates, a translation and rotation must be applied according to

$$\mathbf{l}_n = r_n \, \mathbf{R}_n \, \mathbf{a}'_n + \mathbf{c}_n, \tag{10.21}$$

where $\mathbf{R}_n$ is the 3×3 rotation matrix from the nth local coordinate system into the global coordinate system. Defining the rotated direction cosine vector

$$\mathbf{a}_n \triangleq \mathbf{R}_n \, \mathbf{a}'_n,$$

allows (10.21) to be rewritten as

$$\mathbf{l}_n = r_n \, \mathbf{a}_n + \mathbf{c}_n.$$

Now assume that there are M_4 orthogonal and bisecting sensor quadruples with the bearing lines

$$\mathbf{l}_n = r_n \mathbf{a}_n + \mathbf{c}_n \; \forall \, n = 1, \ldots, M_4.$$

The point of nearest intersection for any pair of sensor lines can then be taken as a possible location for the source. Figure 10.2 illustrates how the nearest intersection point for two bearing lines is determined. In particular, given the bearing lines,

$$\mathbf{l}_m = r_m \, \mathbf{a}_m + \mathbf{c}_m, \tag{10.22}$$

$$\mathbf{l}_n = r_n \, \mathbf{a}_n + \mathbf{c}_n, \tag{10.23}$$

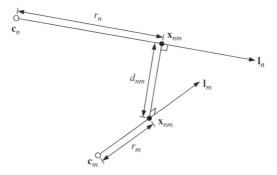

Figure 10.2 Determining the point of nearest intersection for two bearing lines

the shortest distance between the lines is measured along a line parallel to their common normal and is given by (Swokowski 1979)

$$d_{mn} = \frac{|(\mathbf{a}_m \times \mathbf{a}_n) \cdot (\mathbf{c}_m - \mathbf{c}_n)|}{|(\mathbf{a}_m \times \mathbf{a}_n)|}.$$

The point on $\mathbf{l}_m$ closest to $\mathbf{l}_n$, which is denoted as $\hat{\mathbf{x}}_{mn}$, and the point on $\mathbf{l}_n$ with closest intersection to $\mathbf{l}_m$, which is denoted as $\hat{\mathbf{x}}_{nm}$, can be found by solving for the local ranges r_m and r_n, then substituting these values into (10.22–10.23). These local ranges can be obtained by subtracting (10.22) from (10.23) at the point of closest intersection, where it holds that $\mathbf{l}_m = \hat{\mathbf{x}}_{mn}$ and $\mathbf{l}_n = \hat{\mathbf{x}}_{nm} = \hat{\mathbf{x}}_{mn} + d_{mn}(\mathbf{a}_m \times \mathbf{a}_n)$. This difference yields the overdetermined matrix equation

$$r_m \mathbf{a}_m - r_n \mathbf{a}_n = \mathbf{c}_n - \mathbf{c}_m - d_{mn}(\mathbf{a}_m \times \mathbf{a}_n), \tag{10.24}$$

which, when solved, provides the desired local ranges.

Each point $\hat{\mathbf{x}}_{nm}$ of nearest intersection represents a potential source location. In order to determine the final estimate of the speaker's position, these potential source locations are weighted and summed together. Denoting the number of unique microphone pairs as M_2, the weight assigned to $\hat{\mathbf{x}}_{nm}$ can be expressed by

$$W_{nm} = \prod_{i=1}^{M_2} \mathcal{N}\left[T_i(\hat{\mathbf{x}}_{nm}); \tau_{i12}, \sigma_{i12}^2\right],$$

where τ_{i12} and σ_{i12}^2, respectively, are the observed time delay and variance for the ith microphone pair, and the predicted TDOA, in this case, is defined as

$$T_i(\hat{\mathbf{x}}_{nm}) = T(\mathbf{m}_{i1}, \mathbf{m}_{i2}, \hat{\mathbf{x}}_{nm}) \, \forall \, n = 1, \ldots, M_2, \tag{10.25}$$

for $T(\mathbf{m}_1, \mathbf{m}_2, \mathbf{x})$ specified in (10.1). The final location estimate is then given by

$$\hat{\mathbf{x}}_{LI} = \frac{\sum_{m=1}^{M} \sum_{n=1,n\neq m}^{M} W_{mn}\hat{\mathbf{x}}_{mn}}{\sum_{m=1}^{M} \sum_{n=1,n\neq m}^{M} W_{mn}}.$$

By associating each potential source location with a probabilistic value, the weighting terms W_{mn} serve as a means for excluding outlier locations stemming from radically errant TDOA estimates and their consequently incorrect bearing lines. Typically, the weighting term for an aberrant $\hat{\mathbf{x}}_{mn}$ is sufficiently small that the potential location estimate plays little or no role in the final location estimate.

Note that the three methods described above all fall into the category of speaker *localization* systems, inasmuch as they return instantaneous estimates of the speaker's position. It should be noted, however, that each technique can readily be extended into a speaker tracking system by applying a KF to *smooth* the time series of instantaneous estimates, as in Brandstein *et al.* (1997) and Strobel *et al.* (2001), for example. As an alternative, instead of splitting the localization and tracking problem into two parts, it is possible to adopt a unified approach whereby the time series of the speaker's position is estimated directly without recourse to any intermediate localization. Such techniques are described in the next section.

10.2   Speaker Tracking with the Kalman Filter

In this section, we present the speaker tracking system based on an extended Kalman filter that was orginally proposed by Klee *et al.* (2005b). Consider two microphones located at $\mathbf{m}_{n1}$ and $\mathbf{m}_{n2}$ comprising the nth microphone pair, and once more define the TDOA as in (10.25), where $\mathbf{x}$ represents the position of an active speaker and, as before, $M = M_2$ denotes the number of unique microphone pairs. Source localization based on the maximum likelihood criterion (Kay 1993) proceeds by minimizing the error function

$$\epsilon(\mathbf{x}) = \sum_{i=n}^{M} \frac{1}{\sigma_n^2} [\hat{\tau}_n - T_n(\mathbf{x})]^2, \tag{10.26}$$

where σ_n^2 denotes the error covariance associated with this observation, and $\hat{\tau}_n$ is the observed TDOA as in (10.4) and (10.5). Solving for that $\mathbf{x}$ minimizing (10.26) would be eminently straightforward were it not for the fact that (10.25) is nonlinear in $\mathbf{x}$. In the coming development, we will find it useful to have a linear approximation. Hence, we take a partial derivative with respect to $\mathbf{x}$ on both sides of (10.25) and write

$$\nabla_{\mathbf{x}} T_n(\mathbf{x}) = \frac{1}{c} \cdot \left[\frac{\mathbf{x} - \mathbf{m}_{n1}}{D_{n1}} - \frac{\mathbf{x} - \mathbf{m}_{n2}}{D_{n2}} \right],$$

where

$$D_{nm} = |\mathbf{x} - \mathbf{m}_{nm}| \,\forall\, n = 1, \ldots, M; \, m = 1, 2,$$

is the distance between the source and the microphone located at $\mathbf{m}_{nm}$.

Although (10.26) implies that we should find the $\mathbf{x}$ that minimizes the instantaneous error criterion, we would be better advised to attempt to minimize such an error criterion over a series of time instants. In so doing, we exploit the fact that the speaker's position cannot change instantaneously; thus, both the present $\hat{\tau}_i(k)$ and past TDOA estimates $\{\hat{\tau}_i(n)\}_{n=1}^{k-1}$ are potentially useful in estimating a speaker's current position $\mathbf{x}(k)$. Let us approximate $T_n(\mathbf{x})$ with a first-order Taylor series expansion about the last position estimate $\hat{\mathbf{x}}(k-1)$ by writing

$$T_n(\mathbf{x}) \approx T_n(\hat{\mathbf{x}}(k-1)) + \mathbf{c}_n^T(k)\, [\mathbf{x} - \hat{\mathbf{x}}(k-1)], \tag{10.27}$$

where the row vector $\mathbf{c}_n^T(k)$ is given by

$$\mathbf{c}_n^T(k) = [\nabla_{\mathbf{x}} T_n(\mathbf{x})]_{\mathbf{x}=\hat{\mathbf{x}}(k-1)}^T = \frac{1}{c} \cdot \left[\frac{\mathbf{x} - \mathbf{m}_{n1}}{D_{n1}} - \frac{\mathbf{x} - \mathbf{m}_{n2}}{D_{n2}} \right]_{\mathbf{x}=\hat{\mathbf{x}}(k-1)}^T \forall\, n = 1, \ldots, M. \tag{10.28}$$

Substituting the linearization (10.27) into (10.26) provides

$$\epsilon(\mathbf{x}; k) \approx \sum_{n=1}^{M} \frac{1}{\sigma_i^2} \left\{ \hat{\tau}_n(k) - T_n(\hat{\mathbf{x}}(k-1)) - \mathbf{c}_n^T(k)\, [\mathbf{x} - \hat{\mathbf{x}}(k-1)] \right\}^2$$

$$= \sum_{n=1}^{M} \frac{1}{\sigma_n^2} \left[\overline{\tau}_n(k) - \mathbf{c}_n^T(k)\mathbf{x} \right]^2 \tag{10.29}$$

where

$$\overline{\tau}_n(k) = \hat{\tau}_n(k) - \left[T_n(\hat{\mathbf{x}}(k-1)) - \mathbf{c}_n^T(k)\hat{\mathbf{x}}(k-1)\right] \ \forall \, n = 1, \ldots, M. \tag{10.30}$$

Let us define

$$\overline{\tau}(k) = \begin{bmatrix} \overline{\tau}_1(k) \\ \overline{\tau}_2(k) \\ \vdots \\ \overline{\tau}_M(k) \end{bmatrix}, \ \hat{\tau}(k) = \begin{bmatrix} \hat{\tau}_1(k) \\ \hat{\tau}_2(k) \\ \vdots \\ \hat{\tau}_M(k) \end{bmatrix}, \ \mathbf{T}(\hat{\mathbf{x}}(k)) = \begin{bmatrix} T_1(\hat{\mathbf{x}}(k)) \\ T_2(\hat{\mathbf{x}}(k)) \\ \vdots \\ T_M(\hat{\mathbf{x}}(k)) \end{bmatrix}, \ \mathbf{C}(k) = \begin{bmatrix} \mathbf{c}_1^T(k) \\ \mathbf{c}_2^T(k) \\ \vdots \\ \mathbf{c}_M^T(k) \end{bmatrix},$$

$$\tag{10.31}$$

so that (10.30) can be expressed in matrix form as

$$\overline{\tau}(k) = \hat{\tau}(k) - \left[\mathbf{T}(\hat{\mathbf{x}}(k-1)) - \mathbf{C}(k)\hat{\mathbf{x}}(k-1)\right]. \tag{10.32}$$

Similarly, defining

$$\mathbf{\Sigma} = \mathrm{diag}(\sigma_1^2, \sigma_2^2, \cdots, \sigma_M^2) \tag{10.33}$$

enables (10.29) to be expressed as

$$\epsilon(\mathbf{x}; t) = \left[\overline{\tau}(k) - \mathbf{C}(k)\mathbf{x}\right]^T \mathbf{\Sigma}^{-1} \left[\overline{\tau}(k) - \mathbf{C}(k)\mathbf{x}\right]. \tag{10.34}$$

Klee *et al.* proposed to recursively minimize the linearized least squares position estimation criterion (10.34) with the (iterated) extended Kalman filter presented in Sections 4.3.2 and 4.3.3. Their algorithm is readily understood as soon as we make an association between the position $\mathbf{x}(k)$ of the speaker and the state $\mathbf{x}_k$ of the *extended Kalman filter* (EKF), as well as between the TDOA vector $\hat{\tau}(k)$ and the observation $\mathbf{y}$. Following Klee *et al.* (2005b), we can make a simple assumption that the speaker is "stationary" such that the state transition matrix $F_{k|k-1} = \mathbf{I}$. We further associate the "linearized" TDOA estimate $\overline{\tau}(k)$ in (10.32) with the modified observation $\overline{\mathbf{y}}(k)$ appearing in (4.49). The nonlinear functional $\mathbf{H}_k(\mathbf{x})$ appearing in (4.44) corresponds to the TDOA model

$$H_k(\mathbf{x}_k) = \mathbf{T}_k(\mathbf{x}_k) = \begin{bmatrix} T_1(\mathbf{x}_k) \\ T_2(\mathbf{x}_k) \\ \vdots \\ T_M(\mathbf{x}_k) \end{bmatrix},$$

where the individual components $T_n(\mathbf{x}_k)$ are given by (10.25). Moreover, we recognize that the linearized observation functional $\overline{\mathbf{H}}_k(\mathbf{x}_{k|k-1})$ in (4.45) required for the EKF filter is given by (10.28) and (10.31) for our acoustic speaker tracking problem. Furthermore, we can equate the TDOA error covariance matrix $\mathbf{\Sigma}$ in (10.33) with the observation noise covariance $\mathbf{V}_k$. Hence, we have all relations needed on the observation side of the EKF filter.

With the above definitions, it is only necessary to specify an appropriate model of the speaker's motion to complete the algorithm. Assuming the process noise components in the three directions are statistically independent, we can write

$$\mathbf{U}_k = \sigma_P^2 \, T_E^2 \begin{bmatrix} 1 & 0 & 0 \\ 0 & 1 & 0 \\ 0 & 0 & 1 \end{bmatrix}, \tag{10.35}$$

where T_E is the time elapsed since the last update of the state estimate, and σ_P^2 is the process noise power. Klee *et al.* (2005b) set σ_P^2 based on a set of empirical trials to achieve the best localization results.

Before performing an update of the speaker's position estimate $\mathbf{x}_{k|k}$, it was first necessary to determine the time T_E that had elapsed since an observation was last received. While the audio sampling is synchronous for all sensors, it could not be assumed that the speaker constantly spoke, nor that all microphones received the direct signal from the speaker's mouth; i.e., the speaker might sometimes turn so that he is no longer facing a given microphone array. As only the direct signal is useful for localization (Armani *et al.* 2003), the TDOA estimates returned by those sensors receiving only the indirect signal reflected from the walls could not be used for position updates. This was accomplished by setting a threshold on the PHAT (10.4), and using for source localization only those microphone pairs returning a peak in the PHAT above the threshold (Armani *et al.* 2003). This implied that no update was made if the speaker was silent.

As the update of the speaker's position estimate was obtained from the standard update formulae (4.57–4.61) of the *iterated extended Kalman filter* (IEKF), it was no longer necessary to invoke one of the closed form approximations for the speaker's position considered in Section 10.1. Using the formalism of the KF had the added advantage that the uncertainty with respect to the speaker's position, which is reflected in the state estimation error covariance matrices $\mathbf{K}_{k|k-1}$ and $\mathbf{K}_k$, was automatically factored into the state update. This uncertainty was weighted against the uncertainty of the current observation specified by $\boldsymbol{\Sigma}$. Finally, the KF formulation provided the additional flexibility afforded by the model of speaker motion, which could potentially be extended to include velocity and even acceleration terms.

To gain an appreciation for the severity of the nonlinearity in this particular Kalman filtering application, Klee *et al.* (2005b) plotted the actual value of $T_n(\mathbf{x}(k))$ against the linearized version. These plots are shown in Figure 10.3 for deviations parallel to the x- and y-axes from the point about which $T_n(\mathbf{x}(k))$ was linearized. The plots correspond to T-array 4 in the smart room at the Universität Karlsruhe (TH) depicted in Figure 14.1. The nominal position of the speaker, about which the functional $\mathbf{H}_k(\mathbf{x})$ was linearized, was $(x, y, z) = (2.95, 4.08, 1.70)$ m in room coordinates, which is approximately in the middle of the room. As is clear from Figure 10.3, for deviations of ± 1 m from the nominal, the linearized TDOA is within 2.33% of the true value for movement along the x-axis, which was perpendicular to the plane of the array, and within 1.38% for movement along the y-axis, which was parallel to the plane of the array.

Klee *et al.* (2005b) investigated the effect of speaker movement on the number of local iterations required by the IEKF. The local iteration compensates for the difference between the original nonlinear least squares estimation criterion (10.26) and the linearized

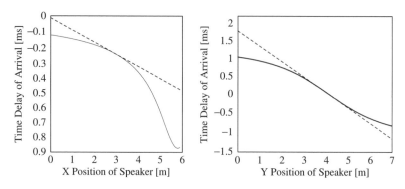

Figure 10.3 Actual (solid) vs linearized (dashed) $T_n(\mathbf{x}_k)$ for movement parallel to the x- and y-axes

criterion (10.29). The difference between the two is only significant during startup or when a significant amount of time has passed since the last update, as in such cases the initial position estimate is far from the true position of the speaker. Once the speaker's position has been acquired to a reasonable accuracy, the linearized model (10.29) matches the original (10.26) quite well. Klee *et al.* (2005b) found that the average number of local iterations increases in proportion to the time since the last update, the distance moved since the last update, and the speaker's velocity. These results correspond well with our expectations, in that significant speaker movement implies that the linearized error criterion (10.29) does not initially match the true criterion (10.26), as shown in Figure 10.3. Hence, several local iterations are required for the position estimate to converge. Five or fewer local iterations were required for convergence in all cases, which was more than sufficient for real-time speaker tracking.

10.2.1 Implementation Based on the Cholesky Decomposition

As noted in Section 4.3.4, the Riccati equation has notoriously poor numerical stability characteristics. In particular, after many iterations wherein the Riccati equation is used to update the predicted state estimation error covariance matrix $\mathbf{K}_{k|k-1}$, a phenomenon known as explosive divergence (Haykin 2002, sect. 11) can occur, whereby $\mathbf{K}_{k|k-1}$ becomes indefinite. As we will now discuss, this undesirable state of affairs can be avoided by replacing a naive implementation of IEKF based on the Riccati equation with a so-called *square-root* implementation. In addition to preventing explosive divergence, such a square-root implementation effectively doubles the precision of the machine on which it is executed (Simon 2006, sects 6.3–6.4).

Using manipulations similar to those leading to (4.65), it is possible to formulate the Riccati equation for the IEKF presented in Section 4.3.3 as

$$\mathbf{K}_{k+1|k} = \mathbf{F}\,\mathbf{K}_{k|k-1}\,\mathbf{F}^T - \mathbf{F}\,\mathbf{G}_k(\boldsymbol{\eta}_i)\,\mathbf{S}_k(\boldsymbol{\eta}_i)\,\mathbf{G}_k^T(\boldsymbol{\eta}_i)\,\mathbf{F}^T + \mathbf{U}_k, \qquad (10.36)$$

where $\mathbf{F} = \mathbf{F}_{k+1|k}$ is the constant transition matrix, $\boldsymbol{\eta}_i$ is the current local iterate, and $\mathbf{S}_k(\boldsymbol{\eta}_i)$ and $\mathbf{G}_k^T(\boldsymbol{\eta}_i)$ are respectively the innovation covariance matrix and Kalman gain calculated at η_i. The Cholesky decomposition (Golub and Van Loan 1990, sect. 4.2) or square-root $\mathbf{K}_{k+1|k}^{1/2}$ of the state estimation error covariance matrix $\mathbf{K}_{k+1|k}$ is the unique lower triangular matrix achieving

$$\mathbf{K}_{k+1|k}^{1/2}\mathbf{K}_{k+1|k}^{T/2} \triangleq \mathbf{K}_{k+1|k}.$$

Square-root implementations of RLS estimators and KFs propagate $\mathbf{K}_{k+1|k}^{1/2}$ forward at each time step instead of $\mathbf{K}_{k+1|k}$. The advantage of this approach is that the Cholesky decomposition exists only for positive definite matrices (Golub and Van Loan 1990, sect. 4.2). Thus, in propagating $\mathbf{K}_{k+1|k}^{1/2}$ instead of $\mathbf{K}_{k+1|k}$, we ensure that the latter remains positive definite. The square-root implementation is derived from the following well-known lemma (Sayed and Kailath 1994).

***Lemma 10.2.1* (matrix factorization)** *Given any two $N \times M$ matrices $\mathbf{A}$ and $\mathbf{B}$ with dimensions $N \leq M$,*

$$\mathbf{AA}^T = \mathbf{BB}^T$$

iff there exists a unitary matrix $\boldsymbol{\theta}$ such that

$$\mathbf{A}\boldsymbol{\theta} = \mathbf{B}.$$

To develop an update strategy based on the Cholesky decomposition, we set $\mathbf{V}_k = \boldsymbol{\Sigma}$ and write

$$\mathbf{A} = \begin{bmatrix} \boldsymbol{\Sigma}^{1/2} & \vdots & \overline{\mathbf{H}}_k(\boldsymbol{\eta}_i)\mathbf{K}_{k|k-1}^{1/2} & \vdots & \mathbf{0} \\ \cdots\cdots\cdots\cdots\cdots\cdots\cdots\cdots\cdots\cdots \\ \mathbf{0} & \vdots & \mathbf{F}\mathbf{K}_{k|k-1}^{1/2} & \vdots & \mathbf{U}_k^{1/2} \end{bmatrix}.$$

We seek a unitary transform $\boldsymbol{\theta}$ that achieves $\mathbf{A}\boldsymbol{\theta} = \mathbf{B}$, such that

$$\mathbf{AA}^T = \begin{bmatrix} \boldsymbol{\Sigma}^{1/2} & \vdots & \overline{\mathbf{H}}_k(\boldsymbol{\eta}_i)\mathbf{K}_{k|k-1}^{1/2} & \vdots & \mathbf{0} \\ \cdots\cdots\cdots\cdots\cdots\cdots\cdots\cdots\cdots\cdots \\ \mathbf{0} & \vdots & \mathbf{F}\mathbf{K}_{k|k-1}^{1/2} & \vdots & \mathbf{U}_k^{1/2} \end{bmatrix} \cdot \begin{bmatrix} \boldsymbol{\Sigma}^{T/2} & \vdots & \mathbf{0} \\ \cdots\cdots\cdots\cdots\cdots\cdots \\ \mathbf{K}_{k|k-1}^{T/2}\overline{\mathbf{H}}_k^T(\boldsymbol{\eta}_i) & \vdots & \mathbf{K}_{k|k-1}^{T/2}\mathbf{F}^T \\ \cdots\cdots\cdots\cdots\cdots\cdots \\ \mathbf{0} & \vdots & \mathbf{U}_k^{T/2} \end{bmatrix}$$

$$= \begin{bmatrix} \mathbf{B}_{11} & \vdots & \mathbf{0} & \vdots & \mathbf{0} \\ \cdots\cdots\cdots\cdots \\ \mathbf{B}_{21} & \vdots & \mathbf{B}_{22} & \vdots & \mathbf{0} \end{bmatrix} \begin{bmatrix} \mathbf{B}_{11}^T & \vdots & \mathbf{B}_{21}^T \\ \cdots\cdots\cdots \\ \mathbf{0} & \vdots & \mathbf{B}_{22}^T \\ \cdots\cdots\cdots \\ \mathbf{0} & \vdots & \mathbf{0} \end{bmatrix} = \mathbf{BB}^T, \tag{10.37}$$

where both $\mathbf{B}_{11}$ and $\mathbf{B}_{22}$ are lower triangular. Performing the required multiplications on the block components of (10.37), we find

$$\mathbf{B}_{11}\mathbf{B}_{11}^T = \mathbf{\Sigma} + \overline{\mathbf{H}}_k(\boldsymbol{\eta}_i)\,\mathbf{K}_{k|k-1}\,\overline{\mathbf{H}}_k^T(\boldsymbol{\eta}_i), \tag{10.38}$$

$$\mathbf{B}_{21}\mathbf{B}_{11}^T = \mathbf{F}\,\mathbf{K}_{k|k-1}\,\overline{\mathbf{H}}_k^T(\boldsymbol{\eta}_i), \tag{10.39}$$

$$\mathbf{B}_{21}\,\mathbf{B}_{21}^T + \mathbf{B}_{22}\,\mathbf{B}_{22}^T = \mathbf{F}\,\mathbf{K}_{k|k-1}\,\mathbf{F}^T + \mathbf{U}_k. \tag{10.40}$$

Comparing (4.57) and (10.38), it is clear that

$$\mathbf{B}_{11}\mathbf{B}_{11}^T = \mathbf{S}_k(\boldsymbol{\eta}_i). \tag{10.41}$$

Hence, as $\mathbf{B}_{11}$ is lower triangular,

$$\mathbf{B}_{11} = \mathbf{S}_k^{1/2}(\boldsymbol{\eta}_i). \tag{10.42}$$

Based on (4.58),

$$\mathbf{G}_k(\boldsymbol{\eta}_i)\,\mathbf{S}_k(\boldsymbol{\eta}_i) = \mathbf{K}_{k|k-1}\,\overline{\mathbf{H}}_k^T(\boldsymbol{\eta}_i).$$

Substituting the last equation for $\mathbf{K}_{k|k-1}\overline{\mathbf{H}}_k^T(\boldsymbol{\eta}_i)$ in (10.39), we find

$$\mathbf{B}_{21}\,\mathbf{B}_{11}^T = \mathbf{F}\,\mathbf{G}_k(\boldsymbol{\eta}_i)\,\mathbf{S}_k(\boldsymbol{\eta}_i) = \mathbf{F}\,\mathbf{G}_k(\boldsymbol{\eta}_i)\,\mathbf{B}_{11}\,\mathbf{B}_{11}^T,$$

where the last equality follows from (10.41). Hence,

$$\mathbf{B}_{21} = \mathbf{F}\,\mathbf{G}_k(\boldsymbol{\eta}_i)\,\mathbf{S}_k^{1/2}(\boldsymbol{\eta}_i). \tag{10.43}$$

Finally, we can rewrite (10.40) as

$$\mathbf{B}_{22}\mathbf{B}_{22}^T = \mathbf{F}\mathbf{K}_{k|k-1}\mathbf{F}^T - \mathbf{B}_{21}\mathbf{B}_{21}^T + \mathbf{U}_{k-1} \tag{10.44}$$

$$= \mathbf{F}\,\mathbf{K}_{k|k-1}\,\mathbf{F}^T - \mathbf{F}\,\mathbf{G}_k^T(\boldsymbol{\eta}_i)\,\mathbf{S}_k(\boldsymbol{\eta}_i)\,\mathbf{G}_k(\boldsymbol{\eta}_i)\,\mathbf{F}^T + \mathbf{U}_k, \tag{10.45}$$

where (10.45) follows from substituting (10.43) into (10.44). The last equation, together with (10.36), implies that

$$\mathbf{B}_{22}\,\mathbf{B}_{22}^T = \mathbf{K}_{k+1|k},$$

or, as $\mathbf{B}_{22}$ is lower triangular,

$$\mathbf{B}_{22} = \mathbf{K}_{k+1|k}^{1/2}. \tag{10.46}$$

In light of (10.42–10.46), we have

$$
\mathbf{A}\theta = \begin{bmatrix} \boldsymbol{\Sigma}^{1/2} & \vdots & \overline{\mathbf{H}}_k(\boldsymbol{\eta}_i)\,\mathbf{K}_{k|k-1}^{1/2} & \vdots & \mathbf{0} \\ \cdots\cdots\cdots\cdots\cdots\cdots\cdots\cdots\cdots \\ \mathbf{0} & \vdots & \mathbf{F}\mathbf{K}_{k|k-1}^{1/2} & \vdots & \mathbf{U}_k^{1/2} \end{bmatrix} \theta
$$

$$
= \begin{bmatrix} \mathbf{S}_k^{1/2}(\boldsymbol{\eta}_i) & \vdots & \mathbf{0} & \vdots & \mathbf{0} \\ \cdots\cdots\cdots\cdots\cdots\cdots\cdots\cdots \\ \mathbf{F}\mathbf{G}_k(\boldsymbol{\eta}_i)\mathbf{S}_k^{1/2}(\boldsymbol{\eta}_i) & \vdots & \mathbf{K}_{k+1|k}^{1/2} & \vdots & \mathbf{0} \end{bmatrix} = \mathbf{B}. \tag{10.47}
$$

Although a new estimate of $\mathbf{K}_{k+1|k}^{1/2}$ is generated with each local iteration, only the final estimate is saved for use in the succeeding time step.

The final position update is accomplished as follows: through forward substitution we can find that $\boldsymbol{\zeta}_k'(\boldsymbol{\eta}_i)$ achieving

$$
\boldsymbol{\zeta}_k(\boldsymbol{\eta}_i) = \mathbf{S}_k^{1/2}(\boldsymbol{\eta}_i)\boldsymbol{\zeta}_k'(\boldsymbol{\eta}_i), \tag{10.48}
$$

where $\boldsymbol{\zeta}_k(\boldsymbol{\eta}_i)$ is defined in (4.60). The preceding development shows that $\mathbf{F}$ is upper triangular for the stationary source model of interest here. Hence, we can find that $\boldsymbol{\zeta}_k''(\boldsymbol{\eta}_i)$ achieving

$$
\mathbf{F}\boldsymbol{\zeta}_k''(\boldsymbol{\eta}_i) = \mathbf{B}_{21}\boldsymbol{\zeta}_k'(\boldsymbol{\eta}_i)
$$

through back substitution on $\mathbf{F}$. Finally, as in (4.61), we update $\boldsymbol{\eta}_i$ according to

$$
\boldsymbol{\eta}_{i+1} = \hat{\mathbf{x}}_{k|k-1} + \boldsymbol{\zeta}_k''(\boldsymbol{\eta}_i), \tag{10.49}
$$

where $\boldsymbol{\eta}_1 = \hat{\mathbf{x}}_{k|k-1}$. The new state estimate $\hat{\mathbf{x}}_{k|k}$ is taken as the final iterate $\boldsymbol{\eta}_f$.

A unitary transform θ that imposes the desired zeros on $\mathbf{A}$ can be readily constructed from a set of Givens rotations. The latter are described in Section B.15.

In the acoustic speaker tracking experiments conducted by Klee *et al.* (2005b), the numerical stability proved adequate using even the KF based directly on the Riccati equation. Instabilities arose, however, when the audio features were supplemented with video information as discussed in Section 10.4.

10.3   Tracking Multiple Simultaneous Speakers

In this section, we consider how the single speaker tracking system based on the KF can be extended to track multiple simultaneously active speakers. The approach presented here was proposed by Gehrig *et al.* (2006), and is based on the generalization of the KF discussed in Section 4.3.6.

As the joint probabilistic data association filter can maintain tracks for multiple active speakers, it is necessary to formulate rules for deciding when a new track should be created, when two tracks should be merged and when a track should be deleted. A new track was always created as soon as a measurement could not be associated with any existing track. But if the time to initialize the filter exceeded a time threshold, the newly created track was immediately deleted. The initialization time of the filter was defined as the time required until the variance of each dimension of $\epsilon_{k+1|k}$ in (4.25) fell below a given threshold. Normally this initialization time was relatively short for a track for which sufficient observations were available, but longer for spurious noises. To merge two or more tracks, a list was maintained with the time-stamp when the two tracks approached one another within a given distance. If, after some allowed interval of overlap, the two tracks did not move apart, then the track with the larger $|K_{k+1|k}|$ was deleted. In all cases, tracks were deleted if their position estimate had not been updated for a given length of time. To detect the active sound source, the track with the smallest error covariance matrix was used, as an active sound source should be associated with enough observations so that the covariance decreases, while error covariance matrices associated with inactive sound sources should increase with time.

10.4   Audio-Visual Speaker Tracking

As mentioned previously, a speaker tracking system based purely on acoustic information is only able to track a speaker when he actually speaks. If the acoustic observations are supplemented with visual information, however, it becomes possible to track a speaker even when he is silent.

An often-cited system for audio-visual source localization was proposed by Strobel *et al.* (2001). The system we present here was first described by Gehrig *et al.* (2005). In the latter algorithm, no explicit position estimates were made by the individual sensors, either audio or video. Rather, as in the work of Welch and Bishop (1997), the observations of the individual sensors were used to *incrementally* update the state of a KF. This combined approach yielded a robust speaker tracking system that functioned reliably both for segments wherein the speaker was silent, which would have been detrimental for an audio-only tracker, and wherein many faces appeared, which would have confused a video-only tracker. After testing the algorithm on a data set consisting of seminars held by actual speakers, Gehrig *et al.* reported that the audio-visual tracker functioned better than a speaker tracking system based either solely on audio or solely on video features.

In order to localize the speaker visually, Gehrig *et al.* minimized a squared-error criterion much like that in (10.26). In this case, the difference between the position produced by a *face detector* and the speaker's predicted position in a video image was minimize. This *two-dimensional* difference was calculated in the camera's image plane. As shown in Figure 10.4, the predicted speaker position **x** was projected onto the image plane **I** of the camera at position **t** with focal length **f**. This resulted in the image point $\hat{\mathbf{x}}$. The difference between $\hat{\mathbf{x}}$ and the position **y** returned by the face detector was then minimized.

The extrinsic parameters **t** and **R** define a camera's translation and rotation with respect to the global three-dimensional coordinate system. To project a point onto the image plane, it was also necessary to have information about the camera's intrinsic parameters: the camera matrix **P** is determined by the focal length f, the sensor pixel sizes p_x and p_y, and the principal point $\begin{bmatrix} c_x & c_y & 1 \end{bmatrix}^T$ according to (Pollefeys 2000)

$$\mathbf{P} = \begin{pmatrix} f/p_x & 0 & c_x \\ 0 & f/p_y & c_y \\ 0 & 0 & 1 \end{pmatrix}. \tag{10.50}$$

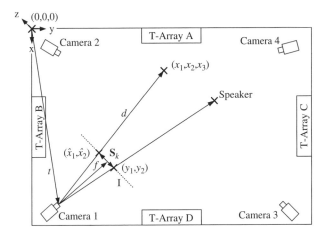

Figure 10.4 Back projection of the speaker's position onto the image plane of a camera

Assuming a simple pinhole camera model, the position estimate **x** could be projected onto the image plane using the *projection equation*,

$$\bar{\mathbf{x}} = \begin{pmatrix} \bar{x}_1 \\ \bar{x}_2 \\ \bar{x}_3 \end{pmatrix} = \mathbf{A}(\mathbf{x} - \mathbf{t}), \tag{10.51}$$

where, for the sake of efficiency,

$$\mathbf{A} = \mathbf{P}\mathbf{R}^T \tag{10.52}$$

can be calculated in advance, as it is not changing. The two-dimensional projection point on the image plane is then specified by

$$f(\mathbf{x}) = \begin{pmatrix} \hat{x}_1 \\ \hat{x}_2 \end{pmatrix} = \begin{pmatrix} \bar{x}_1/\bar{x}_3 \\ \bar{x}_2/\bar{x}_3 \end{pmatrix}. \tag{10.53}$$

As in (10.28) for the audio features, a linearization is required for this nonlinear projection function. Hence, we take the partial derivative of $f(\mathbf{x})$ with respect to $\mathbf{x}$

$$\mathbf{C} = \nabla_{\mathbf{x}} f(\mathbf{x}) \tag{10.54}$$

where

$$c_{mn} = \frac{a_{mn} - a_{3n}\hat{x}_m}{\bar{x}_3} \qquad \text{for } 1 \le m \le 2, 1 \le n \le 3 \tag{10.55}$$

and $\{a_{mn}\}$ are the elements of A.

Face detectors used for visual speaker tracking can be constructed based on the concept of boosted classifier cascades presented by Lienhart and Maydt (2002) and Jones and

Viola (2003). In order to be able to detect faces from different views, two separate cascades – one for frontal and one for profile faces – can be trained, thus covering a range of $\pm 90°$ horizontal head rotation. In order to reduce the rate of false detections, an adaptive background model of the scene was maintained and detections that were not supported by the foreground–background segmentation were ignored.

Performing face detection on an entire video image is computationally expensive. In order to keep this computational expense within reason, each camera was first sent the most recent position estimate from the KF. Thereafter, this three-dimensional position estimate was projected onto the camera's image plane, and the face detector was used to search for a face only within a relatively small neighborhood about this point. If a face was discovered within this region, the innovation vector, given by the difference between the projected position estimate and the location of the detected face, was calculated and returned to the KF. This two-dimensional innovation vector was then used to update the three-dimensional speaker position.

As all video data arrived asynchronously, there was at most a two-dimensional innovation vector available from a single camera to update the three-dimensional speaker position at any given time instant. As noted by Welch and Bishop (1997), this implies that the state of the KF was not *observable* on the basis of the data obtained from any single video sensor. Nonetheless, the state of the KF could be updated from a single observation. Moreover, the true state of the KF becomes observable when estimates from all sensors, both audio and video, are sequentially combined, subject only to very mild restrictions on the positions of the sensors with respect to the speaker and on the update rate (Welch 1996).

10.5 Speaker Tracking with the Particle Filter

The first direct applications of Bayesian filtering techniques to the problem of acoustic source localization without an intermediate position estimate were reported by Vermaak and Blake (2001) and Ward and Williamson (2002). The material presented in this section is taken largely from Ward et al. (2003), who formulated the speaker tracking problem as such. Assuming that the data arriving at each of M sensors of a microphone array is transformed into the frequency domain with a short-time Fourier transform, let

$$\mathbf{Y}_k(\omega) \triangleq \begin{bmatrix} Y_{k,1}(\omega) & Y_{k,2}(\omega) & \cdots & Y_{k,M}(\omega) \end{bmatrix}^T \qquad (10.56)$$

denote the stacked vector of frequency domain samples at time k and angular frequency ω. In the sequel, we will suppress the frequency variable ω when convenient. The vector $\mathbf{Y}_k(\omega)$ is often referred to as the frequency domain *snapshot*. Moreover, let θ denote a generic *localization parameter* and let $\mathbf{h}(\theta, \mathbf{Y}_k)$ denote a generic *localization function*. In Sections 10.5.1 and 10.5.2, we will consider two possible localization functions. We now use the localization function to define the observation at time k,

$$\mathbf{y}_k(\theta) \triangleq \mathbf{h}(\theta, \mathbf{Y}_k), \qquad (10.57)$$

and let $\mathbf{y}_{1:k}(\theta)$ denote all observations up to and including time k. In order to estimate the speaker's current position, Ward et al. (2003) proposed to use a PF, such as those

Algorithm 10.1   Algorithm for speaker tracking with a particle filter

Form an initial set of particles $\{\mathbf{x}_0^{(m)} \, \forall \, m = 1, \ldots, N\}$ and give them uniform weights $\{w_0^{(m)} = 1/N \, \forall \, m = 1, \ldots, N\}$. Then as each new frame of data is received:

1. Resample the particles from the previous frame $\{\mathbf{x}_{k-1}^{(m)} \, \forall \, m = 1, \ldots, N\}$ to form the resampled set of particles $\{\tilde{\mathbf{x}}_{k-1}^{(m)} \, \forall \, i = 1, \ldots, N\}$.
2. Predict the new set of particles $\{\mathbf{x}_k^{(m)} \, \forall \, m = 1, \ldots, N\}$ by propagating the resampled set $\{\mathbf{x}_{k-1}^{(m)} \, \forall \, m = 1, \ldots, N\}$ according to the source dynamics.
3. Transform the raw data into localization measurements through application of the localization function:

$$\mathbf{y}_k(\boldsymbol{\theta}) = \mathbf{h}(\theta, \mathbf{Y}_k).$$

4. Form the likelihood function

$$p(\mathbf{y}_k|\mathbf{x}) = F(\mathbf{y}_k, \mathbf{x}).$$

5. Weight the new particles according to the likelihood function:

$$w_k^{(m)} = p(\mathbf{y}_k|\mathbf{x}_k^{(m)}),$$

and normalize such that $\sum_m w_k^{(m)} = 1$.
6. Compute the current source location estimate $\hat{\mathbf{x}}_k$ as the weighted sum of the particle locations

$$\mathcal{E}\{\mathbf{x}_k\} = \sum_{m=1}^{N} w_k^{(m)} \mathbf{x}_k^{(m)}.$$

7. Store the particles and their respective weights $\{\mathbf{x}_k^{(m)}, w_k^{(m)} \, \forall \, m = 1, \ldots, N\}$ for the next iteration.

---

described in Section 4.4, to track the filtering density $p(\mathbf{x}_k|\mathbf{y}_{1:k})$. In principle, this is achieved through the iterative application of two steps, namely, *prediction* (4.6) and *correction* (4.7).

The only remaining element that must be specified for a complete speaker tracking system based on the PF is a *likelihood function*, which is defined through the relation

$$p(\mathbf{y}_k|\mathbf{x}_k) \triangleq F(\mathbf{y}_k, \mathbf{x}_k).$$

With these definitions, the operation of the complete speaker tracking system can be summarized as in Algorithm 10.1.

10.5.1 Localization Based on Time Delays of Arrival

For speaker tracking based on TDOA information, the localization function corresponds to (10.4), or one of the other methods for estimating TDOAs, mentioned in Section 10.1. In order to estimate the filtering density $p(\mathbf{x}_k|\mathbf{y}_{1:k})$, a set $\{\hat{\tau}_{i,l}\}_{l=1}^{L}$ of possible time delays for the ith microphone pair is determined by choosing the L largest peaks in (10.4). The locations of these peaks on the time axis are, by assumption, distributed according to the likelihood function, which in this case is defined as the Gaussian mixture model,

$$F_i(\mathbf{y}_t, \mathbf{x}_k) = w_0 + \sum_{j=1}^{J} w_j \, \mathcal{N}(\hat{\tau}_{i,l}; T_i(\mathbf{x}_k), \sigma^2),$$

where J is the number of mixture weights, w_j is the weight of the jth mixture, and σ^2 is a predetermined observation variance. The predicted delay $T_i(\mathbf{x}_k)$ is defined in (10.25). The weight $w_0 < 1$ is the probability that none of the observations corresponds to the true source location. Under the assumption that the observations across all sensor pairs are independent, the complete likelihood function can be expressed as

$$F(\mathbf{y}_t, \mathbf{x}_k) = \prod_{i=1}^{M} F_i(\mathbf{y}_t, \mathbf{x}_k).$$

10.5.2 Localization Based on Steered Beamformer Response Power

For localization based on *steered beamformer response power* (SBRP), the localization function is defined as

$$f(\mathbf{x}, \mathbf{Y}_k) \triangleq \int W(\omega) \left| \sum_{m=1}^{M} H_m(\mathbf{x}, \omega) \, \mathbf{Y}_m(\omega) \right|^2 d\omega$$

where

$$H_m(\mathbf{x}, \omega) = a_m \, e^{j\omega(|\mathbf{x}-\mathbf{x}_m|-d_{ref})/c}$$

is the complex-valued beamformer weighting term on the mth sensor, with $a_m \in \mathbb{R}$ the gain applied to the mth sensor output; $\mathbf{x}_m$ is the sensor location; and d_{ref} is the distance to some reference point, which is typically chosen as the center of the sensor array. Setting $\alpha_m = 1/M$ is equivalent to the delay-and-sum beamformer discussed in Section 13.1.3. Chen *et al.* (2002) report that setting all $\alpha_m = 1$ works nearly as well as the optimal solution, whereby α_m is chosen according to the level of the signal at the mth sensor.

The position of the speaker is then determined according to

$$\hat{\mathbf{x}}_k = \mathrm{argmax}_{\mathbf{x}} = f(\mathbf{x}, \mathbf{Y}_k). \tag{10.58}$$

This implies that a multidimensional search over possible speaker positions is required in order to find the position with maximal steered response, which can potentially be very computationally demanding. Speaker position estimation based on SBRP, however, does not require the estimation of intermediate time delays.

For SBRP-based speaker tracking, the likelihood function can be expressed as

$$F_i(\mathbf{y}_t, \mathbf{x}_k) = w_0 + \sum_{j=1}^{J} w_j \, \mathcal{N}(\mathbf{x}_k; \theta^{(j)}, \sigma^2),$$

where $\theta^{(j)}$ is the jth potential speaker position returned by the localization function (10.58).

10.6 Summary and Further Reading

In this chapter, we have introduced one of the principal supporting technologies required for a complete DSR system, namely, algorithms for determining the physical positions of one or more speakers in a room. Speaker localization and tracking – whether based on acoustic features, video features, or both – are important technologies, because the beamforming algorithms discussed in Chapter 13 all assume that the position of the desired speaker is *known*. Moreover, the accuracy of a speaker tracking system has a very significant influence on the recognition accuracy of the entire system.

A class of algorithms was developed wherein the position estimate is obtained from the intersection of several spheres. The first algorithm in this class was proposed by Schau and Robinson (1987), and later came to be known as spherical intersection. Perhaps the best-known. algorithm from this class is the spherical interpolation method of Smith and Abel (1987). Both methods provide closed-form estimates suitable for real-time implementation. The latter method was recently extended by Huang *et al.* (2004).

Brandstein *et al.* (1997) proposed another closed-form approximation for a speaker's position known as linear intersection. Their algorithm proceeds by first calculating a bearing line to the source for each pair of sensors. Thereafter, the point of nearest approach is calculated for each pair of bearing lines, yielding a potential source location. The final position estimate is obtained from a weighted average of these potential source locations.

The first application of Bayesian filtering techniques, namely, PFs, to the acoustic speaker tracking problem was by Vermaak and Blake (2001), which was quickly followed by Ward *et al.* (2003). Oddly enough, a variant of the more conventional extended KF was not used for acoustically tracking single speakers until some time later by Klee *et al.* (2005a), who found that the KF-based tracking approach described in Section 10.2 provided performance superior to the three conventional techniques described in Section 10.1. The use of the EKF for audio-visual tracking was described in Strobel *et al.* (2001) and Gehrig *et al.* (2005). The results obtained with the latter system are also described in Section 14.6. As described in Section 14.5, the use of probabilistic data association filters for tracking multiple simultaneous speakers was investigated by Gehrig *et al.* (2006).

A system for tracking varying numbers of speakers based on PFs was recently proposed by Quinlan and Asano (2008). A combination of the EKF discussed in Section 10.3 with the PF described in Section 10.5 has recently been proposed by Zhong and Hopgood (2008). Other recent work in acoustic speaker tracking is by Cirillo *et al.* (2008). Brutti *et al.* (2005) address a problem related to tracking a speaker's position, namely, that of determining the orientation of the speaker's head.

10.7 Principal Symbols

Symbol	Description
$\mathbf{\Sigma}$	observation error covariance matrix
$\mathbf{A}$	prearray
$\mathbf{B}$	postarray
c	speed of sound
$d_{mn} = D_m - D_n$	range difference between mth and nth microphones
D_n	distance from speaker to nth microphone
$\mathbf{G}_k$	Kalman gain
$\mathbf{F}_k$	state transition matrix
$\mathbf{H}_k(\mathbf{x})$	observation functional
$\mathbf{K}_k$	filtered state estimation error covariance matrix
$\mathbf{K}_{k\mid k-1}$	predicted state estimation error covariance matrix
$\mathbf{m}_n$	position of nth microphone
M_2	number of unique microphone pairs
M_4	number of unique microphone quadruples
N	total number of microphones
$\mathbf{P}$	camera matrix
R_n	distance from origin to nth microphone
R_s	distance from origin to speaker
$\mathbf{S}_k$	innovation covariance matrix
$T(\mathbf{m}_1, \mathbf{m}_2, \mathbf{x})$	time delay of arrival between mth and nth microphones
$\mathbf{U}_k$	process noise
$w_k^{(m)}$	weight of mth particle at kth time step
$\mathbf{x}$	position of speaker
$\mathbf{x}_k^{(m)}$	mth particle at kth time step

11

Digital Filter Banks

As we will learn in Chapter 13, beamforming in the subband domain offers many advantages in terms of computational complexity and speed of convergence with respect to beamforming in the discrete-time domain. Hence, we are led to consider *filter banks* (FBs), which provide the means for transforming an input signal into the subband domain, and back again. As the transformation from time domain to the subband domain with M subbands would entail an M-fold increase in the amount of data to be processed, we will typically *decimate* the subband signals. Such decimation can lead to aliasing if care is not taken in the design of the FB prototype.

The initial work on digital FBs was all but exclusively devoted to subband coding applications (Jayant and Noll 1984). In such applications, the goal is not to alter the contents of the individual subbands, but to digitally transmit or store them in such a way as to achieve the maximum possible fidelity with the minimum number of bits. Typically this entails allocating different numbers of bits to different subbands. There are two important properties of any FB used for digital encoding of waveforms. The first is *perfect reconstruction* (PR), which implies that the output of the FB is a (possibly) delayed version of the input, and that the signal is neither modified nor distorted in any other way. The second is *maximal decimation*, which implies that if a FB has M subbands, the signals at the outputs of each of the M bands can be decimated by a factor of M without losing the PR property. Clearly the requirement for maximal decimation is directly related to the goal of obtaining the highest fidelity encoding with the fewest possible bits.

Some time later it was realized that FBs were also potentially useful for adaptive filtering applications (Haykin 2002). By this time, however, the design of digital FBs was largely considered to be a "solved problem". Hence, researchers in this area were somewhat slow to realize that adaptive filtering imposed very different requirements on a FB than subband coding. As a result, much of the important work on FB design for adaptive filtering and beamforming applications has appeared only in the last several years.

For these reasons, we will present here FB designs based largely on the work of De Haan *et al.* (2003). For such designs, aliasing is not eliminated through cancellation. Rather, the analysis and synthesis prototypes are designed so as to minimize a weighted

Distant Speech Recognition   Matthias Wölfel and John McDonough
© 2009 John Wiley & Sons, Ltd

combination of the residual aliasing distortion and the total response error of the FB. As the subbands are then independent, the performance of the FB does not radically degrade when different scale factors and phase shifts are applied to the subbands. In exchange for this independence, however, it is necessary to give up the computational efficiency of maximal decimation, as these designs are highly dependent on the use of *oversampling* to reduce the residual aliasing distortion to an acceptable level.

In Section 11.1, we introduce the concepts and notation to be used in the balance of the chapter. In particular, we describe the basic idea of modulating a single prototype in order to obtain the impulse responses for an entire bank of filters. Section 11.2 describes how the FBs described in the prior section can be efficiently implemented with the polyphase representation. As a digital FB converts a single time series into the complex output of an entire bank of filters, it produces a large increase in the data rate. Hence, the output of each filter is typically first decimated by retaining only every Dth sample, then subsequently expanded by inserting lengths of zeros. The effects of these operations in the frequency domain are described in Section 11.3. Section 11.4 describes how the processing blocks of the entire FB can be arranged such that all components run at their lowest possible rate. In Section 11.5, we present the Nyquist(M) criterion, which dictates how a single FB prototype can be designed in order to ensure that the sum of all modulated versions thereof sum to a constant in the frequency domain. Having discussed the principles of digital FBs, Sections 11.6 through 11.7.3 describe three prototype design techniques, while Section 11.8 presents metrics comparing the performance of each of these designs.

11.1 Uniform Discrete Fourier Transform Filter Banks

Let us begin our discussion by formally defining a FB as a collection of M filters with a common input and a common output. Such a system is shown in Figure 11.1. The set of transfer functions $\{H_m(z)\}$ comprise the *analysis filter bank*, which splits the input $x[n]$ into M subband signals $\{X_m[n]\}_{m=0}^{M-1}$. The set $\{G_m(z)\}$ of transfer functions comprise the *synthesis filter bank*, which recombines the M subband signals $\{Y_m[n]\}_{m=0}^{M-1}$ into a single output $\hat{x}[n]$. Each $Y_m[n]$ is obtained by multiplying $X_m[n]$ with a complex constant, which is determined with an adaptive filtering or beamforming operation. We consider then a class of FBs wherein the impulse response of each filter is obtained by modulating a prototype impulse response $h_0[n]$ according to

$$h_m[n] = h_0[n]\, e^{j2\pi nm/M} \ \forall\, m = 0, \ldots, M-1, \tag{11.1}$$

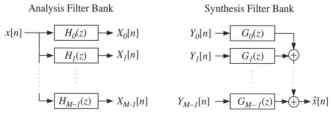

Figure 11.1 Analysis and synthesis filter banks with individual impulse response $\{H_m(z)\}$ and $\{G_m(z)\}$, respectively

which implies that the prototype $h_0[n]$ is modulated as discussed in Section 3.1.2 to obtain the impulse responses $h_m[n]$ of all other filters. The similarity of (3.84) and (11.1) is readily apparent; to make the two equivalent, we need only set $h_0[n] = w[-n]$. Applying the z-transform of both sides of (11.1) we obtain

$$H_m(z) \triangleq H_0(zW_M^m), \tag{11.2}$$

where $W_M = e^{-j2\pi/M}$ is the Mth root of unity. Typically the subscript of W_M will be suppressed when it is clear from the context. Equation (11.2) implies that $H_m(e^{j\omega})$ is a *shifted version* of the frequency response of $H_0(e^{j\omega})$ according to

$$H_m(e^{j\omega}) = H_0(e^{j(\omega - 2\pi m/M)}), \tag{11.3}$$

which follows from (3.26). Likewise, the similarity of (3.85) and (11.2) is quite evident. Making them equivalent requires only that we specify $H_0(z) = W(z^{-1})$.

Similarly, for the synthesis bank, we will assume that the impulse responses of the individual filters are related by

$$g_m[n] = g_0[n]\,e^{j2\pi nm/M},$$

so that we can write

$$G_m(z) \triangleq G_0(zW_M^m). \tag{11.4}$$

A particularly simple prototype impulse response is given by

$$h_0[n] = \begin{cases} 1, & 0 \le n \le M - 1, \\ 0, & \text{otherwise,} \end{cases} \tag{11.5}$$

or, in the z-transform domain,

$$H_0(z) = 1 + z^{-1} + \cdots + z^{-(M-1)} = \frac{1 - z^M}{1 - z}. \tag{11.6}$$

The frequency response of the simple filter $H_0(z)$ can be obtained by substituting $z = e^{j\omega}$ into the last equation in (11.6), then factoring the terms $e^{j\omega M/2}$ and $e^{j\omega/2}$ out of the numerator and denominator, respectively. This provides

$$H_0(e^{j\omega}) = \frac{\sin(M\omega/2)}{\sin(\omega/2)}\, e^{-j\omega(M-1)/2}, \tag{11.7}$$

the magnitude of which is plotted in Figure 11.2. While this response is undoubtedly lowpass in nature, the suppression in the stopband is poor in that the first sidelobe is only 13 dB below the main lobe. We will shortly see how this stopband attenuation can be improved by replacing the simple prototype in (11.5) with a prototype designed to satisfy a number of criteria. In the meantime, we will demonstrate that the analysis bank

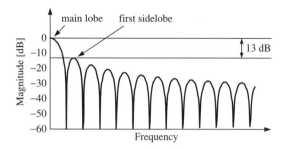

Figure 11.2 Magnitude of the frequency response $H_0(e^{j\omega})$ in (11.7)

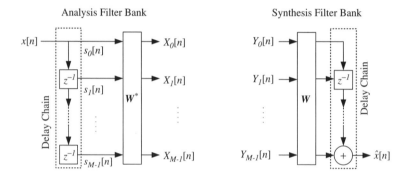

Figure 11.3 The uniform DFT filter bank

in Figure 11.1 can be implemented as shown in Figure 11.3, where **W** denotes the *discrete Fourier transform matrix*, which is defined as

$$[\mathbf{W}]_{mn} \triangleq W_M^{mn} = e^{j2\pi mn/M}, \tag{11.8}$$

and $\mathbf{W}^*$ denotes the *inverse discrete Fourier transform matrix*. To begin, let us define the sequences $s_i[n] = x[n - i]$ at the outputs of the delay chain of the analysis filter bank in Figure 11.3. From the definition (11.8) of **W**, it then follows that

$$X_m[n] = \sum_{i=0}^{M-1} s_i[n]\, W_M^{-mi}. \tag{11.9}$$

Apart from a factor of M, we see that $X_m[n]$ is the inverse *discrete Fourier transform* (DFT) of $s_i[n]$ over the time index i. Now the z-transform of $X_m[n]$ can be expressed as

$$X_m(z) = \sum_{i=0}^{M-1} S_i(z)\, W^{-mi} = \sum_{i=0}^{M-1} z^{-i}\, W^{-mi}\, X(z) = \sum_{i=0}^{M-1} \left(z\, W^m\right)^{-i} X(z). \tag{11.10}$$

The first equation in (11.10) follows from (11.9) and the linearity of the z-transform. The second equation follows from $s_i[n] = x[n - i]$. The final equality in (11.10) can be rewritten as

$$X_m(z) = H_m(z)\, X(z),$$

where

$$H_m(z) = \sum_{i=0}^{M-1} \left(z W^m\right)^{-i} = H(z W^m),$$

which is the desired result.

A FB in which the responses of the filters are related as in (11.2) and (11.4) is known as a *uniform DFT filter bank* (Vaidyanathan 1993, sect. 4.1.2). A physical meaning can readily be attached to the output $X_m[n]$ of the individual filters. In light of (11.9), we can, for fixed n, write

$$X_m[n + M - 1] = \sum_{i=0}^{M-1} x[n + M - 1 - i]\, W^{-mi} = W^m \sum_{l=0}^{M-1} x[n + l]\, W^{ml}, \qquad (11.11)$$

where the second equation follows from the change of variables $l = M - 1 - i$ and the fact that $W^M = 1$. It is clear that $X_m[n]$ is W^m times the mth point of the DFT of the length-M sequence

$$x[n], x[n + 1], \ldots, x[n + M - 1].$$

Thus, as was depicted in Figure 3.4, a stationary window $w[m]$ is used to isolate a segment of $x[m]$, which has been shifted n samples to the left, for the calculation of a M-point DFT. For the simple prototype in (11.6), the window would be rectangular. Other windows could be used, however, to reduce the size of the large sidelobes present in Figure 11.2. In this case, the window would be implemented as a set of unequal scale factors applied to the outputs of the delay chain in Figure 11.3. Note that the nonzero portion of $w[m]$ may also be larger than M, in which case the windowed sequence $w[m] x[n + m]$ must be *time-aliased* prior to the inverse DFT. Such a filter bank design is more useful for beamforming, which requires fine frequency resolution, as opposed to automatic speech recognition, which requires minimal frequency resolution but relatively good time resolution. As previously mentioned, frequency resolution can, in most cases, only be increased at the expense of time resolution and vice versa. Finally, upon comparing the definition (3.80) of the short-time Fourier transform with (11.11) after setting $L = M$, it becomes evident that

$$X_m[n + M - 1] = W^m\, \overline{X}_m[n],$$

provided that $w[n]$ in the former relation is rectangular. The last equation demonstrates the equivalence of the short-time Fourier transform and the output of an uniform DFT filter bank.

11.2 Polyphase Implementation

The *polyphase representation* provides an important insight in that it simplifies many important theoretical results and can be used to achieve an efficient implementation of digital FBs. To begin the discussion of such an implementation, consider a filter

$$H(z) = \sum_{n=-\infty}^{\infty} h[n] z^{-n}.$$

We can develop an equivalent representation by separating $H(z)$ into odd and even components, such that

$$H(z) = \underbrace{\sum_{n=-\infty}^{\infty} h[2n] z^{-2n}}_{E_0(z^2)} + z^{-1} \underbrace{\sum_{n=-\infty}^{\infty} h[2n+1] z^{-2n}}_{E_1(z^2)},$$

where $E_0(z)$ and $E_1(z)$ are the *Type 1 polyphase components*. In a more general form, we can decompose $H(z)$ into M bands by writing

$$H(z) = \sum_{n=-\infty}^{\infty} h[nM] z^{-nM} + z^{-1} \sum_{n=-\infty}^{\infty} h[nM+1] z^{-nM}$$

$$\cdots + z^{-(M-1)} \sum_{n=-\infty}^{\infty} h[nM+M-1] z^{-nM}.$$

The latter is equivalent to the *Type 1 polyphase representation*,

$$H(z) = \sum_{m=0}^{M-1} z^{-m} E_m(z^M) \qquad (11.12)$$

where

$$E_m(z) \triangleq \sum_{n=-\infty}^{\infty} e_m[n] z^{-n},$$

with

$$e_m[n] \triangleq h[nM+m] \,\forall\, 0 \le m \le M-1.$$

A useful variation of (11.12) can be written as the *Type 2 polyphase representation*

$$H(z) = \sum_{m=0}^{M-1} z^{-(M-1-m)} R_m(z^M) \qquad (11.13)$$

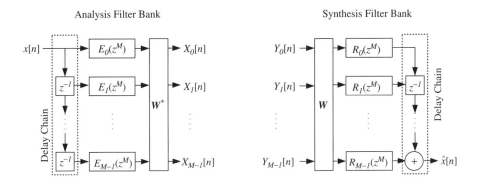

Figure 11.4 Polyphase implementation of the analysis and synthesis filter banks

where the Type 2 polyphase components $R_m(z)$ are permutations of $E_m(z)$, such that $R_m(z) = E_{M-1-m}(z)$ (Vaidyanathan 1993, sect. 4.3).

For a prototype $H_0(z)$ which has been decomposed into polyphase components as in (11.12), the response (11.2) of the mth filter can be expressed as

$$H_m(z) = H_0(zW^m) = \sum_{l=0}^{M-1} (z^{-1}W^{-m})^l E_l(z^M), \qquad (11.14)$$

which holds because $(zW^m)^M = z^M$. The output of $H_m(z)$ can then be expressed as

$$X_m(z) = H_m(z)X(z) = \sum_{l=0}^{M-1} W^{-lm}\left(z^{-l}E_l(z^M)X(z)\right),$$

which implies the analysis FB can be implemented as shown on the left side of Figure 11.4. Similar development can be used to demonstrate that the synthesis bank can be implemented as shown on the right side of Figure 11.4.

11.3 Decimation and Expansion

Having defined the modulated analysis and synthesis FBs, we now introduce two all important operations and consider their effects in the frequency domain. The first operation is that of decimation. An M-fold *decimator* with input $x[n]$ has output

$$y_D[n] = x[nM],$$

for integer M. In the frequency domain, the output of the decimator can be expressed as

$$Y_D(e^{j\omega}) = \frac{1}{M}\sum_{m=0}^{M-1} X(e^{j(\omega-2\pi m)/M}), \qquad (11.15)$$

as we now show. The z-transform of $y_D[n]$ can be written as

$$Y_D(z) = \sum_{n=-\infty}^{\infty} y_D[n]\, z^{-n} = \sum_{n=-\infty}^{\infty} x[nM]\, z^{-n}.$$

We define the intermediate sequence

$$x_1[n] = \begin{cases} x[n], & n = \text{multiple of } M, \\ 0, & \text{otherwise}, \end{cases} \tag{11.16}$$

so that $y_D[n] = x[nM] = x_1[nM]$. Then

$$Y_D(z) = \sum_{n=-\infty}^{\infty} x_1[nM]\, z^{-n} = \sum_{m=-\infty}^{\infty} x_1[m]\, z^{-m/M},$$

which holds because $x_1[n]$ is zero whenever n is other than a multiple of M. Hence,

$$Y_D(z) = X_1(z^{1/M}). \tag{11.17}$$

Now we need only express $X_1(z)$ in terms of $X(z)$. Note that (11.16) can be rewritten as

$$x_1[n] = C_M[n]\, x[n], \tag{11.18}$$

where $C_M[n]$ is the *comb* sequence defined as

$$C_M[n] = \begin{cases} 1, & n = \text{multiple of } M, \\ 0, & \text{otherwise}. \end{cases}$$

Equation (3.59) indicates that the comb sequence can then be expressed as

$$C_M[n] = \frac{1}{M} \sum_{m=0}^{M-1} W_M^{-mn}. \tag{11.19}$$

Substituting (11.19) into (11.18) yields

$$X_1(z) = \sum_{n=-\infty}^{\infty} x_1[n]\, z^{-n} = \sum_{n=-\infty}^{\infty} \left(\frac{1}{M} \sum_{m=0}^{M-1} W_M^{-mn} \right) x[n]\, z^{-n}.$$

Regrouping terms on the right-hand side of the last equation, we find,

$$X_1(z) = \frac{1}{M} \sum_{m=0}^{M-1} \sum_{n=-\infty}^{\infty} x[n]\, W^{-mn}\, z^{-n} = \frac{1}{M} \sum_{m=0}^{M-1} \sum_{n=-\infty}^{\infty} x[n]\, (zW^m)^{-n}.$$

The inner summation above is equal to $X(zW^m)$ so that from (11.17),

$$Y_{\mathrm{D}}(z) = \frac{1}{M} \sum_{m=0}^{M-1} X(z^{1/M} W^m), \tag{11.20}$$

which is clearly equivalent to (11.15) when z is replaced by $e^{j\omega}$.

The effect of the decimator in the frequency domain, which we now summarize, is illustrated in Figure 11.5. Firstly, the decimator stretches the original spectrum $X(e^{j\omega})$ by a factor of M to form $X(e^{j\omega/M})$. Secondly, $M-1$ copies of the stretched spectrum are created by shifting $X(e^{j\omega/M})$ by increments of 2π. Then these shifted versions of the spectrum are summed together and divided by M.

An L–fold *expander* takes as input $x[n]$

$$y_{\mathrm{E}}[n] = \begin{cases} x[n/L], & \text{if } n \text{ is an integer-multiple of } L, \\ 0, & \text{otherwise,} \end{cases} \tag{11.21}$$

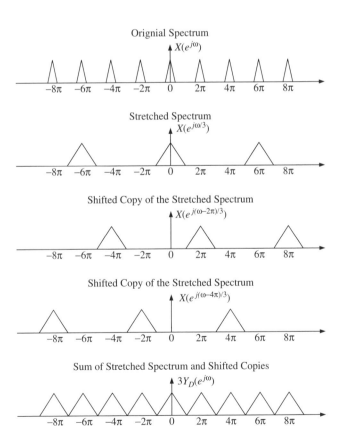

Figure 11.5   Effect of decimation in the frequency domain

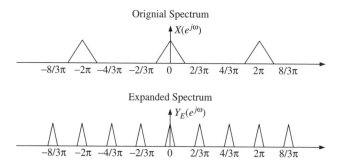

Figure 11.6 Effect of expansion in the frequency domain

for integer L. We will now analyze the effects of expansion in the frequency domain. Based on (11.21), we can write

$$Y_E(z) = \sum_{n=-\infty}^{\infty} y_E[n] z^{-n} = \sum_{n=\text{multiple of } L} y_E[n] z^{-n} \tag{11.22}$$

$$= \sum_{m=-\infty}^{\infty} y_E[mL] z^{-mL} = \sum_{m=-\infty}^{\infty} x[m] z^{-mL}. \tag{11.23}$$

The last equality clearly implies

$$Y_E(z) = X(z^L). \tag{11.24}$$

The effect of the expander in the frequency domain is illustrated in Figure 11.6. From the figure it is clear that the expander simply scales the frequency axis, thereby causing the images, which were previously centered at $\omega = 0, \pm 2\pi, \pm 4\pi, \ldots$, to move down within the range $-\pi \leq \omega < \pi$.

11.4   Noble Identities

Figure 11.7 depicts the *noble identities* in schematic form. As we will shortly learn, these identities are very useful for positioning the decimation and expansion blocks of a FB such that all components run at the lowest possible rate. To prove these identities, we

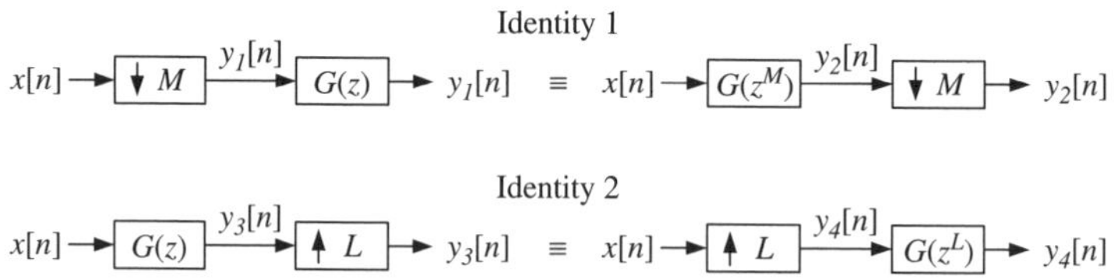

Identity 1

$$x[n] \rightarrow \boxed{\downarrow M} \xrightarrow{y_1[n]} \boxed{G(z)} \rightarrow y_1[n] \quad \equiv \quad x[n] \rightarrow \boxed{G(z^M)} \xrightarrow{y_2[n]} \boxed{\downarrow M} \rightarrow y_2[n]$$

Identity 2

$$x[n] \rightarrow \boxed{G(z)} \xrightarrow{y_3[n]} \boxed{\uparrow L} \rightarrow y_3[n] \quad \equiv \quad x[n] \rightarrow \boxed{\uparrow L} \xrightarrow{y_4[n]} \boxed{G(z^L)} \rightarrow y_4[n]$$

Figure 11.7   The noble identities for multirate systems

begin by defining $y_2'[n] \leftrightarrow Y_2'(z)$ as the output of the $G(z^M)$ block on the right-hand side of Identity 1 in Figure 11.7. We can write

$$Y_2'(z) = G(z^M) X(z).$$

Then based on (11.20),

$$Y_2(z) = \frac{1}{M} \sum_{m=0}^{M-1} Y_2'(z^{1/M} W^m) = \frac{1}{M} \sum_{m=0}^{M-1} G((z^{1/M} W^m)^M) X(z^{1/M} W^m)$$

$$= G(z) \cdot \frac{1}{M} \sum_{m=0}^{M-1} X(z^{1/M} W^m), \tag{11.25}$$

where the final equality follows from $W^{mM} = 1$. Comparing (11.20) and (11.25), it becomes evident that $y_1[n] \leftrightarrow Y_1(z)$ is equivalent to $y_2[n] \leftrightarrow Y_2(z)$, thereby completing the proof of Identity 1.

Now let $y_4'[n] \leftrightarrow Y_4'(z)$ denote the output of the expansion block on the right-hand side of Identity 2, whereupon it follows

$$Y_4(z) = G(z^L) Y_4'(z) = G(z^L) X(z^L), \tag{11.26}$$

where the last expression on the right-hand side of (11.26) follows from (11.24). From the last expression, it is clear that $y_3[n] \leftrightarrow Y_3(e^{j\omega})$ is equivalent to $y_4[n] \leftrightarrow Y_4(e^{j\omega})$, thereby completing the proof.

Without further modification, an analysis bank with M subbands such as that shown in Figure 11.4 would represent an M-fold increase in the data rate. Hence, as mentioned previously the output of the analysis bank is typically *decimated* by some factor D. Were the decimation applied only to the subband outputs, the data rate of each of the polyphase filters would still be equivalent to the input rate of the entire system. To reduce the amount of computation, the first noble identity can be used to push the decimation block all the way to left in the analysis bank, so that each polyphase component runs at the rate M/D. To reconstruct the signal in the synthesis bank, an expansion block is applied to the subband samples from the analysis blank. In order to minimize the data rate in this case, the second noble identity is used to push the expansion block all the way to the right. The resulting analysis and synthesis banks are show in Figure 11.8.

11.5 Nyquist(M) Filters

Suppose that a filter function $H(z)$ has been represented in Type 1 polyphase form, and the zeroth polyphase component is constant, such that

$$H(z) = c + z^{-1} E_1(z^M) + \cdots + z^{-(M-1)} E_{M-1}(z^M). \tag{11.27}$$

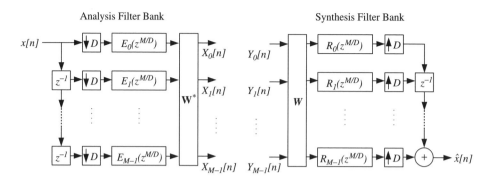

Figure 11.8 Analysis and synthesis banks after pushing the decimation and expansion blocks all the way to the left or right, respectively

A filter with this property is said to be a *Nyquist(M)* or *Mth band filter* (Vaidyanathan 1993, sect. 4.6.1), and its impulse response clearly satisfies

$$h[Mn] = \begin{cases} c, & n = 0, \\ 0, & \text{otherwise.} \end{cases} \tag{11.28}$$

The definition in (11.27) can be generalized by assuming that

$$H(z) = cz^{-m_\mathrm{d}M} + z^{-1} E_1(z^M) + \cdots + z^{-(M-1)} E_{M-1}(z^M), \tag{11.29}$$

in which case, the impulse response of $H(z)$ must then satisfy

$$h[Mn] = \begin{cases} c, & n = m_\mathrm{d}, \\ 0, & \text{otherwise} \end{cases} \tag{11.30}$$

where m_d is the input delay.

As shown in Section B.14, if $H(z)$ satisfies (11.27) with $c = 1/M$, then

$$\sum_{m=0}^{M-1} H(zW_M^m) = Mc = 1, \tag{11.31}$$

where $W_M = e^{-j2\pi/M}$, as before. Hence, all M uniformly shifted versions of $H(e^{j\omega})$ add up to a constant. Similarly, if $H(z)$ satisfies (11.29) then

$$\sum_{m=0}^{M-1} H(zW_M^m) = z^{-m_\mathrm{d}M}, \tag{11.32}$$

in which case, in the absence of decimation, the output of analysis FB would be equivalent to the input delayed by $m_\mathrm{d}M$ samples.

11.6   Filter Bank Design of De Haan *et al.*

This section considers the design technique of De Haan, which is a typical example of a "late" FB design. In this method, a separate analysis and synthesis prototype are designed so as to minimize a weighted combination of the total response error and aliasing distortion.

Figure 11.9 shows a schematic of a direct form implementation of uniform DFT analysis and synthesis filter banks. The outputs $V_m(z)$ of the uniform analysis filters can be expressed as

$$V_m(z) = H_m(z)X(z) = H(zW_M^m)X(z) \; \forall \, m = 0, \dots, M-1.$$

As discussed in Section 11.3, the decimators then expand and copy the output of the synthesis filters according to

$$X_m(z) = \frac{1}{D}\sum_{d=0}^{D-1} V_m(z^{1/D}W_D^d) = \frac{1}{D}\sum_{d=0}^{D-1} H(z^{1/D}W_M^m W_D^d)X(z^{1/D}W_D^d). \qquad (11.33)$$

The last equation indicates that $X_m(z)$ consists of the sum of a stretched output of the mth FB and $D-1$ aliasing terms.

At this point, the "fixed" subband weights F_m can be applied to the decimated signals to achieve the desired adaptive filtering effect:

$$Y_m(z) = F_m X_m(z). \qquad (11.34)$$

The expanders then compress the signals $Y_m(z)$ according to

$$U_m(z) = Y_m(z^D) = \frac{1}{D}F_m \sum_{d=0}^{D-1} H(zW_M^m W_D^d)X(zW_D^d). \qquad (11.35)$$

In the last step, the signals $U_m(z)$ are processed by the synthesis filters $G_m(z)$ in order to suppress the spectral images created by aliasing, and the outputs of the synthesis filters

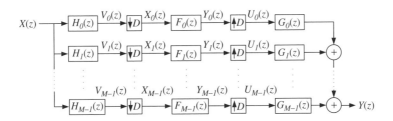

Figure 11.9   Schematic of a direct form implementation of uniform DFT analysis and synthesis filter banks

are summed together

$$Y(z) = \sum_{m=0}^{M-1} U_m(z) \, G_m(z). \tag{11.36}$$

Then the final relation between input and output can be expressed as

$$Y(z) = \frac{1}{D} \sum_{d=0}^{D-1} X(z W_D^d) \sum_{m=0}^{M-1} F_m \, H(z W_M^m W_D^d) \, G(z W_M^m). \tag{11.37}$$

Upon defining

$$A_{m,d}(z) \triangleq \frac{1}{D} F_m \, H(z W_M^m W_D^d) \, G(z W_M^m), \tag{11.38}$$

the output relation (11.37) can be written more conveniently as

$$Y(z) = \sum_{d=0}^{D-1} A_d(z) \, X(z W_D^d), \tag{11.39}$$

where

$$A_d(z) \triangleq \sum_{m=0}^{M-1} A_{m,d}(z). \tag{11.40}$$

Note that the transfer function $A_0(z)$ produces the desired signal, while the remaining transfer functions $A_d(z) \, \forall \, d = 1, \ldots, D - 1$ give rise to residual aliasing in the output signal.

11.6.1   Analysis Prototype Design

To design a suitable prototype for the passband, a number of approaches might be adopted. One possibility would be to specify some desirable behavior in the passband $\Omega_p = [-\omega_p, \omega_p]$ for some $\omega_p > 0$. For example, we might define the *desired passband frequency response* as constant in magnitude with a linear phase delay:

$$H_d(e^{j\omega}) = e^{-j\omega \tau_H} \text{ for } \omega \in \Omega_p, \tag{11.41}$$

where τ_H is the delay in samples introduced by the analysis bank. If a relation such as (11.41) were to hold also for the response of the entire FB, then the output of the FB, in the absence of aliasing distortion, would simply be a delayed version of the input. In a popular PR design (Vaidyanathan 1993, sect. 8), a *para-unitary constraint* is imposed on the filter prototype through the imposition of a *lattice structure* on its components. This para-unitary constraint ensures that the FB will have the desired PR property. Subject to this PR constraint, the energy in the stopband of the prototype is minimized in order to provide good frequency resolution.

The design method of De Haan *et al.* (2003) begins with the specification of a desired response as in (11.41), then defines the *passband response error*,

$$\epsilon_p \triangleq \frac{1}{2\omega_p} \int_{-\omega_p}^{\omega_p} \left| H(e^{j\omega}) - H_d(e^{j\omega}) \right|^2 d\omega. \tag{11.42}$$

Then, instead of minimizing the energy of the filter prototype in the stopband, De Haan *et al.* sought to directly minimize the *inband-aliasing distortion*, defined as

$$\epsilon_i \triangleq \frac{1}{2\pi D^2} \int_{-\pi}^{\pi} \sum_{d=1}^{D-1} \left| H(e^{j\omega/D} W_D^d) \right|^2 d\omega. \tag{11.43}$$

The design of the analysis prototype $h[n]$ is then based on minimizing the objective function

$$\epsilon_h \triangleq \epsilon_p + \epsilon_i. \tag{11.44}$$

By defining, respectively, the *stacked prototype* and *stacked delay chain*

$$\mathbf{h} \triangleq \begin{bmatrix} h[0] & h[1] & \cdots & h[L_h - 1] \end{bmatrix}^T, \tag{11.45}$$

$$\boldsymbol{\phi}_{\mathbf{h}}(z) \triangleq \begin{bmatrix} 1 & z^{-1} & \cdots & z^{-(L_h-1)} \end{bmatrix}^T, \tag{11.46}$$

where L_h is the length of $\mathbf{h}$, it can be readily demonstrated that the passband response error can be expressed as (De Haan 2001)

$$\epsilon_p = \mathbf{h}^T \mathbf{A} \mathbf{h} - 2 \mathbf{h}^T \mathbf{b} + 1, \tag{11.47}$$

where

$$\mathbf{A} = \frac{1}{2\omega_p} \int_{-\omega_p}^{\omega_p} \boldsymbol{\phi}_{\mathbf{h}}(e^{j\omega}) \boldsymbol{\phi}_{\mathbf{h}}^H(e^{j\omega}) d\omega, \tag{11.48}$$

$$\mathbf{b} = \frac{1}{2\omega_p} \int_{-\omega_p}^{\omega_p} \mathrm{Re}\left\{ e^{j\omega\tau_H} \boldsymbol{\phi}_{\mathbf{h}}(e^{j\omega}) \right\} d\omega. \tag{11.49}$$

Typically $L_h = m_h M$ for integer $m_h > 1$. Based on (11.48–11.49), the components of $\mathbf{A}$ and $\mathbf{b}$ can be expressed as

$$A_{m,l} = \mathrm{sinc}(\omega_p(l - m)), \tag{11.50}$$

$$b_m = \mathrm{sinc}(\omega_p(\tau_H - m)), \tag{11.51}$$

where

$$\mathrm{sinc}\, x = \frac{\sin x}{x}.$$

Equations (11.50–11.51) can be readily verified, as we now demonstrate. Let us rewrite (11.42) as

$$
\epsilon_p = \frac{1}{2\omega_p} \int_{-\omega_p}^{\omega_p} \left\{ H(e^{j\omega}) H(e^{-j\omega}) - H(e^{j\omega}) e^{j\omega\tau_H} - H(e^{-j\omega}) e^{-j\omega\tau_H} + 1 \right\} d\omega
$$

$$
= \frac{1}{2\omega_p} \int_{-\omega_p}^{\omega_p} \left\{ \mathbf{h}^T \boldsymbol{\phi}(e^{j\omega}) \boldsymbol{\phi}^T(e^{-j\omega}) \mathbf{h} - 2\operatorname{Re}\left[\mathbf{h}^T \boldsymbol{\phi}(e^{j\omega}) e^{j\omega\tau_H} \right] + 1 \right\} d\omega,
$$

from which (11.47–11.49) follow. The (m, n)th component of $\mathbf{A}$ is readily calculated as

$$
A_{m,n} = \frac{1}{2\omega_p} \int_{-\omega_p}^{\omega_p} e^{j\omega(n-m)} d\omega = \frac{1}{2\omega_p} \frac{1}{j(n-m)} \left[e^{j\omega_p(n-m)} - e^{-j\omega_p(n-m)} \right]
$$

$$
= \frac{1}{2\omega_p} \frac{1}{j(n-m)} \cdot 2j \sin \omega_p(n-m). \tag{11.52}
$$

Then (11.50) follows readily from (11.52). The nth component of $\mathbf{b}$ is also straightforward to calculate as

$$
b_n = \frac{1}{2\omega_p} \int_{-\omega_p}^{\omega_p} e^{j\omega(\tau_H - n)} d\omega = \frac{1}{2\omega_p} \frac{1}{j(\tau_H - n)} \cdot 2j \sin \omega_p(\tau_H - n),
$$

from which (11.51) follows.

Similarly, the inband aliasing term (11.43) can be expressed as

$$
\epsilon_i = \frac{1}{2\pi} \sum_{d=1}^{D-1} \mathbf{h}^T \left[\int_{-\pi}^{\pi} \boldsymbol{\phi}_\mathbf{h}\left(e^{j\omega/D} W_D^d \right) \boldsymbol{\phi}_\mathbf{h}^H \left(e^{j\omega/D} W_D^d \right) d\omega \right] \mathbf{h}. \tag{11.53}
$$

The last equation can be rewritten as

$$
\epsilon_i = \mathbf{h}^T \mathbf{C} \mathbf{h}, \tag{11.54}
$$

where

$$
\mathbf{C} = \frac{1}{2\pi} \sum_{d=1}^{D-1} \int_{-\pi}^{\pi} \boldsymbol{\phi}_\mathbf{h}\left(e^{j\omega/D} W_D^d \right) \boldsymbol{\phi}_\mathbf{h}^H \left(e^{j\omega/D} W_D^d \right) d\omega. \tag{11.55}
$$

The components of $\mathbf{C}$ can then be expressed as

$$
C_{m,l} = \frac{\varphi[l - m] \sin\left(\frac{\pi(l-m)}{D} \right)}{\pi(l - m)/D}, \tag{11.56}
$$

where

$$
\varphi[m] = D \sum_{n=-\infty}^{\infty} \delta[m - nD] - 1. \tag{11.57}
$$

Combining all terms above, De Haan *et al.* (2003) then minimized the objective function

$$\epsilon_{\mathbf{h}} = \epsilon_{\mathrm{p}} + \epsilon_{\mathrm{i}} = \mathbf{h}^T (\mathbf{A} + \mathbf{C})\mathbf{h} - 2\mathbf{h}^T \mathbf{b} + 1. \tag{11.58}$$

The optimal prototype **h** must thus satisfy

$$(\mathbf{A} + \mathbf{C})\mathbf{h} = \mathbf{b}. \tag{11.59}$$

11.6.2 Synthesis Prototype Design

We now derive the optimal synthesis prototype. Let us denote the stacked synthesis prototype as

$$\mathbf{g} = \begin{bmatrix} g[0] & g[1] & \cdots & g[L_{\mathbf{g}} - 1] \end{bmatrix}^T,$$

where $L_{\mathbf{g}}$ is the length of **g**. As in the case of the analysis prototype, the length of the synthesis prototype is chosen to be an integer multiple $m_{\mathbf{g}}$ of M, such that $L_{\mathbf{g}} = m_{\mathbf{g}} M$. In order to design the synthesis prototype, De Haan *et al.* (2003) took as an objective function

$$\epsilon_{\mathbf{g}}(\mathbf{h}) = \epsilon_{\mathrm{t}}(\mathbf{h}) + \epsilon_{\mathrm{r}}(\mathbf{h}), \tag{11.60}$$

where the *total response error* is defined as

$$\epsilon_{\mathrm{t}}(\mathbf{h}) \triangleq \frac{1}{2\pi} \int_{-\pi}^{\pi} \left| A_0(e^{j\omega}) - e^{-j\omega\tau_{\mathrm{T}}} \right|^2 d\omega, \tag{11.61}$$

the total analysis–synthesis FB delay is denoted as τ_{T}, and the *residual aliasing distortion* is

$$\epsilon_{\mathrm{r}}(\mathbf{h}) \triangleq \frac{1}{2\pi} \sum_{d=1}^{D-1} \sum_{m=0}^{M-1} \int_{-\pi}^{\pi} \left| A_{m,d}(e^{j\omega}) \right|^2 d\omega. \tag{11.62}$$

Note the functional dependence of (11.60–11.62) on the analysis prototype **h**. Through manipulations similar to those used in deriving the quadratic objective criterion for the analysis FB, it can be shown that

$$\epsilon_{\mathrm{t}}(\mathbf{h}) = \mathbf{g}^T \mathbf{E} \mathbf{g} - 2\mathbf{g}^T \mathbf{f} + 1, \tag{11.63}$$

where the components of **E** and **f** are given by

$$E_{m,l} = \frac{M^2}{D^2} \sum_{n=-\infty}^{\infty} h^*[nM - m] \, h[nM - l], \tag{11.64}$$

$$f_m = \frac{M}{D} h[\tau_{\mathrm{T}} - m]. \tag{11.65}$$

Similarly, the quadratic form for the residual aliasing distortion is

$$\epsilon_r(\mathbf{h}) = \mathbf{g}^T \mathbf{P} \mathbf{g}, \tag{11.66}$$

where the components of $\mathbf{P}$ are given by

$$P_{m,l} = \frac{M}{D^2} \sum_{n=-\infty}^{\infty} h^*[n+l]\, h[n+m]\, \varphi[m-l], \tag{11.67}$$

$$\varphi[m] = D \sum_{n=-\infty}^{\infty} \delta[m-nD] - 1. \tag{11.68}$$

De Haan *et al.* (2003) introduce a weighting factor v to emphasize either the total response error $(0 < v < 1)$ or residual aliasing distortion $(v > 1)$, such that the final error metric is given by

$$\epsilon_g(\mathbf{h}) = \epsilon_t(\mathbf{h}) + v\epsilon_r(\mathbf{h}) = \mathbf{g}^T (\mathbf{E} + v\mathbf{P})\mathbf{g} - 2\mathbf{g}^T \mathbf{f} + 1. \tag{11.69}$$

In this case, the optimal synthesis prototype $\mathbf{g}$ must satisfy

$$(\mathbf{E} + v\mathbf{P})\mathbf{g} = \mathbf{f}. \tag{11.70}$$

11.7 Filter Bank Design with the Nyquist(M) Criterion

We now present a variation on the FB design of De Haan *et al.* Consider again the Nyquist(M) filters discussed in Section 11.5. A moment's thought will reveal that (11.32) represents a much stronger condition than that aimed at by the minimization of (11.42) or (11.61), inasmuch as (11.32) implies that the response error will vanish, not just for the passband of a single filter, but for the entire working spectrum, including the transition bands between the passbands of adjacent filters. Hence, we are led to consider replacing the terms ϵ_p and $\epsilon_t(\mathbf{h})$ in the optimization criteria (11.44) and (11.60), respectively, with constraints of the form (11.30). This section presents the details of such an approach.

11.7.1 Analysis Prototype Design

Under the constraint (11.30), design of the optimal analysis prototype $\mathbf{h}$ reduces to minimizing the inband aliasing distortion (11.54). To exclude the trivial solution $\mathbf{h} = \mathbf{0}$ from this optimization problem, we impose the additional constraint

$$\mathbf{h}^T \mathbf{h} = 1, \tag{11.71}$$

which is readily achieved through the method of *undetermined Lagrange multipliers*. We posit the modified objective function

$$f(\mathbf{h}) = \mathbf{h}^T \mathbf{C} \mathbf{h} + \lambda (\mathbf{h}^T \mathbf{h} - 1), \tag{11.72}$$

where λ is a *Lagrange multiplier*. Upon setting

$$\nabla_{\mathbf{h}} f(\mathbf{h}) = \mathbf{0},$$

we find

$$\mathbf{C}\mathbf{h} + \lambda\mathbf{h} = \mathbf{0},$$

which implies

$$\mathbf{C}\mathbf{h} = -\lambda\mathbf{h}. \tag{11.73}$$

Hence, $\mathbf{h}$ is clearly an eigenvector of $\mathbf{C}$. Moreover, in order to ensure that $\mathbf{h}$ minimizes (11.54), it must be that eigenvector associated with the *smallest* eigenvalue of $\mathbf{C}$. Note that, in order to ensure that $\mathbf{h}$ satisfies either (11.28) or (11.30), we must delete those rows and columns of $\mathbf{C}$ corresponding to the components of $\mathbf{h}$ that are identically zero. We then solve the eigenvalue problem (11.73) for the remaining components of $\mathbf{h}$, and finally reassemble the complete prototype by appropriately concatenating the zero and nonzero components. This is similar to the construction of the *eigenfilter* described by Vaidyanathan (1993, sect. 4.6.1).

11.7.2 Synthesis Prototype Design

As with the analysis prototype, we can now impose the Nyquist(M) constraint on the complete *analysis–synthesis prototype* $(h * g)[n]$ such that

$$(h * g)[Mn] = \begin{cases} c, & n = m_{\mathrm{d}}, \\ 0, & \text{otherwise,} \end{cases} \tag{11.74}$$

in which case the total response error (11.61) must be identically equal to zero. Subject to this constraint, we wish to minimize the residual aliasing distortion (11.69). Satisfaction of (11.74) clearly reduces to a set of linear constraints of the form

$$\mathbf{g}^T \mathbf{h}_m = \begin{cases} c, & \text{for } m = m_{\mathrm{d}}, \\ 0, & \text{otherwise,} \end{cases} \qquad \forall \, m = -m + 1, \ldots, m - 1,$$

where $\mathbf{h}_m$ is obtained by shifting a time-reversed version of $\mathbf{h}$ by mM samples and padding with zeros as needed. All such constraints can be expressed in matrix form as

$$\mathbf{g}^T \mathbf{H} = \mathbf{c}^T, \tag{11.75}$$

where

$$\mathbf{H} = \begin{bmatrix} \mathbf{h}_{-m_{\mathrm{d}}+1} & \cdots & \mathbf{h}_{m_{\mathrm{d}}} & \cdots & \mathbf{h}_{m_{\mathrm{d}}-1} \end{bmatrix},$$
$$\mathbf{c}^T = \begin{bmatrix} 0 & \cdots & c & \cdots & 0 \end{bmatrix}.$$

For the constrained minimization problem at hand, we again draw upon the method of undetermined Lagrange multipliers and formulate the objective function

$$f(\mathbf{g}) = \mathbf{g}^T \mathbf{Pg} + (\mathbf{g}^T \mathbf{H} - \mathbf{c}^T)\boldsymbol{\lambda} \tag{11.76}$$

where $\boldsymbol{\lambda} = \left[\lambda_{-m_d+1} \cdots \lambda_{m_d} \cdots \lambda_{m_d-1}\right]^T$. Then setting

$$\nabla_{\mathbf{g}} f(\mathbf{g}) = 2\mathbf{Pg} + \mathbf{H}\boldsymbol{\lambda} = 0, \tag{11.77}$$

we find

$$\mathbf{g} = -\frac{1}{2}\mathbf{P}^{-1}\mathbf{H}\boldsymbol{\lambda}. \tag{11.78}$$

The values of the multipliers $\{\lambda_m\}$ can be determined by substituting (11.78) into (11.75) and solving, such that

$$\boldsymbol{\lambda} = -2\left(\mathbf{H}^H \mathbf{P}^{-1}\mathbf{H}\right)^{-1}\mathbf{c}. \tag{11.79}$$

By substituting (11.79) into (11.78), we finally obtain a synthesis prototype

$$\mathbf{g} = \mathbf{P}^{-1}\mathbf{H}\left(\mathbf{H}^T \mathbf{P}^{-1}\mathbf{H}\right)^{-1}\mathbf{c}. \tag{11.80}$$

11.7.3 Alternative Design

The optimal prototypes can be obtained by solving (11.73) and (11.80) if the matrices $\mathbf{C}$ and $\mathbf{P}$ are not singular. As the decimation factor D is reduced to control residual aliasing distortion or as the lengths of the analysis and synthesis prototypes are increased, however, it can happen that $\mathbf{C}$ and $\mathbf{P}$ become very ill-conditioned, to the point where they are *numerically* rank deficient. If $\mathbf{C}$ is in fact singular, we can define its null space $\mathbf{C}_{\text{null}}$, which consists of column vectors $\mathbf{q} \in \mathbb{R}^n$ such that $\mathbf{Cq} = 0$. In order to obtain a basis for the null space, the *singular value decomposition* (SVD) (Golub and Van Loan 1996b, sect. 2.5.3) can be used. Under the SVD, $\mathbf{C}$ is decomposed as

$$\mathbf{C} = \mathbf{U}\boldsymbol{\Sigma}\mathbf{V}^T, \tag{11.81}$$

where $\mathbf{U}$ and $\mathbf{V}$ are orthonormal matrices, and $\boldsymbol{\Sigma}$ is diagonal. The bases of the null space of $\mathbf{C}$ can be obtained from the columns of $\mathbf{U}$, corresponding to the singular values below a given threshold. Such a threshold can be chosen according to

$$\sigma = \max(m, n) \times \max_i(\sigma_i) \times \epsilon,$$

where m and n, respectively, are the number of rows and columns of $\mathbf{C}$, σ_i is the ith singular value, and ϵ is the machine-dependent floating-point precision.

Obviously, the inband-aliasing distortion can be driven to null by an analysis prototype which is a linear combination of the basis vectors of the null space $\mathbf{h} = \mathbf{C}_{\text{null}} \mathbf{x}$. The free parameters $\mathbf{x}$ are determined so as to minimize the passband response error (11.47). Such a solution can be expressed as

$$\mathbf{h} = \mathbf{C}_{\text{null}} (\mathbf{C}_{\text{null}}^T \mathbf{A} \mathbf{C}_{\text{null}})^{-1} \mathbf{C}_{\text{null}}^T \mathbf{b}, \tag{11.82}$$

where the rows and columns of $\mathbf{C}_{\text{null}}$, $\mathbf{A}$ and $\mathbf{b}$, corresponding to the components of $\mathbf{h}$ that are identically zero, are deleted, and $\mathbf{h}$ is reassembled in order to maintain the Nyquist(M) constraint, as before.

For the synthesis prototype design, we can also eliminate residual aliasing distortion (11.66) in a similar manner. Defining the null space of $\mathbf{P}$ to be $\mathbf{P}_{\text{null}}$, we can express the synthesis prototype as $\mathbf{g} = \mathbf{P}_{\text{null}} \mathbf{y}$. Then by substituting this solution into (11.75), we obtain

$$\mathbf{y} = (\mathbf{H}^T \mathbf{P}_{\text{null}})^+ \mathbf{c}, \tag{11.83}$$

where $(\cdot)^+$ indicates the pseudo-inverse of the argument. If the dimensionality of $\mathbf{P}_{\text{null}}$ is greater than or equal to $2m - 1$, we can find a synthesis prototype $\mathbf{g} = \mathbf{P}_{\text{null}} \mathbf{y}$ that achieves zero total response error and zero residual aliasing distortion. Finally, the synthesis prototype can be expressed as a function of the null space $\mathbf{P}_{\text{null}}$, such that

$$\mathbf{g} = \mathbf{P}_{\text{null}} (\mathbf{H}^T \mathbf{P}_{\text{null}})^+ \mathbf{c}. \tag{11.84}$$

In practice, as the inband-aliasing distortion decreases, $\mathbf{P}$ becomes practically singular. Hence, with the method described here, both inband and residual aliasing distortion can be eliminated.

11.8 Quality Assessment of Filter Bank Prototypes

In this section, we will assess the quality of the prototypes obtained using the De Haan and Nyquist(M) methods. We will evaluate these designs primarily in terms of three figures of merit, namely, stopband suppression, response error and aliasing distortion, both inband and residual. We will also present other figures indicating the conditioning of the matrices used in calculating the prototypes.

As we will ultimately use the filter banks designed here for a beamforming application, the outputs of all subbands should be statistically independent. Hence, stopband attenuation is an important indicator of the quality of a prototype, as statistical independence increases with stopband attenuation. Figures 11.10, 11.11 and 11.12 show the frequency responses of the analysis, synthesis and composite analysis–synthesis prototypes, respectively. Each figure presents the frequency responses of the cosine modulated PR FB (Vaidyanathan 1993, sect. 8), de Haan FB and the Nyquist(M) FB, where the number of subbands is $M = 8$ and the decimation factor is $D = 4$. From these figures, it is clear that the FBs designed with the Nyquist(M) constraint provide the best suppression of stopband energy, followed by de Haan prototype and then the PR filter prototype. As previously mentioned, when arbitrary magnitude scalings and phase shifts are applied to the output of each subband in the course of beamforming, the cosine modulated design no longer retains the PR property. Hence, it is necessary to minimize the stopband energy and aliasing error of each filter individually.

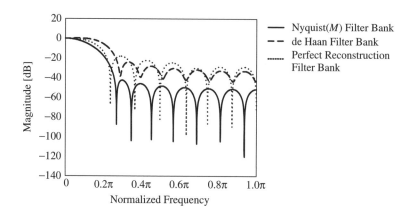

Figure 11.10 Frequency response of analysis filter bank prototypes with $M = 8$ subbands, decimation factor $D = 4$ and filter length $L_h = 16$

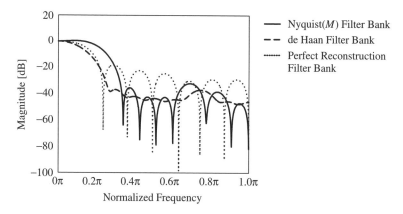

Figure 11.11 Frequency response of synthesis filter bank prototypes with $M = 8$ subbands, decimation factor $D = 4$, and filter length $L_h = L_g = 16$

Other important figures of merit for assessing the quality of filter bank prototypes are inband and residual aliasing distortion (De Haan 2001). Moreover, it is useful to know the relationship between the aliasing distortion and the number M of subbands, as this can be helpful for designing a subband domain beamforming system, as described in Chapter 13. Figure 11.13 shows the inband and residual aliasing distortions plotted against the number of subbands M, where the decimation factor is $D = M/2$. Decreasing the inband aliasing distortion reduces the residual aliasing distortion. It is thus important to suppress the inband aliasing distortion. It is apparent from the figure that the Nyquist(M) filter prototype provides smaller inband aliasing distortion than that designed with de Haan's algorithm. This is quite likely due to the fact that the Nyquist(M) design algorithm minimizes the inband aliasing distortion directly, while de Haan's method minimizes a linear combination of the passband response error and inband aliasing distortion. It is also apparent from the figure that the Nyquist(M) FB can maintain the residual aliasing

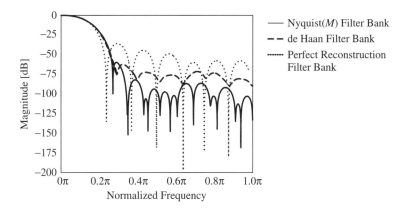

Figure 11.12 Frequency response of composite analysis–synthesis filter bank prototypes with $M = 8$ subbands, decimation factor $D = 4$ and filter length $L_h = L_g = 16$

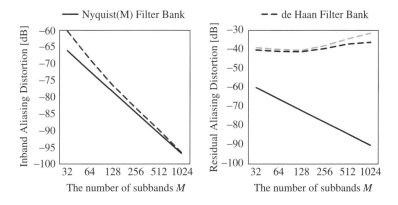

Figure 11.13 Inband ϵ_i and residual ϵ_r aliasing distortion as a function of the number M of subbands for the Nyquist(M) and De Haan filter banks, where the filter length is set to $L_h = L_g = 2M$

distortion at a much lower level than the conventional method. Moreover, the figure indicates that the residual aliasing distortion of the Nyquist(M) FB decreases monotonically with an increasing number of subbands M, while the residual aliasing distortion of the de Haan FB is more or less insensitive to the number of subbands. Once more, De Haan's algorithm minimizes a linear combination of the total response error and residual aliasing distortion (11.62). Hence, the additional term of the total response error $\epsilon_t(\mathbf{h})$ prevents the residual aliasing error $\epsilon_r(\mathbf{h})$ from being perfectly suppressed. On the other hand, in the design technique proposed in Kumatani *et al.* (2008d), only the residual aliasing distortion is minimized, while zero total response error is maintained through the application of a constraint. As a result, the residual aliasing distortion of the Nyquist(M) prototype decreases monotonically as M increases.

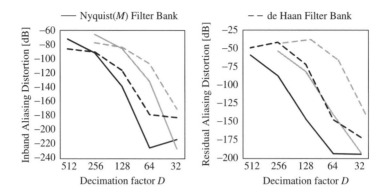

Figure 11.14 Inband ϵ_p and residual $\epsilon_g(\mathbf{h})$ aliasing distortion as a function of decimation factor D. The number of subbands is $M = 256$ (gray) or $M = 512$ (black) and the filter lengths are $L_\mathbf{h} = L_\mathbf{g} = 2M$

The aliasing errors can be also reduced by decreasing the decimation factor D, although this increases the computational complexity associated with adaptive filtering or beamforming. Figure 11.14 presents the inband and residual aliasing distortions as a function of decimation factor D for both the de Haan and the Nyquist(M) FBs, where the number of subbands is either $M = 256$ or 512, and the filter lengths are $L_\mathbf{h} = L_\mathbf{g} = 2M$. In designing the de Haan FBs, the weighting factor in (11.70) was $v = 100.0$. It is clear from the figure that the Nyquist(M) design has lower aliasing distortion than the de Haan design in most cases.

In calculating the inband aliasing distortion for a given decimation factor D, Kumatani *et al.* (2008d) observed that the matrix $\mathbf{C}$ became singular when the number of subbands and the decimation factor were set to $M = 256$ and $D \leq 32$ or when $M = 512$ and $D \leq 64$. In such cases, Kumatani *et al.* computed the nullspace of $\mathbf{C}$ and then used the alternative solution for the design of the analysis and synthesis prototypes instead of that based on the eigendecomposition. Due to this necessity of using the alternative design method, it is important to know when the matrices $\mathbf{C}$ and $\mathbf{P}$ in (11.56) and (11.67), respectively, are ill-conditioned or numerically singular. A matrix $\mathbf{C}$ can always be represented with the singular value decomposition

$$\mathbf{C} = \mathbf{U}\mathbf{\Lambda}\mathbf{V}^T,$$

where both $\mathbf{U}$ and $\mathbf{V}$ are unitary matrices, and $\mathbf{\Lambda}$ is a diagonal matrix. The singular values are to be found on the main diagonal of $\mathbf{\Lambda}$, typically sorted from highest to lowest. The *condition number* of $\mathbf{C}$ is then defined as the ratio of the largest to the smallest singular values.

The logarithms of the condition numbers of $\mathbf{C}$ and $\mathbf{P}$ are plotted in Figure 11.15 for $M = 512$ and $L_\mathbf{h} = L_\mathbf{g} = 2M$. Practically speaking, a matrix is ill-conditioned when its condition number approaches the reciprocal of the floating point precision of the machine used for the calculations. The latter is shown as the threshold in Figure 11.15. As indicated in the figure, the conditioning degrades with decreasing decimation factor. The condition numbers reach the threshold of machine precision for $D \leq 64$.

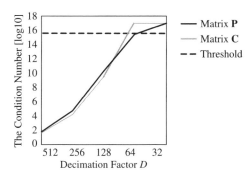

Figure 11.15 Logarithm of the condition number of **C** and **P** as a function of decimation factor D. The number of subbands is $M = 512$ and the filter lengths are $L_\mathbf{h} = L_\mathbf{g} = 2M$

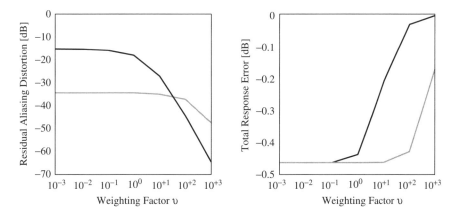

Figure 11.16 Residual aliasing distortion $\epsilon_\mathbf{g}(\mathbf{h})$ and total response error $\epsilon_\mathbf{t}(\mathbf{h})$ as a function of weighting factor v. The number of subbands is $M = 512$ and the filter lengths are $L_\mathbf{h} = L_\mathbf{g} = 2M$. The decimation factor is $D = 256$ (gray) or $D = 512$ (black)

One might intuitively consider that the residual aliasing distortion would decrease monotonically with decreasing decimation factor. Figure 11.14 shows, however, that the residual aliasing distortion of de Haan's FB has a peak at $D = M/2$. In order to investigate this phenomenon further, Kumatani *et al.* calculated the residual aliasing distortions with $D = 256$ and $D = 512$. Figure 11.16 shows the residual aliasing distortions as a function of the weighting factor v in (11.70). It is apparent from the figure that the residual aliasing distortion of $D = 512$ is smaller than that of $D = 256$ in the case of $v \geq 100.0$.

The total response errors as a function of the weighting factor v are also plotted in Figure 11.16. It is clear from the figure that the residual aliasing distortion can be reduced by setting a large weighting factor v, but only at the expense of the total response error. As stated previously, the total response error is zero in the Nyquist(M) filter bank.

11.9 Summary and Further Reading

In this chapter we have discussed digital FBs, which are arrays of bandpass filters that separate an input signal into many narrowband components. A very good if somewhat dated reference for the theory and design of digital filter banks is Vaidyanathan (1993), which describes clearly with many examples the terms and concepts necessary for the comprehension of the vast majority of the FB literature. Vaidyanathan (1993, sect. 11) also describes the relationship between digital FBs and wavelets. A useful summary of more recent work in FB design and its application to audio coding is presented by Schuler (2004). Unfortunately, both of the summaries mentioned concentrate all but exclusively on maximally decimated FB designs which achieve perfect reconstruction through the process of aliasing cancellation. As mentioned throughout this chapter, while such designs are ideal for subband coding applications, they are poorly suited for beamforming and adaptive filtering. Aliasing cancellation functions through the design of a FB prototype such that the aliasing that is perforce present in *one* subband is cancelled by the aliasing present in all *other* subbands. Hence, the subbands are not truly *independent* of one another. In particular, if arbitrary scale factors are applied to the samples of each subband, or, even worse, if some subbands are suppressed entirely, the aliasing cancellation effect will be impaired or destroyed, and strong aliasing components will be present in the resynthesized output of the FB. De Haan (2001) and De Haan *et al.* (2003) address FB design particularly for adaptive filtering, as opposed to subband coding, applications. Kumatani *et al.* (2008d) presented a variation of the de Haan filter bank with better characteristics in terms of total response error and residual aliasing distortion. The essential points of De Haan *et al.* (2003) were presented in Section 11.6, while those of Kumatani *et al.* (2008d) were presented in Sections 11.7 and 11.8.

11.10 Principal Symbols

Symbol	Description
ϵ_i	inband aliasing distortion
ϵ_p	passband response error
ϵ_r	residual aliasing distortion
ϵ_t	total response error
ϕ_h	delay chain
ω	angle frequency
Ω_p	passband
$\mathbf{A}$	matrix to calculate passband response error
$\mathbf{b}$	vector to calculate passband response error
$\mathbf{C}$	matrix to calculate inband aliasing error
D	decimation factor
$\mathbf{g}$	uniform DFT filter bank synthesis prototype

Symbol	Description
$G_m(e^{j\omega})$	frequency response of mth synthesis filter
$h[n]$	impulse response
$\mathbf{h}$	uniform DFT filter bank analysis prototype
$H_m(e^{j\omega})$	frequency response of mth analysis filter
$H_{\mathrm{d}}(e^{j\omega})$	desired passband frequency response
k	frame index
m	subband index
m_{d}	processing delay
M	number of subbands
n	sample index
$\mathbf{P}$	matrix to calculate residual aliasing error
T	sampling interval
$\mathcal{T}\{\cdot\}$	transformation operator
$u[n]$	step sequence
$w[n]$	window function
$W_N = e^{-j2\pi/N}$	Nth root of unity
$x, x[n]$	input signal, input sequence
$y, y[n]$	output signal, output sequence

12

Blind Source Separation

Blind source separation (BSS) is a term used to describe the first class of techniques by which signals from multiple sensors may be combined into one signal for speech recognition or enhancement. The *source separation* in BSS alludes to the fact that two or more sources are typically mixed in the signals reaching the several microphones. For our purposes, we will assume that at least one of these mixed sources is the desired speech, which is to be separated from the other sources, noise, and/or interference, then recognized automatically. This class of methods is known as *blind* because neither the relative positions of the sensors, nor the positions of the sources are assumed to be known.

We will define *independent component analysis* (ICA) as a subcategory of BSS which attempts to separate different sources based only on their statistical characteristics. As discussed in Section 12.2.2, separating signals using only knowledge of their statistics is possible only for non-Gaussian signals; hence, the primary assumption of ICA is that information-bearing signals are *not* Gaussian signals. The pdfs of such non-Gaussian signals can only be adequately characterized through the use of so-called *higher order statistics* (HOS), which are by definition the statistical moments of a pdf of greater than second order. Section 12.2.3 discusses several optimization criteria that are typically applied in ICA, including *mutual information*, *negentropy*, and *kurtosis*. While mutual information can be meaningfully defined for both non-Gaussian and Gaussian random variables, negentropy and kurtosis are both measures of non-Gaussianity, and inherently assume that the signal of interest is non-Gaussian and hence can only be properly specified through HOS. In many cases, the use of such HOS leads to better separation and automatic recognition performance, but often at the cost of the additional complexity required to estimate the structure of the non-Gaussian pdf.

Buchner *et al.* (2004) have proposed to divide BSS algorithms into three major categories:

- those based on *non-whiteness*,
- those based on *nonstationarity*, and
- those based on *non-Gaussianity*.

All known BSS algorithms use one or more of these characteristics. The subset of ICA algorithms typically exploit the non-whiteness and non-Gaussianity of the desired sources in order to achieve separation. The members of the larger class of BSS algorithms exploit only the *second-order statistics* (SOS) of the desired sources, in addition to their inherent nonstationarity and/or non-whiteness.

The balance of this chapter is organized as follows. In Section 12.1, we begin by explaining the necessity of measuring channel quality in order to determine a subset of sensors whose signals can be combined in order to reduce word error rate with respect to that achievable with any single sensor. In Section 12.2 we consider the ICA problem in general. In particular, Section 12.2.1 first presents the definition of instantaneous ICA, wherein two or more sources are mixed additively without memory. While this formulation is simplistic, it is well-suited to frequency- or subband-domain implementations of ICA algorithms. We also consider the ambiguities inherent in ICA algorithms. The definition and implications of statistical independence are discussed in Section 12.2.2. We will also present several parametric pdfs which will prove useful for modeling non-Gaussian random variables. As mentioned above, the assumption of non-Gaussianity is a crucial component of ICA. This discussion of the representation of non-Gaussian random variables along with criteria for determining the degree of deviation from Gaussianity will prove useful not only for the present development, but also for that in Section 13.5, where beamforming algorithms based on measures of non-Gaussianity will be discussed. The three most popular optimization criteria in the ICA field, namely, mutual information, negentropy, and kurtosis, are presented in Section 12.2.3. We will also discuss the important relation between mutual information and negentropy. Section 12.2.4 will describe how the parameters of an ICA system can be updated using the well-known natural gradient method. BSS based solely on second-order statistics will be presented in Section 12.3. In order to enable separation based on SOS, it will prove necessary to exploit one or both of the nonstationarity and non-whiteness of the desired signal. The final section of this chapter will summarize the development here, and provide references for further reading.

12.1 Channel Quality and Selection

ICA, BSS, and beamforming entail combining the signals from several microphones in some advantageous way. Improving recognition accuracy through such a combination depends critically on using only those channels that are reliable, as the use of unrealiable channels can actually degrade performance. Hence, such unrealiable channels must be excluded from the combination, or their contribution must at very least be discounted. To decide if a particular input stream should be included or how much it should contribute to the combination, reliable measures for channel quality are required. The worst case is that wherein a given channel provides no information about the desired source or sources. In this case, including the uninformative channel in the combination would potentially introduce additional noise or other distortions into the combination without providing any additional information. Anguera *et al.* (2005) have also shown that the robust selection of the reference channel in array processing has a decisive effect on the effectiveness of beamforming techniques. An ideal channel selection criterion should be reliable, highly correlated with word error rate, and function in an unsupervised fashion. Moreover, the selection criterion should take into account as much knowledge as possible about the front-end and acoustic models used by the recognition system.

Several different objective functions have been successfully applied to the channel selection problem for distant speech recognition.

Signal-to-Noise Ratio

The *signal-to-noise ratio* (SNR) is possibly the most widely used objective function for measuring channel quality. This is mainly due to its simplicity and low computational complexity. The chief advantage of SNR is that it provides a good indication of channel quality with relatively little computation. SNR has, however, a number of inherent drawbacks as a measure of channel quality. Firstly, SNR requires speech activity detection in order to distinguish between speech and silence regions. Secondly, additional knowledge about the signal is commonly ignored, even though some knowledge of the human hearing apparatus can easily be integrated. Thirdly, different classes of interest, such as phoneme classes, are also typically ignored. Finally, SNR cannot be applied in a number of domains, most particularly, it cannot be applied in the normalized cepstral domain, which is that most often used for ASR feature extraction.

Decoder-Based Methods

Decoder-based measures use information provided by the recognition system in order to perform channel selection. Two primary approaches have been proposed in the literature:

- *Maximum likelihood* – This method chooses the channel achieving the highest likelihood as measured with the hidden Markov model used for ASR (Shimizu *et al.* 2000).
- *Difference in feature compensation* – Comparing first best word hypothesis of different recognition passes with uncompensated and unsupervised compensated feature vectors on a channel indicates how much a system output has changed by adaptation. A channel is considered to be good or reliable if unsupervised adaptation does not lead to a significant change in the feature vectors extracted from it (Obuchi 2004).

The principal advantage of decoder-based methods is the close coupling between the channel selection criteria and the recognition system, which provides more reliable estimates. The chief disadvantage of decoder-based methods is that they require that the feature stream from each sensor be decoded separately in order to avoid performance degradations due to channel mismatch. This in turn can lead to a drastic increase in computation time. Comparing differences on a word level has the additional disadvantage that short utterances do not provide sufficient granularity to distinguish channel quality.

Class Separability

To consider all possible information available in the recognition front-end, class separability-based measures can be used for channel selection, where a channel with higher class separability is considered to be better. One such measure of class separability is given in (5.59).

The classes used to determine class separability are not known *a priori*. Thus, these classes must be estimated on the observed data; e.g., by split and merge training. It seems to be helpful to exclude those segments or classes containing silence regions, as the separation between the different phonemes is the main focus here. The choice of an ideal number of classes need not necessarily be equivalent to the number of phonemes, but might depend on the amount of available data. For very short utterances containing, for example, 60 or fewer frames of speech, better results may be obtained when the number of classes is smaller than the number of phonemes (Wölfel 2007).

12.2 Independent Component Analysis

Many BSS algorithms are based on the use of *higher order statistics*, which, as we will show in Section 12.2.3, implies the use of non-Gaussian pdfs to model the statistical behavior of random variables. Collectively, such algorithms are perhaps best known as belonging to the field of independent component analysis. In this section we first define ICA and discuss its inherent assumptions. We also present three optimization criteria that are commonly used in the ICA field. These ICA optimization criteria will prove useful both for the development in this chapter, as well as that in Section 13.5, where we will consider nonconventional beamforming algorithms.

Before proceding, we will say a word about the notation used here. To wit, in the balance of the chapter, we will use upper case letters to denote random variables or vectors, and lower case letters to denote the particular values these variables assume for a given trial.

12.2.1 Definition of ICA

Here we will define the simplest form of ICA, namely, instantaneous ICA. Consider a random vector of N independent components

$$\mathbf{S} = \begin{bmatrix} S_1 & S_2 & \ldots & S_N \end{bmatrix},$$

as well as a *mixing matrix* $\mathbf{A}$. The *instantaneous ICA model* is by definition

$$\mathbf{X} \triangleq \mathbf{A}\,\mathbf{S}, \qquad (12.1)$$

where

$$\mathbf{X} = \begin{bmatrix} X_1 & X_2 & \ldots & X_N \end{bmatrix},$$

is a random vector of mixtures. Clearly the components $\{X_n\}$ are not independent, as each X_n is a linear combination of the independent components $\{S_n\}$. The ICA model is generative in that it describes how the observable components $\{X_n\}$ are generated through the process of mixing the independent components $\{S_n\}$. The latter are said to be *latent components* in that they cannot be observed directly. Moreover, the mixing matrix $\mathbf{A}$ is assumed to be unknown. Only the vector $\mathbf{X}$ is observed, and both $\mathbf{A}$ and $\mathbf{S}$ must be estimated on the basis of this observation under conditions that are as general as possible.

There are two fundamental assumptions inherent in ICA:

1. The components of $\mathbf{S}$ are independent statistically. The nature and implication of such independence will be discussed in 12.2.2.

2. The components of **S** have non-Gaussian pdfs. This non-Gaussianity will be the primary
   cue used to determine the identity of the independent components.

Note that these assumptions are not universally made in the more general field of BSS.
In BSS, the desired components are often taken to be merely uncorrelated, and not fully
independent.

For the sake of simplicity, we assume that **A** is square and invertible. Under these
assumptions, the goal of ICA can be stated as that of finding a *demixing matrix* **B**, which
ideally would be equivalent to $\mathbf{A}^{-1}$, such that

$$\mathbf{Y} = \mathbf{B}\mathbf{X} = \mathbf{B}\mathbf{A}\mathbf{X} = \mathbf{X}. \tag{12.2}$$

Were we unable to achieve (12.2) identically, we might still hope to realize a solution
which would remove all statistical dependence in the extracted outputs. Such a solution
would attempt to adapt the demixing matrix such that

$$\lim_{k \to \infty} \mathbf{B}(k)\,\mathbf{A} = \mathbf{\Phi}\,\mathbf{D}, \tag{12.3}$$

where **Φ** is an $N \times N$ permutation matrix with a single unity entry in every row and
column and **D** is a diagonal nonsingular scaling matrix. Then each Y_n would be a scaled
version of some S_m, where $n \neq m$ in general.

In the context of distant speech recognition, we might well associate the components of
S with the voices of several speakers who speak simultaneously, as in the so-called cocktail
party problem (Cherry 1953). While ICA can be posed as a problem in the separation of
convolutive mixtures, the simpler model (12.1) will be sufficient for our purposes here and
in Chapter 13. This is largely due to the fact that our primary interest is in beamforming
methods implemented in the frequency or subband domain based on insights from the
ICA field. In this case, all source separation problems devolve to instantaneous separation
problems. As mentioned previously, we will investigate such beamforming methods in
Sections 13.5.2 through 13.5.4.

It should be noted that the field of BSS, as opposed to that of ICA narrowly understood,
does not always make the latter assumption described above. That is, it is not always
strictly assumed that the signals of interest are non-Gaussian. BSS techniques based on
second-order methods will be discussed in Section 12.3.

Inherent in the ICA model of (12.1) are two ambiguities:

1. The *variances* of the independent components **S** cannot be uniquely determined.
2. As implied by (12.3), the *order* of the independent components **S** also cannot be
   uniquely determined.

The first ambiguity follows from the fact that two or more independent random
variables remain independent if they are multiplied by some scalar. Hence, there is no
basis for determining the variance of the independent components. The most frequent
approach for circumventing this problem is to assume that each component S_n has unit
variance, such that $\sigma_n^2 = \mathcal{E}\{S_n^2\} = 1$. This assumption is typically built into the demixing
matrix **B**.

The second ambiguity follows from the observation that, as both $\mathbf{S}$ and $\mathbf{A}$ are unknown, any of the independent components could well be the first one, which is clear from (12.3). Formally, a permutation matrix $\mathbf{\Phi}$ and its inverse could be substituted into the model (12.1) to yield $\mathbf{X} = \mathbf{A}\mathbf{\Phi}^{-1}\mathbf{\Phi}\mathbf{S}$. The components of the matrix vector product $\mathbf{\Phi}\mathbf{S}$ are the same independent components found in $\mathbf{S}$, but in a different order. Hence, the matrix $\mathbf{A}\mathbf{\Phi}^{-1}$ could be viewed as yet another conceivable demixing matrix to be estimated by an ICA algorithm.

12.2.2 Statistical Independence and its Implications

In this section, we consider the definition of *statistical independence*, as well as the implications thereof. In particular, we will discuss that this independence can be applied to the problem of separating two or more independent components from an additive combination. By definition, two random variables Y_1 and Y_2 are said to be statistically independent (Papoulis 1984) iff

$$p_{Y_1,Y_2}(y_1, y_2) = p_{Y_1}(y_1)\, p_{Y_2}(y_2). \tag{12.4}$$

Given two functions h_1 and h_2, it follows from definition (12.4) that if Y_1 and Y_2 are statistically independent, then

$$\mathcal{E}\{h_1(Y_1)\, h_2(Y_2)\} = \mathcal{E}\{h_1(Y_1)\}\, \mathcal{E}\{h_2(Y_2)\}. \tag{12.5}$$

This fact can readily be proven from the definition of the expectation operator, namely,

$$\mathcal{E}\{h_1(Y_1)\, h_2(Y_2)\} \triangleq \int\int h_1(y_1)\, h_2(y_2)\, p(y_1, y_2)\, dy_1\, dy_2, \tag{12.6}$$

$$= \int h_1(y_1)\, p(y_1)\, dy_1 \int h_2(y_2)\, p(y_2)\, dy_2, \tag{12.7}$$

where (12.7) follows from (12.6) upon substituting (12.4) into the latter.

Two random variables Y_1 and Y_2 are, by definition, *statistically uncorrelated* iff

$$\mathcal{E}\{Y_1\, Y_2\} - \mathcal{E}\{Y_1\}\, \mathcal{E}\{Y_2\} = 0. \tag{12.8}$$

If Y_1 and Y_2 are independent, they are also uncorrelated, which is readily seen by setting $h_1(Y_1) = Y_1$ and $h_2(Y_2) = Y_2$ in (12.5).

While independent implies uncorrelated, the converse does not hold. The property (12.5) of independent random variables can be used to demonstrate why uncorrelated and independent are not generally the same. Assume that the random variables Y_1 and Y_2 are discrete-valued, and distributed such that the pairs $(y_1, y_2) = (0, 1), (0, -1), (1, 0), (-1, 0)$ all have probability $1/4$. Then Y_1 and Y_2 are uncorrelated, as is apparent from (12.8). They are not independent, however, which is readily seen by setting $h_1(Y) = h_2(Y) = Y^2$ and noting

$$\mathcal{E}\{Y_1^2\, Y_2^2\} = 1 \neq \frac{1}{4} = \mathcal{E}\{Y_1^2\}\, \mathcal{E}\{Y_2^2\}.$$

Hence, the property (12.5) does not hold, from which it follows that Y_1 and Y_2 are not independent.

As another example illustrating an instance when uncorrelated does not imply independent, consider a random vector $\mathbf{S}$ of N zero-mean independent components all having the same variance σ_S^2. Clearly, the covariance matrix of $\mathbf{S}$ can be expressed as

$$\mathbf{\Sigma}_\mathbf{S} \triangleq \sigma_S^2 \mathbf{I}, \tag{12.9}$$

where $\mathbf{I}$ is the identity matrix. Defining $\mathbf{X} = \mathbf{VS}$ where $\mathbf{V}$ is a $N \times N$ unitary transform, we see that the covariance matrix of $\mathbf{X}$ can be expressed as

$$\mathbf{\Sigma}_\mathbf{X} = \mathcal{E}\{\mathbf{XX}^T\} = \mathbf{V}\mathcal{E}\{\mathbf{SS}^T\}\mathbf{V}^T = \sigma_S^2 \mathbf{I},$$

where the latter equality follows from (12.9) and $\mathbf{VV}^T = \mathbf{I}$. Hence, the components of $\mathbf{X}$ are once more uncorrelated. They are by no means independent, however, inasmuch as they consist of linear combinations of the independent components $\mathbf{S}$.

As explained by Hyvärinen and Oja (2000), the entire field of ICA is based on the assumption that all signals of real interest are *not* Gaussian-distributed. Briefly, their reasoning is founded on two points:

1. The *central limit theorem* states that the pdf of the sum of independent random variables will approach Gaussianity in the limit as more and more components are added, *regardless* of the pdfs of the individual components. This implies that the sum of several random variables will be closer to Gaussian than any of the individual components. Thus, if the original independent components comprising the sum are sought, one must look for components with pdfs that are the *least* Gaussian.
2. As discussed in Section 12.2.2, entropy is the basic measure of information in *information theory* (Gallager 1968). It is well known that a Gaussian random variable has the highest entropy of all random variables with a given variance (Gallager 1968, Thm. 7.4.1), which holds also for complex Gaussian random variables (Neeser and Massey 1993, Thm. 2). Hence, a Gaussian random variable is, in some sense, the least *predictable* of all random variables, which is why the Gaussian pdf is most often associated with *noise*. Information-bearing signals contain structure that makes them more predictable than Gaussian random variables. Hence, if an information-bearing signal is sought, one must once more look for a signal that is *not* Gaussian.

An illustration of the central limit theorem is provided in Figure 12.1. In this figure, the pdfs of the random variable given by the sum,

$$Y_N = \sum_{n=1}^{N} X_n,$$

are plotted for several values of N, where each X_n is a zero-mean Laplacian random variable with a variance of $\sigma_n^2 = \sigma_N^2/N$. The variance of each X_n is scaled in this fashion

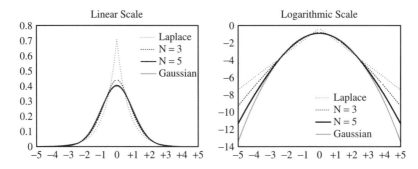

Figure 12.1 Plot of the Gaussian pdf and the pdf obtained by summing together N Laplacian random random variables for several values of N. As predicted by the central limit theorem, the sum becomes ever more Gaussian with increasing N

to ensure that the final sum has variance

$$\sigma_N^2 = \sum_{n=1}^{N} \sigma_n^2 = 1,$$

regardless of the number N of terms in the sum. The pdf $p_{Y_N}(y)$ of Y_N is readily obtained from the relation

$$p_{Y_N}(y) = p_{X_1}(x) * p_{X_2}(x) * \cdots * p_{X_N}(x),$$

where each $p_{X_n}(x)$ is the pdf of the random variable X_n, and $p_{X_n}(x) * p_{X_{n+1}}(x)$ denotes the convolution of the pdfs $p_{X_n}(x)$ and $p_{X_{n+1}}(x)$ according to (Papoulis 1984, sect. 6.2)

$$p_{X_n}(x) * p_{X_{n+1}}(x) \triangleq \int_{-\infty}^{\infty} p_{X_n}(z)\, p_{X_{n+1}}(x-z)\, dz.$$

From Figure 12.1 it is clear that adding more Laplacian random variables together brings the pdf of the sum ever closer to the Gaussian. The left side of the figure, which is plotted in a linear space, indicates that the sharp peak of the Laplacian is "rounded off" through the addition of several component random variables, and that the probability mass formerly in the peak moves into the intermediate regions around the mean. The right side of Figure 12.1, which is the same plot as that on the left but on a logarithmic scale, indicates that probability mass is also transferred out of the tail of the pdf through the addition of more and more Laplacian r.v.s, inasmuch as the tail of $p_{Y_N}(y)$ for increasing N approaches that of the Gaussian pdf.

The fact that the pdf of speech is super-Gaussian has often been reported in the literature; see, for example, Martin (2005) and Kumatani *et al.* (2007). Noise, on the other hand, is typically Gaussian-distributed. As discussed in Section 13.5.2, when speech is corrupted by noise, reverberation, or the speech of another speaker, its pdf becomes more nearly Gaussian. Hence, it is possible to remove or suppress the damaging effects of these distortions by adjusting the demixing matrix of an ICA system or active weight vector of a beamformer so as to produce output signals that are maximally non-Gaussian.

In order to optimize the parameters of an ICA system or beamformer with respect to the negentropy or mutual information criteria considered in Section 12.2.3, it is necessary to know or estimate the form of the relevant pdf. Alternatively, we can assume that the required pdf has a known parametric form, and then estimate the parameters of this model from a set of training data. The *generalized Gaussian* (GG) pdf, which is well known and finds frequent application in the ICA field, is one such parametric model. The GG pdf for a real-valued r.v. Y with zero-mean value can be expressed as

$$p_{GG}(y) = \frac{1}{2\,\Gamma(1 + 1/f)\,\hat{\sigma}\,A(f)} \exp\left\{-\left|\frac{y}{\hat{\sigma}\,A(f)}\right|^f\right\}, \qquad (12.10)$$

where $\Gamma(.)$ denotes the *Gamma function*, f is the *shape factor*, which controls how fast the tail of the pdf decays, and

$$A(f) \triangleq \left[\frac{\Gamma(1/f)}{\Gamma(3/f)}\right]^{1/2}. \qquad (12.11)$$

Note that the GG with $f = 1$ corresponds to the Laplace pdf, and that setting $f = 2$ yields the conventional Gaussian pdf, whereas in the case of $f \to +\infty$ the GG pdf converges to a uniform distribution. The GG and several other pdfs are generalized in Section B.5 for the important case of circular, complex random variables.

Plots of the Gaussian and three super-Gaussian univariate pdfs are provided on the left of Figure 12.2. From the figure, it is clear that the Laplace, K_0, and Γ pdfs exhibit the "spikey" and "heavy-tailed" characteristics that are typical of super-Gaussian pdfs. This implies that they have a sharp concentration of probability mass at the mean, relatively little probability mass as compared with the Gaussian at intermediate values of the argument, and a relatively large amount of probability mass in the tail; i.e., far from the mean. The right side of Figure 12.2 shows the GG pdf with the same scaling factor $\hat{\sigma}^2 = 1$ and various shape factors $f = 0.5, 1, 2$, and 4. The shape factors f can be estimated from training data using the method described in Section 13.5.2. As mentioned previously,

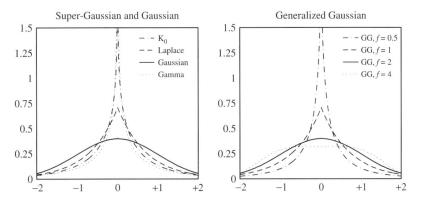

Figure 12.2 Plot of the likelihood of Gaussian, super-Gaussian and GG pdfs. Note that the GG pdf with $f = 1$ and $f = 2$ is equivalent to the Laplacian and Gaussian pdfs, respectively.

$f = 2$ corresponds to the conventional Gaussian, and $f = 1$ yields the Laplacian pdf. From the figure it is clear that a smaller shape parameter generates a pdf with a spikier peak and a heavier tail. In Section 13.5.2, we will present empirical evidence that speech is not only super-Gaussian, but also *super-Laplacian*.

12.2.3 ICA Optimization Criteria

In this section, we consider three of the optimization criteria that are commonly used in the ICA field, as well as some relations between them. These optimization criteria are equivalent to, or – under appropriate conditions – can be equated to, measures of *non-Gaussianity*, which stands to reason given the logic of Section 12.2.2. Such measures are often referred to as *contrast functions* in the ICA literature (Douglas 2001a), although this term seems to have different meanings for different authors. Comon (1994), who was among the first to use the term, defined a contrast function by beginning with the observation that if $\mathbf{X}$ is a vector of N *independent* random components, then the pdf of $\mathbf{X}$ can be expressed as

$$\hat{p}_{\mathbf{X}}(\mathbf{x}) = \prod_{i=1}^{N} p_{X_i}(x_i). \tag{12.12}$$

Hence, one natural way of determining whether the components are truly independent is to measure statistically the *distance* between the actual pdf $p_{\mathbf{X}}(\mathbf{x})$ of $\mathbf{X}$ and that pdf $\hat{p}_{\mathbf{X}}(\mathbf{x})$ in (12.12) obtained by assuming that the components of $\mathbf{X}$ are truly independent. Several possible measures for the distance between pdfs of two random vectors $\mathbf{X}$ and $\mathbf{Y}$ have appeared in the literature (Basseville 1989). Perhaps the best known of these is the *Kullback divergence*, which is defined as

$$D(p_{\mathbf{X}} \| p_{\mathbf{Y}}) \triangleq \int p_{\mathbf{X}}(\mathbf{x}) \log \frac{p_{\mathbf{X}}(\mathbf{x})}{p_{\mathbf{Y}}(\mathbf{x})} \, d\mathbf{x}. \tag{12.13}$$

The Kullback divergence satisfies

$$D(p_{\mathbf{X}} \| p_{\mathbf{Y}}) \geq 0, \tag{12.14}$$

with equality iff $p_{\mathbf{X}}(\mathbf{x}) = p_{\mathbf{Y}}(\mathbf{x})$ almost everywhere. The latter property is due to the convexity of the logarithm (Blahut 1987).

Comon (1994) defined a contrast function formally as a mapping Ψ from the set of densities $\{p_{\mathbf{X}}(\mathbf{x}) \, \forall \mathbf{x} \in \mathbb{C}^N\}$ to $\mathbb{R}$ satisfying the following three requirements:

- $\Psi(p_{\mathbf{X}})$ does not change if the components of $\mathbf{X}$ are permuted such that

$$\Psi(p_{\mathbf{\Phi}\mathbf{X}}) = \Psi(p_{\mathbf{X}}) \, \forall \text{ permutations } \mathbf{\Phi}.$$

- $\Psi(p_{\mathbf{X}})$ is invariant to scale changes, such that

$$\Psi(p_{\mathbf{\Lambda}\mathbf{X}}) = \Psi(p_{\mathbf{X}}) \, \forall \text{ diagonal, invertible } \mathbf{\Lambda}.$$

- If $\mathbf{X}$ has independent components, then

$$\Psi(p_{\mathbf{AX}}) \leq \Psi(p_{\mathbf{X}}) \, \forall \text{ invertible } \mathbf{A}.$$

Although their definition was not so formal, Hyvärinen and Oja (2000) presented a contrast function simply as a measure of non-Gaussianity, such as the negentropy and kurtosis criteria discussed subsequently. These authors have all explored the connection between such contrast functions and mutual information, which we will shortly consider. We will first, however, introduce the concept of entropy, which will play a pivotal role in the definitions of both mutual information and negentropy.

Entropy

As mentioned previously, *entropy* is the basic measure of pure information in the field of information theory (Gallager 1968, sect. 2.2). Although entropy cannot be used directly as a contrast function for source separation or beamforming, it plays a prominent role in the definitions of mutual information and negentropy, both of which are well-known contrast functions in the ICA field. Hence, we will define entropy here before proceding to consider the latter criteria. For a continuous-valued random variable Y, entropy is defined as

$$H(Y) \triangleq -\mathcal{E}\left\{\log p_Y(Y)\right\} = -\int p_Y(y) \log p_Y(y) \, dy, \qquad (12.15)$$

where $p_Y(y)$ is the pdf of Y. As we will soon see, entropy as defined in (12.15) is closely related to mutual information, and also plays a decisive role in the definition of negentropy.

The entropy of a random variable indicates how much information a single observation of the variable provides. As a larger entropy implies more information is conveyed with each observation, we associate a large entropy with a lack of predictability, as mentioned previously. As noted in Section 12.2.2, a Gaussian variable has the largest entropy among all random variables of equal variance.

It is instructive to consider the entropy of a zero-mean Gaussian random variable Y. The pdf of Y can be expressed as

$$p_Y(y) = \frac{1}{\sqrt{2\pi\sigma_Y^2}} e^{-y^2/2\sigma_Y^2},$$

where $\sigma_Y^2 = \mathcal{E}\{Y^2\}$ is the variance of Y. The expectation in (12.15) can be evaluated according to

$$H(Y) = -\mathcal{E}\{\log p_Y(Y)\} = \mathcal{E}_Y\left\{\frac{1}{2}\log 2\pi\sigma_Y^2 + \frac{1}{2}\frac{Y^2}{\sigma_Y^2}\right\}$$

$$= \frac{1}{2}\log 2\pi\sigma_Y^2 + \frac{1}{2}\int_{-\infty}^{\infty} \frac{y^2}{\sigma_Y^2} p_Y(y) \, dy. \qquad (12.16)$$

$$= \frac{1}{2}\left(1 + \log 2\pi\sigma_Y^2\right), \qquad (12.17)$$

where (12.17) follows from (12.16) given that the integral in the latter is unity.

Mutual information

The *mutual information* (MI) between two random variables X and Y is a measure of how much information X and Y have in commmon. Formally, the MI of X and Y is defined by

$$I(X; Y) \triangleq \mathcal{E} \left\{ \log \frac{p(X, Y)}{p(X)p(Y)} \right\} \qquad (12.18)$$

$$= \mathcal{E} \{\log p(X, Y)\} - \mathcal{E} \{\log p(X)\} - \mathcal{E} \{\log p(Y)\}. \qquad (12.19)$$

At one extreme, if X and Y are statistically independent, then $I(X; Y) = 0$. At the other extreme, if X and Y are identical, then knowing Y determines X, and vice versa. In this case, $I(X; Y) = H(X) = H(Y)$. From (12.19) it is obvious that MI is symmetric, such that $I(X; Y) = I(Y; X)$.

From the definitions (12.15) and (12.19), it is apparent that MI can be equivalently expressed as a function of the marginal entropies $H(X)$, $H(Y)$ along with the *joint entropy* $H(X, Y)$ according to Gallager (1968; sect. 2.2)

$$I(X; Y) = H(X) + H(Y) - H(X, Y), \qquad (12.20)$$

where, by definition,

$$H(X, Y) \triangleq -\mathcal{E}\{\log p(X, Y)\}. \qquad (12.21)$$

Alternatively, $I(X; Y)$ satisfies the relation

$$I(X; Y) = H(X) - H(X|Y) = H(Y) - H(Y|X),$$

where $H(X|Y)$ is the *conditional entropy* of X given Y, which is by definition

$$H(X|Y) \triangleq -\mathcal{E}\{p(X|Y)\}.$$

The quantity $H(X|Y)$ is known as *equivocation* because it indicates how much remains unknown about X once Y is known. A similar expression holds for $H(Y|X)$. Moreover, upon comparing the definitions (12.13) and (12.18), it is clear that $I(X; Y)$ can be expressed as the Kullback divergence between the joint probability density $p_{X,Y}(X, Y)$ and the marginal density of the two random variables $p_X(X) \, p_Y(Y)$ as

$$I(X; Y) = D(p_{X,Y} \parallel p_X \, p_Y).$$

Given this interpretation in terms of the Kullback divergence, mutual information can be viewed as the natural measure of the statistical independence of two or more random variables. Note that for N random variables $Y_1, \ldots, Y_N$, definition (12.19) can be readily extended as

$$I(Y_1; \ldots; Y_N) \triangleq \mathcal{E} \left\{ \log \frac{p(Y_1, \ldots, Y_N)}{p(Y_1), \ldots, p(Y_N)} \right\}. \tag{12.22}$$

For the development in this chapter, however, we will be concerned exclusively with the MI of two random variables.

As implied by (12.14) and (12.18), MI is always non-negative, and identically zero iff the two random variables are statistically independent. The latter property is clear from (12.19), inasmuch as statistical independence implies $p(y_1, y_2) = p(y_1) \, p(y_2)$ by definition. For these reasons, MI is in some sense the most natural contrast function. Subsequently, we will explore the relationship between MI and negentropy, which is another contrast function typically used in the ICA field.

We will now calculate the mutual information for the zero-mean Gaussian random variables Y_1 and Y_2. For jointly Gaussian random variables,

$$p_{Y_1, Y_2}(y_1, y_2) = \frac{1}{\sqrt{|2\pi \mathbf{\Sigma_Y}|}} \exp \left[-\frac{1}{2} \mathbf{y}^T \mathbf{\Sigma_Y}^{-1} \mathbf{y} \right],$$

where $\mathbf{y} = [y_1 \quad y_2]^T$. The covariance matrix $\mathbf{\Sigma_Y} \triangleq \mathcal{E}\{\mathbf{y}\mathbf{y}^T\}$ appearing in the last equation can be written in the form (Anderson 1984, sect. 2.3)

$$\mathbf{\Sigma_Y} = \begin{bmatrix} \sigma_1^2 & \sigma_1 \sigma_2 \rho_{12} \\ \sigma_1 \sigma_2 \rho_{12} & \sigma_2^2 \end{bmatrix}, \tag{12.23}$$

where

$$\rho_{12} = \frac{\epsilon_{12}}{\sigma_1 \sigma_2}, \qquad \sigma_i^2 = \mathcal{E}\{Y_i^2\}, \ \forall \, i = 1, 2, \quad \text{and} \quad \epsilon_{12} = \mathcal{E}\{Y_1 Y_2\}.$$

Hence, the joint entropy $H(Y_1, Y_2)$ defined in (12.21) can be evaluated as

$$H(Y_1, Y_2) \triangleq -\mathcal{E}\{\log p(Y_1, Y_2)\} = \mathcal{E} \left\{ \frac{1}{2} \log |2\pi \mathbf{\Sigma_Y}| + \frac{1}{2} \mathbf{y}^T \mathbf{\Sigma_Y}^{-1} \mathbf{y} \right\}$$

$$= \frac{1}{2} \log |2\pi \mathbf{\Sigma_Y}| + \frac{1}{2} \int_{\mathbf{y}} \mathbf{y}^T \mathbf{\Sigma_Y}^{-1} \mathbf{y} \, p(\mathbf{y}) \, d\mathbf{y}. \tag{12.24}$$

Due to the *whitening* (Fukunaga 1990, sect. 2.3) provided by the term $\mathbf{\Sigma_Y}^{-1}$, the integral in (12.24) decouples into two integrals of the same form as that in (12.16). Hence, when (12.17) and (12.24) are substituted back into (12.20), the integral terms cancel out, and what remains is

$$I(Y_1; Y_2) = -\frac{1}{2} \log \left[4\pi^2 \sigma_1^2 \sigma_2^2 (1 - \rho_{12}^2) \right] + \frac{1}{2} \log 2\pi \sigma_1^2 + \frac{1}{2} \log 2\pi \sigma_2^2,$$

or, upon cancelling common terms,

$$I(Y_1; Y_2) = -\frac{1}{2} \log \left(1 - \rho_{12}^2\right). \tag{12.25}$$

From (12.25) it is clear that minimizing the MI between two zero-mean Gaussian r.v.s is equivalent to minimizing the squared magnitude of their cross-correlation coefficient, and that

$$I(Y_1; Y_2) = 0 \leftrightarrow \rho_{12}^2 = 0.$$

This is another illustration of the necessity of using non-Gaussian pdfs in order to account for the higher order moments of the independent components. That is, under a Gaussian assumption, uncorrelated and independent are equivalent.

Negentropy

We next consider a second contrast function that is frequently used in the ICA field. The *negentropy* J of a random variable Y is defined as

$$J(Y) \triangleq H(Y_{\text{gauss}}) - H(Y) \tag{12.26}$$

where Y_{gauss} is a Gaussian random variable with the same variance σ_Y^2 as Y. The entropy $H(Y_{\text{gauss}})$ of this Gaussian random variable is given by (12.17). Given that a Gaussian random variable has the highest entropy of all random variables with the same variance, negentropy is non-negative, and identically zero if and only if Y has a Gaussian distribution.

Note that the differential entropy of the GG pdf for the real-valued random variable Y can be expressed as

$$H_{\text{gg}}(Y) = -\int_{-\infty}^{+\infty} p_{\text{gg}}(y) \log p_{\text{gg}}(y) \, dy$$

$$= \log 2 + \log \Gamma \left(1 + \frac{1}{f}\right) + \log \hat{\sigma} + \log A(f) + \frac{1}{f}, \tag{12.27}$$

where $p_{\text{gg}}(y)$ is defined in (12.10).

Negentropy is the optimal estimator of non-Gaussianity in the statistical sense, and possesses the interesting and useful property of being invariant for invertible linear transformations; see Comon (1994) and Hyvärinen (1999). While negentropy finds frequent application in ICA, it is not used exclusively because it requires knowledge of the fine structure of the pdf of the desired signal. This is indeed a hindrance if an algorithm is sought, such as *fast ICA*, that works with any conceivable signal (Hyvärinen and Oja 2000). For the purposes of the present volume, however, we will confine our attention solely to speech signals and their subband samples. Hence, as we will discover in Section 13.5.2, it is very worth-while to estimate the pdf of speech, which is highly non-Gaussian, from a set of training data. Thereafter, negentropy can be used as the optimization criterion for either source separation or beamforming. Alternatively, it is possible to use kurtosis as a measure of non-Gaussianity. As discussed in the next section, kurtosis offers the advantage of simplicity in that the exact structure of the pdf need not be known in order to perform parameter optimization, as with negentropy. Rather, it is only necessary to calculate the second- and fourth-order moments.

Before leaving this section, we will develop an interesting connection between MI and negentropy. Based on (12.22), the MI of a set of random variables $\{X_1, X_2, \ldots, X_N\}$ for some $N \geq 2$ can be expressed as

$$I(X_1, X_2, \ldots, X_N) \triangleq \sum_{i=1}^{N} H(X_i) - H(\mathbf{X});$$

where

$$\mathbf{X} = \begin{bmatrix} X_1 & X_2 & \cdots & X_N \end{bmatrix}^T.$$

Note that for an invertible linear transformation $\mathbf{Y} = \mathbf{WX}$,

$$I(Y_1, Y_2, \ldots, Y_N) = \sum_{i=1}^{N} H(Y_i) - H(\mathbf{X}) - \log |\mathbf{W}|;$$

where $|\mathbf{W}|$ indicates the determinant of $\mathbf{W}$ (Cover and Thomas 1991). If all Y_i are constrained to be uncorrelated and to have unit variance, then

$$\mathcal{E}\{\mathbf{YY}^T\} = \mathbf{W}\,\mathcal{E}\{\mathbf{XX}^T\}\,\mathbf{W}^T = \mathbf{I},$$

from which it follows that

$$|\mathbf{I}| = 1 = |\mathbf{W}\mathcal{E}\{\mathbf{XX}^T\}\mathbf{W}^T| = |\mathbf{W}|\,|\mathcal{E}\{\mathbf{XX}^T\}|\,|\mathbf{W}^T|.$$

The last equation implies that $|\mathbf{W}|$ is constant. Moreover, for Y_i of unit variance, entropy and negentropy are the same up to a sign change and an additive constant. Hence,

$$I(Y_1, Y_2, \ldots, Y_N) = C - \sum_{i=1}^{N} J(Y_i);$$

where C is a constant that does not depend on $\mathbf{W}$. Therefore, under the assumption of uncorrelated random variables, *minimizing* MI is equivalent to *maximizing* the negentropy of each individual Y_i.

As mentioned previously, the difficulty of applying negentropy directly as an optimization criterion lies in the requirement of knowing the individual pdfs of the random components. If MI is used as an optimization criterion, the problem becomes even more difficult, because the joint pdf of all components must be known in addition to the marginal pdfs of the individual components. Moreover, the functional form of the joint pdf changes as the number N of indepedent components increases. Several approaches for obtaining the required estimates of the pdfs are possible. Two such approaches are presented in Chapter 13 in the context of acoustic beamforming, which has goals similar to BSS and ICA, but begins from different assumptions, namely, that the geometry of the sensor array is known and the positions of the desired speakers are known or can be reliably estimated. The first approach, which is described in Sections 13.5.2, is based on the use of the generalized Gaussian as a parametric model for the pdf of subband samples of speech. Due to the simplicity of the GG pdf, the scale and shape factors specifying its form can be readily estimated from a training set. The second approach, which is described in Sections 13.5.4, is based on the use of the Meier G-function to model a certain class of super-Gaussian pdfs. The representation of such pdfs with the Meier G-function is convenient in that it enables higher order variates to be estimated in a straightforward fashion as soon as the univariate pdf and the covariance matrix of the components are known.

There are still other possible approaches for modeling super-Gaussian pdfs in terms of higher order cumulants. Comon (1994) used a fourth-order Edgeworth expansion (Abramowitz and Stegun 1965, 1972) of a non-Gaussian pdf in order to approximate the MI (12.14) in terms of higher order cumulants, noting that the latter are more accessible than the fine structure of the pdf itself. Amari *et al.* (1996), on the other hand, preferred to use the Gram–Charlier expansion (Stuart and Ord 1994) to model the pdfs of all non-Gaussian random variables.

Kurtosis

Here we present yet another contrast function from the field of independent component analysis. The *excess kurtosis* of a random variable Y with zero mean is defined as

$$\text{kurt}(Y) \triangleq \mathcal{E}\{Y^4\} - 3(\mathcal{E}\{Y^2\})^2. \tag{12.28}$$

Much like the negentropy criterion considered in the last section, kurtosis is a measure of the *non-Gaussianity* of Y (Hyvärinen and Oja 2000). The Gaussian pdf has zero kurtosis, pdfs with *positive kurtosis* are *super-Gaussian*, and those with *negative kurtosis* are *sub-Gaussian*. As shown in Table 2.2, of the three super-Gaussian pdfs in Figure 12.2, the Γ pdf has the highest kurtosis, followed by the K_0, then by the Laplace pdf. This fact manifests itself in Figure 12.2, where it is clear that as the kurtosis increases, the pdf becomes more and more spiky and heavy-tailed. As explained in Section 13.5.2, the kurtosis of the GG pdf can be controlled by adjusting the shape factor f. Although kurtosis is widely used as a measure of non-Gaussianity, it is relatively sensitive to outliers, in that the value calculated for kurtosis can be strongly influenced by a few samples with relatively low observation probability (Hyvärinen and Oja 2000).

Note that the definition of kurtosis is not entirely unproblematic, inasmuch as there is not one but three such definitions that are often used in the literature. We are already acquainted with the first definition of kurtosis, namely (12.28), which has the desirable property of being identically zero for Gaussian random variables. Another common definition is

$$\text{kurt}(Y) \triangleq \frac{\mathcal{E}\{Y^4\}}{(\mathcal{E}\{Y^2\})^2},$$

which assumes a value of 3 for Gaussian random variables. The third definition is

$$\text{kurt}(Y) \triangleq \frac{\mathcal{E}\{Y^4\}}{(\mathcal{E}\{Y^2\})^2} - 3,$$

which is sometimes referred to as *normalized kurtosis*. The latter definition also assumes a value of zero for Gaussian random variables.

12.2.4 Parameter Update Strategies

In this section, we consider strategies for recursively updating the demixing matrix under various optimization criteria. We will begin by considering the straightforward stochastic gradient descent, and thereafter the natural gradient method. The latter method will lead to faster convergence in many cases.

Let us define the vector

$$\mathbf{f}(\mathbf{y}) \triangleq [f(y_1) \ f(y_2) \ \cdots \ f(y_N)],$$

where the *score function* is by definition

$$f(y) \triangleq -\frac{\partial \log p_S(y)}{\partial y},$$

and $p_S(y)$ is the pdf of the independent components $\{S_n\}$. For convenience, let us assume that each of the independent components $\mathbf{S} = [S_1 \ S_2 \ \cdots \ S_N]^T$ is identically distributed, and that there is no permutation problem such that $\mathbf{\Phi} = \mathbf{I}$ in (12.3). Then a contrast function can be defined as

$$\hat{\mathcal{J}}(\mathbf{B}) \triangleq -\log \left[|\mathbf{B}| \prod_{i=1}^{N} p_s(y_i(k)) \right], \tag{12.29}$$

where $|\mathbf{B}|$ denotes the determinant of the demixing matrix $\mathbf{B}$. So defined, the expected value of $\hat{\mathcal{J}}(\mathbf{B})$ is equivalent to the Kullback divergence $D(p_Y \,||\, \hat{p}_Y)$ up to a constant, where $\hat{p}_Y$ is the pdf of $\mathbf{Y}$ under the assumption of independent components as in (12.12). A stochastic gradient descent procedure can be readily derived from this contrast function, with the update rule

$$\mathbf{B}(k+1) = \mathbf{B}(k) - \mu(k) \frac{\partial \hat{\mathcal{J}}(\mathbf{B}(k))}{\partial \mathbf{B}}$$
$$= \mathbf{B}(k) - \mu(k) \left[\mathbf{B}^{-T}(k) - \mathbf{f}(\mathbf{y}(k))\mathbf{x}^T(k) \right], \tag{12.30}$$

where $\mu(k) > 0$ is the step size.

In ICA, as with so many other optimization problems, gradient descent leads to simple optimization schemes, but very slow convergence. If the optimization surface is quadratic or nearly so, this speed of convergence can be improved through an update based on Newton's rule. Unfortunately, this is not the case in ICA, where all too often the optimization criterion is not quadratic. A modification of (12.30), dubbed *natural gradient* by Amari (1998) and *relative gradient* by Cardoso (1998), leads to much faster convergence. This modification can be expressed as

$$\mathbf{B}(k+1) = \mathbf{B}(k) - \mu(k)\frac{\partial \hat{\mathcal{J}}(\mathbf{B}(k))}{\partial \mathbf{B}}\mathbf{B}^T(k)\mathbf{B}(k)$$

$$= \mathbf{B}(k) + \mu(k)\left[\mathbf{I} - \mathbf{f}(\mathbf{y}(k))\,\mathbf{y}^T(k)\right] \qquad (12.31)$$

To gain some insight into the behavior of (12.31), let us define the combined system matrix $\mathbf{C}(k)$ as

$$\mathbf{C}(k) \triangleq \mathbf{B}(k)\,\mathbf{A}. \qquad (12.32)$$

Based on a comparision of (12.3) and (12.32), we would clearly hope that

$$\lim_{k\to\infty}\mathbf{C}(k) \to \mathbf{\Phi}\mathbf{D}.$$

Post-multiplying both sides of (12.31) by $\mathbf{A}$ and recognizing that $\mathbf{y}(k) = \mathbf{C}(k)\,\mathbf{s}(k)$, we can write the update in (12.31) as

$$\mathbf{C}(k+1) = \mathbf{C}(k) + \mu(k)\left[\mathbf{I} - \mathbf{f}(\mathbf{C}(k)\,\mathbf{s}(k))\,\mathbf{s}^T(k)\mathbf{C}^T(k)\right]\mathbf{C}(k).$$

The last equation depends only on the combined system matrix $\mathbf{C}(k)$, the signal vector $\mathbf{s}(k)$ of the source, and the step size $\mu(k)$. The effect of the mixing matrix $\mathbf{A}$ has been absorbed as an initial condition into $\mathbf{C}(0) = \mathbf{B}(0)\,\mathbf{A}$. Hence, as long as $\mathbf{C}(k)$ can "escape" from bad initial conditions, the evolutionary behavior of the combined system $\mathbf{C}(k)$ is not limited by $\mathbf{A}$.

The uniform performance provided by (12.31) is due to the equivariance property (Cardoso 1998) achieved by the natural/relative gradient. Moreover, the computational complexity of the parameter update is actually reduced, in that it is no longer necessary to calculate $\mathbf{B}^{-T}(k)$ as in (12.30). Further details regarding the natural gradient and its properties can be found in Douglas and Amari (2000).

12.3 BSS Algorithms based on Second-Order Statistics

The algorithms described in this section can more properly be said to belong to the field of BSS than that of ICA, inasmuch as they rely solely on the use of second-order statistics. While these algorithms are relatively simple, they are in many cases not as powerful as their counterparts based on HOS. The following development is based on that in Douglas (2001b, sect. 7.2.4).

Now rather than requiring that the sources are spatially-independent, we will make the weaker assumption that the sources are only spatially-uncorrelated, such that

$$\mathcal{E}\{S_i(k)\,S_j(k+l)\} = 0 \ \forall \ i \neq j, l. \qquad (12.33)$$

This condition is weaker than spatial independence, and opens up the possibility for separating sources that are nearly Gaussian-distributed. Each source is required to be temporally correlated, such that the normalized cross-correlation matrix

$$\hat{\mathbf{R}}_{SS} \triangleq \left[\mathcal{E}\{\mathbf{S}(k)\,\mathbf{S}^T(k)\} \right]^{-1} \mathcal{E}\{\mathbf{S}(k)\,\mathbf{S}^T(k+l)\} \tag{12.34}$$

has N unique eigenvalues for some $l \neq 0$. This condition is identical to the requirement that the normalized cross-correlation coefficients

$$\rho_i(l) \triangleq \frac{\mathcal{E}\{S_i(k)\,S_i(k-l)\}}{\sqrt{\mathcal{E}\{S_i^2(k)\}\,\mathcal{E}\{S_i^2(k-l)\}}}$$

are distinct $\forall\, 1 \leq i \leq N$ and at least one value of $l \neq 0$. This additional constraint on the correlation statistics is required to satisfy certain identifiability conditions, without which this formulation of the BSS problem will not function properly (Douglas 2001b).

In order to understand how (12.33–12.34) yield a separating solution, consider the corresponding normalized cross-correlation matrix of the input signals,

$$\hat{\mathbf{R}}_{XX}(l) \triangleq \left[\mathcal{E}\{\mathbf{X}(k)\,\mathbf{X}^T(k)\} \right]^{-1} \mathcal{E}\{\mathbf{X}(k)\,\mathbf{X}^T(k+l)\} \tag{12.35}$$

$$= \left[\mathbf{A}\,\mathcal{E}\{\mathbf{S}(k)\,\mathbf{S}^T(k)\}\mathbf{A}^T \right]^{-1} \mathbf{A}\,\mathcal{E}\{\mathbf{S}(k)\,\mathbf{S}^T(k+l)\}\mathbf{A}^T. \tag{12.36}$$

Without loss of generality, we may assume that each source signal has unit variance, such that

$$\mathcal{E}\{\mathbf{S}(k)\,\mathbf{S}^T(k)\} = \mathbf{I}.$$

It then follows that

$$\hat{\mathbf{R}}_{XX}(l) = \mathbf{A}^{-T}\mathcal{E}\{\mathbf{S}(k)\,\mathbf{S}^T(k+l)\}\mathbf{A}^T = \mathbf{A}^{-T}\,\hat{\mathbf{R}}_{SS}(l)\,\mathbf{A}^T. \tag{12.37}$$

Let us define the eigendecomposition of $\hat{\mathbf{R}}_{XX}(l)$ as

$$\hat{\mathbf{R}}_{XX}(l) = \mathbf{Q}\,\mathbf{\Lambda}\,\mathbf{Q}^{-1},$$

which implies that

$$\mathbf{\Lambda}(l) = \mathbf{\Phi}\hat{\mathbf{R}}_{SS}(l)\,\mathbf{\Phi}^T \quad \text{and} \quad \mathbf{Q} = \mathbf{A}^{-T}\,\mathbf{\Phi}^T.$$

Then the demixing matrix $\mathbf{B}$ can be calculated according to

$$\mathbf{B} = \mathbf{Q}^T.$$

Hence, all that is required is the eigendecomposition of a normalized cross-correlation matrix.

In practice, most BSS algorithms based on temporal correlation identify the demixing by solving a system of N^2 nonlinear equations for the N^2 entries of $\mathbf{B} = [\mathbf{b}_1 \cdots \mathbf{b}_N]^T$

such that

$$\mathbf{b}_i^T \, \mathcal{E}\{\mathbf{X}(k)\mathbf{X}^T(k)\}\mathbf{b}_j = \delta_{ij} \; \forall \; 1 \le i \le j \le N, \tag{12.38}$$

$$\mathbf{b}_i^T \, \mathcal{E}\{\mathbf{X}(k)\mathbf{X}^T(k+l)\}\mathbf{b}_j = 0 \; \forall \; 1 \le i < j \le N, \tag{12.39}$$

where

$$\delta_{ij} = \begin{cases} 1, & \text{for } i = j, \\ 0, & \text{otherwise.} \end{cases}$$

With regard to the diagonal components of (12.39), it will in general hold that $\mathbf{b}_i^T \, \mathcal{E}\{\mathbf{X}(k)\mathbf{X}^T(k+l)\}\mathbf{b}_j \ne 0, 1$. Algorithms for solving a system such as (12.38–12.39) are known as *joint diagonalization* procedures (Fukunaga 1990, sect. 10.2), inasmuch as they search for a matrix whose rows are the eigenvectors of at least two different matrices. Alternatively, a preprocessing stage can be used to calculate a pre-whitened sequence $\mathbf{V}(k)$ whose eigenvectors are the orthogonal separating matrix

$$\mathbf{W}^T \mathcal{E}\{\mathbf{V}(k) \, \mathbf{V}^T(k)\}\mathbf{W} = \mathbf{I}. \tag{12.40}$$

BSS using temporal decorrelation has the advantage of requiring only second-order statistics, which implies that it is more robust when confronted with limited data from which the data-dependent quantities must be estimated.

Consider once more the mixing and demixing matrix mode in (12.2), where we will now assume that the number of independent sources N is not necessarily equal to the number of sensors M. Let us define the *coherence function* between outputs i and j as

$$C_{i,j}(\omega, t) \triangleq \frac{S_{i,j}(\omega, t)}{\sqrt{S_{i,i}(\omega, t) \, S_{j,j}(\omega, t)}}, \tag{12.41}$$

where

$$S_{i,j}(\omega, t) \triangleq \mathcal{E}\{Y_i(e^{j\omega}, t) \, Y_j^*(e^{j\omega}, t)\} \tag{12.42}$$

is the cross-power spectral density at time k. Fancourt and Parra (2001) proposed to perform separation on convolutive mixtures by minimizing the sum of the squared magnitudes between all $N \times (N-1)/2$ distinct pairs of outputs of the form

$$J = \sum_k \sum_{i,j} |C_{i,j}(\omega, k)|^2. \tag{12.43}$$

While the optimization criterion (12.41–12.43) bears a certain similarity to that presented previously, it should be borne in mind that the criterion (12.41–12.43) is formulated in the frequency domain. Such a formulation has the advantage that it is no longer necessary to check that the random components are uncorrelated for all possible time lags (Fancourt and Parra 2001).

Equations (12.41–12.43) can be expressed in matrix form according to

$$J = \sum_t |\mathbf{C}_{\mathbf{YY}}(\omega, k)| = \text{trace}\left[\mathbf{C}_{\mathbf{YY}}^H(\omega, k) \cdot \mathbf{C}_{\mathbf{YY}}(\omega, k)\right], \tag{12.44}$$

where $\mathbf{C_{YY}}(\omega, k)$ is the $N \times N$ matrix of coherence functions whose components are $C_{Y_i Y_j}$. The matrix of coherence functions can then be expressed as

$$\mathbf{C_{YY}}(\omega, k) = \mathbf{\Lambda_{YY}^{-1/2}}(\omega, k) \times \mathbf{S_{YY}}(\omega, k) \times \mathbf{\Lambda_{YY}^{-1/2}}(\omega, k) \tag{12.45}$$

in terms of a $N \times N$ matrix of cross-spectral densities between the outputs, $\mathbf{S_{YY}}$ whose components are $S_{i,j}$ and a diagonal matrix $\mathbf{\Lambda_{YY}^{-1/2}}$, whose diagonal elements are $S_{i,i}$. Substituting (12.45) into (12.44) provides

$$J = \sum_t \text{trace} \left[\mathbf{\Lambda_{YY}^{-1}}(\omega, k) \times \mathbf{S_{YY}}(\omega, k) \times \mathbf{\Lambda_{YY}^{-1}}(\omega, k) \times \mathbf{S_{YY}}(\omega, k) \right]. \tag{12.46}$$

The cross-power spectral densities required to optimize (12.46) can be obtained from the recursive update

$$\mathbf{S_{YY}}(\omega, k) = \mu \mathbf{S_{YY}}(\omega, k - T) + (1 - \mu)\mathbf{Y}(\omega, k) \times \mathbf{Y}^H(\omega, k),$$

where $0 < \mu < 1$ is a *forgetting factor*, and T is the sampling interval. Fancourt and Parra showed that the separation weights can be recursively updated according to

$$\Delta \mathbf{W}(\omega, k) = -v\mathbf{\Lambda_{YY}^{-1}} \times [\mathbf{S_{YY}}(\omega, k) - \mathbf{\Lambda_{YY}}] \times \mathbf{\Lambda_{YY}^{-1}} \times \mathbf{S_{YX}}(\omega, k), \tag{12.47}$$

where v is a step size and $\mathbf{S_{YX}}(\omega, k)$ is the $N \times M$ matrix of cross-power spectral densities between the *output* and the *inputs*, which can be calculated as

$$\mathbf{S_{YX}}(\omega, k) = \mu \mathbf{S_{YX}}(\omega, k - T) + (1 - \mu)\mathbf{Y}(\omega, k) \times \mathbf{X}^H(\omega, k).$$

12.4 Summary and Further Reading

In this chapter, we discussed two classes of algorithms, both of which use several microphones in order to separate speech mixed with other speech or noise. This is done with knowledge neither of the relative locations of the microphones, nor of the positions of the desired speakers. Independent component analysis is based on the specific assumption that the source signal of interest has a non-Gaussian pdf, which – in the case of human speech – is clearly true. In order to perform separation, ICA algorithms estimate a demixing matrix that returns two or more signals that are as non-Gaussian as possible. We examined several optimization criteria that are commonly used in ICA, namely, mutual information, negentropy, and kurtosis. As discussed in this chapter, the latter two criteria are measures of deviation from Gaussianity, and hence are useful in searching for non-Gaussian sources. Mutual information is a measure of statistical independence, and thus can be used to search for statistically-independent sources. Mutual information can also be directly related to negentropy under the assumption that two or more sources are uncorrelated.

Blind source separation does not uniformly assume that the desired sources are non-Gaussian, and thus is somewhat more general than ICA. Instead, the separation criteria under BSS can be based on assumptions of non-whiteness or nonstationarity.

Moreover, BSS does not rely exclusively on the use of higher-order statistics of the source pdfs; many BSS algorithms are based only on the use of second-order statistics.

While negentropy is effective as an optimization criterion for ICA, it has the disadvantage of requiring a specific assumption as regards the pdf of the desired source. Several proposals to alleviate this requirement have appeared in the ICA literature. Hyvärinen and Oja (2000) used an approximation of negentropy based on higher order moments, namely,

$$J(Y) \approx \frac{1}{12} (\mathcal{E}\{Y^3\})^2 + \frac{1}{48} \text{kurt}^2(Y).$$

Hyvärinen (1998) also developed new approximations of negentropy based on the maximum-entropy principle. These approximations can in general be expressed in the form

$$J(Y) \approx \sum_{i=1}^{P} k_i \left[\mathcal{E}\{G_i(Y)\} - \mathcal{E}\{G_i(v)\} \right]^2$$

where $\{k_i\}$ is a set of positive constants, $\{G_i\}$ is a set of some nonquadratic constants, and v is a Gaussian random variable of zero mean and unit variance.

Cardoso (1998) as well as Hyvärinen and Oja (2000) describe the connection between MI and other contrast functions such as *maximum likelihood* and *infomax*. There are several useful tutorials and summaries of research within the BSS and ICA fields. A very accessible and readable treatment is given by Hyvärinen and Oja (2000). A very comprehensive – if somewhat less accessible – summary of the state-of-the-art in BSS, including an extensive discussion of the similarities and differences of several extant algorithms, is given by Buchner *et al.* (2004). Another useful tutorial is Douglas (2001a).

12.5   Principal Symbols

Symbol	Description
ω	angular frequency
Φ	permutation matrix
$\mathbf{A}$	mixing matrix
$\mathbf{B}$	demixing matrix
$\mathbf{D}$	diagonal scaling matrix
$\mathbf{S}$	vector of independent components
$\mathbf{X}$	vector of random mixtures
$D(p_{\mathbf{X}} \| p_{\mathbf{Y}})$	Kullback divergence between $p_{\mathbf{X}}$ and $p_{\mathbf{Y}}$
$H(Y)$	entropy of Y
$J(Y)$	negentropy of Y
$p_{\text{GG}}(y)$	generalized Gaussian probability density function
$I(Y_1, Y_2)$	mutual information between Y_1 and Y_2

13

Beamforming

In this chapter, we investigate a class of techniques – known collectively as *beamforming* – by which signals from several sensors can be combined to emphasize a desired source and suppress interference from other directions. Beamforming begins with the assumption that the positions of all sensors are *known*, and that the position of the desired source is known or can be estimated. The simplest of beamforming algorithms, the *delay-and-sum beamformer*, uses only this geometrical knowledge to combine the signals from several sensors. More sophisticated *adaptive beamformers* attempt to minimize the total output power of the array under the constraint that the desired source must be unattenuated. The conventional adaptive beamforming algorithms attempt to minimize a quadratic optimization criterion related to signal-to-noise ratio under a distortionless constraint in the look direction. Recent research has revealed, however, that such quadratic criteria are not optimal for acoustic beamforming of human speech. Hence, we also present beamformers based on non-conventional optimization criteria that have appeared more recently in the literature.

Any reader well acquainted with the conventional array processing literature will certainly have already seen the material in Sections 13.1 through 13.4. The interaction of propagating waves with the sensors of a beamformer are described in Section 13.1.1, as are the effects of sensor spacing and beam steering on the spatial sensitivity of the array. The *beam pattern*, which is a plot of array sensitivity versus direction of arrival of propagating wave, is defined and described in Section 13.1.2. The simplest beamformer, namely the delay-and-sum beamformer, is presented in Section 13.1.3, and the effects of beam steering are discussed in Section 13.1.4. Quantitative measures of beamforming performance are presented in Section 13.2, the most important of which are directivity, as presented in Section 13.2.1, and array gain, as presented in Section 13.2.2. These measures will be used to evaluate the conventional beamforming algorithms described later in the chapter.

In Section 13.3, we take up the discussion of the conventional beamforming algorithms. The *minimum variance distortionless response* (MVDR) is presented in Section 13.3.1, and its performance is analyzed in Sections 13.3.2 and 13.3.3. The beamforming algorithms based on the MVDR design, including the minimum mean square error and maximum signal-to-noise ratio beamformers, have the advantage of being tractable to analyze in simple acoustic environments. As discussed in Section 13.3.4, the superdirective beam-former, which is based on particular assumptions about the ambient noise field, has proven useful in real acoustic environments. The *minimum mean-square error* (MMSE) beamformer is presented in Section 13.3.5 and its relation to the MVDR beamformer is discussed. The maximum signal-to-noise ratio design is then presented in Section 13.3.6. The *generalized sidelobe canceller* (GSC), which is to play a decisive role in the latter sections of this chapter, is presented in Section 13.3.7. As discussed in Section 13.3.8, diagonal loading is a very simple technique for adding robustness into adaptive beam-forming designs.

Section 13.4, the last about the conventional beamforming algorithms, discusses implementations of adaptive beamforming algorithms that are suitable for online operation. Firstly, a convergence analysis of designs based on stochastic gradient descent is presented in Section 13.4.1, thereafter the various *least mean-square* (LMS) error designs, are presented in Section 13.4.2. These designs provide a complexity that is linear with the number N of sensors in the array, but can be slow to converge under unfavorable acoustic conditions. The *recursive least square* (RLS) error design, whose complexity increases as N^2, is discussed in Section 13.4.3. In return for this greater complexity, the RLS designs can provide better convergence characteristics. The RLS algorithms are known to be susceptible to numerical instabilities. A way to remedy this problem, namely the square-root implementation, is discussed in Section 13.4.4.

Recent research has revealed that the optimization criteria used in conventional array processing are not optimal for acoustic beamforming applications. In Section 13.5 of this chapter we discuss nonconventional optmization criteria for beamforming. A beamformer that maximizes the likelihood of the output signal with respect to a *hidden Markov model* (HMM) such as those discussed in Chapters 7 and 8 is discussed in Section 13.5.1.

Section 13.5.2 presents a nonconventional beamforming algorithm based on the optimization of a negentropy criterion subject to a distortionless constraint. The negentropy criterion provides an indication of how non-Gaussian a random variable is. Human speech is a highly non-Gaussian signal, but becomes more nearly Gaussian when corrupted with noise or reverberation. Hence, in adjusting the active weight vectors of a GSC so as to provide a maximally non-Gaussian output subject to a distortionless constraint, the harmful effects of noise and reverberation on the output of the array can be minimized. A refinement of the *maximum negentropy beamformer* (MNB) is presented in Section 13.5.3, whereby a HMM is used to capture the nonstationarity of the desired speaker's speech.

It happens quite often when two or more people speak together, that they will speak simultaneously, thereby creating regions of overlapping or simultaneous speech. Thus, the recognition of such simultaneous speech is an area of active research. In Section 13.5.4, we present a relatively new algorithm for separating overlapping speech into different output streams. This algorithm is based on the construction of two beamformers in GSC configuration, one pointing at each active speaker. To provide optimal separation performance, the active weight vectors of both GSCs are optimized jointly to provide two output streams with *minimum mutual information* (MinMI). This approach is also motivated in large part by research within the ICA field. The geometric source separation algorithm is presented in Section 13.5.5, which under the proper assumptions can be shown to be related to the MinMI beamformer.

Section 13.6 discusses a technique for automatically inferring the geometry of a micro-phone array based on a diffuse noise assumption.

In the final section of the chapter, we present our conclusions and recommendations for further reading.

13.1 Beamforming Fundamentals

Here we consider the fundamental concepts required to describe the interaction of propa-gating sound waves with sensor arrays. In this regard, the discussion here is an extension of that in Section 2.1. The exposition in this section is based largely on Van Trees (2002, sect. 2.2), and will make extensive use of the basic signal processing concepts developed in Chapter 3.

13.1.1 Sound Propagation and Array Geometry

To begin, consider an arbitrary array of N sensors. We will assume for the moment that the locations $\mathbf{m}_n$, for $n = 0, 1, \ldots, N - 1$ of the sensors are known. These sensors produce a set of signals denoted by the vector

$$\mathbf{f}(t, \mathbf{m}) = \begin{bmatrix} f(t, \mathbf{m}_0) \\ f(t, \mathbf{m}_1) \\ \vdots \\ f(t, \mathbf{m}_{N-1}) \end{bmatrix}.$$

For the present, we will also work in the continuous-time domain t. This is done only to avoid the *granularity* introduced by a discrete-time index. But this will cease to be an issue when we move to the subband domain, as the phase shifts and scaling factors to be applied in the subband domain are continuous-valued, regardless of whether or not this is so for the signals with which we begin. The output of each sensor is processed with a *linear time-invariant* (LTI) filter with impulse response $h_n(\tau)$ and filter outputs are then summed to obtain the final output of the beamformer:

$$y(t) = \sum_{n=0}^{N-1} \int_{-\infty}^{\infty} h_n(t - \tau) \, f_n(\tau, \mathbf{m}_n) \, d\tau.$$

In matrix notation, the sensor weights of the delay-and-sum beamformer can be expressed as

$$y(t) = \int_{-\infty}^{\infty} \mathbf{h}^T(t - \tau) \, \mathbf{f}(\tau, \mathbf{m}) \, d\tau, \qquad (13.1)$$

where

$$\mathbf{h}(t) = \begin{bmatrix} h_0(t) \\ h_1(t) \\ \vdots \\ h_{N-1}(t) \end{bmatrix}.$$

Moving to the frequency domain by applying the continuous-time Fourier transform (3.48) enables (13.1) to be rewritten as

$$Y(\omega) = \int_{-\infty}^{\infty} y(t)\, e^{-j\omega t}\, dt = \mathbf{H}^T(\omega)\, \mathbf{F}(\omega, \mathbf{m}), \qquad (13.2)$$

where

$$\mathbf{H}(\omega) = \int_{-\infty}^{\infty} \mathbf{h}(t) e^{-j\omega t}\, dt, \qquad (13.3)$$

$$\mathbf{F}(\omega, \mathbf{m}) = \int_{-\infty}^{\infty} \mathbf{f}(t, \mathbf{m}) e^{-j\omega t}\, dt, \qquad (13.4)$$

are, respectively, the vectors of frequency responses of the filters and spectra of the signals produced by the sensors.

In building an actual beamforming system, we will not, of course, work with continuous-time Fourier transforms as implied by (13.2). Rather, the output of each microphone will be sampled then processed with an analysis filter bank such as was described in Chapter 11 to yield a set of subband samples. The N samples for each center frequency $\omega_m = 2\pi m/M$, where M is the number of subband samples, will then be gathered together and the inner product (13.2) will be calculated, whereupon all M beamformer outputs can then be transformed back into the time domain by a synthesis bank. We are justified in taking this approach by the reasoning presented in Section 11.1, where it was explained that the output of the analysis bank can be interpreted as a short-time Fourier transform of the sampled signals subject only to the condition that the signals are sampled often enough in time to satisfy the Nyquist criterion. Beamforming in the subband domain has the considerable advantage that the active sensor weights can be optimized for each subband independently, which provides a tremendous computational savings with respect to a time-domain filter-and-sum beamformer with filters of the same length on the output of each sensor.

Although the filter frequency responses are represented as constant with time in (13.2–13.4), in subsequent sections we will relax this assumption and allow $\mathbf{H}(\omega)$ to be adapted in order to maximize or minimize an optimization criterion. We will in this case, however, make the assumption that is standard in adaptive filtering theory, namely, that $\mathbf{H}(\omega)$ changes sufficiently slowly such that (13.2) is valid for the duration of a single subband snapshot (Haykin 2002). This implies, however, that the system is no longer actually linear.

We will typically use spherical coordinates (r, θ, ϕ) to describe the propagation of sound waves through space. The relation between these spherical coordinates and the Cartesian coordinates (x, y, z) is illustrated in Figure 13.1. So defined, $r > 0$ is the *radius* or *range*, the *polar angle* θ assumes values on the range $0 \le \theta \le \pi$, and the *azimuth* assumes values on the range $0 \le \phi \le 2\pi$. Letting ϕ vary over its entire range is normal for circular arrays, but with the linear arrays considered in Section 13.1.3, it is typical for the sensors to be shielded acoustically from the rear so that, effectively, no sound propagates in the range $\pi \le \phi \le 2\pi$.

In the classical array-processing literature, it is quite common to make a *plane wave* assumption, which implies that the source of the wave is so distant that the locus of points

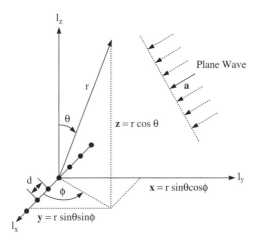

Figure 13.1 Relation between the spherical coordinates (r, θ, ϕ) and Cartesian coordinates (x, y, z)

with the same phase or *wavefront* is a plane. Such an assumption is seldom justified in acoustic beamforming through air, as the aperture of the array is typically of the same order of magnitude as the distance from the source to the sensors. Nonetheless, such an assumption is useful in introducing the conventional array-processing theory, our chief concern in this section, because it simplifies many important concepts. It is often useful in practice as well, in that it is not always possible to reliably estimate the distance from the source to the array, in which case the plane wave assumption is the only possible choice.

Consider then a plane wave shown in Figure 13.1 propagating in the direction

$$\mathbf{a} = \begin{bmatrix} a_x \\ a_y \\ a_z \end{bmatrix} = \begin{bmatrix} -\sin\theta\cos\phi \\ -\sin\theta\sin\phi \\ -\cos\theta \end{bmatrix}.$$

The first simplification this produces is that the *same signal* $f(t)$ arrives at each sensor, but not at the *same time*. Hence, we can write

$$\mathbf{f}(t, \mathbf{m}) = \begin{bmatrix} f(t - \tau_0) \\ f(t - \tau_1) \\ \vdots \\ f(t - \tau_{N-1}) \end{bmatrix}, \tag{13.5}$$

where the *time delay of arrival* (TDOA) τ_n appearing in (13.5) can be calculated through the inner product,

$$\tau_n = \frac{\mathbf{a}^T \mathbf{m}_n}{c} = -\frac{1}{c}[m_{n,x} \cdot \sin\theta\cos\phi + m_{n,y} \cdot \sin\theta\sin\phi + m_{n,z} \cdot \cos\theta], \tag{13.6}$$

c is the *velocity of sound*, and $\mathbf{m}_n = [m_{n,x} \quad m_{n,y} \quad m_{n,z}]$. Each τ_n represents the difference in arrival time of the wavefront at the nth sensor with respect to the origin.

If we now define the *direction cosines*

$$\mathbf{u} \triangleq -\mathbf{a}, \tag{13.7}$$

then τ_n can be expressed as

$$\tau_n = -\frac{1}{c} [u_x m_{n,x} + u_y m_{n,y} + u_z m_{n,z}] = -\frac{\mathbf{u}^T \mathbf{m}_n}{c}. \tag{13.8}$$

The time-delay property (3.50) of the continuous-time Fourier transform implies that under the signal model (13.5), the nth component of $\mathbf{F}(\omega)$ defined in (13.4) can be expressed as

$$F_n(\omega) = \int_{-\infty}^{\infty} f(t - \tau_n) e^{-j\omega t} \, dt = e^{-j\omega \tau_n} F(\omega), \tag{13.9}$$

where $F(\omega)$ is the Fourier transform of the original source. From (13.7) and (13.8) we infer

$$\omega \tau_n = \frac{\omega}{c} \mathbf{a}^T \mathbf{m}_n = -\frac{\omega}{c} \mathbf{u}^T \mathbf{m}_n. \tag{13.10}$$

For plane waves propagating in a locally homogeneous medium, the *wave number* is defined as

$$\mathbf{k} = \frac{\omega}{c} \mathbf{a} = \frac{2\pi}{\lambda} \mathbf{a}, \tag{13.11}$$

where λ is the wavelength corresponding to the angular frequency ω. Based on (13.7), we can now express the wavenumber as

$$\mathbf{k} = -\frac{2\pi}{\lambda} \begin{bmatrix} \sin \theta \cos \phi \\ \sin \theta \sin \phi \\ \cos \theta \end{bmatrix} = -\frac{2\pi}{\lambda} \mathbf{u}.$$

Assuming that the speed of sound is constant implies that

$$|\mathbf{k}| = \frac{\omega}{c} = \frac{2\pi}{\lambda}. \tag{13.12}$$

Physically, the wavenumber represents both the direction of propagation and frequency of the plane wave. As indicated by (13.11), the vector $\mathbf{k}$ specifies the direction of propagation of the plane wave. Equation (13.12) implies that the magnitude of $\mathbf{k}$ determines the frequency of the plane wave.

Together (13.10) and (13.11) imply that

$$\omega \tau_n = \mathbf{k}^T \mathbf{m}_n. \tag{13.13}$$

Hence, the Fourier transform of the propagating wave whose nth component is (13.9) can be expressed in vector form as

$$\mathbf{F}(\omega) = F(\omega)\,\mathbf{v_k}(\mathbf{k}), \tag{13.14}$$

where the *array manifold vector*, defined as

$$\mathbf{v_k}(\mathbf{k}) \triangleq \begin{bmatrix} e^{-j\mathbf{k}^T\,\mathbf{m}_0} \\ e^{-j\mathbf{k}^T\,\mathbf{m}_1} \\ \vdots \\ e^{-j\mathbf{k}^T\,\mathbf{m}_{N-1}} \end{bmatrix}, \tag{13.15}$$

represents a complete "summary" of the interaction of the array geometry with a propagating wave. As mentioned previously, beamforming is typically performed in the discrete-time Fourier transform domain, through the use of digital filter banks. This implies that the time-shifts must be specified in *samples*, in which case the array manifold vector must be represented as

$$\mathbf{v}_{\mathrm{DT}}(\mathbf{x}, \omega_m) \triangleq \begin{bmatrix} e^{-j\omega_m\,\tau_0/T_{\mathrm{s}}} \\ e^{-j\omega_m\,\tau_1/T_{\mathrm{s}}} \\ \vdots \\ e^{-j\omega_m\,\tau_{N-1}/T_{\mathrm{s}}} \end{bmatrix}, \tag{13.16}$$

where the subband center frequencies are $\{\omega_m\}$, the propagation delays $\{\tau_n\}$ are calculated according to (13.8), and T_{s} is the sampling interval defined in Section 3.1.4.

13.1.2 Beam Patterns

In Section 3.1.1 we demonstrated that the complex exponential sequence $f[n] = e^{j\omega n}$ is an eigensequence for any digital LTI system. It can be similarly shown that

$$f(t) = e^{j\omega t} \tag{13.17}$$

is an eigenfunction for any analog LTI system. This implies that if the complex exponential (13.17) is taken as the input to a single-input, single-output LTI system, the output of the system always has the form

$$y(t) = G(\omega)\,e^{j\omega t},$$

where, as discussed in Section 3.1, $G(\omega)$ is the frequency response of the system. For the analysis of multiple-input, single-output systems used in array processing, we consider eigenfunctions of the form

$$f_n(t, \mathbf{m}_n) = \exp\left[j(\omega t - \mathbf{k}^T\,\mathbf{m}_n)\right], \tag{13.18}$$

which is in fact the definition of a plane wave. For the entire array, we can write

$$\mathbf{f}(t, \mathbf{m}) = e^{j\omega t}\, \mathbf{v_k}(\mathbf{k}). \qquad (13.19)$$

The response of the array to a plane wave input can be expressed as

$$y(t, \mathbf{k}) = \Upsilon(\omega, \mathbf{k})\, e^{j\omega t},$$

where the *frequency–wavenumber response function* (Van Trees 2002, sect. 2.2) is defined as

$$\Upsilon(\omega, \mathbf{k}) \triangleq \mathbf{H}^T(\omega)\, \mathbf{v_k}(\mathbf{k}),$$

and $\mathbf{H}(\omega)$ is the Fourier transform of $\mathbf{h}(t)$ defined in (13.3). Just as the frequency response $H(\omega)$ defined in (3.13) specifies the response of conventional LTI system to a sinusoidal input, the frequency–wavenumber response function specifies the response of an array to a plane wave input with wavenumber $\mathbf{k}$ and angular frequency ω. Observe that the notation $\Upsilon(\omega, \mathbf{k})$ is redundant in that the angular frequency ω is uniquely specified by the wavenumber $\mathbf{k}$ through (13.12). We retain the argument ω, however, to stress the frequency-dependent nature of the frequency–wavenumber response function.

The *beam pattern* indicates the sensitivity of the array to a plane wave with wavenumber $\mathbf{k} = \frac{2\pi}{\lambda}\, \mathbf{a}(\theta, \phi)$, and is defined as

$$B(\omega : \theta, \phi) \triangleq \Upsilon(\omega, \mathbf{k})\big|_{\mathbf{k} = \frac{2\pi}{\lambda}\, \mathbf{a}(\theta, \phi)},$$

where $\mathbf{a}(\theta, \phi)$ is a unit vector with spherical coordinate angles θ and ϕ. The primary difference between the frequency–wavenumber response function and the beam pattern is that the arguments in the beam pattern must correspond to the physical angles θ and ϕ.

13.1.3 Delay-and-Sum Beamformer

In a *delay-and-sum beamformer*[1] (DSB), the impulse response of the filter on each sensor is a shifted impulse:

$$h_n(t) = \frac{1}{N}\, \delta(t + \tau_n),$$

where $\delta(t)$ is the Dirac delta function. The time shifts $\{\tau_n\}$ are calculated according to (13.13), such that the signals from each sensor in the array upon which a plane wave with wavenumber $\mathbf{k}$ and angular frequency ω impinges are added *coherently*. As we will shortly see, this has the effect of enhancing the desired plane wave with respect to plane waves propagating in other directions, provided certain conditions are met. If the signal is

---

[1] Many authors (Van Trees 2002) refer to the delay-and-sum beamformer as the *conventional beamformer*. In this volume, however, we will reserve the term "conventional" to refer to the conventional adaptive beamformer algorithms – namely, the minimum variance distortionless response, MMSE, and maximum signal-to-noise ratio beamformers – discussed in Section 13.3.

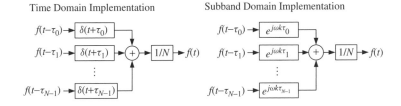

Figure 13.2 Time and subband domain implementations of the delay-and-sum beamformer

narrowband with a center frequency of ω_c, then, as indicated by (3.50), a time delay of τ_n corresponds to a linear phase shift, such that the complex weight applied to the output of the nth sensor can be expressed as

$$w_n^* = H_n(\omega_c) = \frac{1}{N} e^{j\omega_c \tau_n}.$$

In matrix form, this becomes

$$\mathbf{w}^H(\omega_c) = \mathbf{H}^T(\omega_c) = \frac{1}{N} \mathbf{v}_{\mathbf{k}}^H(\mathbf{k}), \qquad (13.20)$$

where the array manifold vector $\mathbf{v}_{\mathbf{k}}(\mathbf{k})$ is defined in (13.15) and (13.16) for the continuous- and discrete-time cases, respectively. The narrowband assumption is justified in that, as mentioned previously, we will apply an analysis filter bank to the output of each sensor to divide it into M narrowband signals. As discussed in Section 11, the filter bank prototype is designed to minimize aliasing distortion, which implies it will have good suppression in the stopband. This assertion is readily verified through an examination of the frequency response plots in Figures 11.10 through 11.12. Both time and subband domain implementations of the DSB are shown in Figure 13.2.

A simple *discrete Fourier transform* (DFT) can also be used for the subband analysis and resynthesis. This approach, however, is suboptimal in that it corresponds to a uniform DFT filter bank with a prototype impulse response whose values are constant. This implies that there will be large sidelobes in the stopband, as shown in Figure 11.2, and that the complex samples at the output of the different subbands will be neither statistically independent nor uncorrelated.

In order to gain an appreciation of the behavior of a sensor array, we now introduce several simplifying assumptions. Firstly, we will consider the case of a uniform linear array with equal intersensor spacing as shown in Figure 13.1. The nth sensor is located at

$$m_{n,x} = \left(n - \frac{N-1}{2} \right) d, \quad m_{n,y} = m_{n,z} = 0 \, \forall n = 0, \ldots, N-1,$$

where d is the intersensor spacing. As a further simplification, assume that plane waves propagate only parallel to the $x-y$ plane, so that the array manifold vector (13.15) can be expressed as

$$\mathbf{v_k}(k_x) = \left[e^{j\left(\frac{N-1}{2}\right)k_x d} \quad \cdots \quad e^{j\left(\frac{N-1}{2}-1\right)k_x d} \quad \cdots \quad e^{-j\left(\frac{N-1}{2}\right)k_x d} \right]^T,$$

where the x-component of $\mathbf{k}$ is by definition

$$k_x \triangleq -\frac{2\pi}{\lambda} \cos\phi = -k_0 \cos\phi,$$

and

$$k_0 \triangleq |\mathbf{k}| = \frac{2\pi}{\lambda}.$$

Let $u_x = \cos\phi$ denote the direction cosine with respect to the x-axis, and let us define

$$\psi \triangleq -k_x \, d = \frac{2\pi}{\lambda} \cos\phi \cdot d = \frac{2\pi}{\lambda} u_x \, d. \tag{13.21}$$

The variable ψ contains the all-important ratio d/λ as well as the *direction of arrival* (DOA) in $u = u_x = \cos\phi$. Hence ψ is a succinct summary of all information needed to calculate the sensitivity of the array. The wavenumber response as a function of k_x can then be expressed as

$$\Upsilon(\omega, k_x) = \mathbf{w}^H \mathbf{v_k}(k_x) = \sum_{n=0}^{N-1} w_n^* \, e^{-j\left(n-\frac{N-1}{2}\right) k_x \, d}. \tag{13.22}$$

The array manifold vector can be represented in the other spaces according to

$$[\mathbf{v}_\phi(\phi)]_n = e^{j(n-\frac{N-1}{2})\frac{2\pi d}{\lambda} \cos\phi},$$

$$[\mathbf{v}_u(u)]_n = e^{j(n-\frac{N-1}{2})\frac{2\pi d}{\lambda} u},$$

$$[\mathbf{v}_\psi(\psi)]_n = e^{j(n-\frac{N-1}{2})\psi},$$

where $[\cdot]_n$ denotes the nth component of the relevant array manifold vector. The representations of the beam pattern given above are useful for several reasons. Firstly, the ϕ–space is that in which the physical wave actually propagates, hence it is inherently useful. As we will learn in Section 13.1.4, the representation in u–space is useful inasmuch as, due to the definition $u \triangleq \cos\phi$, steering the beam in this space is equivalent to simply shifting the beam pattern. Finally, the ψ–space is useful because the definition (13.21) directly incorporates the all-important ratio d/λ, whose significance will be discussed in Section 13.1.4.

Based on (13.21), the beam pattern can also be expressed as a function of ϕ, u, or ψ:

$$B_\phi(\phi) = \mathbf{w}^H \mathbf{v}_\phi(\phi) = e^{-j(\frac{N-1}{2})\frac{2\pi d}{\lambda} \cos\phi} \sum_{n=0}^{N-1} w_n^* \, e^{jn\frac{2\pi d}{\lambda} \cos\phi},$$

$$B_u(u) = \mathbf{w}^H \mathbf{v}_u(u) = e^{-j\left(\frac{N-1}{2}\right)\frac{2\pi d}{\lambda}u} \sum_{n=0}^{N-1} w_n^* e^{jn\frac{2\pi d}{\lambda}u}, \tag{13.23}$$

$$B_\psi(\psi) = \mathbf{w}^H \mathbf{v}_\psi(\psi) = e^{-j\left(\frac{N-1}{2}\right)\psi} \sum_{n=0}^{N-1} w_n^* e^{jn\psi}.$$

Now we introduce a further simplifying assumption, namely, that all sensors are uniformly weighted, such that

$$w_n = \frac{1}{N} \,\forall\, n = 0, 1, \ldots, N-1.$$

In this case, the beam pattern in ψ-space can be expressed as

$$B_\psi(\psi) = \frac{1}{N} e^{-j\left(\frac{N-1}{2}\right)\psi} \sum_{n=0}^{N-1} e^{jn\psi}. \tag{13.24}$$

Using the identity

$$\sum_{n=0}^{N-1} x^n = \frac{1-x^N}{1-x}$$

it is possible to rewrite (13.24) as

$$
\begin{aligned}
B_\psi(\psi) &= \frac{1}{N} e^{-j\left(\frac{N-1}{2}\right)\psi} \left(\frac{1-e^{jN\psi}}{1-e^{j\psi}}\right) \\
&= \frac{1}{N} e^{-j\left(\frac{N-1}{2}\right)\psi} \cdot \frac{e^{jN\psi/2}}{e^{j\psi/2}} \cdot \frac{e^{-jN\psi/2}-e^{jN\psi/2}}{e^{-j\psi/2}-e^{j\psi/2}} \\
&= \operatorname{sinc}_N\left(\frac{\psi}{2}\right) \,\forall\, -\frac{2\pi d}{\lambda} \le \psi \le \frac{2\pi d}{\lambda},
\end{aligned}
\tag{13.25}
$$

where

$$\operatorname{sinc}_N(x) \triangleq \frac{1}{N}\frac{\sin(Nx)}{\sin x}. \tag{13.26}$$

From the final equality in (13.25), which is plotted against both linear and decibel axes in Figure 13.3, it is clear that $B_\psi(\psi)$ is periodic with period 2π for odd N. Moreover, $B_\psi(\psi)$ assumes its maximum values when both numerator and denominator of (13.26) are *zero*, in which case it can be shown to assume a value of unity through the application of L'Hospital's rule.

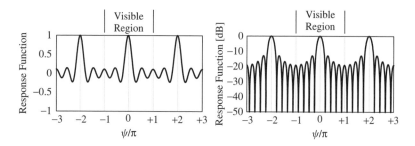

Figure 13.3 Comparison between a beam pattern on a linear and logarithmic scale, $\psi = \frac{2\pi}{\lambda} d \cos \phi$, $N = 20$

Substituting the relevant equality from (13.21), the beam pattern can be expressed in ϕ-space as

$$B_\phi(\phi) = \text{sinc}_N \left(\frac{\pi d}{\lambda} \cos \phi \right) \, \forall \, 0 \le \phi \le \pi. \tag{13.27}$$

In u-space this becomes

$$B_u(u) = \text{sinc}_N \left(\frac{\pi d}{\lambda} u \right) \, \forall \, -1 \le u \le 1. \tag{13.28}$$

A comparison of the beam pattern in different spaces is provided in Figure 13.4. Note that in each of (13.25), (13.27) and (13.28), we have indicated the allowable range on the argument of the beam pattern. As shown in Figure 13.4, this range is known as the *visible region*, because this is the region in which waves may actually propagate. It is often useful, however, to assume that ψ, ϕ, and u can vary over the entire real axis. In this case, every point outside of the range outside of the visible region is said to lie in the *virtual region*. Clearly, the beam patterns as plotted in the k_x-, ψ- and u_x-spaces are just scaled replicas, just as we would expect given the linear relationships between these variables manifest in (13.21). The beam pattern plotted in ϕ-space, on the other hand, has a noticeably narrower main lobe and significantly longer sidelobes due to the term $\cos \phi$ appearing in (13.21).

The portion of the visible region where the array provides maximal sensitivity is known as the *main lobe*. A *grating lobe* is a sidelobe with the same height as the main lobe. As mentioned previously, such lobes appear when the numerator and denominator of (13.26) are both zero, which for $\text{sinc}_N (\psi/2)$ occurs at intervals of

$$\psi = 2\pi m,$$

for odd N. In direction cosine or u-space, the beam pattern (13.23) is specified by $B_u(u) = \text{sinc}_N (\pi d u / \lambda)$ and the grating lobes appear at intervals of

$$u = \frac{\lambda}{d} m \, \forall \, m = 1, 2, \ldots. \tag{13.29}$$

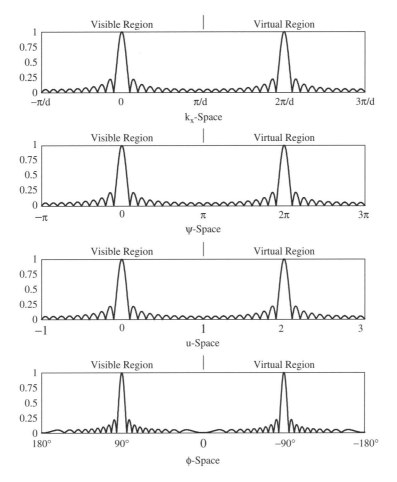

Figure 13.4 Beam pattern plots in k_x-, ψ-, u- and ϕ-spaces for a linear array with $d = \lambda/2$ and $N = 20$

The grating lobes are harmless as long as they remain in the virtual region. If the spacing between the sensors of the array is chosen to be too large, however, the grating lobes can move into the visible region. The effect is illustrated in Figure 13.5. The quantity that determines whether a grating lobe enters the visible region is the ratio d/λ. For a uniformly-weighted, uniform linear array, we must require $d/\lambda < 1$ in order to ensure that no grating lobe enters the visible region. We will shortly find, however, that *steering* can cause grating lobes to move into the visible region even when this condition is satisfied.

13.1.4 Beam Steering

Steering of the beam pattern is typically accomplished at the digital rather than physical level so that the array "listens" to a source emanating from a known or estimated position. For a plane wave, recall that the sensor inputs are given by (13.19). We would like the output to be time-aligned to the "target" wavenumber $\mathbf{k} = \mathbf{k}_T$, which is known as the

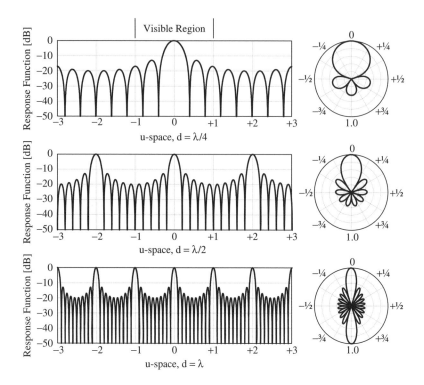

Figure 13.5 Effect of element spacing on beam patterns in linear and polar coordinates for $N = 10$

main response axis or *look direction*. As noted before, steering can be accomplished with time delays, or phase shifts. We will, however, universally prefer the latter based on our use of filter banks to carve up the sensor outputs into narrowband signals. The steered sensor inputs can then be expressed as

$$f_s(t, \mathbf{m}) = e^{j\omega t} \mathbf{v_k}(\mathbf{k} - \mathbf{k}_T),$$

and the steered frequency wavenumber response as

$$\Upsilon(\omega, \mathbf{k}|\mathbf{k}_T) = \Upsilon(\omega, \mathbf{k} - \mathbf{k}_T).$$

Hence, in wavenumber space, steering is equivalent to a simple shift, which is the principal advantage of plotting beam patterns in this space.

When DSB is steered to $\mathbf{k} = \mathbf{k}_T$, the sensor weights become

$$\mathbf{w} = \frac{1}{N}\, \mathbf{v_k}(\mathbf{k}_T). \tag{13.30}$$

The *delay-and-sum beam pattern*, which by definition is

$$B_{\mathrm{dsb}}(\mathbf{k} : \mathbf{k}_T) \triangleq \left. \frac{1}{N} \mathbf{v_k}^H(\mathbf{k}_T) \mathbf{v_k}(\mathbf{k}) \right|_{\mathbf{k}=\mathbf{a}(\theta,\phi)}, \tag{13.31}$$

is that beam pattern obtained when a DSB is steered to wavenumber $\mathbf{k}_T$ and evaluated at wavenumber $\mathbf{k} = \mathbf{a}(\theta, \phi)$. For a linear array, the delay-and-sum beam pattern can be expressed as

$$B_{\text{dsb}}(\psi : \psi_T) = \frac{1}{N} \mathbf{v}_\psi^H(\psi_T) \mathbf{v}_\psi(\psi) = \frac{1}{N} \frac{\sin\left(N\frac{\psi - \psi_T}{2}\right)}{\sin\left(\frac{\psi - \psi_T}{2}\right)},$$

or alternatively in u-space as

$$B_{\text{dsb}}(u : u_T) = \frac{1}{N} \mathbf{v}_u^H(u_T) \mathbf{v}_u(u) = \frac{1}{N} \frac{\sin\left[\frac{\pi N d}{\lambda}(u - u_T)\right]}{\sin\left[\frac{\pi d}{\lambda}(u - u_T)\right]}.$$

The *broadside angle* $\overline{\phi} = \phi - \pi/2$ is, by definition, measured with respect to the y-axis and has the same sense as ϕ. The effect of array steering with respect to $\overline{\phi}$ is illustrated in Figure 13.6. Based on the fact that steering corresponds to a simple shift in u-space, we can readily develop a requirement for excluding grating lobes from the visible region. Replacing u with $u - u_T$ in (13.29), we find that the position of the first grating lobe can be expressed as

$$u - u_{\text{T}} = \frac{\lambda}{d};$$

hence, keeping grating lobes out of the visible region requires

$$|u - u_{\text{T}}| \le \frac{\lambda}{d}.$$

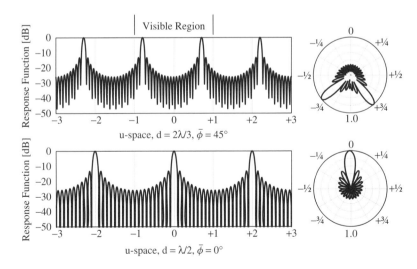

Figure 13.6 Effect of steering on the grating lobes for $N = 20$ plotted in linear and polar coordinates

This inequality can be further overbounded as

$$|u - u_\mathrm{T}| \leq |u| + |u_\mathrm{T}| \leq 1 + |\sin \overline{\phi}_\mathrm{max}| \leq \frac{\lambda}{d},$$

where $\overline{\phi}_\mathrm{max}$ is the maximum broadside angle to which the pattern is to be steered. Thus the condition for excluding grating lobes from the visible region can be expressed as

$$\frac{d}{\lambda} \leq \frac{1}{1 + |\sin \overline{\phi}_\mathrm{max}|}.$$

If the array is to be steered over the full range, which is to say, over the entire half plane, then we must require

$$d \leq \frac{\lambda}{2}. \tag{13.32}$$

This inequality bears a remarkable similarity to the Nyquist criterion described in Section 3.1.4 for the sampling of continuous-time signals, and for good reason. A microphone array is perhaps best thought of as *sampling* a wave propagating through space. The wave is sampled, however, both in *time* and in *space*. If not sampled often enough in either dimension, which means at least twice on every wavelength, aliasing occurs. The entrance of grating lobes into the visible region is nothing more than the manifestation of *spatial aliasing*.

Although it is perhaps not immediately obvious, the inequality (13.32) illustrates why designing a microphone array for acoustic beamforming applications is such a daunting task. Consider that speech is typically sampled at 16 kHz for audio applications. This implies that it must be bandlimited to 8 kHz prior to analog to digital conversion. In Section 2.1.2, we learned that sound propagates with a speed of roughly 343 m/s at sea level through air. This implies that the length of the shortest wave, which corresponds to the highest frequency, can be calculated as

$$\lambda_\mathrm{min} = \frac{34,300 \,\mathrm{cm}\,\mathrm{s}^{-1}}{8,000 \,\mathrm{s}^{-1}} \approx 4.28 \,\mathrm{cm}.$$

Hence, based on (13.32), in order to properly sample the wave with the shortest length using a uniform linear array, we must have an intersensor spacing of approximately 2 cm. Now consider that the lower limit on the bandwidth of human speech is typically taken as 300 Hz.[2] This implies that the length of the longest wave is

$$\lambda_\mathrm{max} = \frac{34,300 \,\mathrm{cm}\,\mathrm{s}^{-1}}{300 \,\mathrm{s}^{-1}} \approx 114.3 \,\mathrm{cm}.$$

In order to achieve good directivity at low frequencies, the *aperture* of the array, which is equivalent to the total distance between the sensors that are furthest apart, should be

---

[2] Even this lower limit of 300 Hz for bandwidth of human speech is too high, inasmuch as the fundamental frequency f_0 for voiced speech of male speakers is in the range of 120 Hz.

greater than the length of the longest wave. Taking the quotient of 114.3 cm divided by 2 cm, we find that our uniform linear array requires more than 50 sensors in order to adequately capture human speech. Moreover, its total length must be more than a meter. This is why a uniformly weighted, uniform linear array is seldom used in practice.

As we will learn in Section 14.1, using such an enormous array is feasible for some applications, such as in smart rooms or smart offices. If the array must be mounted on a humanoid robot, a laptop computer, a rearview mirror, or a PDA, however, such vast stretches of space are simply unavailable. In such applications, perhaps only four sensors can be used, and the total available space is perhaps 10 or 20 cm, or even less. Beamforming is still possible in such cases, but the performance provided by the simple DSB we have described in this section is insufficient. This is the primary reason for considering the superdirective and adaptive designs described in subsequent sections of this chapter.

Early work on the optimal placement of the elements of a microphone array was described by Silverman (1987). Alvarado (1990) investigated optimal spacing for linear microphone arrays. Rabinkin *et al.* (1996) demonstrated that the performance of microphone array systems is affected by the microphone placement. In Rabinkin *et al.* (1997) a method to evaluate the microphone array configuration was derived and an outline for optimum microphone placement under practical considerations was described.

One final note before closing this section: The development based on the definition of the array manifold vector in (13.15) is useful for demonstrating theoretically the basic concepts of beamforming. This definition is also sometimes useful practically for building far-field data capture systems. In acoustic beamforming it may happen, however, that the distance from the source to any one of the sensors of an array is of the same order as the aperture of the array itself. In such a case, it is better to use a spherical wave assumption. The latter implies that the array manifold vector must be redefined as

$$\mathbf{v}_{\mathbf{x}}(\mathbf{x}) \triangleq \begin{bmatrix} e^{-j\omega|\mathbf{x}-\mathbf{m}_0|/c} \\ e^{-j\omega|\mathbf{x}-\mathbf{m}_1|/c} \\ \vdots \\ e^{-j\omega|\mathbf{x}-\mathbf{m}_{N-1}|/c} \end{bmatrix},$$

where $\mathbf{x}$ is the position of the desired source in Cartesian coordinates. As before, if we are working in the discrete-time domain, the time delays must be expressed in samples such that

$$\mathbf{v}_{\mathrm{DT}}(\mathbf{x}) \triangleq \begin{bmatrix} e^{-j\omega_m|\mathbf{x}-\mathbf{m}_0|/(c\,T_s)} \\ e^{-j\omega_m|\mathbf{x}-\mathbf{m}_1|/(c\,T_s)} \\ \vdots \\ e^{-j\omega_m|\mathbf{x}-\mathbf{m}_{N-1}|/(c\,T_s)} \end{bmatrix},$$

where T_s is once more the sampling interval and $\{\omega_m\}$ is the set of center frequencies of the filter bank. This clearly follows from the fact that the array manifold vector is nothing more than a vector of linear phase delays.

13.2 Beamforming Performance Measures

As our principal interest here is in *distant speech recognition* (DSR), we will in all cases take *word error rate* (WER) as the most important measure of system performance. Nonetheless, it is useful to consider other simpler measures to gauge the performance of the beamforming component of a DSR system that do not require running the entire system in order to evaluate. The two most important such measures of array performance are *directivity* and *array gain*. In addition to being inherently useful, these measures are frequently reported in the array-processing literature. In this section, we introduce both measures and mention their similarities and differences.

13.2.1 Directivity

The *directivity* is formally defined as the maximum sensitivity of the array divided by its average sensitivity over its working range, which is typically either a sphere or hemisphere. In order to derive a suitable expression for calculating directivity, we may begin by defining the *power pattern* as

$$P(\theta, \phi) \triangleq |B(\omega : \theta, \phi)|^2. \tag{13.33}$$

The differential element of area on the surface of a sphere or hemisphere of radius ρ is given by

$$\Delta A = \rho \sin \theta \, d\theta \, d\phi.$$

Moreover, the total area of such a sphere is $4\pi r^2$ or $2\pi r^2$ for a hemisphere. The directivity D is, as mentioned above, defined as the maximum radiation intensity divided by the average radiation intensity. Assuming then that only sound radiating in the front half plane of a hemisphere of radius $r = 1$ is captured, we can express the directivity as

$$D \triangleq \frac{P(\theta_T, \phi_T)}{\frac{1}{2\pi} \int_0^\pi d\theta \int_0^\pi d\phi \, \sin \theta \cdot P(\theta, \phi)},$$

where (θ_T, ϕ_T) are the coordinates of the look direction. Assuming the sensor weights are normalized such that $P(\theta_T, \phi_T) = 1$, then

$$D = \left\{ \frac{1}{2\pi} \int_0^\pi d\theta \int_0^{2\pi} d\phi \, \sin \theta \cdot P(\theta, \phi) \right\}^{-1}.$$

We will now calculate the directivity of a uniformly weighted uniform linear array, such as that considered in Section 13.1.3. For such a linear array $B(\theta, \phi) = B(\phi)$, which implies that

$$D = \left\{ \frac{1}{2} \int_0^\pi |B(\phi)|^2 \sin \phi d\phi \right\}^{-1}.$$

Equivalently, in u-space, we set $u = \cos\phi$, and can write

$$D = \left\{ \frac{1}{2} \int_{-1}^{1} |B_u(u)|^2 du \right\}^{-1}.$$

Substituting for $B_u(u)$ from (13.23), we find

$$D = \left\{ \frac{1}{2} \int_{-1}^{1} \sum_{n=0}^{N-1} w_n^* e^{jn(\frac{2\pi d}{\lambda})(u-u_T)} \sum_{m=0}^{N-1} w_m e^{-jm(\frac{2\pi d}{\lambda})(u-u_T)} du \right\}^{-1}.$$

Rearranging and integrating the latter expression provides

$$D = \left\{ \sum_{n=0}^{N-1} \sum_{m=0}^{N-1} w_m^* w_n^* e^{j(\frac{2\pi d}{\lambda})(m-n)u_T} \operatorname{sinc}\left[\frac{2\pi d}{\lambda}(n-m) \right] \right\}^{-1}.$$

Upon defining the *sinc matrix* as

$$[\mathbf{S}(r)]_{nm} \triangleq \operatorname{sinc} 2\pi r(n-m),$$

and assuming that the steering component is included in $\mathbf{w}_s$, we can write

$$D = \mathbf{w}_s^H \mathbf{S}(d/\lambda)\, \mathbf{w}_s.$$

For the special case of a standard linear array with $d = \lambda/2$, we can further simplify as

$$D = \left\{ \sum_{n=0}^{N-1} |w_n|^2 \right\}^{-1} = (\mathbf{w}^H \mathbf{w})^{-2} = |\mathbf{w}|^{-1},$$

where $|\mathbf{w}| = (\mathbf{w}^H \mathbf{w})^{1/2}$. For a uniformly weighted array, $w_n = 1/N$, so that

$$\sum_{n=0}^{N-1} |w_n|^2 = \frac{1}{N},$$

or, finally,

$$D = N. \tag{13.34}$$

Hence, for the special case of a uniformly weighted, uniform linear array with $d = \lambda/2$, the directivity does *not* depend on the look direction. This, however, does not hold for other inter-element spacings. It can be shown that uniform weighting maximizes the directivity of the standard linear array (Van Trees 2002, sect. 2.6.1).

The *directivity index* is expressed in dB and given by:

$$D_{\text{dB}} \triangleq 10 \log_{10} N. \tag{13.35}$$

The last equation implies that each doubling of the number of elements in a uniformly weighted, uniform linear array increases the directivity index by approximately 3 dB.

13.2.2 Array Gain

In the conventional beamforming algorithms, the sensor weights are typically set so as to maximize the *signal-to-noise ratio* (SNR), although this maximization is performed independently in each subband; see the discussion of this point in Section 13.3.6. The improvement in SNR obtained through the interaction of the propagating wave, the array geometry, and the sensor weights is measured by the *array gain*, which by definition is the ratio of two SNRs, namely, the SNR at the input of any given sensor, and the final SNR at the output of the beamformer. In this section, we will calculate the array gain of the DSB, which will serve as a baseline for gauging the effectiveness of the more elaborate beamformer designs considered in Section 13.3.

In the remainder of this chapter, we consider the desired signal as well as any interference as *random variables*, and consequently characterize both by their statistical properties. In particular, we will initially assume that both source and interference are zero-mean circular Gaussian complex random processes. This assumption will be modified in Section 13.5, however, as it clearly does not hold for human speech. Moreover, in the balance of this section and Section 13.3, we will assume that the second-order characteristics of both the sources and interferences are *known*. In Section 13.4, we will investigate techniques whereby these second-order statistical characteristics can be deduced directly from the data impinging on a sensor array.

Let $\mathbf{X}(\omega) \in \mathbb{C}^N$ denote a subband domain snapshot, which is a vector of N complex subband samples, one per microphone, for frequency ω. We will assume that $\mathbf{X}(\omega)$ is generated by the snapshot model

$$\mathbf{X}(\omega) = \mathbf{F}(\omega) + \mathbf{N}(\omega), \tag{13.36}$$

where $\mathbf{F}(\omega)$ denotes the subband domain snapshot of the desired signal and $\mathbf{N}(\omega)$ denotes that of the noise or other interference impinging on the sensors of the array. We will assume that $\mathbf{F}(\omega)$ and $\mathbf{N}(\omega)$ are uncorrelated and that the signal vector $\mathbf{F}(\omega)$ can be expressed as in (13.14).

We now introduce the notation necessary for specifying the second-order statistics of random variables and vectors. In general, for some complex scalar random variable $Y(\omega)$, we will define

$$\Sigma_Y(\omega) \triangleq \mathcal{E}\{|Y(\omega)|^2\}.$$

Similarly, for a complex random vector $\mathbf{X}(\omega)$, let us define the *spatial spectral matrix* as

$$\mathbf{\Sigma}_X(\omega) \triangleq \mathcal{E}\{\mathbf{X}(\omega)\mathbf{X}^H(\omega)\}.$$

By way of illustrating the concept of array gain with a concrete example, we will now calculate the array gain for a DSB. Let us begin by assuming that the component of the desired signal reaching each component of a sensor array is $F(\omega)$ and the component of the noise or interference reaching each sensor is $N(\omega)$. This implies that the SNR at the input of the array can be expressed as

$$\text{SNR}_{\text{in}}(\omega) \triangleq \frac{\Sigma_F(\omega)}{\Sigma_N(\omega)}. \tag{13.37}$$

Defining $\mathbf{w}^H(\omega) = \mathbf{H}^T(\omega)$ as in (13.20) enables (13.2) to be rewritten as

$$Y(\omega) = \mathbf{w}^H(\omega)\,\mathbf{X}(\omega) = Y_F(\omega) + Y_N(\omega), \tag{13.38}$$

where $Y_F(\omega) \triangleq \mathbf{w}^H(\omega)\,\mathbf{F}(\omega)$ and $Y_N(\omega) \triangleq \mathbf{w}^H(\omega)\,\mathbf{N}(\omega)$ are, respectively, the signal and noise components in the output of the beamformer. Hence, the variance of the output of the beamformer can be calculated according to

$$\Sigma_Y(\omega) = \mathcal{E}\{|Y(\omega)|^2\} = \Sigma_{Y_F}(\omega) + \Sigma_{Y_N}(\omega), \tag{13.39}$$

where

$$\Sigma_{Y_F}(\omega) = \mathbf{w}^H(\omega)\,\boldsymbol{\Sigma}_\mathbf{F}(\omega)\,\mathbf{w}(\omega), \tag{13.40}$$

is the signal component of the beamformer output, and

$$\Sigma_{Y_N}(\omega) = \mathbf{w}^H(\omega)\,\boldsymbol{\Sigma}_\mathbf{N}(\omega)\,\mathbf{w}(\omega), \tag{13.41}$$

is the noise component. Equation (13.39) follows directly from the assumption that $\mathbf{F}(\omega)$ and $\mathbf{N}(\omega)$ are uncorrelated. Expressing the snapshot of the desired signal once more as in (13.14), we find that the spatial spectral matrix $\mathbf{F}(\omega)$ of the desired signal can be written as

$$\boldsymbol{\Sigma}_\mathbf{F}(\omega) = \Sigma_F(\omega)\,\mathbf{v_k}(\mathbf{k}_s)\,\mathbf{v_k}^H(\mathbf{k}_s), \tag{13.42}$$

where $\Sigma_F(\omega) = \mathcal{E}\{|F(\omega)|^2\}$. Substituting (13.42) into (13.40), we can calculate the output signal spectrum as

$$\Sigma_{Y_F}(\omega) = \mathbf{w}^H(\omega)\,\mathbf{v_k}(\mathbf{k}_s)\,\Sigma_F(\omega)\,\mathbf{v_k}^H(\mathbf{k}_s)\,\mathbf{w}(\omega) = \Sigma_F(\omega), \tag{13.43}$$

where the final equality follows from the definition (13.20) of the DSB. Substituting (13.20) into (13.41) it follows that the noise component present at the output of the DSB is given by

$$\Sigma_{Y_N}(\omega) = \frac{1}{N^2}\mathbf{v}^H(\mathbf{k}_s)\,\boldsymbol{\Sigma}_\mathbf{N}(\omega)\,\mathbf{v}(\mathbf{k}_s) \tag{13.44}$$

$$= \frac{1}{N^2}\mathbf{v}^H(\mathbf{k}_s)\boldsymbol{\rho}_\mathbf{N}(\omega)\mathbf{v}(\mathbf{k}_s)\Sigma_N(\omega), \tag{13.45}$$

where the *normalized spatial spectral matrix* $\boldsymbol{\rho}_N(\omega)$ is defined through the relation

$$\boldsymbol{\Sigma}_\mathbf{N}(\omega) \triangleq \Sigma_N(\omega)\,\boldsymbol{\rho}_\mathbf{N}(\omega). \tag{13.46}$$

Hence, the SNR at the output of the beamformer is given by

$$\text{SNR}_{\text{out}}(\omega) \triangleq \frac{\Sigma_{Y_F}(\omega)}{\Sigma_{Y_N}(\omega)} = \frac{\Sigma_F(\omega)}{\mathbf{w}^H(\omega)\,\boldsymbol{\Sigma}_\mathbf{N}(\omega)\mathbf{w}(\omega)}. \tag{13.47}$$

Then based on (13.37) and (13.47), we can calculate the array gain of the DSB as

$$A_{\text{dsb}}(\omega, \mathbf{k}_s) = \frac{\Sigma_{Y_F}(\omega)}{\Sigma_{Y_N}(\omega)} \bigg/ \frac{\Sigma_F(\omega)}{\Sigma_N(\omega)} = \frac{N^2}{\mathbf{v}^H(\mathbf{k}_s)\,\boldsymbol{\rho}_{\mathbf{N}}(\omega)\,\mathbf{v}(\mathbf{k}_s)}. \qquad (13.48)$$

In the presence of isotropic or uncorrelated noise, $\boldsymbol{\rho}_{\mathbf{N}}(\omega)$ reduces to the identity matrix and the denominator of (13.48) reduces to N. Hence, comparing (13.34) and (13.48), it becomes apparent that the directivity D is equivalent to the array gain in the presence of isotropic or uncorrelated noise. For other cases directivity and array gain differ. The directivity metric takes into account solely the sensitivity of the array as represented by the power pattern (13.33). The array gain, on the other hand, considers both the sensitivity of the array as well as the acoustic environment in which it operates as characterized by the noise snapshot $\mathbf{N}(\omega)$ in (13.36). Hence, the array gain is more representative of the performance of a beamformer in a particular acoustic environment. The directivity is useful, however, as a "rough" indicator of the performance of a beamformer without knowledge of the specific acoustic environment.

13.3 Conventional Beamforming Algorithms

In this section, we investigate the class of conventional beamforming algorithms. The reader should please note that this class of conventional beamforming algorithms, all of which are adaptive, is distinct from the DSB described in Section 13.1.3, which is a fixed design. The conventional beamforming algorithms are designated as such because they all seek to minimize a quadratic optimization criterion. While each of the four conventional algorithms minimizes a different criterion, the various criteria lead to the same matrix processing element; the four conventional algorithms are then distinguished only by the scalar processing element or *postfilter* applied to the output of the matrix processor. These algorithms were designed not for speech processing, but for other signal processing applications such as radar, sonar, and radio astronomy (Van Trees 2002, sect. 1.2).

13.3.1 Minimum Variance Distortionless Response Beamformer

Although data-independent designs such as the DSB can give substantial reductions in the WER of a DSR system, further reductions can be achieved through the use of algorithms that adapt to a particular acoustic environment. In this section, we will investigate the first such design. We will maintain the signal model of (13.14) and (13.36). Whenever it becomes cumbersome, we will suppress the functional dependence of $\mathbf{X}(\omega)$ on ω. It must be borne in mind, however, that this frequency dependence is present whether or not it is explicitly indicated. Moreover, although ω is represented as continuous-valued, we will typically calculate snapshots $\mathbf{X}(\omega)$ for a discrete set of filter bank center frequencies $\{\omega_m\}$.

Many adaptive beamforming algorithms impose a *distortionless constraint*, which implies that, in the absence of noise, the output of the beamformer is equivalent to the desired source signal. In particular, a plane wave arriving along the main response axis $\mathbf{k}_s$ under a distortionless constraint will be neither amplified nor attenuated by the beamformer, so that

$$Y(\omega) = F(\omega), \qquad (13.49)$$

where the beamformer output $Y(\omega)$ is specified in (13.38), and $F(\omega)$ is the Fourier transform of the original source signal. Substituting (13.14) into (13.36), and the latter into (13.38), it follows that

$$Y(\omega) = F(\omega)\, \mathbf{w}^H(\omega)\, \mathbf{v}(\mathbf{k}_s) = F(\omega).$$

Hence, the distortionless constraint can be expressed as

$$\mathbf{w}^H(\omega)\, \mathbf{v}(\mathbf{k}_s) = 1. \tag{13.50}$$

Clearly setting

$$\mathbf{w}^H(\omega) = \frac{1}{N}\mathbf{v}^H(\mathbf{k}_s),$$

as is the case for the DSB, will satisfy (13.50). Thus, the DSB satisfies the distortionless constraint, as due other quiescent weight vector designs, such as the Dolph–Chebyshev design (Van Trees 2002, sect. 4.1.8).

Now let us characterize the noise snapshot model as a zero-mean process with spatial spectral matrix

$$\mathbf{\Sigma_N}(\omega) = \mathcal{E}\{\mathbf{N}(\omega)\mathbf{N}^H(\omega)\} = \mathbf{\Sigma}_c(\omega) + \sigma_w^2\mathbf{I},$$

where $\mathbf{\Sigma}_c$ and $\sigma_w^2\mathbf{I}$ are the spatially correlated and uncorrelated portions, respectively, of the noise covariance matrix. Whereas spatially correlated interference is due to the propagation of some interfering signal through space, uncorrelated noise is typically due to the self-noise of the sensors. As we will learn in Section 13.3.8, however, adding an additional uncorrelated portion to the spatial spectral matrix adds robustness to the beamformer in the presence of various types of mismatch.

The beamformer output will be as specified in (13.38). When noise is present, we can write

$$Y(\omega) = F(\omega) + Y_N(\omega),$$

where, according to (13.36), $Y_N(\omega) = \mathbf{w}^H(\omega)\mathbf{N}(\omega)$ is the component of the noise remaining in the output of the beamformer. The power spectrum of the noise at the output of the beamformer is given by (13.44). In addition to satisfying the distortionless constraint, we wish also to minimize this output variance, and thereby minimize the influence of the noise. To solve the constrained optimization problem, we can apply the method of Lagrange multipliers. As we are dealing here with complex vectors, we must use the method for taking derivatives described in Appendix B.16. That is, we first define the "symmetric" objective function

$$F \triangleq \mathbf{w}^H(\omega)\, \mathbf{\Sigma_N}(\omega)\, \mathbf{w}(\omega) + \lambda[\mathbf{w}^H(\omega)\mathbf{v}(\mathbf{k}_s) - 1] + \lambda^*[\mathbf{v}(\mathbf{k}_s)^H\mathbf{w} - 1], \tag{13.51}$$

where λ is a complex Lagrange multiplier, to incorporate the constraint into the objective function. Taking the complex gradient with respect to $\mathbf{w}^H$, equating this gradient to zero, and solving yields

$$\mathbf{w}_{\mathrm{mvdr}}^H(\omega) = -\lambda \, \mathbf{v}^H(\mathbf{k}_s) \, \mathbf{\Sigma}_{\mathbf{N}}^{-1}(\omega). \tag{13.52}$$

Applying now the distortionless constraint (13.50), we find

$$\lambda = -\left[\mathbf{v}^H(\mathbf{k}_s) \, \mathbf{\Sigma}_{\mathbf{N}}^{-1}(\omega) \, \mathbf{v}(\mathbf{k}_s)\right]^{-1}.$$

Thus, the optimal sensor weights are given by

$$\mathbf{w}_{\mathrm{o}}^H(\omega) = \Lambda(\omega) \, \mathbf{v}^H(\mathbf{k}_s) \, \mathbf{\Sigma}_{\mathbf{N}}^{-1}(\omega) = \mathbf{w}_{\mathrm{mvdr}}^H(\omega), \tag{13.53}$$

where

$$\Lambda(\omega) \triangleq \left[\mathbf{v}^H(\mathbf{k}_s) \, \mathbf{\Sigma}_{\mathbf{N}}^{-1}(\omega) \, \mathbf{v}(\mathbf{k}_s)\right]^{-1}. \tag{13.54}$$

This solution is known as the MVDR and was first derived by Capon (1969). Shown in Figure 13.7 is a schematic of the MVDR beamformer. The quantity $\Lambda(\omega)$ is equivalent to the spectral power of the noise component present in $Y(\omega)$, as can be seen from the following chain of equalities:

$$\Sigma_{Y_N}(\omega) = \mathbf{w}_{\mathrm{mvdr}}^H(\omega) \, \mathbf{\Sigma}_{\mathbf{N}}(\omega) \, \mathbf{w}_{\mathrm{mvdr}}(\omega) \tag{13.55}$$

$$= \mathbf{v}^H(\mathbf{k}_s) \, \mathbf{\Sigma}_{\mathbf{N}}^{-1}(\omega) \, \mathbf{\Sigma}_{\mathbf{N}}(\omega) \, \mathbf{\Sigma}_{\mathbf{N}}^{-1}(\omega) \, \mathbf{v}(\mathbf{k}_s) \cdot \Lambda^2(\omega) = \Lambda(\omega). \tag{13.56}$$

Equation (13.56) follows from (13.55) upon substituting (13.53) into the latter.

Note that (13.53) implies that the sensor weights for each subband are designed independently. This is one of the chief advantages of subband domain adaptive beamforming. In particular, the transformation into the subband domain has the effect of a divide and conquer optimization scheme; i.e., a single optimization problem over MN free parameters, where M is the number of subbands and N is the number of sensors, is converted into M optimization problems, each with N free parameters. Each of the M optimization problems is solved independently, which is a direct result of the statistical independence of the subband samples produced by the high stopband suppression of each filter in the digital filter bank. If a time signal is then required at the output of the beamformer, a synthesis filter of the type discussed in Section 11.1 can be used to transform the beamformed set of subband samples back into the time domain.

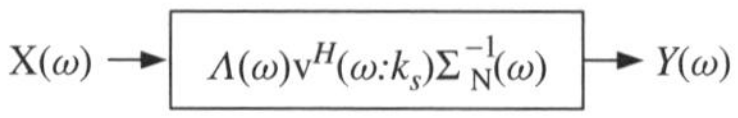

$$X(\omega) \longrightarrow \boxed{\Lambda(\omega)\mathbf{v}^H(\omega\!:\!k_s)\Sigma_{\mathbf{N}}^{-1}(\omega)} \longrightarrow Y(\omega)$$

Figure 13.7 Minimum variance distortionless response processor

13.3.2 Array Gain of the MVDR Beamformer

We will now calculate the array gain of the MVDR beamformer, and compare it to that of the conventional beamformer. As $\mathbf{w}_{\text{mvdr}}^H(\omega)$ satisfies the distortionless constraint, the power spectrum of the desired signal at the output of the beamformer can be expressed as

$$\Sigma_{Y_F}(\omega) = \Sigma_F(\omega),$$

where $\Sigma_F(\omega)$ is the power spectrum of the desired signal $F(\omega)$ at the input of each sensor. Hence, based on (13.56), the output SNR can be written as

$$\frac{\Sigma_F(\omega)}{\Sigma_{Y_N}(\omega)} = \frac{\Sigma_F(\omega)}{\Lambda(\omega)}.$$

If we assume that the noise spectrum at the input of each sensor is the same, then the input SNR is $\Sigma_F(\omega)/\Sigma_N(\omega)$. As discussed in Section 13.2, the array gain at a particular frequency is the ratio of the SNR at the output of the array to the SNR at the input, and can be expressed as

$$A_{\text{mvdr}}(\omega, \mathbf{k}_s) = \frac{\Sigma_F(\omega)}{\Lambda(\omega)} \bigg/ \frac{\Sigma_F(\omega)}{\Sigma_N(\omega)} = \frac{\Sigma_N(\omega)}{\Lambda(\omega)} \tag{13.57}$$

$$= \Sigma_N(\omega)\, \mathbf{v}^H(\mathbf{k}_s)\, \mathbf{\Sigma}_N^{-1}(\omega)\, \mathbf{v}(\mathbf{k}_s), \tag{13.58}$$

where the final equality follows from (13.54). Hence, based on (13.46) we may rewrite (13.58) as

$$A_{\text{mvdr}}(\omega, \mathbf{k}_s) = \mathbf{v}^H(\mathbf{k}_s)\, \boldsymbol{\rho}_N^{-1}(\omega)\, \mathbf{v}(\mathbf{k}_s). \tag{13.59}$$

If the noises at all sensors are spatially uncorrelated, then $\boldsymbol{\rho}_N(\omega)$ is the identity matrix and the MVDR beamformer reduces to the DSB. From (13.48) and (13.59), it can be seen that, in this case, the array gain is

$$A_{\text{mvdr}}(\omega, \mathbf{k}_s) = A_{\text{dsb}}(\omega, \mathbf{k}_s) = N. \tag{13.60}$$

In all other cases,

$$A_{\text{mvdr}}(\omega, \mathbf{k}_s) > A_{\text{dsb}}(\omega, \mathbf{k}_s).$$

13.3.3 MVDR Beamformer Performance with Plane Wave Interference

It is instructive to analyze the performance of the conventional and MVDR beamformers in the presence of both a plane wave interfering signal emanating from a particular direction and spatially uncorrelated noise. Hence, we consider here a desired signal with array manifold vector $\mathbf{v}(\mathbf{k}_s)$ and a single plane-wave interfering signal with manifold vector $\mathbf{v}(\mathbf{k}_1)$, in addition to uncorrelated sensor noise with power σ_w^2. In this case, the spatial spectral matrix $\boldsymbol{\Sigma_N}(\omega)$ of the noise can be expressed as

$$\boldsymbol{\Sigma_N}(\omega) = \sigma_w^2 \mathbf{I} + M_1(\omega) \mathbf{v}(\mathbf{k}_1) \mathbf{v}^H(\mathbf{k}_1), \tag{13.61}$$

where $M_1(\omega)$ is the spectrum of the interfering signal. Applying the matrix inversion lemma, as described in Appendix B.2, to (13.61) provides

$$\boldsymbol{\Sigma_N^{-1}} = \frac{1}{\sigma_w^2} \left[\mathbf{I} - \frac{M_1}{\sigma_w^2 + N M_1} \mathbf{v}_1 \mathbf{v}_1^H \right], \tag{13.62}$$

where we have suppressed ω and $\mathbf{k}$ for convenience, and defined $\mathbf{v}_1 \triangleq \mathbf{v}(\mathbf{k}_1)$. The noise spectrum at each element of the array can be expressed as

$$\Sigma_N = \sigma_w^2 + M_1. \tag{13.63}$$

Substituting (13.62) into (13.53), we find

$$\mathbf{w}_{mvdr}^H = \frac{\Lambda}{\sigma_w^2} \mathbf{v}_s^H \left[\mathbf{I} - \frac{M_1}{\sigma_w^2 + N M_1} \mathbf{v}_1 \mathbf{v}_1^H \right]. \tag{13.64}$$

Let us define the *spatial correlation coefficient* between the desired signal and the interference as

$$\rho_{s1} \triangleq \frac{\mathbf{v}_s^H \mathbf{v}_1}{N}, \tag{13.65}$$

and note that

$$\rho_{s1} = B_{dsb}(\mathbf{k}_1 : \mathbf{k}_s),$$

where $B_{dsb}(\mathbf{k}_1 : \mathbf{k}_s)$ is the delay-and-sum beam pattern (13.31) aimed at $\mathbf{k}_s$, the wavenumber of the desired signal, and evaluated at $\mathbf{k}_1$, the wavenumber of the interference. With this definition (13.64) can be rewritten as

$$\mathbf{w}_{mvdr}^H = \frac{\Lambda}{\sigma_w^2} \left[\mathbf{v}_s^H - \rho_{s1} \frac{N M_1}{\sigma_w^2 + N M_1} \mathbf{v}_1^H \right]. \tag{13.66}$$

The normalizing coefficient (13.54) then reduces to

$$\Lambda = \left\{ \frac{1}{\sigma_w^2} N \left[1 - \frac{N M_1}{\sigma_w^2 + N M_1} |\rho_{s1}|^2 \right] \right\}^{-1}. \tag{13.67}$$

The beamformer just derived is represented schematically in Figure 13.8. It is clear that the upper and lower branches of this MVDR beamformer correspond to conventional beamformers pointing at the desired signal and the interference, respectively. The necessity of the bottom branch is readily apparent if we reason as follows: The path labeled $\hat{\mathbf{N}}_1(\omega)$ is the minimum mean-square estimate of the interference plus noise. This noise estimate

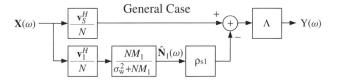

Figure 13.8   Optimum MVDR beamformer in the presence of a single interferer

is scaled by ρ_{s1} and subtracted from the output of the DSB in the upper path, in order to remove that portion of the noise and interference captured by the upper path.

Observe that in the case where $N M_1 \gg \sigma_{\mathrm{w}}^2$, we may rewrite (13.66) as

$$\mathbf{w}_{\mathrm{mvdr}}^H = \frac{\Lambda}{\sigma_{\mathrm{w}}^2} \mathbf{v}_{\mathrm{s}}^H \mathbf{P}_I^\perp,$$

where $\mathbf{P}_I^\perp = \mathbf{I} - \mathbf{v}_1 \mathbf{v}_1^H$ is the projection matrix onto the space orthogonal to the interference, as discussed in Appendix B.17. This case is shown schematically in Figure 13.8, which indicates that the beamformer is placing a perfect null on the interference.

The array gain of the MVDR beamformer in the presence of plane wave interference can be calculated by substituting (13.63) and (13.67) into (13.57), which provides

$$A_{\mathrm{mvdr}} = N(1 + \sigma_{\mathrm{I}}^2) \left[\frac{1 + N\sigma_{\mathrm{I}}^2(1 - |\rho_{\mathrm{s1}}|^2)}{1 + N\sigma_{\mathrm{I}}^2} \right],$$

where the *interference-to-noise ratio* (INR), defined as

$$\sigma_{\mathrm{I}}^2 \triangleq \frac{M_1}{\sigma_{\mathrm{w}}^2},$$

is the ratio of spatially correlated to uncorrelated noise. Beam patterns corresponding to several values of σ_{I}^2 and u_{I}, the direction cosine of the interference, are shown in Figure 13.9. Observe that the suppression of the interference is not perfect when either σ_{I}^2 is verly low, or u_{I} is very small such that the interference moves within the main lobe region of the delay-and-sum beam pattern.

The array gain of the DSB in the presence of a single interferer is readily obtained by substituting (13.61) and (13.63) into (13.48), whereupon we find

$$A_{\mathrm{dsb}}(\omega, \mathbf{k}_{\mathrm{s}}) = \frac{N^2(\sigma_{\mathrm{w}}^2 + M_1)}{\mathbf{v}_{\mathrm{s}}^H (\sigma_{\mathrm{w}}^2 \mathbf{I} + M_1 \mathbf{v}_1 \mathbf{v}_1^H) \mathbf{v}_{\mathrm{s}}} = \frac{N(1 + \sigma_{\mathrm{I}}^2)}{1 + \sigma_{\mathrm{I}}^2 N |\rho_{\mathrm{s1}}|^2}.$$

The array gains for both DSB and optimal MVDR beamformer at various INR levels are plotted in Figure 13.10. From the figure several facts are evident: Firstly, when $1 - |\rho_{\mathrm{s1}}|^2$

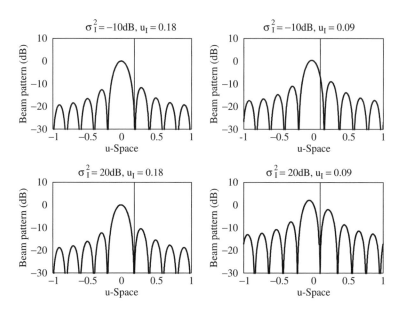

Figure 13.9   MVDR beam patterns for single plane wave interference

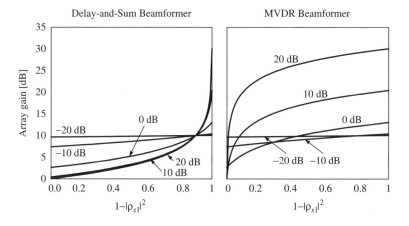

Figure 13.10   Array gains for conventional and MVDR beamformers as a function of $(1 - |\rho_{s1}|^2)$ for a 10-element array. The curves are labeled with the corresponding value of σ_I^2

approaches zero, which corresponds to the case wherein the interference moves *inside* the main lobe, the array gains of both the DSB and MVDR beamformers drop to zero. Secondly, for moderate to high values of σ_I^2 and $1 - |\rho_{s1}|^2 > 0.2$, the MVDR beamformer provides a substantially higher array gain than the DSB. Thirdly, the MVDR beam-former is clearly more effective at suppressing correlated than uncorrelated noise, which is apparent from the fact that the array gain it provides increases with increasing σ_I^2. Finally, for very low values of σ_I^2, the array gain provided by both beamformers approaches 10 dB regardless of $1 - |\rho_{s1}|^2$, which is exactly what is to be expected given that the MVDR becomes a DSB for $\boldsymbol{\rho}_N(\omega) = \mathbf{I}$; i.e., when the noise is completely uncorrelated. That the array gain for both beamformers should be $10\,\text{dB}$ is evident from (13.60).

13.3.4 Superdirective Beamformers

In this section and the one following, we develop two variants on the MVDR beamformer. The first variant will involve a specific assumption with respect to the noise field in which the MVDR beamformer is to operate, and will lead to a particular solution for the sensor weights. The second variant will not place any restrictions on the sensor weights, but will instead involve the use of a postfilter at the output of the MVDR beamformer. The latter design will be shown to be equivalent to the MMSE processor.

Consider the frequency-dependent beam pattern for a linear DSB with an intersensor spacing of $d = 4$ cm and $N = 10$ elements shown on the left side of Figure 13.11. As is clear from the figure, the directivity of the linear DSB at low frequencies is poor due to the fact that the wavelength is much longer than the aperture of the array. The beam pattern for very low frequencies is nearly flat, indicating that the directivity is effectively zero.

The superdirective beamformer design provides a remedy for such low-frequency directivity problems. Let us define the *cross-correlation coefficient* between the inputs of the mth and nth sensors as

$$\rho_{mn}(\omega) \triangleq \frac{\mathcal{E}\{X_m(\omega)X_n^*(\omega)\}}{\sqrt{\mathcal{E}\{|X_m(\omega)|^2\}\,\mathcal{E}\{|X_n(\omega)|^2\}}}, \tag{13.68}$$

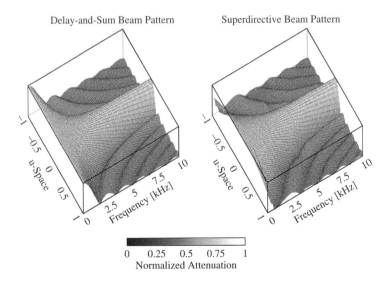

Figure 13.11 Frequency-dependent delay-and-sum and superdirective beam patterns for a linear array with an intersensor spacing of $d = 4$ cm and $N = 10$ elements

where ω is the angular frequency. The superdirective design is then obtained by replacing the spatial spectral matrix Σ_N in (13.53) and (13.54) with the *coherence matrix* Γ_N corresponding to a diffuse noise field. The (m, n)th component of the latter can be expressed as

$$\Gamma_{N,m,n}(\omega) = \text{sinc}\left(\frac{\omega \, d_{m,n}}{c}\right) = \rho_{mn}(\omega), \qquad (13.69)$$

where $d_{m,n}$ is the distance between the mth and nth elements of the array. The frequency-dependent beam pattern obtained with this superdirective design is shown on the right side of Figure 13.11. The name 'superdirective' implies that the beamformer has a higher directivity index, as defined in (13.35), than a DSB with the same array geometry (Bitzer and Simmer 2001).

As an alternative to the assumption of a spherically isotropic noise field, it is also common to assume a *cylindrically isotropic noise field* in the design of superdirective beamformers. As explained in Elko (2001), the cylindrically isotropic coherence function can be expressed as

$$\Gamma_{N,m,n}(\omega) = J_0\left(\frac{\omega \, d_{m,n}}{c}\right), \qquad (13.70)$$

where J_0 denotes the zeroth-order Bessel function[3] of the first kind (Cron and Sherman 1962). This design is well suited to modeling babble noise, such as is encountered in the *cocktail party problem*. In particular, this design has been used in the design of hearing aids (Doerbecker 1997). The coherence functions corresponding to the spherically and cylindrically isotropic noise fields are plotted in Figure 13.12. From the figure, it is clear that the cylindrically isotropic noise field retains a higher correlation over greater distances than the spherically isotropic field.

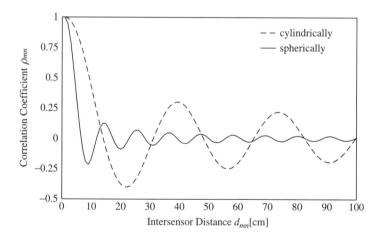

Figure 13.12   Spherically and cylindrically isotropic coherence functions for 1000 Hz

---

[3] For a brief discussion of Bessel functions, see Appendix B.12.

13.3.5 Minimum Mean Square Error Beamformer

As we will see in this section, the performance of the MVDR beamformer can be enhanced by applying a frequency dependent weighting to the output of the beamformer. This has the effect of introducing a final filtering operation, or a *postfilter*, on the beamformer output. Let us again consider the same single plane-wave model as in (13.14) and (13.36), and once more assume that $F(\omega)$ and $N(\omega)$ are uncorrelated. The spatial spectral matrix of $X(\omega)$ can be expressed as

$$\mathbf{\Sigma_X}(\omega) = \Sigma_F(\omega)\,\mathbf{v}(\mathbf{k}_s)\,\mathbf{v}^H(\mathbf{k}_s) + \mathbf{\Sigma_N}(\omega).$$

Let $D(\omega)$ denote the snapshot of the desired signal, which is equivalent to the source snapshot $F(\omega)$. We now define the matrix processor

$$\hat{D}(\omega) = \mathbf{w}^H(\omega)\,\mathbf{X}(\omega).$$

The *mean-square error* (MSE) is defined as

$$\zeta(\mathbf{w}(\omega)) \triangleq E\left\{\left|D(\omega) - \mathbf{w}^H(\omega)\,\mathbf{X}(\omega)\right|^2\right\}$$
$$= E\left\{(D(\omega) - \mathbf{w}^H(\omega)\,\mathbf{X}(\omega))(D^*(\omega) - \mathbf{X}^H(\omega)\,\mathbf{w}(\omega))\right\}.$$

In order to minimize the MSE, we take the complex gradient of ζ with respect to $\mathbf{w}(\omega)$ and equate the result to zero, to find

$$\mathcal{E}\left\{D(\omega)\,\mathbf{X}^H(\omega)\right\} - \mathbf{w}^H(\omega)\,E\left\{\mathbf{X}(\omega)\mathbf{X}^H(\omega)\right\} = \mathbf{0},$$

so that

$$\mathbf{\Sigma}_{D\,\mathbf{X}^H}(\omega) = \mathbf{w}^H_{\text{mmse}}(\omega)\,\mathbf{\Sigma_X}(\omega).$$

Hence, the MMSE solution is

$$\mathbf{w}^H_{\text{mmse}}(\omega) = \mathbf{\Sigma}_{D\,\mathbf{X}^H}(\omega)\,\mathbf{\Sigma}_\mathbf{X}^{-1}(\omega). \tag{13.71}$$

From the signal model, and the assumption that noise and signal are uncorrelated we find

$$\mathbf{\Sigma}_{D\,\mathbf{X}^H}(\omega) = \mathcal{E}\{D(\omega)D^*(\omega)\mathbf{v}^H(\mathbf{k}_s) + D(\omega)\,\mathbf{N}(\omega)\} = \Sigma_F(\omega)\,\mathbf{v}^H(\mathbf{k}_s),$$

where $D(\omega) = F(\omega)$ by assumption. This implies that (13.71) can be specialized according to

$$\mathbf{w}^H_{\text{mmse}}(\omega) = \Sigma_F(\omega)\,\mathbf{v}^H(\mathbf{k}_s)\,\mathbf{\Sigma}_\mathbf{X}^{-1}(\omega).$$

The spatial spectral matrix of the subband snapshot $\mathbf{X}$ can be expressed as

$$\boldsymbol{\Sigma}_{\mathbf{X}}(\omega) = \Sigma_F(\omega)\,\mathbf{v}(\mathbf{k}_s)\,\mathbf{v}^H(\mathbf{k}_s) + \boldsymbol{\Sigma}_{\mathbf{N}}(\omega).$$

This latter equation can be rewritten by applying the matrix inversion lemma described in Appendix B.2, with

$$\mathbf{A} = \boldsymbol{\Sigma}_{\mathbf{N}}(\omega),\ \ \mathbf{B} = \mathbf{v}(\mathbf{k}_s),\ \ \mathbf{C} = \Sigma_F(\omega),\ \ \mathbf{D} = \mathbf{v}^H(\mathbf{k}_s),$$

whereupon we find,

$$\boldsymbol{\Sigma}_{\mathbf{X}}^{-1} = \boldsymbol{\Sigma}_{\mathbf{N}}^{-1} - \Sigma_F\,\boldsymbol{\Sigma}_{\mathbf{N}}^{-1}\,\mathbf{v}\left(1 + \Sigma_F \mathbf{v}^H\,\boldsymbol{\Sigma}_{\mathbf{N}}^{-1}\,\mathbf{v}\right)^{-1}\mathbf{v}^H\boldsymbol{\Sigma}_{\mathbf{N}}^{-1}. \tag{13.72}$$

The dependence on ω in (13.72) has been suppressed out of convenience. Defining $\Lambda(\omega)$ as in (13.54) and substituting into (13.72), we learn

$$\mathbf{w}_{\mathrm{mmse}}^H(\omega) = \frac{\Sigma_F(\omega)}{\Sigma_F(\omega) + \Lambda(\omega)} \cdot \Lambda(\omega)\,\mathbf{v}^H(\mathbf{k}_s)\,\boldsymbol{\Sigma}_{\mathbf{N}}^{-1}(\omega). \tag{13.73}$$

Comparing (13.53) and (13.73), it is clear that the MMSE beamformer consists of a MVDR beamformer followed by a frequency-dependent scalar multiplicative factor. Recall now that $\Sigma_F(\omega)$ is the power spectral density of the signal at the input of the beamformer, which, due to the distortionless constraint (13.50), is also the power spectral density of the signal at the output of the MVDR beamformer. Also recall that, in light of (13.56), $\Lambda(\omega)$ is the power spectral density of the noise at the output of the MVDR beamformer. Hence, upon comparing the ratio in (13.73) with (4.16), it becomes apparent that the multiplicative factor mentioned above is equivalent to a Wiener postfilter. The MMSE beamformer is shown schematically in Figure 13.13.

While (13.73) is optimal in the mean square sense, it is not sufficient to design a MMSE beamformer. This follows from the fact that the spectra of both the desired signal $D(\omega)$ and disturbance $\Lambda(\omega)$ at the output of the beamformer must be *known*. In practice they can only be estimated, and forming this estimate is the art in Wiener postfilter design. One of the earliest and best-known proposals for estimating these quantities was by Zelinski (1988). A good survey of current techniques is given by Simmer *et al.* (2001). Other approaches to postfilter design based on the particle filter and other techniques were discussed in Chapter 6.

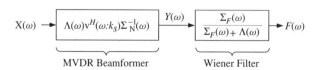

Figure 13.13 MMSE processor

13.3.6 Maximum Signal-to-Noise Ratio Beamformer

At this point, we have derived the optimal sensor weights for both the MVDR and MMSE beamformers, and discovered that the MMSE beamformer differs from the MVDR beamformer only through a scalar factor in the subband domain, which is equivalent to a postfilter in the time domain. In Section 13.2.2, we also defined the array gain as the *ratio* of SNR at the input and output of a beamformer. Given that the array gain is an often-used metric for the quality of a beamformer, we might well ask what beamformer design maximizes the array gain? Inasmuch as the SNR at the input to the beamformer is fixed, maximizing the array gain is equivalent to maximizing SNR at the output of the beamformer.

Van Trees (2002, sect. 6.2.3) demonstrates that maximizing output SNR results in the sensor weights

$$\mathbf{w}^H_{\text{max. SNR}} = \mathbf{v}^H(\mathbf{k}_s)\, \Sigma_{\mathbf{N}}^{-1}(\omega) \qquad (13.74)$$

Comparing (13.53) and (13.73) with (13.74), it is clear that the matrix processor in the maximum SNR beamformer is equivalent to that in both the MVDR and MMSE beamformers. The only difference between the sensor weights produced by the three optimization criteria is in the final scalar factor or postfilter. This fact indicates that the MVDR beamformer creates a one-dimensional signal subspace in which all subsequent processing occurs.

13.3.7 Generalized Sidelobe Canceler

While the distortionless constraint is nearly always the constraint first chosen, it is not the *only* reasonable constraint. Quite often additional constraints are applied in order to prevent the beam from becoming too narrow or too broad, or to place a null on a known source of interference (Van Trees 2002, sect. 6.7.1). If there are M_c linear constraints in total, the constraint equation can be written as

$$\mathbf{w}^H \mathbf{C} = \mathbf{g}^H, \qquad (13.75)$$

where $\mathbf{w}^H$ and $\mathbf{g}^H$ are $1 \times N$ and $1 \times M_c$ vectors respectively, and $\mathbf{C}$ is a $N \times M_c$ matrix. In such a case, the first column of $\mathbf{C}$ is typically $\mathbf{v}(\mathbf{k}_s)$ and the first element of $\mathbf{g}$ is typically unity, so the processor remains distortionless.

Hereafter we will, as a matter of convenience, suppress the dependence on the frequency ω in all quantities. In order to derive the optimal sensor weights for the beamformer with multiple constraints, we define the function

$$J \triangleq \mathbf{w}^H \, \Sigma_{\mathbf{N}} \, \mathbf{w} + \left(\mathbf{w}^H \mathbf{C} - \mathbf{g}^H\right) \lambda + \lambda^H \left(\mathbf{C}^H \mathbf{w} - \mathbf{g}\right), \qquad (13.76)$$

where the vector of Lagrange multipliers λ is of length M_c due to the M_c constraints. Taking the complex gradient of J with respect to $\mathbf{w}^H$ and setting to zero provides

$$\Sigma_{\mathbf{N}} \mathbf{w} + \mathbf{C} \lambda = \mathbf{0},$$

or

$$\mathbf{w} = -\boldsymbol{\Sigma}_{\mathbf{N}}^{-1}\mathbf{C}\boldsymbol{\lambda}.$$

Applying the linear constraint, we find

$$-\boldsymbol{\lambda}^H\mathbf{C}^H\boldsymbol{\Sigma}_{\mathbf{N}}^{-1}\mathbf{C} = \mathbf{g}^H.$$

Hence, the final result is

$$\mathbf{w}_{\text{lcmv}}^H = \mathbf{g}^H\left(\mathbf{C}^H\boldsymbol{\Sigma}_{\mathbf{N}}^{-1}\mathbf{C}\right)^{-1}\mathbf{C}^H\boldsymbol{\Sigma}_{\mathbf{N}}^{-1}, \tag{13.77}$$

which is known as the *linear constraint minimum variance* (LCMV) solution.

A useful beamforming structure can be obtained by dividing the total N-dimensional weight space into *constraint* and *orthogonal* spaces. The constraint space is determined by the range space of the columns of $\mathbf{C}$, an $N \times M_c$ matrix. Then let us define the *blocking matrix* $\mathbf{B}$ as an $N \times (N - M_c)$ matrix with linearly independent columns such that

$$\mathbf{C}^H\mathbf{B} = \mathbf{0}, \tag{13.78}$$

where $\mathbf{0}$ is an $M_c \times (N - M_c)$ matrix of zeros. The orthogonal space is defined by the columns of $\mathbf{B}$.

Let us assume that the optimal weights correspond to the LCMV solution (13.77), and partition $\mathbf{w}_{\text{lcmv}}^H$ into two components,

$$\mathbf{w}_o^H = \mathbf{w}_c^H - \mathbf{w}_p^H,$$

where $\mathbf{w}_c^H$ and $\mathbf{w}_p^H$ are the projections of $\mathbf{w}_o^H$ onto the constraint and orthogonal spaces, respectively. As explained in Appendix B.17, the projection matrix onto the constraint space is

$$\mathbf{P_C} = \mathbf{C}\left(\mathbf{C}^H\mathbf{C}\right)^{-1}\mathbf{C}^H, \tag{13.79}$$

and $\mathbf{w}_c^H$ can be expressed as

$$\mathbf{w}_c^H = \mathbf{w}_{\text{lcmv}}^H\mathbf{P_C}. \tag{13.80}$$

Substituting (13.77) and (13.79) into (13.80) provides

$$\mathbf{w}_c^H = \mathbf{w}_o^H\mathbf{P_C} = \mathbf{g}^H\left(\mathbf{C}^H\boldsymbol{\Sigma}_{\mathbf{N}}^{-1}\mathbf{C}\right)^{-1}\mathbf{C}^H\boldsymbol{\Sigma}_{\mathbf{N}}^{-1}\left[\mathbf{C}\left(\mathbf{C}^H\mathbf{C}\right)^{-1}\mathbf{C}^H\right]. \tag{13.81}$$

That component of $\mathbf{w}_o$ lying in the constraint space is known as the *quiescent weight vector*, and will be denoted as $\mathbf{w}_q$. Canceling the common terms in (13.81), we find

$$\mathbf{w}_c^H = \mathbf{g}^H\left(\mathbf{C}^H\mathbf{C}\right)^{-1}\mathbf{C}^H = \mathbf{w}_q^H. \tag{13.82}$$

Observe that $\mathbf{w}_c^H$ is determined solely by the constraints and is independent of $\boldsymbol{\Sigma}_X$, much as we might expect.

Let $\mathbf{P}_C^\perp$ denote the projection operator described in Appendix B.17 for the space orthogonal to the constraints, which can be expressed as

$$\mathbf{P}_C^\perp = \mathbf{B} \left(\mathbf{B}^H \mathbf{B} \right)^{-1} \mathbf{B}^H . \tag{13.83}$$

Thus, the second component of $\mathbf{w}_{\text{lcmv}}^H$ can be expressed as

$$\mathbf{w}_p^H = \mathbf{w}_{\text{lcmv}}^H \mathbf{P}_C^\perp = \mathbf{w}_o^H \mathbf{B} \left(\mathbf{B}^H \mathbf{B} \right)^{-1} \mathbf{B}^H .$$

Substituting (13.77) into the last equation provides

$$\mathbf{w}_p^H = \mathbf{g}^H \left(\mathbf{C}^H \boldsymbol{\Sigma}_N^{-1} \mathbf{C} \right)^{-1} \mathbf{C}^H \boldsymbol{\Sigma}_N^{-1} \cdot \mathbf{B} \left(\mathbf{B}^H \mathbf{B} \right)^{-1} \mathbf{B}^H .$$

The formulation leads to the implementation shown in Figure 13.14.

Now note that

$$Y_b = \mathbf{w}_p^H \mathbf{X}$$

is obtained by multiplying $\mathbf{X}$ by a matrix completely contained in the $\mathbf{B}$ subspace. Defining the *active weight vector* $\mathbf{w}_a$ through the relation

$$\mathbf{w}_p^H \triangleq \mathbf{w}_a^H \mathbf{B}^H ,$$

leads to the processor shown in Figure 13.15, whose output can be expressed as

$$Y = (\mathbf{w}_q - \mathbf{B}_a \mathbf{w}_a)^H \mathbf{X} . \tag{13.84}$$

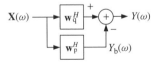

Figure 13.14   Partitioned linearly constrained minimum power beamformer

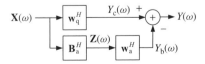

Figure 13.15   Generalized sidelobe canceler

This final configuration is known as the GSC. The active weight vector is known as such because it is this vector that is adapted during the execution of the adaptive beamforming algorithms that will be discussed in Section 13.4.

The GSC is useful because it converts a constrained optimization problem into an unconstrained optimization problem. It is also more computationally efficient than the direct form given that the optimization is performed in a space with dimensionality $N - M_c$ instead of N.

Recall that we seek to minimize the output power subject to the constraint (13.75). Substituting (13.82) into (13.75) we learn that the upper path of the GSC exactly satisfies this constraint. The lower path is orthogonal to the constraint space, and hence has no impact on whether or not the complete weight vector satisfies (13.75). Subject to (13.75), we wish to minimize the total output power

$$P_o = \left(\mathbf{w}_q - \mathbf{B}\mathbf{w}_a\right)^H \Sigma_X \left(\mathbf{w}_q - \mathbf{B}\mathbf{w}_a\right).$$

Taking the gradient of P_o with respect to $\mathbf{w}_a$ and equating to zero gives

$$\left(\mathbf{w}_q^H - \mathbf{w}_a^H \mathbf{B}^H\right) \Sigma_X \mathbf{B} = \mathbf{0},$$

so that the active weight vector can be expressed as

$$\hat{\mathbf{w}}_a^H = \mathbf{w}_q^H \Sigma_X \mathbf{B} \left(\mathbf{B}^H \Sigma_X \mathbf{B}\right)^{-1}.$$

Equation (13.78) implies that $\mathbf{B}$ lies entirely in the null space of $\mathbf{C}$. Although $\mathbf{B}$ must satisfy this constraint, its specification is not unique. There remains the question of how such a blocking matrix may be derived. One common technique is to first form the projection operator for the space orthogonal to the constraints according to

$$\mathbf{P}_C^{\perp} = \mathbf{I} - \mathbf{C} \left(\mathbf{C}^H \mathbf{C}\right)^{-1} \mathbf{C}^H, \tag{13.85}$$

then to use a modified Gram–Schmidt orthogonalization procedure (Golub and Van Loan 1996b) to construct a basis of $N - M_c$ independent vectors for this orthogonal space. Another way to obtain a minimal orthonormal basis is to perform a *singular value decomposition* (SVD) on $\mathbf{P}_C^{\perp}$ to obtain

$$\mathbf{P}_C^{\perp} = \mathbf{U}\boldsymbol{\Lambda}\mathbf{V}^T.$$

Because $\mathbf{P}_C^{\perp}$ is rank deficient, M_c of the singular values on the main diagonal of $\boldsymbol{\Lambda}$ will be very close to zero. In order to obtain a minimal basis for the range space of the columns of $\mathbf{P}_C^{\perp}$, we need only take the $N - M_c$ singular vectors from the columns of $\mathbf{U}$ associated with the largest singular values on the main diagonal of $\boldsymbol{\Lambda}$. An even simpler technique for determining a basis for the null space of $\mathbf{C}$ is to perform a SVD directly on $\mathbf{C}$ to obtain

$$\mathbf{C} = \mathbf{U}_C \boldsymbol{\Lambda}_C \mathbf{V}_C^T.$$

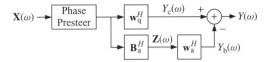

Figure 13.16 Generalized sidelobe canceler with phase presteering

Thereafter the columns of $\mathbf{U_C}$ associated with the *smallest* $N - M_c$ singular values on the main diagonal of $\mathbf{\Lambda_C}$ can be used as the desired basis for the null space of $\mathbf{C}$.

Another common technique for designing $\mathbf{B}$ and $\mathbf{w}_q$ stems from the need to track a *moving* source. In this case, the modified GSC structure shown in Figure 13.16 is used, wherein beam steering is performed as a preprocessing step by applying appropriate phase shifts to the beamformer inputs. The presteering vector is exactly that given in (13.20). This implies that the quiescent weight vector and blocking matrix are real-valued and have a very simple structure, namely (Griffiths and Jim 1982),

$$\mathbf{w}_q^T = \begin{bmatrix} 1 & 1 & \cdots & 1 \end{bmatrix}, \tag{13.86}$$

$$\mathbf{B}^T = \begin{bmatrix} 1 & -1 & 0 & 0 & \cdots & 0 & 0 \\ 0 & 1 & -1 & 0 & \cdots & 0 & 0 \\ 0 & 0 & 1 & -1 & \cdots & 0 & 0 \\ \vdots & \vdots & \vdots & \vdots & \ddots & \vdots & \vdots \\ 0 & 0 & 0 & 0 & \cdots & 1 & -1 \end{bmatrix}. \tag{13.87}$$

If an orthonormal specification of $\mathbf{B}$ is required, the Gram–Schmidt procedure or alternatively the SVD can be applied to the rows of (13.87). The advantage of such a design is that $\mathbf{w}_q$ and $\mathbf{B}$ need not change when the speaker moves, rather, it is only necessary to adjust the presteering phase shifts.

The GSC was used by Owsley (1971), as well as Applebaum and Chapman (1976). Griffiths and Jim (1982) analyzed its behavior and coined the term generalized sidelobe canceller. Other early work with the GSC was due to Er and Cantoni (1983) and Cox *et al.* (1987).

13.3.8 Diagonal Loading

A practical beamforming algorithm must typically contend with various forms of mismatch between the conditions for which the algorithm was designed, and those under which it must operate in a realistic acoustic environment. The most common form of mismatch is due to *steering errors*, whereby there is some inaccuracy in the estimate of the true speakers' positions. Another common form of mismatch is due to differences among the individual sensors of the microphone array, both in their gain or phase responses. Yet another form of mismatch stems from imprecise knowledge of the exact locations of the sensors. This latter form of mismatch can arise, for example, when these locations must be inferred using techniques such as that described in Section 13.6.

The sensitivity of the MVDR beamformer to all such forms of mismatch increases as $|\mathbf{w}|^2$ increases (Van Trees 2002, sect. 6.6.4). This suggests that the robustness of the entire system can be increased by limiting the total size of the sensor weight vector through the use of a *quadratic constraint*

$$|\mathbf{w}|^2 \leq T_{\mathrm{o}},$$

where T_{o} is a design parameter. It must hold that,

$$T_{\mathrm{o}} \geq \frac{1}{N},$$

as $1/N$ is the minimum value that $|\mathbf{w}|^2$ can assume and still satisfy the distortionless constraint (13.50). We can then seek to minimize the variance of the output of the beamformer subject to both a distortionless and a quadratic constraint according to

minimize $\qquad\qquad\qquad \mathbf{w}^H \boldsymbol{\Sigma}_{\mathbf{X}} \mathbf{w},$ $\qquad\qquad\qquad$ (13.88)

such that $\qquad\qquad\qquad \mathbf{w}^H \mathbf{v}_{\mathrm{m}} = 1,$ $\qquad\qquad\qquad$ (13.89)

and $\qquad\qquad\qquad\qquad \mathbf{w}^H \mathbf{w} = T_{\mathrm{o}},$ $\qquad\qquad\qquad$ (13.90)

where we use an equality constraint in (13.90) for simplicity. Once more applying the method of undetermined Lagrange multipliers, the function to be minimized in solving the constrained optimization problem above can be expressed as

$$F \triangleq \mathbf{w}^H \boldsymbol{\Sigma}_{\mathbf{X}} \mathbf{w} + \lambda_1 \left(\mathbf{w}^H \mathbf{w} - T_{\mathrm{o}} \right) + \lambda_2 \left(\mathbf{w}^H \mathbf{v}_{\mathrm{m}} - 1 \right) + \lambda_2^* \left(\mathbf{v}_{\mathrm{m}}^H \mathbf{w} - 1 \right). \qquad (13.91)$$

Differentiating the last equation with respect to $\mathbf{w}$ and setting the result to zero provides

$$\mathbf{w}^H \boldsymbol{\Sigma}_{\mathbf{X}} + \lambda_1 \mathbf{w}^H + \lambda_2^* \mathbf{v}_{\mathrm{m}}^H = \mathbf{0}.$$

Hence, the optimal sensor weights can be expressed as

$$\mathbf{w}^H = -\lambda_2^* \mathbf{v}_m^H \left(\lambda_1 \mathbf{I} + \boldsymbol{\Sigma}_{\mathbf{X}} \right)^{-1}. \qquad (13.92)$$

Upon substituting the right-hand side of (13.92) into the constraint (13.89), we arrive at the solution for λ_2^* as

$$\lambda_2^* = -\frac{1}{\mathbf{v}_m^H \left(\lambda_1 \mathbf{I} + \boldsymbol{\Sigma}_{\mathbf{X}} \right)^{-1} \mathbf{v}_m}. \qquad (13.93)$$

Finally, substituting (13.93) into (13.92) provides

$$\mathbf{w}^H = \frac{\mathbf{v}_{\mathrm{m}}^H \left(\boldsymbol{\Sigma}_{\mathbf{X}} + \lambda_1 \mathbf{I} \right)^{-1}}{\mathbf{v}_{\mathrm{m}}^H \left(\boldsymbol{\Sigma}_{\mathbf{X}} + \lambda_1 \mathbf{I} \right)^{-1} \mathbf{v}_{\mathrm{m}}}. \qquad (13.94)$$

From (13.94) it is clear that the quadratic constraint effectively adds extra weight to the main diagonal of $\boldsymbol{\Sigma}_{\mathbf{X}}$; i.e., $\mathbf{w}^H$ is designed for a higher level of uncorrelated noise than is actually present. For simplicity, let us specify the level $\lambda_1 = \sigma_L^2$ of diagonal loading directly and write

$$\mathbf{w}^H = \frac{\mathbf{v}_m^H \left(\Sigma_X + \sigma_L^2 \mathbf{I} \right)^{-1}}{\mathbf{v}_m^H \left(\Sigma_X + \sigma_L^2 \mathbf{I} \right)^{-1} \mathbf{v}_m}. \tag{13.95}$$

Equation (13.95) can be rewritten as

$$\mathbf{w}_{dl-mvdr}^H = \frac{\mathbf{v}_m^H \left(\mathbf{I} + \Sigma_X / \sigma_L^2 \right)^{-1}}{\mathbf{v}_m^H \left(\mathbf{I} + \Sigma_X / \sigma_L^2 \right)^{-1} \mathbf{v}_m}.$$

Clearly, in the limit of large σ_L^2,

$$\lim_{\sigma_L^2 \to \infty} \mathbf{w}_{dl-mvdr}^H = \frac{\mathbf{v}_m^H}{\mathbf{v}_m^H \mathbf{v}_m} = \frac{1}{N} \mathbf{v}_m^H = \mathbf{w}_{dsb}^H,$$

where the final equality follows from (13.30). In other words, the diagonally loaded MVDR beamformer approaches the DSB in the limit of very large diagonal loading. This comes as no surprise given that uniform weighting is optimal for spatially uncorrelated noise, as discussed in Section 13.2.1.

13.4 Recursive Algorithms

In prior sections, we assumed that the second-order characteristics of both sources and interferences were known. Henceforth, we will remove this assumption. Instead we consider means by which such statistics can be estimated from the actual data. As we are considering algorithms for recursively updating the weight vectors of a beamformer, we will introduce a time index denoted either as k or K; e.g., the active weight vector at time k is $\mathbf{w}_a(k)$. We will adopt the convention that K represents the most recent time step, but k is still required to indicate summations over the snapshots of prior time steps. It must be borne in mind that the frequency dependence of $\mathbf{w}_a(k)$, although not explicitly indicated, remains nonetheless.

The conventional beamforming algorithms described here are in one way or another based on the notion of minimizing the output power of the beamformer subject to a distortionless constraint. Such algorithms were initially proposed *not* for acoustic beamforming applications, but for arrays of antennas and other sensors that can be assumed to operate, at least approximately, in a free field. Hence, all such algorithms assume that the desired signal comes from one or more *known* directions. This assumption is unwarranted for beamforming in realistic acoustic environments inasmuch as hard surfaces such as walls and tables cause reflections. As mentioned previously, this phenomenon leads to the well-known signal cancellation problem, whereby the desired signal is canceled by the action of the beamformer (Widrow *et al.* 1982). Several methods for combatting this problem for the conventional beamforming algorithms are mentioned in Section 13.7. Algorithms that are *not* based on the concept of minimizing the variance of the beamformer output subject to a distortionless constraint, and thus are not prone to the signal cancellation problem, are discussed in Section 13.5.

13.4.1 Gradient Descent Algorithms

In the sequel, instead of solving for the optimal beamforming weights directly, we will use a *gradient descent* procedure (Bertsekas 1995). This implies that at each iteration, a small step will be taken in the downhill direction of the MSE optimization criterion. This will in fact yield a set of very simple algorithms, but at the possible price of slow convergence if the conditions of the acoustic environment are not favorable. The favorability of the environment will be determined by the distribution of the eigenvalues in the spatial spectral matrix of the input of the beamformer.

The MSE can be expressed as a function of the vector $\mathbf{w}$ of beamforming weights according to

$$\xi(\mathbf{w}) \triangleq \mathcal{E}\left\{ \left(D - \mathbf{w}^H \mathbf{X}\right)\left(D^* - \mathbf{X}^H \mathbf{w}\right)\right\}$$
$$= \Sigma_D - \mathbf{w}^H \mathbf{p} - \mathbf{p}^H \mathbf{w} + \mathbf{w}^H \Sigma_\mathbf{X} \mathbf{w}, \tag{13.96}$$

where

$$\Sigma_D \triangleq \mathcal{E}\left\{D^* D\right\}, \tag{13.97}$$

$$\mathbf{p} \triangleq \mathcal{E}\left\{\mathbf{X} D^*\right\} = \Sigma_{\mathbf{X} D^*}, \tag{13.98}$$

$$\Sigma_\mathbf{X} \triangleq \mathcal{E}\left\{\mathbf{X}\mathbf{X}^H\right\}. \tag{13.99}$$

For our present purposes, we will assume that the quantities in (13.97–13.99) are known. The gradient of $\xi(\mathbf{w})$ with respect to $\mathbf{w}^H$ is

$$\nabla_{\mathbf{w}^H} \xi = -\mathbf{p} + \Sigma_\mathbf{X} \mathbf{w}. \tag{13.100}$$

Setting the right-hand side of the last equation to zero gives the familiar Wiener–Hopf equation,

$$\Sigma_\mathbf{X} \mathbf{w}_o = \mathbf{p}. \tag{13.101}$$

The corresponding MMSE is

$$\xi_0 = \Sigma_D - \mathbf{w}_o^H \Sigma_\mathbf{X} \mathbf{w}_o = \Sigma_D - \mathbf{p}^H \mathbf{w}_o.$$

In keeping with the gradient descent rule, we will not use (13.101) to solve for the optimal sensor weights. Rather, the weights $\mathbf{w}(K)$ will be iteratively updated according to

$$\mathbf{w}(K) = \mathbf{w}(K-1) + \alpha(-\nabla_{\mathbf{w}^H} \xi),$$

where α denotes the *step size*. Substituting the gradient (13.100) into the last equation provides

$$\mathbf{w}(K) = \mathbf{w}(K-1) + \alpha\left[\mathbf{p} - \Sigma_\mathbf{X}\,\mathbf{w}(K-1)\right] \forall K = 1, 2, \ldots. \tag{13.102}$$

Now let us define the *weight-error vector*

$$\mathbf{w}_e(K) \triangleq \mathbf{w}(K) - \mathbf{w}_o.$$

Substituting for $\mathbf{w}(K) = \mathbf{w}_e(K) + \mathbf{w}_o$ into (13.102), we find

$$\mathbf{w}_e(K) = (\mathbf{I} - \alpha \mathbf{\Sigma}_X) \mathbf{w}_e(K-1). \tag{13.103}$$

We now seek to analyze the convergence characteristics of the simple gradient algorithm. Toward this end, let us perform an eigendecomposition on $\mathbf{\Sigma}_X$, such that,

$$\mathbf{\Sigma}_X = \mathbf{U}\mathbf{\Lambda}\mathbf{U}^H. \tag{13.104}$$

Given the unitary property of $\mathbf{U}^H$, substituting (13.104) into (13.103) provides

$$\mathbf{U}^H \mathbf{w}_e(K) = (\mathbf{I} - \alpha \mathbf{\Lambda}) \mathbf{U}^H \mathbf{w}_e(K-1). \tag{13.105}$$

Now let us define the vector of *independent coordinates*

$$\mathbf{v}(K) \triangleq \mathbf{U}^H \mathbf{w}_e(K). \tag{13.106}$$

Substituting (13.106) into (13.105), we learn

$$\mathbf{v}(K) = (\mathbf{I} - \alpha \mathbf{\Lambda}) \mathbf{v}(K-1). \tag{13.107}$$

Assuming $\mathbf{w}(0) = \mathbf{w}_e(K) + \mathbf{w}_o = \mathbf{0}$, it follows from (13.106) that

$$\mathbf{v}(0) = -\mathbf{U}^H \mathbf{w}_o.$$

Let $v_n(K)$ denote the nth component of $\mathbf{v}(K)$. The term $(\mathbf{I} - \alpha \mathbf{\Lambda})$ in (13.107) is diagonal, which implies that the components of $\mathbf{v}(K)$ can be treated independently, according to

$$v_n(K) = (1 - \alpha \lambda_n) \, v_n(K-1) \, \forall n = 0, 1, \ldots, N-1, \, K \geq 0, \tag{13.108}$$

where $1 - \alpha \lambda_n$ is clearly the nth component on the main diagonal of $\mathbf{I} - \alpha \mathbf{\Lambda}$.
    The solution to (13.108) is

$$v_n(K) = (1 - \alpha \lambda_n)^K \, v_n(0) \, \forall n = 0, 1, \ldots, N-1, \, K \geq 0. \tag{13.109}$$

As $\mathbf{\Sigma}_X$ is conjugate symmetric and positive definite, all of its eigenvalues are real and positive. Hence, $v_n(K)$ is a *geometric series*. By defining the *time constant*

$$\tau_n = \frac{-1}{\ln(1 - \alpha \lambda_n)}, \tag{13.110}$$

we can create a continuous version of (13.109) according to

$$v_n(t) = e^{-t/\tau_n} v_n(0).$$

For small step sizes α, it is possible to approximate (13.110) as

$$\tau_n \approx \frac{1}{\alpha \lambda_n}. \tag{13.111}$$

The convergence of (13.109) requires that

$$|1 - \alpha \lambda_n| < 1 \,\forall\, n = 0, 1, \ldots, N - 1.$$

In order to ensure convergence, α must thus be chosen to satisfy

$$0 < \alpha < \frac{2}{\lambda_{\max}},$$

where $\lambda_{\max}$ is the largest eigenvalue of $\boldsymbol{\Sigma}_{\mathbf{X}}$. Hence, the maximum admissible step size α is determined by the largest eigenvalue $\lambda_{\max}$. Based on (13.111), the corresponding mode will have a time constant of $(\alpha \lambda_{\max})^{-1}$. The time constant of mode corresponding to the smallest eigenvalue, however, will have a time constant of $(\alpha \lambda_{\min})^{-1}$. Hence, if the ratio of the biggest to the smallest eigenvalue is very large, the time required for convergence of all modes may be unacceptable. This is the primary weakness of LMS estimation, and the price to be paid for its simplicity.

13.4.2 Least Mean Square Error Estimation

The last section discussed weight vector update formulae based on the notion of *steepest descent*. In this section, we consider *least mean square error* (LMSE) estimation algorithms. Every LMSE estimation algorithm is a *stochastic* version of a steepest descent algorithm. The principal advantage of the LMSE algorithms, with respect to the recursive least squares algorithms described in Section 13.4.3, is their computational simplicity. The principal disadvantage of the LMSE algorithms is their slow rate of convergence.

In the sequel, we will derive four LMSE beamforming algorithms. The first is the original LMSE algorithm proposed by Widrow *et al.* (1967), which is based on a MMSE criterion and involves an unconstrained optimization. The second algorithm was proposed by Griffiths (1969), and assumes that both the DOA as well as the power of the desired signal are known. Unlike the LMS algorithm proposed by Widrow *et al.*, however, direct knowledge of the desired signal is not required. The last two algorithms are based on the imposition of linear constraints on the weight vectors of the beamformer. The algorithm proposed by Frost (1972) partitions the weight vector into two components, namely, a quiescent weight vector $\mathbf{w}_q$ which satisfies a distortionless constraint, and a second component which is constrained to lie in the null space of the constraints. The final algorithm is the LMS implementation of the GSC discussed in Section 13.3.7, which was originally proposed by Griffiths and Jim (1982). In keeping with the GSC formalism, the Griffiths and Jim beamformer performs a LMS update on the active weight vector $\mathbf{w}_a$, while leaving the quiescent weight vector $\mathbf{w}_q$ unchanged. A recent tutorial on these algorithms, together with an extensive list of references, can be found in Glentis *et al.* (1999).

Widrow LMS Algorithm

For a vector $\mathbf{w}$ of sensor weights, consider again the MSE defined in (13.96) through (13.99). While we previously assumed that both $\mathbf{p}$ and Σ_X were known, an LMS algorithm requires *estimates* of the expectations in (13.98) and (13.99). Simple choices for such estimates are given by the *instantaneous values*

$$\hat{\mathbf{p}}(K) = \mathbf{X}(K)D^*(K), \tag{13.112}$$

and

$$\hat{\Sigma}_X = \mathbf{X}(K)\,\mathbf{X}^H(K). \tag{13.113}$$

Substituting (13.112) and (13.113) into (13.100), the estimate of the gradient is found to be

$$\hat{\nabla}\xi(K) = -\mathbf{X}(K)\,D^*(K) + \mathbf{X}(K)\,\mathbf{X}^H(K)\hat{\mathbf{w}}(K).$$

The weight vector update formula is then,

$$\hat{\mathbf{w}}(K) = \hat{\mathbf{w}}(K-1) + \alpha(K)\mathbf{X}(K)\left[D^*(K) - \mathbf{X}^H(K)\,\hat{\mathbf{w}}(K-1)\right], \tag{13.114}$$

where $\alpha(K)$ is the step size at time K. The notation $\hat{\mathbf{w}}(K)$ indicates that the algorithm is based on an estimate of the gradient, instead of the actual gradient. Equation (13.114) can also be rewritten as

$$\hat{\mathbf{w}}^H(K) = \hat{\mathbf{w}}^H(K-1) + \alpha(K)\mathbf{X}^H(K)e_p(K), \tag{13.115}$$

where

$$e_p(K) \triangleq D(K) - \tilde{Y}_p(K), \quad\text{and}\quad \tilde{Y}_p(K) \triangleq \hat{\mathbf{w}}^H(K-1)\,\mathbf{X}(K). \tag{13.116}$$

As mentioned previously, this formulation of the LMS beamformer is due to Widrow *et al.* (1967).

In order to include a diagonal loading term $\sigma_L^2\mathbf{I}$, as discussed in Section 13.3.8, we must modify (13.115) as

$$\hat{\mathbf{w}}^H(K) = \hat{\mathbf{w}}^H(K-1) + \alpha(K)\mathbf{X}^H(K)D(K) - \alpha(K)\left\{\hat{\mathbf{w}}^H(K-1)\left[\sigma_L^2\mathbf{I} + \mathbf{X}(K)\mathbf{X}(K)^H\right]\right\}.$$

The latter can be rewritten as

$$\hat{\mathbf{w}}^H(K) = \left[1 - \alpha(K)\,\sigma_L^2\right]\hat{\mathbf{w}}^H(K-1) + \alpha(K)\mathbf{X}^H(K)\,e_p(K),$$

where $e_p(K)$ is defined in (13.116). Equivalently,

$$\hat{\mathbf{w}}^H(K) = \beta_L(K)\,\hat{\mathbf{w}}^H(K-1) + \alpha(K)\mathbf{X}^H(K)\,e_p(K), \tag{13.117}$$

where

$$\beta_L(K) = 1 - \alpha(K)\,\sigma_L^2.$$

From (13.115), (13.116), and (13.117), it is evident that the algorithm proposed by Widrow *et al.*, although simple, requires knowledge of the desired signal $D(\omega)$, which is a definite drawback.

Griffiths LMS algorithm

The algorithm due to Griffiths (1969) assumes that both the DOA and power of the desired signal are known, but the *desired signal $D(k)$* itself is not required. Hence, the Griffiths algorithm is based on more realistic assumptions than that of Widrow *et al.* In order to develop this algorithm, let us once more assume that $\boldsymbol{\Sigma}_{\mathbf{X}D^*}$ as defined in (13.98) is known. Moreover, we shall assume that the desired signal $D(K)$ and noise $\mathbf{N}(K)$ are uncorrelated. For the narrowband case, we will once more adopt the signal model in (13.14) and (13.36) and write

$$\mathbf{X}(K) = F(K)\,\mathbf{v}_s + \mathbf{N}(K),$$

where the desired signal is given by

$$D(K) = F(K).$$

Upon defining the signal power $\Sigma_F = \mathcal{E}\{|D(\omega)|^2\}$, we can write

$$\boldsymbol{\Sigma}_{\mathbf{X}D^*} = \mathcal{E}\{[\mathbf{v}_s\,F(K) + \mathbf{N}(K)]D^*(K)\} = \Sigma_F\,\mathbf{v}_s.$$

The Griffiths update formula is then,

$$\hat{\mathbf{w}}^H(K) = \hat{\mathbf{w}}^H(K-1) + \alpha(K)\left[\Sigma_F\,\mathbf{v}_s^H - \mathbf{X}^H(K)\,\tilde{Y}_p(K)\right]. \tag{13.118}$$

Although Σ_F appears in the equation above, it does not impose a hard constraint. With diagonal loading, the upate formula becomes

$$\hat{\mathbf{w}}^H(K) = \beta_L(K)\,\hat{\mathbf{w}}^H(K-1) + \alpha(K)\left[\Sigma_F\mathbf{v}_s^H - \mathbf{X}^H(K)\,\tilde{Y}_p(K)\right]. \tag{13.119}$$

From (13.118) and (13.119), it is apparent that the Griffiths algorithm does not require direct knowledge of the desired signal $D(\omega)$, but only of the array manifold vector $\mathbf{v}_s$ and signal power Σ_F, as maintained at the outset of this discussion.

Frost LMS Algorithm

The third version of the LMS algorithm, due to Frost (1972), is based on the linear constraint (13.75). Such a constraint can be satisfied by forming a modified objective function as in (13.76), and taking the partial derivative with respect to $\mathbf{w}^*$, which yields

$$\frac{\partial J}{\partial \mathbf{w}^*} = \mathbf{\Sigma}_{\mathbf{N}} \mathbf{w}(K-1) + \mathbf{C} \boldsymbol{\lambda}(K-1).$$

The LMS update can now be specified as

$$\mathbf{w}(K) = \mathbf{w}(K-1) - \alpha \left[\mathbf{\Sigma}_{\mathbf{N}} \mathbf{w}(K-1) + \mathbf{C} \boldsymbol{\lambda}(K-1) \right].$$

We can solve for $\boldsymbol{\lambda}(K-1)$ by requiring that $\mathbf{w}(K)$ satisfies (13.75). After some straightforward manipulations, the solution reduces to

$$\mathbf{w}^H(K) = \mathbf{w}^H(K-1)(\mathbf{I} - \alpha \mathbf{\Sigma}_{\mathbf{X}}) \mathbf{P}_{\mathbf{C}}^{\perp} + \mathbf{w}_{\mathrm{q}}^H, \tag{13.120}$$

where the quiescent weight vector is given by

$$\mathbf{w}_{\mathrm{q}} = \mathbf{C} \left(\mathbf{C}^H \mathbf{C} \right)^{-1} \mathbf{g}, \tag{13.121}$$

and, as described in Appendix B.17, $\mathbf{P}_{\mathbf{C}}^{\perp}$ is the perpendicular projection operator onto the space orthogonal to the constraints. The weight update equation is then obtained by substituting (13.113) into (13.120), from which we find

$$\hat{\mathbf{w}}^H(K) = \left[\hat{\mathbf{w}}^H(K-1) - \alpha(K) \mathbf{X}^H(K) \tilde{Y}_p(K) \right] \mathbf{P}_{\mathbf{C}}^{\perp} + \mathbf{w}_{\mathrm{q}}^H,$$

where $\tilde{Y}_p(K)$ is defined in (13.116).

For the simple MVDR beamformer, the quiescent weight vector can be expressed as

$$\mathbf{w}_{\mathrm{q}}^H = \frac{1}{N} \mathbf{v}_{\mathrm{s}}^H,$$

and

$$\mathbf{P}_{\mathbf{C}}^{\perp} = \mathbf{I} - \mathbf{v}_{\mathrm{s}} \left(\mathbf{v}_{\mathrm{s}}^H \mathbf{v}_{\mathrm{s}} \right)^{-1} \mathbf{v}_{\mathrm{s}}^H.$$

For the more general linearly constrained minimum power case, $\mathbf{w}_{\mathrm{q}}$ is still given by (13.121), and

$$\mathbf{P}_{\mathbf{C}}^{\perp} = \mathbf{I} - \mathbf{C} \left(\mathbf{C}^H \mathbf{C} \right)^{-1} \mathbf{C}^H.$$

GSC–LMS Algorithm

Here we formulate the GSC version of the LMS algorithm. Recall that, as shown in Figure 13.15, $Y_{\mathrm{c}}(K)$ and $Y_{\mathrm{b}}(K)$ in the GSC correspond to $D(K)$ and $Y(K)$, respectively, in the MMSE algorithm. Hence, we can specialize (13.115) by replacing $\mathbf{X}^H(K)$ with the output $\mathbf{Z}^H(\omega) = \mathbf{B}^H \mathbf{X}(K)$ of the blocking matrix, such that

$$\hat{\mathbf{w}}_{\mathrm{a}}^H(K) = \hat{\mathbf{w}}_{\mathrm{a}}^H(K-1) + \alpha(K)\,\mathbf{Z}^H(K)\,e(K), \tag{13.122}$$

where

$$e(K) = Y_{\mathrm{c}}(K) - \tilde{Y}_{\mathrm{b}}(K)$$
$$= \left[\mathbf{w}_{\mathrm{q}} - \mathbf{B}\hat{\mathbf{w}}_{\mathrm{a}}(K-1)\right]^H \mathbf{X}(K). \tag{13.123}$$

The total weight vector can then be expressed as

$$\hat{\mathbf{w}}^H(K) = [\mathbf{w}_{\mathrm{q}} - \hat{\mathbf{w}}_{\mathrm{a}}(K)\,\mathbf{B}]^H,$$

and the total beamformer output is given by

$$Y(K) = \hat{\mathbf{w}}^H(K)\,\mathbf{X}(K) = [\mathbf{w}_{\mathrm{q}} - \hat{\mathbf{w}}_{\mathrm{a}}(K)\,\mathbf{B}]^H \mathbf{X}(K).$$

A natural choice for the initial condition is

$$\hat{\mathbf{w}}_{\mathrm{a}} = \mathbf{0}.$$

If the columns of $\mathbf{B}$ are orthogonal, such that

$$\mathbf{B}^H \mathbf{B} = \mathbf{I},$$

then the adaptive performance of the GSC implementation will be identical to that of the direct form implementation. As previously discussed, the algorithm described above is the narrowband complex version of the GSC–LMS algorithm originally proposed by Griffiths and Jim (1982).

Step Size

It remains to choose the step size $\alpha(K)$. Van Trees (2002, sect. 7.7.2.2) describes a technique for setting the step size proposed by Goodwin and Sin (1984) and Söderström and Stoica (1989) known as the normalized LMS algorithm. The step size chosen is based on the sample-dependent estimate

$$\alpha(K) = \frac{\gamma}{\beta + \mathbf{X}^H(K)\,\mathbf{X}(K)}$$

with $\beta > 0$ and $0 < \gamma < 2$. A second version can be expressed as

$$\alpha(K) = \frac{\gamma}{\sigma_x^2(K)},$$

where

$$\sigma_x^2(K) = \beta\sigma_x^2(K-1) + (1-\beta)\mathbf{X}^H(K)\,\mathbf{X}(K)$$

with $0 < \beta < 1$ is known as the *power-normalized LMS* algorithm. In this case, β is most often set close to 1, such as $\beta \geq 0.99$. The constant γ usually assumes values in the range $0.005 \leq \gamma \leq 0.05$. If γ is too small, the convergence will be slow, while if too large, there will be stability problems.

13.4.3 Recursive Least Squares Estimation

Here we consider adaptive beamforming algorithms in which an estimate of the inverse of the spatial spectral covariance matrix $\hat{\boldsymbol{\Sigma}}_{\mathbf{X}}^{-1}$ is *recursively* updated with each new block of data. The required recursion will follow directly from the matrix inversion lemma. We will discuss recursive implementations of the MVDR, MMSE and GSC beamformers. Of these, the GSC formulation will prove the most useful and practical, as it does not require the *desired* signal to be available. In addition, we will find that the recursion reduces the computational complexity from $\mathcal{O}(N^3)$ to $\mathcal{O}(N^2)$. While $\mathcal{O}(N^2)$ requires more computation than the $\mathcal{O}(N)$ needed by the LMS algorithms discussed in Section 13.4.2, this additional computational complexity will potentially be offset by a faster rate of convergence.

In addition to introducing the RLS algorithms themselves, we seek here to accomplish two further objectives. Firstly, we will investigate the relations between the *Kalman filter* (KF) described in Section 4.3.1 and the RLS algorithms described here. Secondly, we will present a square-root implementation of the MMSE algorithm similar to that presented previously for the *iterated extended Kalman filter* (IEKF) described in Section 10.2.1. The square-root implementation considered here, however, will propagate the Fisher information matrix instead of the state error covariance matrices. Such square-root implementations were first introduced in the 1960s in connection with the Apollo space program to cope with the very limited precision of the computers of that day (Kaminski *et al.* 1971). While deficiencies related to finite precision are seldom problematic for modern workstations, they can raise their ugly heads as soon as an algorithm is ported to an embedded system, which, it must be conceded, is the wave of the future.

In the following, we will not be overly careful in distinguishing between the calculation and update of $\hat{\boldsymbol{\Sigma}}_{\mathbf{X}}^{-1}$ and the update of $\hat{\boldsymbol{\Sigma}}_{\mathbf{N}}^{-1}$. This is due to the fact that, for acoustic beamforming applications, updating $\hat{\boldsymbol{\Sigma}}_{\mathbf{X}}^{-1}$ or $\hat{\boldsymbol{\Sigma}}_{\mathbf{N}}^{-1}$ are largely equivalent, in that the adaptation of the sensor weight vectors and hence the update of the spatial spectral covariance matrix must be halted whenever the desired source is active, which is done to prevent signal cancellation. Van Trees (2002) actually distinguishes between the MVDR solution, and the minimum power distortionless response solution based on whether the adaptation continues when the desired source is active. For the reason mentioned above, we will not find it necessary to introduce such a distinction here. Note that several authors, including Herbold and Kellermann (2002); Herbold *et al.* (2007); Hoshuyama *et al.* (1999) and Warsitz *et al.* (2008), have investigated adapting the blocking matrix during periods when the desired speaker is silent in order to prevent leakage of the desired signal into the blocking matrix.

MVDR Estimation

In this section, we will derive the MVDR form of the RLS estimator. To begin this derivation, let us express the output of the beamformer as

$$Y(k) = D(k) + Y_N(k) \, \forall \, k = 1, 2, \ldots, K,$$

where $D(k)$ is the desired signal and $N(k)$ represents the corrupting influence of noise and reverberation. In the least squares approach, we seek to minimize

$$\zeta_{Y_N}(K) = \sum_{k=1}^{K} \mu^{K-k} |Y_N(k)|^2$$

for real $0 < \mu < 1$. As before, we will apply the distortionless constraint (13.50). The latter implies minimizing $\zeta_{Y_N}(K)$ is equivalent to minimizing

$$\zeta_Y(K) = \sum_{k=1}^{K} \mu^{K-k} |Y(k)|^2.$$

As before, the output of the beamformer is given by

$$Y(k) = \mathbf{w}^H(k) \, \mathbf{X}(k),$$

where $\mathbf{w}(k)$ is now represented as a function of k inasmuch as it will be adapted whenever new data is received.

As before, we will apply the method of Lagrange multipliers in order to incorporate the distortionless constraint (13.50) into the optimization objective function by writing

$$F \triangleq \mathbf{w}^H(K) \, \mathbf{\Phi}(K) \, \mathbf{w}(K) + \lambda \left[\mathbf{w}^H(K) \, \mathbf{v}_s - 1 \right] + \lambda^* \left[\mathbf{v}_s^H \, \mathbf{w}(K) - 1 \right],$$

where $\mathbf{v}_s = \mathbf{v}(\mathbf{k}_s)$ and

$$\mathbf{\Phi}(K) = \sum_{k=1}^{K} \mu^{K-k} \mathbf{X}(k) \, \mathbf{X}^H(k) \tag{13.124}$$

is the *exponentially-weighted sample spectral matrix*. Proceeding exactly as in Section 13.3.1, we arrive at

$$\hat{\mathbf{w}}_{\mathrm{mvdr}}^H(K) = \Lambda(K) \, \mathbf{v}_s^H \, \mathbf{\Phi}^{-1}(K), \tag{13.125}$$

where

$$\Lambda(K) \triangleq \left[\mathbf{v}_s^H \, \mathbf{\Phi}^{-1}(K) \, \mathbf{v} \right]^{-1}.$$

Then using manipulations equivalent to those leading to (13.55–13.56), we find

$$\zeta_Y(K) = \left[\mathbf{v}_s^H \, \mathbf{\Phi}^{-1}(K) \mathbf{v}_s\right]^{-1} = \Lambda(K).$$

Comparing the relations above to (13.53–13.54), it is evident that the least squares distortionless response beamformer is equivalent to the MVDR beamformer with $\mathbf{\Sigma_N}$ replaced by $\mathbf{\Phi}(K)$.

To implement the least square error beamformer efficiently, we must calculate $\mathbf{\Phi}^{-1}(K)$ from $\mathbf{\Phi}^{-1}(K-1)$. From (13.124), it is clear

$$\mathbf{\Phi}(K) = \mu \, \mathbf{\Phi}(K-1) + \mathbf{X}(K) \, \mathbf{X}^H(K). \tag{13.126}$$

Applying the matrix inversion lemma described in Appendix B.2 to (13.126), we find

$$\mathbf{\Phi}^{-1}(K) = \mu^{-1} \, \mathbf{\Phi}^{-1}(K-1) - \frac{\mu^{-2} \, \mathbf{\Phi}^{-1}(K-1) \, \mathbf{X}(K) \, \mathbf{X}^H(K) \, \mathbf{\Phi}^{-1}(K-1)}{1 + \mu^{-1} \, \mathbf{X}^H(K) \, \mathbf{\Phi}^{-1}(K-1) \, \mathbf{X}(K)}. \tag{13.127}$$

Let us define the *precision matrix* as

$$\mathbf{P}(K) \triangleq \mathbf{\Phi}^{-1}(K), \tag{13.128}$$

and the *Kalman gain* vector as

$$\mathbf{g}_{\mathrm{rls}}(K) \triangleq \frac{\mu^{-1} \, \mathbf{P}(K-1) \, \mathbf{X}(K)}{1 + \mu^{-1} \, \mathbf{X}^H(K) \, \mathbf{P}(K-1) \, \mathbf{X}(K)}. \tag{13.129}$$

Then substituting (13.128) and (13.129) into (13.127) we arrive at the *Riccati equation*,

$$\mathbf{P}(K) = \mu^{-1} \, \mathbf{P}(K-1) - \mu^{-1} \mathbf{g}_{\mathrm{rls}}(K) \, \mathbf{X}^H(K) \, \mathbf{P}(K-1). \tag{13.130}$$

Substituting (13.130) into (13.125) provides

$$\hat{\mathbf{w}}_{\mathrm{mvdr}}^H(K) = \mu^{-1} \Lambda(K) \, \mathbf{v}_s^H \, \mathbf{P}(K-1) \left[\mathbf{I} - \mathbf{X}(K) \mathbf{g}_{\mathrm{rls}}^H(K)\right]$$

$$= \hat{\mathbf{w}}_{\mathrm{mvdr}}^H(K-1) \left\{ \frac{\Lambda(K)}{\mu \, \Lambda(K-1)} \left[\mathbf{I} - \mathbf{X}(K) \mathbf{g}_{\mathrm{rls}}^H(K)\right] \right\},$$

where the second equality follows from $\hat{\mathbf{w}}_{\mathrm{mvdr}}^H(K-1) = \Lambda(K-1) \, \mathbf{v}_s^H \, \mathbf{P}(K-1)$. The term in brackets is a $N \times N$ matrix used to update $\hat{\mathbf{w}}(K-1)$. Clearly, the last equation provides a recursive solution for $\hat{\mathbf{w}}(K)$. The algorithm is typically initialized by setting

$$\mathbf{P}(0) = \frac{1}{\sigma_i^2} \mathbf{I},$$

Algorithm 13.1   MVDR–RLS beamformer

Initialize: $\mathbf{P}(0) = \frac{1}{\sigma_0^2}\mathbf{I}$, $\hat{\mathbf{w}}(0) = \frac{\mathbf{v}_s}{N}$.

Compute: At each snapshot $K = 1, 2, \ldots$, compute

$$\mathbf{g}(K) = \frac{\mu^{-1}\mathbf{P}(K-1)\mathbf{X}(K)}{1 + \mu^{-1}\mathbf{X}^H(K)\mathbf{P}(K-1)\mathbf{X}(K)},$$

$$\mathbf{P}(K) = \mu^{-1}\mathbf{P}(K-1) - \mu^{-1}\mathbf{g}(K)\mathbf{X}^H(K)\mathbf{P}(K-1),$$

$$\Lambda(K) = \left[\mathbf{v}_s^H \mathbf{P}(K)\mathbf{v}_s\right]^{-1},$$

and

$$\hat{\mathbf{w}}_{\text{mvdr}}^H(K) = \frac{\Lambda(K)}{\mu\,\Lambda(K-1)}\hat{\mathbf{w}}_{\text{mvdr}}^H(K-1)\left[\mathbf{I} - \mathbf{X}(K)\mathbf{g}_{\text{rls}}^H(K)\right],$$

where the output of the array is given by

$$\tilde{Y}(K) \triangleq \hat{\mathbf{w}}_{\text{mvdr}}^H(K-1)\mathbf{X}(K).$$

---

where σ_i^2 is the initial snapshot variance. Moreover, the sensor weights can be initialized with the weights of the delay-and-sum beamformer,

$$\hat{\mathbf{w}}(0) = \mathbf{w}_q = \frac{\mathbf{v}_s}{N},$$

or some other weights satisfying the distortionless constraint with a better sidelobe pattern (Van Trees 2002, sect. 4.1.8). The MVDR–RLS beamformer is summarized in Algorithm 13.1.

MMSE Estimation

In order to formulate the MMSE beamformer, let us denote the desired response as $D(k)$, and *MMSE innovation* for frame k as

$$s_{\text{mmse}}(k) \triangleq D(k) - \mathbf{w}^H(k)\mathbf{X}(k) \,\forall\, k = 1, \ldots, K. \tag{13.131}$$

Hence, we seek to minimize

$$\zeta_\mu(K) = \sum_{k=1}^{K} \mu^{K-k} |s_{\text{mmse}}(k)|^2$$

$$= \sum_{k=1}^{K} \mu^{K-k} [D(k) - \mathbf{w}^H(K)\mathbf{X}(k)][D^*(k) - \mathbf{X}^H(k)\mathbf{w}(K)].$$

Taking the gradient of $\zeta_\mu(K)$ with respect to $\mathbf{w}^H(K)$ and setting the result to zero provides

$$\mathbf{\Phi}(K)\hat{\mathbf{w}}_{\text{mmse}}(K) = \mathbf{\Phi}_{\mathbf{X}D^*}(K), \tag{13.132}$$

where $\mathbf{\Phi}(K)$ is as defined in (13.124) and

$$\mathbf{\Phi}_{\mathbf{X}D^*}(K) \triangleq \sum_{k=1}^{K} \mu^{K-k} \mathbf{X}(k) D^*(k) = \mathbf{X}(K) D^*(K) + \mu \mathbf{\Phi}_{\mathbf{X}D^*}(K-1). \tag{13.133}$$

Based on (13.132), the optimal weights can be expressed as

$$\hat{\mathbf{w}}_{\text{mmse}}(K) = \mathbf{\Phi}^{-1}(K) \mathbf{\Phi}_{\mathbf{X}D^*}(K) = \mathbf{P}(K) \mathbf{\Phi}_{\mathbf{X}D^*}(K). \tag{13.134}$$

Comparing (13.71) with (13.134), it becomes evident that the RLS version of the MMSE beamformer is equivalent to that considered before with $\mathbf{\Sigma_X}$ replaced by $\mathbf{\Phi}(K)$, and $\mathbf{\Sigma}_{\mathbf{X}D^*}(\omega)$ replaced with $\mathbf{\Phi}_{\mathbf{X}D^*}(K)$; i.e., the ensemble averages have been replaced by time averages. The final output is

$$Y(K) = \hat{\mathbf{w}}_{\text{mmse}}^H(K) \mathbf{X}(K).$$

The prior partial results (13.129) and (13.130) for the RLS case are still applicable for MMSE estimation. Substituting (13.130) and (13.133) into (13.134) provides

$$\hat{\mathbf{w}}_{\text{mmse}}^H(K) = \hat{\mathbf{w}}_{\text{mmse}}^H(K-1) + \mathbf{g}_{\text{rls}}^H(K) s_{\text{mmse}}(K), \tag{13.135}$$

where $s_{\text{mmse}}(K)$ is defined in (13.131). Clearly, $s_{\text{mmse}}(K)$ is the error between $D(K)$ and the beamformer output when the *current* input sample $\mathbf{X}(K)$ is applied to the *prior* weight vector $\hat{\mathbf{w}}_{\text{mmse}}(K-1)$. The final MMSE–RLS beamforming algorithm is illustrated in Algorithm 13.2.

The reader will note that we make use of the same terminology, namely, Kalman gain and Riccati equation, to describe the function of the RLS estimator that was previously used in Chapter 4 to characterize the operation of the KF. This is so because a RLS estimator can be likened to a KF, with the following simplications. Firstly, the RLS estimator is used to estimate a deterministic set of parameters rather than a stochastic state vector, and hence has no process noise. Secondly, as the RLS estimator assumes that the parameters to be estimated are fixed and not evolving in time, the transition matrix $\mathbf{F}_{k|k-1}$ appearing in the state equation (4.1) of the KF can be assumed to be the identity matrix.

That the KF subsumes the RLS estimator was first unequivocally demonstrated by Sayed and Kailath (1994) and can be established in straightforward fashion, as we now show. For simplicity, we will assume a forgetting factor of $\mu = 1$, although the result holds for any $0 < \mu < 1$; see Haykin (2002, sect. 10.8). In light of the simplifications mentioned above, we will begin by rewriting the process and observation equations (4.1–4.2) as

$$\mathbf{x}_k = \mathbf{x}_{k-1}, \tag{13.136}$$

$$y_k = \mathbf{h}_k^T \mathbf{x}_k + v_k, \tag{13.137}$$

Algorithm 13.2   MMSE–RLS beamformer

Initialize: $\mathbf{P}(0) = \frac{1}{\sigma_0^2} \mathbf{I}$, $\hat{\mathbf{w}}(0) = \frac{\mathbf{v}_s}{N}$. At each snapshot $K = 1, 2, \ldots$, compute

$$\mathbf{g}_{\mathrm{rls}}(K) = \frac{\mu^{-1} \mathbf{P}(K-1) \mathbf{X}(K)}{1 + \mu^{-1} \mathbf{X}^H(K) \mathbf{P}(K-1) \mathbf{X}(K)}$$

and

$$\mathbf{P}(K) = \mu^{-1} \mathbf{P}(K-1) - \mu^{-1} \mathbf{g}_{\mathrm{rls}}(K) \mathbf{X}^H(K) \mathbf{P}(K-1)$$

Compute $s_{\mathrm{rls}}(K)$ from

$$s_{\mathrm{rls}}(K) = D(K) - \hat{\mathbf{w}}_{\mathrm{lse}}^H(K-1) \mathbf{X}(k)$$

Compute $\hat{\mathbf{w}}_{\mathrm{lse}}^H(K)$ from

$$\hat{\mathbf{w}}_{\mathrm{lse}}^H(K) = \hat{\mathbf{w}}_{\mathrm{lse}}^H(K-1) + \mathbf{g}_{\mathrm{rls}}^H(K) s_{\mathrm{rls}}(K)$$

Compute the beamformer output from

$$Y(K) = \hat{\mathbf{w}}_{\mathrm{lse}}^H(K) \mathbf{X}(K)$$

where $\mathbf{h}_k$ is the observation vector, and y_k and v_k are the scalar observation and observation noise, respectively. The process noise is zero and $\mathbf{F}_{k|k-1} = \mathbf{I}$ implies that (4.32) reduces to

$$\mathbf{K}_{k|k-1} = \mathbf{K}_{k-1}.$$

If we now assume that the scalar observation noise v_k has unit variance, and consider (4.26) and (4.31), the Kalman gain vector can be expressed as

$$\mathbf{g}_k = \frac{\mathbf{K}_{k-1}\mathbf{h}_k}{1 + \mathbf{h}_k^T \mathbf{K}_{k-1}\mathbf{h}_k}. \tag{13.138}$$

The numerators of (13.129) and (13.138) are seen to be equivalent upon setting

$$\mathbf{K}_{k-1} = \mathbf{P}(k-1),$$

$$\mathbf{h}_k = \mathbf{X}(k),$$

whereupon the identity,

$$\mathbf{g}_k = \mathbf{g}_{\mathrm{rls}}(K),$$

becomes apparent. Finally, comparing (4.29) to (13.135), we associate

$$\hat{\mathbf{x}}_{k|k} = \hat{\mathbf{w}}_{\text{rls}}(k)$$

and

$$s_k = s^*_{\text{mmse}}(k),$$

where s_k is the one-dimensional or scalar innovation vector in the KF, and $s_{\text{mmse}}(K)$ is MMSE innovation as defined in (13.131).

GSC Estimation

As explained in Section 13.3.7, a beamformer in GSC configuration will attempt to minimize the total output power of the beamformer under the distortionless constraint. This implies that the difference between the lower and upper branches of the beamformer will be minimized in a mean square sense. Hence, the derivation of the GSC form of the RLS beamformer is eminently straightforward. Beginning with the solution for the optimal MMSE weights given above, we need only replace the desired signal $D(k)$ with the output $Y_c(K)$ of the upper branch of the GSC. Denoting once more the output of the blocking matrix as $\mathbf{Z}(K)$, as in Figure 13.15, it remains then only to estimate that active weight vector $\mathbf{w}_a^H(K)$ which minimizes the difference between $Y_c(K)$ and $\mathbf{w}_a^H \mathbf{Z}(K)$ in a least squares sense.

Based on (13.124) and (13.133), let us write

$$\Phi_{\mathbf{Z}}(K) = \sum_{k=1}^{K} \mu^{K-k} \mathbf{Z}(k)\,\mathbf{Z}^H(k) = \mathbf{B}^H\,\Phi_{\mathbf{X}}(K)\,\mathbf{B}$$

and

$$\Phi_{\mathbf{Z}Y_c^*}(K) = \sum_{k=1}^{K} \mu^{K-k} \mathbf{Z}(k)\,Y_c^*(k) = \mathbf{B}^H\,\Phi_{\mathbf{X}}(K)\,\mathbf{w}_q.$$

Then the GSC–RLS beamformer can be readily adapted from its MMSE–RLS counterpart. The final sequence of steps is shown in Algorithm 13.3. In light of (13.134), the optimal weights are given by

$$\hat{\mathbf{w}}_{\text{gsc}-\text{rls}}(K) = \Phi_{\mathbf{Z}}^{-1}(K)\,\Phi_{\mathbf{Z}Y_c^*}(K). \tag{13.139}$$

The GSC–RLS beaformer enjoys a computational advantage over the direct form implementation inasmuch it operates on a vector of length $N - M_c$ instead of length N, where N and M_c are the numbers of sensors and constraints, respectively.

13.4.4 *Square-Root Implementation of the RLS Beamformer*

As we have done in Section 10.2.1 for the IEKF, we will now develop a square-root implementation of the MMSE version of the RLS beamformer. The development, however, can

Algorithm 13.3 GSC–RLS beamformer

Initialize: $\mathbf{P}_{\mathbf{Z}}(0) = \frac{1}{\sigma_o^2}\mathbf{I}$, $\hat{\mathbf{w}}_a(0) = \mathbf{0}$. At each snapshot $K = 1, 2, \ldots$, compute

$$\mathbf{g}_{\text{gsc}}(K) = \frac{\mu^{-1}\mathbf{P}_{\mathbf{Z}}(K-1)\mathbf{Z}(K)}{1 + \mu^{-1}\mathbf{Z}^H(K)\mathbf{P}_{\mathbf{Z}}(K-1)\mathbf{Z}(K)}$$

and

$$\mathbf{P}_{\mathbf{Z}}(K) = \mu^{-1}\mathbf{P}_{\mathbf{Z}}(K-1) - \mu^{-1}\mathbf{g}_{\text{gsc}}(K)\mathbf{Z}^H(K)\mathbf{P}_{\mathbf{Z}}(K-1).$$

Compute $s_{\text{rls}}(K)$ from

$$s_{\text{rls}}(K) = Y_c(K) - \hat{\mathbf{w}}_a^H(K-1)\mathbf{Z}(k)$$

Compute $\hat{\mathbf{w}}_{\text{lse}}^H(K)$ from

$$\hat{\mathbf{w}}_a^H(K) = \hat{\mathbf{w}}_a^H(K-1) + \mathbf{g}_{\text{rls}}^H(K)s_{\text{rls}}(K)$$

Compute the beamformer output from

$$Y(K) = Y_c(\omega) - \hat{\mathbf{w}}_a^H(K)\mathbf{Z}(K)$$

be readily extended to the case of the GSC RLS beamformer. As mentioned previously, the square-root implementation is based on the Cholesky decomposition, which exists only for symmetric positive-definite matrices. Hence, a square-root implementation is immune to the explosive divergence phenomenon to which direct implementations are subject, whereby the covariance matrices, which must be updated at each time step, become indefinite. As noted in Section 4.3.4, square-root implementations effectively double the numerical precision of the direction form implementation (Simon 2006, sect. 6.3–6.4), although they require somewhat more computation.

The covariance form of the RLS estimator propagates $\mathbf{P}^{1/2}(K)$, the square-root of the estimation error covariance matrix. Unlike the square-root implementation of the IEKF presented in Section 10.2.1, the square-root implementation presented here will be based on the propagation of the square-root of the *Fisher information matrix* $\mathbf{P}^{-1}(K)$, which is in fact equivalent to the spatial sample spectral matrix $\mathbf{\Phi}(K)$ as indicated in (13.128). As we will see, this strategy will prove conducive to applying diagonal loading, which, as discussed in Section 13.3.8, is typically done in adaptive beamforming in order to improve robustness.

Let us begin by rewriting (13.130) as

$$\mathbf{g}(K)\mathbf{X}^H(K)\mathbf{P}(K-1) = \mathbf{P}(K-1) - \mu\mathbf{P}(K). \qquad (13.140)$$

Similarly, we can rewrite (13.127) to show

$$\mathbf{g}(K) + \mu^{-1}\mathbf{g}(K)\mathbf{X}^H(K)\mathbf{P}(K-1)\mathbf{X}(K) = \mu^{-1}\mathbf{P}(K-1)\mathbf{X}(K),$$

from which it follows

$$\mathbf{g}(K) = \mu^{-1}\mathbf{P}(K-1)\mathbf{X}(K) - \mu^{-1}\mathbf{g}(K)\mathbf{X}^H(K)\mathbf{P}(K-1)\mathbf{X}(K). \qquad (13.141)$$

Substituting (13.140) into (13.141) and canceling terms, we find

$$\mathbf{g}(K) = \mu\mathbf{P}(K)\mathbf{X}(K). \qquad (13.142)$$

Substituting (13.142) into (13.140) then provides

$$\mathbf{P}(K-1) = \mu\mathbf{P}(K)\mathbf{X}(K)\mathbf{X}^H(K)\mathbf{P}(K-1) + \mu\mathbf{P}(K). \qquad (13.143)$$

Substituting (13.131) and (13.142) into (13.135) yields

$$\mathbf{w}^H(K) = \mathbf{w}^H(K-1) + \mu\mathbf{X}^H(K)\mathbf{P}(K)[D(K) - \mathbf{w}^H(K-1)\mathbf{X}(K)] \qquad (13.144)$$

$$= \mathbf{w}^H(K-1)[\mathbf{I} - \mu\mathbf{X}(K)\mathbf{X}^H(K)\mathbf{P}(K)] + \mu\mathbf{X}^H(K)\mathbf{P}(K)D(K). \qquad (13.145)$$

From (13.143), it is clear

$$\mathbf{I} - \mu\mathbf{X}(K)\,\mathbf{X}^H(K)\,\mathbf{P}(K) = \mu\mathbf{P}^{-1}(K-1)\,\mathbf{P}(K). \qquad (13.146)$$

Then substituting (13.146) into (13.145) provides

$$\mathbf{w}^H(K) = \mu\mathbf{w}^H(K-1)\,\mathbf{P}^{-1}(K-1)\,\mathbf{P}(K) + \mu\mathbf{X}^H(K)\,\mathbf{P}(K)\,D(K).$$

Upon premultiplying the last equation by $\mathbf{P}^{-1}(K)$ and substituting (13.128), we arrive at the first equation in the information RLS recursion, namely,

$$\mathbf{w}^H(K)\boldsymbol{\Phi}(K) = \mu\mathbf{w}^H(K-1)\,\boldsymbol{\Phi}(K-1) + \mu\mathbf{X}^H(K)D(K). \qquad (13.147)$$

The second equation in the recursion is (13.126), which we repeat here,

$$\boldsymbol{\Phi}(K) = \mu\boldsymbol{\Phi}(K-1) + \mathbf{X}(K)\mathbf{X}^H(K). \qquad (13.148)$$

Let us express $\boldsymbol{\Phi}(K)$ in factored form as

$$\boldsymbol{\Phi}(K) = \boldsymbol{\Phi}^{H/2}(K)\,\boldsymbol{\Phi}^{1/2}(K).$$

where $\boldsymbol{\Phi}^{H/2}(K)$ is *lower* triangular. Then (13.147) and (13.148) can be expressed as

$$\mathbf{w}^H(K)\boldsymbol{\Phi}^{H/2}(K)\boldsymbol{\Phi}^{1/2}(K) = \mu\mathbf{w}^H(K-1)\boldsymbol{\Phi}^{H/2}(K-1)\boldsymbol{\Phi}^{1/2}(K-1) + \mu D(K)\mathbf{X}^H(K),$$

$$\qquad (13.149)$$

$$\boldsymbol{\Phi}^{H/2}(K)\boldsymbol{\Phi}^{1/2}(K) = \mu\boldsymbol{\Phi}^{H/2}(K-1)\boldsymbol{\Phi}^{1/2}(K-1) + \mathbf{X}(K)\mathbf{X}^H(K). \qquad (13.150)$$

Hence, let us construct the pre-array

$$
\mathbf{A} = \begin{bmatrix} \mu^{1/2}\mathbf{\Phi}^{H/2}(K-1) & \mu^{1/2}\mathbf{X}(K) \\ \mu^{-1/2}\hat{\mathbf{w}}^H(K-1)\,\mathbf{\Phi}^{H/2}(K-1) & D(K) \end{bmatrix}.
$$

Once more, we derive a unitary transform $\mathbf{\Theta}(K)$ that achieves

$$
\mathbf{A}\mathbf{\Theta}(K) = \begin{bmatrix} \mu^{1/2}\mathbf{\Phi}^{H/2}(K-1) & \mu^{1/2}\mathbf{X}(K) \\ \mu^{-1/2}\hat{\mathbf{w}}^H(K-1)\,\mathbf{\Phi}^{H/2}(K-1) & D(K) \end{bmatrix}\mathbf{\Theta}(K)
$$

$$
= \begin{bmatrix} \mathbf{B}_{11}^H(K) & \mathbf{0} \\ \mathbf{b}_{21}^H(K) & b_{22}^*(K) \end{bmatrix} = \mathbf{B}. \tag{13.151}
$$

The construction of such a unitary transform $\mathbf{\Theta}(K)$ through a set of Givens rotations is described in Appendix B.15. By multiplying out the submatrices of $\mathbf{A}$ and $\mathbf{B}$, we arrive at

$$
\mathbf{B}_{11}^H(K)\mathbf{B}_{11}(K) = \mu\mathbf{\Phi}^{H/2}(K-1)\mathbf{\Phi}^{1/2}(K-1) + \mathbf{X}(K)\mathbf{X}^H(K), \tag{13.152}
$$

$$
\mathbf{b}_{21}^H(K)\mathbf{B}_{11}(K) = \mathbf{w}^H(K-1)\mathbf{\Phi}^{H/2}(K-1)\mathbf{\Phi}^{1/2}(K-1) + \mu D(K)\mathbf{X}^H(K). \tag{13.153}
$$

Comparing (13.152) to (13.150) and (13.153) to (13.149), it is apparent that

$$
\mathbf{B}_{11}^H(K) = \mathbf{\Phi}^{H/2}(K),
$$

$$
\mathbf{b}_{21}^H(K) = \hat{\mathbf{w}}^H(K)\mathbf{\Phi}^{H/2}(K).
$$

Hence, $\mathbf{B}_{11}^H(K)$ is exactly the Cholesky factor needed for the next iteration. Moreover, we can solve for the optimal $\hat{\mathbf{w}}^H(K)$ through backward substitution on

$$
\hat{\mathbf{w}}^H(K)\mathbf{B}_{11}^H(K) = \mathbf{b}_{21}^H(K),
$$

as described in Appendix B.15.

In Section 13.3.8, we discussed the fact that additional *diagonal loading* is often applied to the spatial spectral covariance matrix $\mathbf{\Sigma}_\mathbf{X}$, as in (13.95). This extra diagonal loading limits the size of $\hat{\mathbf{w}}(K)$ and thereby improves the robustness of the beamformer. As $\mathbf{\Sigma}_\mathbf{X}$ is replaced by $\mathbf{\Phi}(K)$ in RLS algorithms, we will now consider a technique whereby diagonal loading can be applied in the square-root implementation considered above. Whenever $\mu < 1$, this loading decays with time, in which case $\hat{\mathbf{w}}(K)$ generally grows larger with increasing K. As we now show, the principal advantage of the information form of the RLS estimator is that it enables this diagonal loading to be easily replenished.

Let $\mathbf{e}_i$ denote the ith unit vector. It is desired to add the loading $\beta^2(K)$ to the ith diagonal component of $\mathbf{\Phi}(K)$, such that

$$
\mathbf{\Phi}_L(K) = \mathbf{\Phi}(K) + \beta^2(K)\,\mathbf{e}_i\,\mathbf{e}_i^T. \tag{13.154}
$$

This can be accomplished by forming the prearray

$$\mathbf{A} = \left[\boldsymbol{\Phi}^{H/2}(K) \vdots \beta(K) \mathbf{e}_i \right],$$

and constructing a unitary transformation $\boldsymbol{\theta}_i$ that achieves

$$\mathbf{A}\boldsymbol{\theta}_i = \left[\boldsymbol{\Phi}_L^{H/2}(K) \vdots \mathbf{0} \right],$$

where $\mathbf{P}_L^{-H/2}(K)$ is the desired Cholesky decomposition of (13.154). The application of each $\boldsymbol{\theta}_i$ requires $\mathcal{O}(N^2)$ operations. Hence, diagonally loading all diagonal components of $\boldsymbol{\Phi}^{H/2}(K)$ is an $\mathcal{O}(N^3)$ procedure. The diagonal loading need not be maintained at an exact level, however, but only within a broad range. Thus, with each iteration of RLS estimation, the diagonal components of $\boldsymbol{\Phi}^{1/2}(K)$ can be successively loaded. In this way, the entire process remains $\mathcal{O}(N^2)$.

13.5 Nonconventional Beamforming Algorithms

While effective in some cases, conventional beamforming makes several assumptions that are clearly unwarranted when beamforming is performed prior to DSR. Firstly, conventional beamforming assumes that both desired as well as interference or noise sources are Gaussian-distributed. Although this may be more or less true of noise sources, it is patently untrue for human speech and subband samples thereof – the source of present interest. In assuming that speech is Gaussian, a great deal of valuable statistical information is discarded – information that could be used for more effective beamforming, as we will discuss in this section. Secondly, conventional beamforming assumes that the desired source and the corrupting noise or interference are statistically independent. This is also untrue. As discussed in Section 2.4.2, the most detrimental distortion with which we must contend in a DSR scenario are echoes and reverberation. As these are nothing more than delayed versions of the desired speech, they are highly correlated with it. Thirdly, conventional beamforming assumes that the desired signal emanates from only one direction, namely, the direct path, while distortions can emanate from any direction. This is also untrue, in that reflections from hard surfaces such as tables and walls, which are invariably present in realistic acoustic environments, are sufficient to ensure that the desired source can also emanate from any direction, not merely from the direct path. Fourthly, as discussed in Section 13.3, conventional adaptive beamformers are most often based either wholly or in part on the minimization of the variance of the beamformer's output subject to a distortionless constraint. While this optimization criterion can effectively null out interfering signals as discussed in Section 13.3.3, it can also lead to signal cancellation (Widrow *et al.* 1982). The most common solution to this problem is to halt the adaptation of the active weight vector whenever the desired source is active (Cohen *et al.* 2003; Herbordt and Kellermann 2003; Nordholm *et al.* 1993). Such a technique is unlikely to be effective against reverberation from a speaker's own voice, however, as if the speaker is not active, this distortion is not present.

Here, we begin to investigate the possibility of modifying the faulty assumptions inherent in conventional beamforming. In Section 13.5.1, we will consider a beamforming algorithm that maximizes the likelihood of the cepstral features, such as used for automatic speech recognition, with respect to an auxiliary HMM. Thereafter, in Section 13.5.2 we consider means of explicitly modeling the non-Gaussian nature of speech with a number of parameterized super-Gaussian probability density functions. Following Kumatani *et al.* (2008a), we will then apply an optimization criterion from the field of ICA, namely, negentropy, to the task of optimizing the weights of a beamformer. These studies are based on the observation that the negentropy of subband samples of human speech *decreases* when the speech is corrupted by noise or reverberation; i.e., the speech becomes more nearly Gaussian. Hence, by using a negentropy optimization criterion during beamforming, it is possible to suppress the effects of noise and reverberation. In Section 13.5.3 we seek to simultaneously model the nonstationary and non-Gaussian nature of speech by using an auxiliary HMM together with a negentropy optimization criterion during adaptive beamforming. Overlapping or competing speech, whereby two or more speakers are simultaneously active, is a common problem in DSR applications. In Section 13.5.4, we minimize the MI between the outputs of two beamformers in GSC configuration in order to effectively separate the voices of two simultaneously active speakers. Section 13.5.5 presents a simpler variant of the MinMI beamformer, known as the *geometric source separation* (GSS) beamformer. The characteristics of the MinMI and GSS beamformers are then compared. In all the beamforming algorithms considered in the chapter thus far, it has been uniformly assumed that the geometry of the sensor array is *known*. It is conceivable, however, to consider another class of nonconventional algorithms whereby the geometry of the sensor array must be *inferred* prior to any beamforming operations. Section 13.6, describes an initial step in that direction, wherein the statistical properties of diffuse noise are used to estimate the distances between all pairs of microphones in an array, and these intersensor distances are thereafter used to infer the complete geometry of the array.

As should be clear from this brief introduction, much of the development in the latter portion of this chapter concerns the incorporation of as much knowledge as possible about the nature of human speech into the beamforming process. This includes knowledge about the statistical characteristics of speech and its subband samples, as well as the nonstationarity thereof. All of this is in addition to the geometric information that is taken for granted in all beamforming algorithms. The foundation of this approach to the beamforming problem was outlined in the invited papers by McDonough and Wölfel (2008) and McDonough *et al.* (2008b).

13.5.1 Maximum Likelihood Beamforming

Seltzer *et al.* (2004) proposed a novel approach to beamforming for DSR applications whereby a HMM is used to calculate the likelihood of a sequence of acoustic features. The weight vectors of the adaptive beamformer are then adjusted in order to maximize this acoustic likelihood. Seltzer *et al.* investigated the performance of such a *maximum likelihood beamformer* (MLB) in a direct form implementation, where the log-likelihood was calculated in the log-spectral domain. This approach was subsequently extended by Raub *et al.* (2004), who considered both a direct form implementation of the MLB, as well as an implementation based on the GSC. Additionally, Raub *et al.* (2004) compared the performance when the active weight vectors of the MLB were optimized, based on an acoustic likelihood calculated in the log-spectral versus the cepstral domain. In what follows, we will present the GSC form of the MLB, where the weight vector optimization is performed in the cepstral domain.

Let $\mathbf{v} = \{v_n\}$ denote a vector of cepstral coefficients associated with a vector $\mathbf{V} = \{V_m\}$ of subband samples. The relationship between the components of $\mathbf{v}$ and $\mathbf{V}$ can be expressed as

$$v_n = \sum_{m=0}^{M-1} T_{nm} \log |V_m|^2 \ \forall n = 0, 1, \ldots, N-1, \tag{13.155}$$

where $T_{nm} = T_{n,m}^{(2)}$ are the components of the Type 2 *discrete cosine transform* (DCT) matrix specified in (B.1), and M and N are the numbers of subbands and cepstral coefficients, respectively. As explained in Section 5.2.2, a nonlinear *mel-warping* is typically applied to the frequency axis prior to the calculation of cepstral coefficients. Hence, let us rewrite (13.155) as

$$v_n = \sum_{m=0}^{M-1} T_{nm} \log |\tilde{V}_m|^2, \tag{13.156}$$

where

$$|\tilde{V}_m|^2 \triangleq \sum_l M_{ml} |V_l|^2 \tag{13.157}$$

are the mel-warped frequency or subband components, and $\mathbf{M} = \{M_{ml}\}$ is the mel-warping matrix defined in (5.5–5.7).

As explained in Section 13.3.7, conventional GSC beamformers attempt to minimize output power subject to a distortionless constraint. For present purpose, we will retain the GSC structure, but consider an optimization metric similar to that proposed by Seltzer *et al.* (2004). Assuming that $\mathbf{V}$ is the output (13.84) of the GSC, the mth component of $\mathbf{V}$ is given by

$$V_m = \left(\mathbf{w}_{q,m} - \mathbf{B}_m \mathbf{w}_{a,m}\right)^H \mathbf{X}_m,$$

where $\mathbf{w}_{q,m}$, $\mathbf{B}_m$, $\mathbf{w}_{a,m}$, and $\mathbf{X}_m$, are, respectively, the quiescent weight vector, blocking matrix, active weight vector and snapshot for the mth subband. This implies

$$|V_m|^2 = (\mathbf{w}_{q,m} - \mathbf{B}_m \mathbf{w}_{a,m})^H \mathbf{X}_m \mathbf{X}_m^H (\mathbf{w}_{q,m} - \mathbf{B}_m \mathbf{w}_{a,m}).$$

Taking a partial derivative with respect to $\mathbf{w}_a^*$ on both sides of (13.156) gives

$$\frac{\partial v_n}{\partial \mathbf{w}_{a,m}^*} = \xi_{nm}(K) \cdot \frac{\partial |V_m|^2}{\partial \mathbf{w}_{a,m}^*}, \tag{13.158}$$

where $\{|\tilde{V}_m|^2\}_m$ are the mel-warped subband components in (13.157), and

$$\xi_{nm}(K) \triangleq \sum_{l=0}^{M-1} \frac{T_{nl} M_{lm}}{|\tilde{V}_l|^2}. \tag{13.159}$$

Moreover, we can write

$$
\begin{aligned}
|V_m|^2 &= (\mathbf{w}_{q,m}^H - \mathbf{w}_{a,m}^H \mathbf{B}_m^H) \mathbf{X}_m\, \mathbf{X}_m^H (\mathbf{w}_{q,m} - \mathbf{B}_m \mathbf{w}_{a,m}) \\
&= \mathbf{w}_{q,m}^H \mathbf{X}_m\, \mathbf{X}_m^H \mathbf{w}_{q,m} - \mathbf{w}_{q,m}^H\, \mathbf{X}_m\, \mathbf{X}_m^H\, \mathbf{B}_m\, \mathbf{w}_{a,m} \\
&\quad - \mathbf{w}_{a,m}^H\, \mathbf{B}_m^H\, \mathbf{X}_m\, \mathbf{X}_m^H\, \mathbf{w}_{q,m} + \mathbf{w}_{a,m}^H\, \mathbf{B}_m^H\, \mathbf{X}_m\, \mathbf{X}_m^H\, \mathbf{B}_m\, \mathbf{w}_{a,m}.
\end{aligned}
$$

Taking care to handle the derivative with respect to the complex vector as indicated in Appendix B.16, we can now evaluate the desired partial derivative as

$$
\begin{aligned}
\frac{\partial\, |V_m|^2}{\partial \mathbf{w}_{a,m}^*} &= -\mathbf{B}_m^H\, \mathbf{X}_m\, \mathbf{X}_m^H\, \mathbf{w}_{q,m} + \mathbf{B}_m^H\, \mathbf{X}_m\, \mathbf{X}_m^H\, \mathbf{B}_m\, \mathbf{w}_{a,m} \\
&= \mathbf{B}_m^H\, \mathbf{X}_m\, \mathbf{X}_m^H \left(\mathbf{B}_m \mathbf{w}_{a,m} - \mathbf{w}_{q,m} \right).
\end{aligned} \tag{13.160}
$$

Substituting (13.160) into (13.158), we arrive at

$$
\frac{\partial v_n}{\partial \mathbf{w}_{a,m}^*} = \xi_{nm}(K) \cdot \mathbf{B}_m^H\, \mathbf{X}_m\, \mathbf{X}_m^H \left(\mathbf{B}_m \mathbf{w}_{a,m} - \mathbf{w}_{q,m} \right). \tag{13.161}
$$

We now seek to develop an LMS-style algorithm for the *maximum likelihood* (ML) beamformer. Our approach will be based on the recursive expectation–maximization algorithm described by Titterington (1984). Consider the auxiliary function

$$
Q(\Lambda|\Lambda(K-1)) = \frac{1}{2} \sum_K [\mathbf{v}(K) - \boldsymbol{\mu}(K)]^T\, \boldsymbol{\Sigma}^{-1}(K)\, [\mathbf{v}(K) - \boldsymbol{\mu}(K)],
$$

where $\boldsymbol{\mu}(K)$ and $\boldsymbol{\Sigma}(K)$ denote the mean and variance associated with the Kth vector $\mathbf{v}(K)$ of cepstral coefficients, as determined by the Viterbi algorithm discussed in Section 7.1.2. Let us then define the auxiliary function

$$
\begin{aligned}
Q(\Lambda|\Lambda(K-1)) &\triangleq \frac{1}{2} [\mathbf{v}(K) - \boldsymbol{\mu}(K)]^T \boldsymbol{\Sigma}^{-1}(K)[\mathbf{v}(K) - \boldsymbol{\mu}(K)] \\
&= \frac{1}{2} [\mathbf{v}^T(K) \boldsymbol{\Sigma}^{-1}(K) \mathbf{v}(K) - 2\mathbf{v}^T(K) \boldsymbol{\Sigma}^{-1}(K) \boldsymbol{\mu}(K) \\
&\quad + \boldsymbol{\mu}^T(K) \boldsymbol{\Sigma}^{-1}(K) \boldsymbol{\mu}(K)].
\end{aligned}
$$

Assuming that $\boldsymbol{\Sigma}(K)$ is diagonal as in (8.7), the partial derivative of $Q(\Lambda|\Lambda(K-1))$ with respect to $v_n(K)$, the nth component of $\mathbf{v}(K)$, can be expressed as

$$
\frac{\partial Q(\Lambda|\Lambda(K-1))}{\partial v_n(K)} = \frac{v_n(K) - \mu_n(K)}{\sigma_n^2(K)}, \tag{13.162}
$$

where $\sigma_n^2(K)$ is the nth diagonal component of $\mathbf{\Sigma}(K)$. From the chain rule it then follows that

$$\frac{\partial Q(\Lambda|\Lambda(K-1))}{\partial \mathbf{w}_{\mathrm{a},m}^*} = \sum_{n=0}^{L-1} \frac{\partial Q(\Lambda|\Lambda(K-1))}{\partial v_n} \cdot \frac{\partial v_n}{\partial \mathbf{w}_{\mathrm{a},m}^*}.$$

Substituting (13.162) into the last equation provides

$$\frac{\partial Q(\Lambda|\Lambda(K-1))}{\partial \mathbf{w}_{\mathrm{a},m}^*} = -v_m(K) \cdot \mathbf{Z}_m(K)\, e_m^*(K), \qquad (13.163)$$

where $\mathbf{Z}_m(K)$ is the output of the blocking matrix for the mth subband, the output of the beamformer for the current snapshot $\mathbf{X}_m(K)$ using the old sensor weights $\hat{\mathbf{w}}_{\mathrm{a},m}(K-1)$ is

$$e_m(K) \triangleq \left[\mathbf{w}_{\mathrm{q},m} - \mathbf{B}_m \hat{\mathbf{w}}_{\mathrm{a},m}(K-1)\right]^H \mathbf{X}_m(K), \qquad (13.164)$$

and

$$v_m(K) \triangleq \sum_{n=0}^{L-1} \frac{v_n(K) - \mu_n(K)}{\sigma_n^2(K)} \cdot \xi_{nm}(K). \qquad (13.165)$$

The LMS update rule can then be expressed as

$$\hat{\mathbf{w}}_{\mathrm{a},m}(K) = \hat{\mathbf{w}}_{\mathrm{a},m}(K-1) - \alpha_m(K) \cdot \frac{\partial Q(\Lambda|\Lambda(K-1))}{\partial \mathbf{w}_{\mathrm{a},m}^*}, \qquad (13.166)$$

where $\alpha_m(K)$ is the step size. Upon substituting (13.163) into (13.166), we find

$$\hat{\mathbf{w}}_{\mathrm{a},m}^H(K) = \hat{\mathbf{w}}_{\mathrm{a},m}^H(K-1) + \alpha_m(K) \cdot v_m(K)\, \mathbf{Z}_m^H(K)\, e_m(K). \qquad (13.167)$$

It is remarkable that (13.164–13.167) differ from the conventional LMS update rule (13.122–13.123) for a GSC beamformer only by the factor $v_m(K)$.

Step Size

With our LMS update rule for the ML beamformer in place, the only thing lacking for a complete algorithm description is some means of setting the step size $\alpha_m(K)$. For the latter purpose, it is possible to adapt the heuristic described in Section 13.4.2, known as the power normalized LMS algorithm. For some constant γ with a typical value $0.005 < \gamma < 0.05$, and some constant β close to unity (e.g., $\beta \geq 0.99$), the step size can be set according to

$$\alpha_m(K) = \frac{\gamma}{\breve{\sigma}_m^2(K)},$$

where

$$\check{\sigma}_m^2(K) = \beta \check{\sigma}_m^2(K-1) + (1-\beta)\, \mathbf{X}_m^H(K)\, \mathbf{X}_m(K)$$

is the average power in the mth subband.

Quadratic Constraint

Van Trees (2002, sect. 7.7.4) describes a simple technique, which was originally proposed by Cox *et al.* (1987), for enforcing the quadratic constraint,

$$|\hat{\mathbf{w}}_{a,m}(K)|^2 \le g^2.$$

Much like the diagonal loading technique described in Section 13.3.8, enforcing such a quadratic constraint adds robustness to the beamforming algorithm by preventing the active weight vector from becoming too large. The algorithm first calculates

$$\tilde{\mathbf{w}}_{a,m}^H(K) = \hat{\mathbf{w}}_{a,m}^H(K-1) + \alpha_m(K) \cdot \nu_m(K)\, \mathbf{Z}_m^H(K)\, e_m(K).$$

Thereafter, the final weight vector is obtained from

$$\hat{\mathbf{w}}_{a,m}(K) = \begin{cases} \tilde{\mathbf{w}}_{a,m}(K), & if\,|\tilde{w}_{a,m}(K)|^2 \le g^2, \\ c_m(K)\, \tilde{\mathbf{w}}_{a,m}(K), & \text{otherwise.} \end{cases}$$

The scale factor $c_m(K)$ is given by

$$c_m(K) = \frac{g}{|\tilde{\mathbf{w}}_{a,m}(K)|}.$$

A schematic of the weight update is shown in Figure 13.17. As shown in the figure, $\mathbf{w}_a^H(K-1)$ is updated through the addition of $\alpha_m(K) \cdot \nu_m(K)\, \mathbf{Z}_m^H(K)\, e_m(K)$, then scaled back to have size g, as indicated by the half circle.

Note that for the HMM beamformers, the weights in the subbands *cannot* be set independently, as the subbands are no longer truly independent due to the calculation of the log-likelihood optimization criterion in the cepstral domain. This implies that the same scale factor $c(K)$ should be used for *all* subbands, such that

$$c(K) = \min_m c_m(K) = \frac{g}{\max_m |\tilde{\mathbf{w}}_{a,m}(K)|}.$$

Then the final weight vector is obtained from

$$\hat{\mathbf{w}}_{a,m}(K) = \begin{cases} \tilde{\mathbf{w}}_{a,m}(K), & if\,|\tilde{w}_{a,m}(K)|^2 \le g^2, \\ c(K)\tilde{\mathbf{w}}_{a,m}(K), & \text{otherwise.} \end{cases}$$

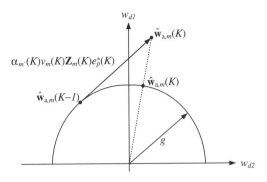

Figure 13.17   Scaling the tentative update vector

13.5.2   *Maximum Negentropy Beamforming*

In Section 12.2.2, we presented theoretical arguments taken from the field of ICA that nearly all information bearing signals are *not* Gaussian-distributed (Hyvärinen and Oja 2000). In this section and those to follow, we seek to exploit this fact in order to perform more effective acoustic beamforming and, in the case of overlapping or simultaneous speakers, better speech separation. Let us begin by presenting *empirical* evidence that human speech is in fact non-Gaussian.

Figure 13.18 shows histograms of the real parts and magnitudes of subband samples of human speech at $f_s = 800$ Hz. These samples were generated with the uniform DFT analysis bank presented in Section 11.1. The speech material, which was recorded with a close-talking microphone, was taken from the development set of the *Speech Separation Challenge* (SSC), Part 2 (Lincoln *et al.* 2005). Superimposed on the speech histograms

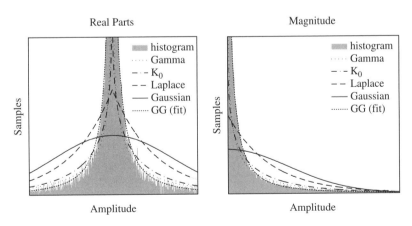

Figure 13.18   Histogram of real parts or magnitude of subband components and the likelihood of pdfs

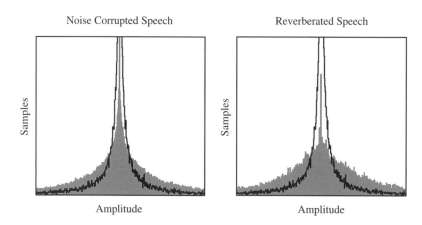

Figure 13.19 Subband domain histograms of clean speech (black line) and speech corrupted with noise or reverberation

in Figure 13.18 are plots of several well-known pdfs, namely, the Γ, K_0 or Bessel, and Laplace pdfs, along with both Gaussian and *generalized Gaussian* (GG) pdfs. Details of these pdfs are discussed in Appendix B.5. Note that the parameters of the GG pdf shown in Figure 13.18 were estimated from the same SSC development set. As noted in Section 12.2.2, the super-Gaussian pdfs exhibit spikey and heavy-tailed characteristics. It is clear from Figure 13.18 that the distribution of clean speech is not Gaussian but super-Gaussian. Figure 13.18 also suggests that, among those pdfs shown, the GG is the most suitable for modeling subband samples of speech.

That human speech and its subband samples are inherently non-Gaussian would not be a particularly useful property were it not for the fact that such subband samples become *more* nearly Gaussian-distributed when corrupted by noise, reverberation, or with competing speech from another speaker. Empirical evidence of the former two points, as first appeared in Kumatani *et al.* (2008b), is presented in Figure 13.19, which shows histograms of clean speech and speech corrupted with both noise and reverberation in the subband domain. It is clear from this figure that the pdf of speech corrupted with noise has less probability mass around the mean and in the tail than the clean speech, but more probability mass in intermediate regions. This indicates that the pdf of the noise-corrupted signal, which is in fact the sum of the speech and noise signals, is closer to Gaussian than that of clean speech. Much the same is true of speech corrupted with reverberation, which stands to reason inasmuch as reverberant speech is the sum of many different, independent portions of the original signal. In the cases of both noise and reverberation, the central limit theorem implies that the corrupted signal should be more nearly Gaussian than the original clean speech, as discussed in Section 12.2.2.

These facts would indeed support the hypothesis that seeking an enhanced speech signal that is maximally non-Gaussian is an effective way to suppress the distorting effects of noise and reverberation. As discussed in Section 12.2.3, two well-known measures of non-Gaussianity are negentropy and kurtosis. Moreover, the former bears a close relation with MI, which is another popular optimization criterion in the ICA field.

Modeling Subband Samples with the Generalized Gaussian PDF

Let the *scale* and *shape factors* of the GG pdf be denoted by $\hat{\sigma}$ and f, respectively. Repeating (B.23–B.24) from Appendix B.5.1, the GG pdf can be generalized for complex circular random variables as

$$p_{\text{gg}}(z) \triangleq \frac{f}{2\pi \, \hat{\sigma}^2 \, B_c^2(f) \, \Gamma(2/f)} \exp\left\{ -\left| \frac{z}{\hat{\sigma} \, B_c(f)} \right|^f \right\}, \qquad (13.168)$$

where

$$B_c(f) \triangleq \left[\frac{\Gamma(2/f)}{\Gamma(4/f)} \right]^{1/2}. \qquad (13.169)$$

Among several methods for estimating the scale and shape factors of the GG pdf that have previously appeared in the literature, the moment and ML methods are arguably the most straightforward. Kumatani *et al.* (2008b) used the moment method (Kokkinakis and Nandi 2005) in order to initialize the parameters of the GG pdf and then updated them with the ML estimate (Varanasi and Aazhang 1989). The shape factors were estimated from training samples offline and held fixed during the adaptation of the active weight vectors. The shape factor for each subband was estimated independently, as the optimal pdf is frequency-dependent. We now describe a somewhat simpler algorithm for estimating the shape and scale factors.

For a set $\mathcal{Y} = \{Y(k)\}_{k=0}^{K-1}$ of real-valued training samples, the log-likelihood function under the GG pdf can be expressed as

$$\log p_{\text{gg}}(\mathcal{Y}; \hat{\sigma}, f) = K \log \frac{f}{2\pi \hat{\sigma}^2 B_c^2(f) \, \Gamma(2/f)} - \frac{1}{B_c^f(f) \, \hat{\sigma}^f} \sum_{k=0}^{K-1} |Y(k)|^f, \qquad (13.170)$$

which follows directly from (B.27). Based on (13.170), we can express the partial derivative with respect to $\hat{\sigma}$ of the log-likelihood of the entire data set as

$$\frac{\partial \log p_{\text{gg}}(\mathcal{Y}; \hat{\sigma}, f)}{\partial \hat{\sigma}} = -\frac{2K}{\hat{\sigma}} + \frac{f}{B_c^f(f) \, \hat{\sigma}^{f+1}} \sum_{k=0}^{K-1} |Y(k)|^f. \qquad (13.171)$$

By equating the right-hand side of (13.171) to zero and solving for $\hat{\sigma}$, we find

$$\hat{\sigma} = \frac{1}{B_c(f)} \left[\frac{f}{2K} \sum_{k=0}^{K-1} |Y(k)|^f \right]^{1/f}. \qquad (13.172)$$

By substituting the optimal value of $\hat{\sigma}$ given in (13.172) back into (13.170), the log-likelihood of the training set can be expressed solely as a function of the shape factor f. Thereafter, the optimal value of f can be obtained with a simple line search based, for example, on Brent's method (Press *et al.* 1992, sect. 10.2).

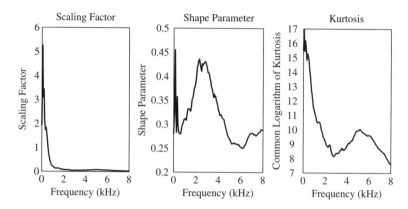

Figure 13.20 The parameters of the GG pdf for each frequency bin versus the scaling parameter $\hat{\sigma}_{|Y|}$, versus the shape parameter f and versus the kurtosis

The scale and shape factors estimated from a set of training data provide insight into the characteristics of human speech. Figure 13.20 shows the scaling parameter $\hat{\sigma}_{|Y|}$ and the shape parameter f for each subband. The training samples used for estimating the GG pdf here were taken from clean speech data in the SSC development set (Lincoln *et al.* 2005).

It is clear from Figure 13.20 that the scale factor $\hat{\sigma}_{|Y|}$, which is approximately the square root of the variance, becomes smaller at higher frequencies. Hence, the figure implies that the lower frequency regions have a higher spectral energy than the higher frequency regions; this is not surpising given that the highest energy phones, namely, the vowels, have most of their spectral content below 1000 Hz. That $f < 2$ for all subbands is a strong indicator of the super-Gaussian nature of speech. Moreover, the figure indicates that the subband samples of speech are more super-Gaussian than the Laplace pdf given that $f < 1$ for all frequencies.

Differential negentropy (12.26) is a measure of the non-Gaussianity of a pdf. Unfortunately, the definition of differential negentropy in (12.26) does not admit a closed-form solution for all choices of the pdf of Y. Hence, as in Rauch *et al.* (2008) we will consider instead the *empirical negentropy*, which can be expressed as

$$J_e(\mathcal{Y}) \triangleq \frac{1}{K} \sum_{k=0}^{K-1} \left[\log \frac{p_{\text{gg}}(Y(k))}{p_{\text{Gauss}}(Y(k))} \right] + \alpha |\mathbf{w}_a|^2, \tag{13.173}$$

where $\alpha |\mathbf{w}_a|^2$ is a regularization term intended to add robustness by penalizing large active weight vectors in the GSC beamformer shown in Figure 13.15. As the optimization of the active weights is to be performed for each subband independently, we will suppress the subband index for the balance of this section. Taking partial derivatives on both sides of (13.173), we obtain

$$\frac{\partial J_e(\mathcal{Y})}{\partial \mathbf{w}_a^*} = \frac{1}{K} \sum_{k=0}^{K-1} \left[\frac{\partial \log p_{\text{gg}}(Y(k))}{\partial \mathbf{w}_a^*} - \frac{\partial \log p_{\text{Gauss}}(Y(k))}{\partial \mathbf{w}_a^*} \right] + \alpha \mathbf{w}_a. \tag{13.174}$$

Based on (B.24), the partial derivative $\partial \log p_{gg}(Y(k))/\partial \mathbf{w}_a^*$ can be expressed as

$$\frac{\partial \log p_{GG}(Y(k))}{\partial \mathbf{w}_a^*} = \frac{f|Y(k)|^{f-2}}{2\left[B_c(f)\hat{\sigma}\right]^f} \mathbf{B}^H \mathbf{X} Y^*(k). \qquad (13.175)$$

The comparable term for the Gaussian pdf in (13.174) can be determined by substituting $f = 2$ into (13.175), whereupon we find

$$\frac{\partial \log p_{Gauss}(Y(k))}{\partial \mathbf{w}_a^*} = \frac{1}{\hat{\sigma}^2} \mathbf{B}^H \mathbf{X} Y^*(k). \qquad (13.176)$$

Substituting (13.175) and (13.176) into (13.174), we arrive at

$$\frac{\partial J_e(Y)}{\partial \mathbf{w}_a^*} = \frac{1}{K} \sum_{k=0}^{K-1} \left\{ \frac{f|Y(k)|^{f-2}}{2\left[B_c(f)\hat{\sigma}\right]^f} - \frac{1}{\hat{\sigma}^2} \right\} \mathbf{B}^H \mathbf{X} Y^*(k) + \alpha \mathbf{w}_a. \qquad (13.177)$$

It is possible to implement a numerical optimization algorithm based on (13.173) and (13.177).

Simulation

As indicated in Section 13.3.1, a conventional MVDR beamformer attempts to minimize output power subject to a distortionless constraint. As discussed in Section 13.3.3, such a beamformer can null out an interfering signal, but is prone to the signal cancellation problem in the presence of an interfering signal which is correlated with the desired signal (Widrow *et al.* 1982). In realistic environments, interference signals are highly correlated with the desired signal, as the desired signal is reflected from hard surfaces such as walls and tables. In such environments, the MNB algorithm would attempt not only to eliminate interfering signals but also *strengthen* those reflections from the desired source. Of course, any reflected signal would be delayed with respect to the direct path signal. Such a delay would, however, manifest itself as a phase shift in the subband domain, and could thus be removed through a suitable choice of $\mathbf{w}_a$. Hence, the MNB offers the possibility of steering both nulls *and* sidelobes; the former toward the undesired signal and its reflections, the latter toward reflections of the desired signal.

In order to verify that the MNB algorithm forms sidelobes directed toward the reflections of a desired signal, Kumatani *et al.* (2008a) conducted experiments with a simulated acoustic environment. As shown in Figure 13.21, these experiments were based on a simple configuration where there was a sound source, a reflective surface, and a linear array of eight microphones with 10-cm intersensor spacing. Actual speech data was used as a source in the simulation, which was based on the *image method* (Allen and Berkley 1979). Figure 13.22 shows beam patterns at $f_s = 800$ Hz and $f_s = 1500$ Hz obtained with a DSB and the MNB with the GG pdf.

Given that a beam pattern shows the sensitivity of an array to plane waves, but the beam patterns in Figure 13.22 were made with a near-field source and reflection, a second set of simulations was also conducted in which the source and reflection were assumed to produce plane waves. The results of this second simulation are shown in Figure 13.23. Once more, it is apparent that the MNB emphasizes the reflection from the desired source.

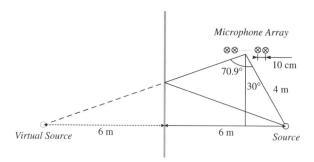

Figure 13.21   Configuration of a source, sensors, and reflective surface for simulation

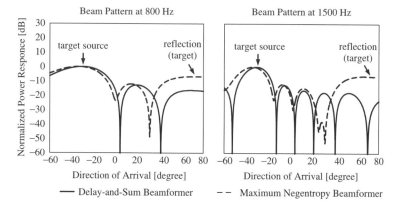

Figure 13.22   Beam patterns produced by the delay-and-sum and the maximum negentropy beamforming algorithms using a spherical wave assumption for $f_s = 800$ and $f_s = 1500$ Hz

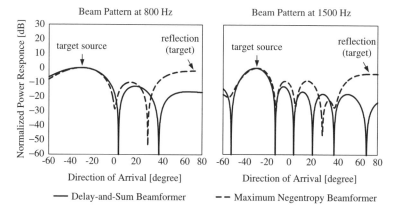

Figure 13.23   Beam patterns produced by the delay-and-sum and the maximum negentropy beamforming algorithms using a plane wave assumption for $f_s = 800$ and $f_s = 1500$ Hz

13.5.3 Hidden Markov Model Maximum Negentropy Beamforming

As explained in Chapter 12, there are three properties of any given signal that can potentially be exploited for *blind source separation* (BSS) or ICA, which we repeat here for convenience: non-whiteness, non-Gaussianity, and nonstationarity. With the formulation of the maximum negentropy beamformer in Section 13.5.2, we have exploited the first two of the these three properties, and to this added the geometric information that comes with knowledge of the geometry of the sensor array as well as the position of the desired source or sources. The maximum likelihood beamformer discussed in Section 13.5.1 accounts for the nonstationarity or time evolution of speech through use of an auxiliary HMM. In this section, we will follow Rauch *et al.* (2008) and combine this time evolution modeling with the other three properties mentioned above, also with the help of an auxiliary HMM. Firstly, we will show how the cepstral mean of an HMM can be used to derive an estimate of the *power spectral density* (PSD) in the subband domain. Indeed, the auxiliary HMM can be adapted to the characteristics of a given speaker's voice using the speaker adaptation techniques based on the all-pass transform presented in Section 9.2.2. Thereafter we will derive the gradient information needed to implement a *conjugate gradients* (Bertsekas 1995, sect. 1.6) algorithm capable of optimizing the active weight vector of a beamformer in GSC configuration. The result will be the *HMM maximum negentropy beamformer* (HMM–MNB).

Reconstructing the Power Spectral Density from the Cepstral Mean

For our negentropy calculations we need an estimate of the variance in the subband domain. Note that if the mean of the power spectrum is known, such an estimate of the variance of the mth subband is readily available as $\sigma_m^2(k) = \mathcal{E}\{|Y_m(k)|^2\}$. This is because the PSD is equivalent to the expected value of the square of the subband magnitude. Our aim is therefore to obtain the mean PSD value. We will now show the relationship between the mean cepstral vector and the mean PSD vector.

Let $\mathbf{Y}(k)$ and $\mathbf{c}(k)$ denote the kth vectors of subband samples and cepstral coefficients, respectively. The relationship between $\mathbf{Y}(k)$ and $\mathbf{c}(k)$ can be expressed as

$$\mathbf{c}(k) = \mathbf{T}_\nu \log |\mathbf{Y}(k)|^2, \qquad (13.178)$$

where $\mathbf{T}_\nu$ is the Type 2 DCT matrix, whose components are give in (B.1), which has been truncated to ν *rows*.[4] In (13.178), the square magnitude and logarithm are calculated individually for each component $Y_m(k)$ of $\mathbf{Y}(k)$. Typically, ν will assume a relatively small value (e.g., $\nu = 13$), which implies that it will model only the spectral envelope due to the resonances of the vocal tract. The more spectral rapid variations due to the harmonic structure of voiced speech must then be modeled by the super-Gaussian pdfs. In this case, no mel warping as in (5.5–5.7) is applied to the subband samples prior to applying the DCT, as this would only decrease the frequency resolution of the filter bank.

---

[4] A brief description of the properties of the DCT is provided in Appendix B.1.

If we calculate the cepstral mean $\overline{\mu}$ over K frames, we have

$$\overline{\mu} \triangleq \frac{1}{K} \sum_{k=0}^{K-1} \mathbf{c}(k) = \frac{1}{K} \mathbf{T}_\nu \sum_{k=0}^{K-1} \log |\mathbf{Y}(k)|^2 . \tag{13.179}$$

In all that follows, we will use k as an index over time and m as an index over subbands. Now let

$$\hat{\mu}(k) \triangleq \mathbf{A}^{(s)} \mu(k) + \mathbf{b}^{(s)}$$

denote the transformed speaker-dependent mean aligned to the kth frame of subband samples by the Viterbi algorithm, where $\mathbf{A}^{(s)}$ and $\mathbf{b}^{(s)}$ are, respectively, a transformation matrix and cepstral bias vector intended to compensate for the unique characteristics of the voice of speaker s. As mentioned previously, for the experiments reported in Rauch *et al.* (2008), $\mathbf{A}^{(s)}$ and $\mathbf{b}^{(s)}$ were determined from sparsely parameterized all-pass transforms described in Section 9.2.2. Let us further define

$$\tilde{\mu}(k) \triangleq \hat{\mu}(k) + \overline{\mu}.$$

With this definition, the true kth cepstral mean $\tilde{\mu}(k)$ is obtained from the sum of a short-term perturbation $\hat{\mu}(k)$ and a long-term average $\overline{\mu}$ that is, obviously, independent of the frame index k.

The inverse $\mathbf{T}^{-1}$ of $\mathbf{T}$ is defined in (B.3). Let $\mathbf{T}_\nu^{-1}$ denote the inverse of $\mathbf{T}$ truncated to ν *columns*. The diagonal covariance matrix $\boldsymbol{\Sigma}_\mathbf{Y}(k)$ of the kth frame of subband components can be approximated as

$$\boldsymbol{\Sigma}_\mathbf{Y}(k) \approx \exp\left(\eta(k)\right), \tag{13.180}$$

where

$$\eta(k) \triangleq \mathbf{T}_\nu^{-1} \tilde{\mu}(k) = \mathbf{T}_\nu^{-1} \hat{\mu}(k) + \overline{\eta}, \tag{13.181}$$

and

$$\overline{\eta} = \frac{1}{K} \mathbf{T}_\nu^{-1} \mathbf{T}_\nu \sum_{k'=0}^{K-1} \log |\mathbf{Y}(k')|^2. \tag{13.182}$$

Clearly $\boldsymbol{\Sigma}_\mathbf{Y}(k)$ is the desired power spectral density. Were ν taken to be equal to the length of the input feature, it would hold that $\mathbf{T}_\nu^{-1}\mathbf{T}_\nu = \mathbf{I}$, where $\mathbf{I}$ is the identity matrix. For smaller values of ν, $\mathbf{T}_\nu^{-1}\mathbf{T}_\nu$ is not diagonal, but remains *diagonally dominated*. This is an important characteristic, for it enables the optimization to be done sequentially over the subbands with little loss in accuracy (Rauch *et al.* 2008). As with the square magnitude and logarithm, the exponential operation in (13.180) is applied component by component.

Figure 13.24 shows an example of the reconstructed log PSD for one frame of a test utterance beamformed with the MNB. The average PSD value is compared to the original PSD and the PSD reconstructed from the 13 original cepstral coefficients, as well as

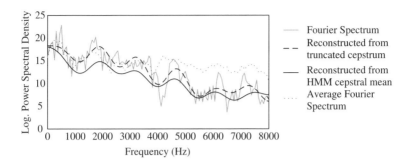

Figure 13.24 Original and reconstructed logarithmic power spectral densities of a test frame

that reconstructed from the 13-coefficient cepstral mean in the HMM state aligned with this frame. In addition to these cepstral envelopes, the long-term spectral mean averaged over the entire utterance is also shown in the figure. We observe that the HMM-based reconstruction approximates the spectral envelope well in the log PSD domain. The PSD obtained by averaging over the entire utterance, on the other hand, only approximates the long-term *spectral tilt*, and hence does not capture the nonstationarity of human speech.

Gradient Calculation

In Kumatani *et al.* (2008b), time averages of the moments of $Y_m(k)$ were used to calculate the differential entropies required to evaluate (12.26). For the initial study on HMM negentropy beamforming, Rauch *et al.* (2008) chose instead to replace the exact differential entropy with the empirical negentropy given in (13.173). In order to use this empirical negentropy as an optimization criterion for an adaptive beamformer, it is necessary to calculate the gradient of empirical negentropy with respect to the active weight vectors $\mathbf{w}_a^*$.

Based on (13.182), we can denote the mth component of $\overline{\eta}$ as $\overline{\eta}_m$ and write

$$\frac{\partial \overline{\eta}_m}{\partial \mathbf{w}_{a,m}^*} = \frac{1}{K} \mathbf{t}_m' \mathbf{t}_m \sum_{k'=0}^{K-1} \frac{1}{|Y_m(k')|^2} \cdot \frac{\partial |Y_m(k')|^2}{\partial \mathbf{w}_{a,m}^*}$$

$$= \frac{1}{K} \mathbf{t}_m' \mathbf{t}_m \mathbf{B}_m^H \sum_{k'=0}^{K-1} \frac{1}{|Y_m(k')|^2} \cdot \mathbf{X}_m(k') \, Y_m^*(k'), \qquad (13.183)$$

where $\mathbf{t}_m'$ and $\mathbf{t}_m$ are, respectively, the mth *row* of $\mathbf{T}_v^{-1}$ and the mth *column* of $\mathbf{T}_v$. In writing (13.183), we account only for the effect of $\mathbf{w}_{a,m}^*$ on $\overline{\eta}_m$, and ignore its effect on any other $\overline{\eta}_n$ for $n \neq m$. Thus we are exploiting the fact that $\mathbf{T}_v^{-1}\mathbf{T}_v$, as previously mentioned, is diagonally dominated. Let $\hat{\sigma}_m^2(k)$ denote the mth diagonal component of $\boldsymbol{\Sigma}_Y(k)$. It then follows that

$$\frac{\partial \hat{\sigma}_m(k)}{\partial \mathbf{w}_{a,m}^*} = \frac{1}{2} \cdot \exp\left(\frac{1}{2}\eta_m\right) \cdot \frac{\partial \overline{\eta}_m}{\partial \mathbf{w}_{a,m}^*} = \frac{1}{2} \cdot \hat{\sigma}_m(k) \cdot \frac{\partial \overline{\eta}_m}{\partial \mathbf{w}_{a,m}^*}.$$

Finally, based on (B.28), we can write

$$\frac{\partial \log p(Y_m(k); f, \hat{\sigma}_m(k))}{\partial \hat{\sigma}_m(k)} \cdot \frac{\partial \hat{\sigma}_m(k)}{\partial \mathbf{w}_{\mathrm{a},m}^*} = \frac{1}{2} \cdot \left[\left(\frac{f \, |Y_m(k)|^f}{B_{\mathrm{c}}^f(f) \, \hat{\sigma}_m^f(k)} - 2 \right) \cdot \frac{\partial \overline{\eta}_m}{\partial \mathbf{w}_{\mathrm{a},m}^*} \right]. \quad (13.184)$$

The partial derivative $\partial J_{\mathrm{e}}(\mathcal{Y}_m)/\partial \mathbf{w}_{\mathrm{a},m}^*$ required for the numerical optimization can be expressed as

$$\frac{\partial J_{\mathrm{e}}(\mathcal{Y})}{\partial \mathbf{w}_{\mathrm{a},m}^*} = \frac{1}{K} \sum_{k=0}^{K-1} \left[\frac{\partial J_{\mathrm{e}}(\mathcal{Y})}{\partial Y_m(k)} \cdot \frac{\partial Y_m(k)}{\partial \mathbf{w}_{\mathrm{a},m}^*} + \frac{\partial J_{\mathrm{e}}(\mathcal{Y})}{\partial \hat{\sigma}_m(k)} \cdot \frac{\partial \hat{\sigma}_m(k)}{\partial \mathbf{w}_{\mathrm{a},m}^*} \right]. \quad (13.185)$$

Based on manipulations similar to those leading to (13.177), the first term can be expressed as

$$\frac{\partial J_{\mathrm{e}}(\mathcal{Y})}{\partial Y_m(k)} \cdot \frac{\partial Y_m(k)}{\partial \mathbf{w}_{\mathrm{a},m}^*} = \frac{1}{K} \left\{ \frac{f \, |Y_m(k)|^{f-2}}{2 \left[B_{\mathrm{c}}(f) \, \hat{\sigma} \right]^f} - \frac{1}{\hat{\sigma}^2} \right\} \mathbf{B}_m^H \mathbf{X}_m(k) \, Y_m^*(k).$$

Following the definition (13.173), the other term in (13.185) can be expressed as

$$\frac{\partial J_{\mathrm{e}}(\mathcal{Y}_m)}{\partial \hat{\sigma}_m(k)} = -\frac{\partial \log p_{\mathrm{Gauss}}(Y_m(k))}{\partial \hat{\sigma}_m(k)} + \frac{\partial \log p_{\mathrm{gg}}(Y_m(k))}{\partial \hat{\sigma}_m(k)}.$$

13.5.4 Minimum Mutual Information Beamforming

Competing or overlapping speech, whereby two or more people speak simultaneously, is a frequently encountered problem in DSR applications (Shriberg *et al.* 2001). The *minimum mutual information beamformer* was proposed by Kumatani *et al.* (2007) in order to address the task of separating the speech of two simultaneously active speakers. Those authors constructed one subband domain beamformer in GSC configuration for each source. In contrast to the conventional beamforming algorithms described in Section 13.3, they then jointly adjusted the active weight vectors of both GSCs to obtain two output signals with MinMI. Assuming that the subband snapshots are Gaussian-distributed, this MinMI criterion reduces to the requirement that the *cross-correlation coefficient* of each of the subband outputs of the two GSCs vanishes. In this section, we describe the MinMI beamformer of Kumatani *et al.* In the next section, we describe the GSS algorithm proposed by Parra and Alvino (2002), which attempts to decorrelate the outputs of two GSC beamformers. We will thereafter compare and contrast the MinMI and GSC beamformers.

Consider a subband beamformer in GSC configuration as described in Section 13.3.7. Assuming there are two such beamformers aimed at different sources, as shown in Figure 13.25, the output of the *i*th beamformer for a given subband can be expressed as

$$Y_i(K) = \left(\mathbf{w}_{\mathrm{q},i} - \mathbf{B}_i \mathbf{w}_{\mathrm{a},i} \right)^H \mathbf{X}(K) \, \forall \, i = 1, 2, \quad (13.186)$$

where *i* is an index over active speakers. As the optimization will be performed independently for each subband, there is no need to retain an explicit subband index. While

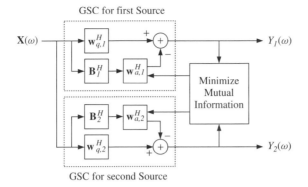

GSC for first Source

GSC for second Source

Figure 13.25 Schematic of generalized sidelobe canceling (GSC) beamformers for each active source

the active weight vector $\mathbf{w}_{\mathrm{a},i}$ is typically chosen to minimize one of the optimization criteria discussed in Section 13.3, here we will, as previously mentioned, develop an optimization procedure to find the $\mathbf{w}_{\mathrm{a},i}$ that *minimizes* the MI $I(Y_1, Y_2)$. Let us define the cross-correlation coefficient ρ_{12} between Y_1 and Y_2 as (Anderson 1984, sect. 2.3)

$$\rho_{12} \triangleq \frac{\epsilon_{12}}{\sigma_1 \sigma_2},\tag{13.187}$$

where

$$\epsilon_{12} \triangleq \mathcal{E}\{Y_1 Y_2^*\} = \left(\mathbf{w}_{\mathrm{q},1} - \mathbf{B}_1 \mathbf{w}_{\mathrm{a},1}\right)^H \mathbf{\Sigma}_{\mathbf{X}} \left(\mathbf{w}_{\mathrm{q},2} - \mathbf{B}_2 \mathbf{w}_{\mathrm{a},2}\right),\tag{13.188}$$

and, based on (13.186), the variance $\sigma_i^2 = \mathcal{E}\left\{Y_i Y_i^*\right\}$ of Y_i can be expressed as

$$\sigma_i^2 = \left(\mathbf{w}_{\mathrm{q},i} - \mathbf{B}_i \mathbf{w}_{\mathrm{a},i}\right)^H \mathbf{\Sigma}_{\mathbf{X}} \left(\mathbf{w}_{\mathrm{q},i} - \mathbf{B}_i \mathbf{w}_{\mathrm{a},i}\right) \ \forall \, i = 1, 2.\tag{13.189}$$

In (13.189), the covariance matrix of the snapshot $\mathbf{X}$ is once more denoted as $\mathbf{\Sigma}_{\mathbf{X}} = \mathcal{E}\{\mathbf{X}\mathbf{X}^H\}$. From (13.187), it then follows that

$$|\rho_{12}|^2 = \frac{|\epsilon_{12}|^2}{\sigma_1^2 \sigma_2^2}.\tag{13.190}$$

It is straightforward to extend the development of Section 12.2.3 for the complex Gaussian random variables whose pdf is given by (B.25), which yields

$$I(Y_1, Y_2) = -\log\left(1 - |\rho_{12}|^2\right).\tag{13.191}$$

This implies that

$$I(Y_1, Y_2) = 0 \leftrightarrow |\rho_{12}| = 0,$$

and minimizing $I(Y_1, Y_2)$ implies minimizing $|\rho_{12}|$.

Minimizing the MI criterion yields a weight vector $\mathbf{w}_{a,i}$ capable of canceling interference without the signal cancellation problems encountered in conventional beamforming. Moreover, the GSC constraint imposed here resolves problems of source permutation and scaling ambiguity encountered in conventional frequency and subband domain BSS algorithms (Buchner *et al.* 2004).

Parameter Optimization: Gaussian pdf

In the absence of a closed-form solution for those $\mathbf{w}_{a,i}$ minimizing $I(Y_1, Y_2)$, we must use a numerical optimization algorithm. Such an optimization algorithm typically requires gradient information. In this section, we will derive the relations necessary to minimize the MinMI criterion under the assumption that the outputs of the beamformers are Gaussian-distributed. Hence, let us apply the chain rule as described in Appendix B.16 to (13.190) by writing

$$
\frac{\partial |\rho_{12}|^2}{\partial \mathbf{w}_{a,1}^*} = \frac{1}{\sigma_1^4 \sigma_2^4} \left(\frac{\partial \epsilon_{12}}{\partial \mathbf{w}_{a,1}^*} \epsilon_{12}^* \sigma_1^2 \sigma_2^2 - \frac{\partial \sigma_1^2}{\partial \mathbf{w}_{a,1}^*} |\epsilon_{12}|^2 \sigma_2^2 \right)
$$

$$
= \frac{1}{\sigma_1^4 \sigma_2^4} \left[-\mathbf{B}_1^H \Sigma_{\mathbf{X}} (\mathbf{w}_{q,2} - \mathbf{B}_2 \mathbf{w}_{a,2}) \epsilon_{12}^* \sigma_1^2 \sigma_2^2 + \mathbf{B}_1^H \Sigma_{\mathbf{X}} (\mathbf{w}_{q,1} - \mathbf{B}_1 \mathbf{w}_{a,1}) |\epsilon_{12}|^2 \sigma_2^2 \right].
$$

The last equation can be simplified to

$$
\frac{\partial |\rho_{12}|^2}{\partial \mathbf{w}_{a,1}^*} = \frac{1}{\sigma_1^4 \sigma_2^4} \mathbf{B}_1^H \Sigma_{\mathbf{X}} \cdot \left[|\epsilon_{12}|^2 \sigma_2^2 (\mathbf{w}_{q,1} - \mathbf{B}_1 \mathbf{w}_{a,1}) - \epsilon_{12}^* \sigma_1^2 \sigma_2^2 (\mathbf{w}_{q,2} - \mathbf{B}_2 \mathbf{w}_{a,2}) \right].
$$
(13.192)

From symmetry it then follows that

$$
\frac{\partial |\rho_{12}|^2}{\partial \mathbf{w}_{a,2}^*} = \frac{1}{\sigma_1^4 \sigma_2^4} \mathbf{B}_2^H \Sigma_{\mathbf{X}} \cdot \left[|\epsilon_{12}|^2 \sigma_1^2 (\mathbf{w}_{q,2} - \mathbf{B}_2 \mathbf{w}_{a,2}) - \epsilon_{12} \sigma_1^2 \sigma_2^2 (\mathbf{w}_{q,1} - \mathbf{B}_1 \mathbf{w}_{a,1}) \right].
$$
(13.193)

To formulate an algorithm for minimizing $I(Y_1, Y_2)$, we need only begin from (13.191) and write

$$
\frac{\partial I(Y_1, Y_2)}{\partial \mathbf{w}_{a,i}^*} = \frac{1}{2\left(1 - |\rho_{12}|^2\right)} \cdot \frac{\partial |\rho_{12}|^2}{\partial \mathbf{w}_{a,i}^*},
$$
(13.194)

which, together with (13.192) and (13.193), is sufficient to calculate the required gradients.

As discussed in Section 13.3.8 conventional beamforming algorithms typically apply a regularization term that penalizes large active weight vectors, and thereby improves robustness by inhibiting the formation of excessively large sidelobes. Such a regularization term can be applied in the present instance by defining the modified optimization criterion

$$
\mathcal{I}(Y_1, Y_2; \alpha) = I(Y_1, Y_2) + \alpha |\mathbf{w}_{a,1}|^2 + \alpha |\mathbf{w}_{a,2}|^2
$$
(13.195)

for some real $\alpha > 0$. Taking the partial derivative on both sides of (13.195) yields

$$\frac{\partial \mathcal{I}(Y_1, Y_2; \alpha)}{\partial \mathbf{w}_{\text{a},i}^*} = \frac{1}{2\left(1 - |\rho_{12}|^2\right)} \cdot \frac{\partial |\rho_{12}|^2}{\partial \mathbf{w}_{\text{a},i}^*} + \alpha \mathbf{w}_{\text{a},i}. \tag{13.196}$$

Parameter Optimization: Super-Gaussian pdfs

While the Gaussian assumption simplifies the parameter optimization problem in that all relevant information is contained in the covariance matrix of the subband samples, it is in fact suboptimal inasmuch as human speech, as discussed in Section 13.5.2, is not Gaussian-distributed. In order to estimate beamforming parameters with an MinMI criterion using non-Gaussian pdfs, we first approximate mutual information with the *empirical mutual information* according to

$$I(Y_1, Y_2) \approx \frac{1}{N} \sum_{k=0}^{K-1} \left[\log p(Y_1(k), Y_2(k)) - \log p(Y_1(k)) - \log p(Y_2(k)) \right], \tag{13.197}$$

where

$$Y_i(k) = (\mathbf{w}_{\text{q},i} - \mathbf{B}_i \mathbf{w}_{\text{a},i})^H \mathbf{X}(k), \ \forall i = 1, 2,$$

for each $\mathbf{X}(k)$ drawn from a *training set* $\mathcal{X} = \{\mathbf{X}(k)\}_{k=0}^{K-1}$. From (13.197), it follows that

$$\frac{\partial I(Y_1, Y_2)}{\partial \mathbf{w}_{\text{a},i}^*} \approx \frac{1}{K} \sum_{k=0}^{K-1} \left[\frac{\partial \log p(Y_1(k), Y_2(k))}{\partial \mathbf{w}_{\text{a},i}^*} - \frac{\partial \log p(Y_1(k))}{\partial \mathbf{w}_{\text{a},i}^*} - \frac{\log p(Y_2(k))}{\partial \mathbf{w}_{\text{a},i}^*} \right].$$

Hence, we need expressions for the partial derivatives of $\log p(Y_1, Y_2)$ and $\log p(Y_i)$ with respect to $\mathbf{w}_{\text{a},i}^*$. Regardless of the precise density, this will require evaluating the intermediate quantities which we now calculate. Based on (13.189), we have,

$$\frac{\partial \sigma_i^2}{\partial \mathbf{w}_{\text{a},i}^*} = -\mathbf{B}_i^H \mathbf{\Sigma}_{\mathbf{X}} (\mathbf{w}_{\text{q},i} - \mathbf{B}_i \mathbf{w}_{\text{a},i}).$$

It is also readily shown that

$$\frac{\partial \sigma_i}{\partial \mathbf{w}_{\text{a},i}^*} = \frac{1}{2\sigma_i} \frac{\partial \sigma_i^2}{\partial \mathbf{w}_{\text{a},i}^*} = -\frac{1}{2\sigma_i} \mathbf{B}_i^H \mathbf{\Sigma}_{\mathbf{X}} (\mathbf{w}_{\text{q},i} - \mathbf{B}_i \mathbf{w}_{\text{a},i}).$$

Beginning from (13.186), we can write

$$|Y_i| = \sqrt{\left(\mathbf{w}_{\text{q},i} - \mathbf{B}_i \mathbf{w}_{\text{a},i}\right)^H \mathbf{X} \mathbf{X}^H \left(\mathbf{w}_{\text{q},i} - \mathbf{B}_i \mathbf{w}_{\text{a},i}\right)}.$$

Hence,

$$\frac{\partial |Y_i|}{\partial \mathbf{w}_{\text{a},i}^*} = -\frac{1}{2|Y_i|} \mathbf{B}_i^H \mathbf{X} \mathbf{X}^H \left(\mathbf{w}_{\text{q},i} - \mathbf{B}_i \mathbf{w}_{\text{a},i}\right).$$

We can express Σ_Y as

$$\Sigma_Y = \begin{bmatrix} \sigma_1^2 & \epsilon_{12} \\ \epsilon_{12}^* & \sigma_2^2 \end{bmatrix},$$ (13.198)

where ϵ_{12} is defined in (13.188). From (13.198) it is clear that

$$|\Sigma_Y| = \sigma_1^2 \sigma_2^2 - \epsilon_{12} \epsilon_{12}^*.$$ (13.199)

Thus,

$$\begin{aligned}
\frac{\partial |\Sigma_Y|}{\partial \mathbf{w}_{a,1}^*} &= \sigma_2^2 \frac{\partial \sigma_1^2}{\partial \mathbf{w}_{a,1}^*} - \epsilon_{12}^* \frac{\partial \epsilon_{12}}{\partial \mathbf{w}_{a,1}^*} \\
&= -\sigma_2^2 \mathbf{B}_1^H \Sigma_X \left(\mathbf{w}_{q,1} - \mathbf{B}_1 \mathbf{w}_{a,1} \right) + \epsilon_{12}^* \mathbf{B}_1^H \Sigma_X \left(\mathbf{w}_{q,2} - \mathbf{B}_2 \mathbf{w}_{a,2} \right) \\
&= \mathbf{B}_1^H \Sigma_X \left[\epsilon_{12}^* \left(\mathbf{w}_{q,2} - \mathbf{B}_2 \mathbf{w}_{a,2} \right) - \sigma_2^2 \left(\mathbf{w}_{q,1} - \mathbf{B}_1 \mathbf{w}_{a,1} \right) \right].
\end{aligned}$$ (13.200)

It then follows from symmetry that

$$\frac{\partial |\Sigma_Y|}{\partial \mathbf{w}_{a,2}^*} = \mathbf{B}_2^H \Sigma_X \left[\epsilon_{12} \left(\mathbf{w}_{q,1} - \mathbf{B}_1 \mathbf{w}_{a,1} \right) - \sigma_1^2 \left(\mathbf{w}_{q,2} - \mathbf{B}_2 \mathbf{w}_{a,2} \right) \right].$$ (13.201)

Based on (13.198) and (13.199), the inverse of Σ_Y can be expressed as

$$\Sigma_Y^{-1} = \frac{1}{|\Sigma_Y|} \begin{bmatrix} \sigma_2^2 & -\epsilon_{12} \\ -\epsilon_{12}^* & \sigma_1^2 \end{bmatrix} = \frac{1}{\sigma_1^2 \sigma_2^2 - |\epsilon_{12}|^2} \begin{bmatrix} \sigma_2^2 & -\epsilon_{12} \\ -\epsilon_{12}^* & \sigma_1^2 \end{bmatrix}.$$

Let us now define

$$s \triangleq \mathbf{Y}^H \Sigma_Y^{-1} \mathbf{Y} = \frac{f_{12}}{|\Sigma_Y|},$$ (13.202)

where

$$f_{12} = \sigma_2^2 |Y_1|^2 - \epsilon_{12}^* Y_1 Y_2^* - \epsilon_{12} Y_1^* Y_2 + \sigma_1^2 |Y_2|^2.$$

From the definition (13.202) it then follows that

$$\begin{aligned}
\frac{\partial s}{\partial \mathbf{w}_{a,1}^*} &= \frac{1}{|\Sigma_Y|^2} \left\{ \left[(\sigma_2^2 Y_1^* - \epsilon_{12}^* Y_2^*) \frac{\partial Y_1}{\partial \mathbf{w}_{a,1}^*} - Y_1^* Y_2 \frac{\partial \epsilon_{12}}{\partial \mathbf{w}_{a,1}^*} + |Y_2|^2 \frac{\partial \sigma_1^2}{\partial \mathbf{w}_{a,1}^*} \right] |\Sigma_Y| - f_{12} \frac{\partial |\Sigma_Y|}{\partial \mathbf{w}_{a,1}^*} \right\} \\
&= \frac{1}{|\Sigma_Y|} \left[(\epsilon_{12}^* Y_2^* - \sigma_2^2 Y_1^*) \mathbf{B}_1^H \mathbf{X} + Y_1^* Y_2 \mathbf{B}_1^H \Sigma_X (\mathbf{w}_{q,2} - \mathbf{B}_2 \mathbf{w}_{a,2}) \right]
\end{aligned}$$

$$-|Y_2|^2 \mathbf{B}_1^H \Sigma_{\mathbf{X}}(\mathbf{w}_{\mathrm{q},1} - \mathbf{B}_1 \mathbf{w}_{\mathrm{a},1})\big] - \frac{f_{12}}{|\Sigma_{\mathbf{Y}}|^2} \frac{\partial |\Sigma_{\mathbf{Y}}|}{\partial \mathbf{w}_{\mathrm{a},1}^*}$$

$$= \frac{1}{|\Sigma_{\mathbf{Y}}|} \mathbf{B}_1^H \big[(\epsilon_{12}^* Y_2^* - \sigma_2^2 Y_1^*) \mathbf{X} + Y_1^* Y_2 \Sigma_{\mathbf{X}}(\mathbf{w}_{\mathrm{q},2} - \mathbf{B}_2 \mathbf{w}_{\mathrm{a},2})$$

$$-|Y_2|^2 \Sigma_{\mathbf{X}}(\mathbf{w}_{\mathrm{q},1} - \mathbf{B}_1 \mathbf{w}_{\mathrm{a},1})\big] - \frac{s}{|\Sigma_{\mathbf{Y}}|} \frac{\partial |\Sigma_{\mathbf{Y}}|}{\partial \mathbf{w}_{\mathrm{a},1}^*}, \tag{13.203}$$

and from symmetry,

$$\frac{\partial s}{\partial \mathbf{w}_{\mathrm{a},2}^*} = \frac{1}{|\Sigma_{\mathbf{Y}}|} \mathbf{B}_2^H \big[(\epsilon_{12} Y_1^* - \sigma_1^2 Y_2^*) \mathbf{X} + Y_1 Y_2^* \Sigma_{\mathbf{X}}(\mathbf{w}_{\mathrm{q},1} - \mathbf{B}_1 \mathbf{w}_{\mathrm{a},1})$$

$$-|Y_1|^2 \Sigma_{\mathbf{X}}(\mathbf{w}_{\mathrm{q},2} - \mathbf{B}_2 \mathbf{w}_{\mathrm{a},2})\big] - \frac{s}{|\Sigma_{\mathbf{Y}}|} \frac{\partial |\Sigma_{\mathbf{Y}}|}{\partial \mathbf{w}_{\mathrm{a},2}^*}. \tag{13.204}$$

We now specialize the calculation of the gradient of $I(Y_1, Y_2)$ for two super-Gaussian pdfs, namely, the Laplace and K_0 pdfs.

Laplace Pdf

Based on the univariate complex Laplace pdf in Table B.3, we can write

$$\log p_{\mathrm{Laplace}}(Y_i) = \log 2 - \frac{1}{2} \log \pi - \log \sigma_i^2 + \log K_0 \left(\frac{2\sqrt{2}|Y_i|}{\sigma_i} \right),$$

and

$$\frac{\partial \log p_{\mathrm{Laplace}}(Y_i)}{\partial \mathbf{w}_{\mathrm{a},i}^*} = -\frac{1}{\sigma_i^2} \frac{\partial \sigma_i^2}{\partial \mathbf{w}_{\mathrm{a},i}^*} + \frac{K_0' \left(\frac{2\sqrt{2}|Y_i|}{\sigma_i} \right)}{K_0 \left(\frac{2\sqrt{2}|Y_i|}{\sigma_i} \right)} \cdot \frac{2\sqrt{2}}{\sigma_i^2} \left(\sigma_i \frac{\partial |Y_i|}{\partial \mathbf{w}_{\mathrm{a},i}^*} - |Y_i| \frac{\partial \sigma_i}{\partial \mathbf{w}_{\mathrm{a},i}^*} \right)$$

$$= -\frac{1}{\sigma_i^2} \frac{\partial \sigma_i^2}{\partial \mathbf{w}_{\mathrm{a},i}^*} + \frac{K_1 \left(\frac{2\sqrt{2}|Y_i|}{\sigma_i} \right)}{K_0 \left(\frac{2\sqrt{2}|Y_i|}{\sigma_i} \right)} \cdot \frac{2\sqrt{2}}{\sigma_i^2} \left(\sigma_i \frac{\partial |Y_i|}{\partial \mathbf{w}_{\mathrm{a},i}^*} - \frac{|Y_i|}{2\sigma_i} \frac{\partial \sigma_i^2}{\partial \mathbf{w}_{\mathrm{a},i}^*} \right), \tag{13.205}$$

where the final equality follows from

$$K_0'(z) = \frac{d K_0(z)}{dz} = K_1(z).$$

Based on the bivariate complex Laplace pdf in Table B.3, we can write

$$\log p_{\mathrm{Laplace}}(\mathbf{Y}) = \log 16 - \frac{3}{2} \log \pi - \log |\Sigma_{\mathbf{Y}}| - \frac{1}{2} \log s + \log K_1(4\sqrt{s}).$$

Hence,

$$
\begin{aligned}
\frac{\partial \log p_{\text{Laplace}}(\mathbf{Y})}{\partial \mathbf{w}_{a,i}^*} &= -\frac{1}{|\Sigma_{\mathbf{Y}}|} \cdot \frac{\partial |\Sigma_{\mathbf{Y}}|}{\partial \mathbf{w}_{a,i}^*} - \frac{1}{2s} \frac{\partial s}{\partial \mathbf{w}_{a,i}^*} + \frac{K_1'(4\sqrt{s})}{K_1(4\sqrt{s})} \cdot \frac{2}{\sqrt{s}} \cdot \frac{\partial s}{\partial \mathbf{w}_{a,i}^*} \\
&= -\frac{1}{|\Sigma_{\mathbf{Y}}|} \cdot \frac{\partial |\Sigma_{\mathbf{Y}}|}{\partial \mathbf{w}_{a,i}^*} - \left[\frac{1}{2s} + \frac{K_0(4\sqrt{s}) + K_2(4\sqrt{s})}{\sqrt{s}\,K_1(4\sqrt{s})} \right] \frac{\partial s}{\partial \mathbf{w}_{a,i}^*},
\end{aligned} \quad (13.206)
$$

where the final equality follows from

$$
K_1'(z) = -\frac{1}{2} \left[K_0(z) + K_2(z) \right].
$$

K_0 Pdf

Based on the univariate complex K_0 pdf in Table B.3, we can write

$$
\log p_{\text{Bessel}}(Y) = -\frac{1}{2} \log \pi - \log \sigma_i - \log |Y| - \frac{2|Y|}{\sigma_i}.
$$

Thus,

$$
\begin{aligned}
\frac{\partial \log p(Y_i)}{\partial \mathbf{w}_{a,i}^*} &= -\frac{1}{\sigma_i} \frac{\partial \sigma_i}{\partial \mathbf{w}_{a,i}^*} - \frac{1}{|Y_i|} \frac{\partial |Y_i|}{\partial \mathbf{w}_{a,i}^*} - \frac{2}{\sigma_i^2} \left(\sigma_i \frac{\partial |Y_i|}{\partial \mathbf{w}_{a,i}^*} - |Y_i| \frac{\partial \sigma_i}{\partial \mathbf{w}_{a,i}^*} \right) \\
&= \left(\frac{2|Y_i|}{\sigma_i^2} - \frac{1}{\sigma_i} \right) \frac{\partial \sigma_i}{\partial \mathbf{w}_{a,i}^*} - \left(\frac{1}{|Y_i|} + \frac{2}{\sigma_i} \right) \frac{\partial |Y_i|}{\partial \mathbf{w}_{a,i}^*}.
\end{aligned} \quad (13.207)
$$

Based on the bivariate complex K_0 pdf in Table B.3, we can write

$$
\log p_{\text{Bessel}}(\mathbf{Y}) = -\frac{3}{2} \log \pi + \log(\sqrt{2} + 4\sqrt{s}) - \log |\Sigma_{\mathbf{Y}}| - \frac{3}{2} \log s - 2\sqrt{2s}.
$$

Hence,

$$
\begin{aligned}
\frac{\partial \log p(\mathbf{Y})}{\partial \mathbf{w}_{a,i}^*} &= \frac{1}{\sqrt{2} + 4\sqrt{s}} \cdot \frac{2}{\sqrt{s}} \cdot \frac{\partial s}{\partial \mathbf{w}_{a,i}^*} - \frac{1}{|\Sigma_{\mathbf{Y}}|} \cdot \frac{\partial |\Sigma_{\mathbf{Y}}|}{\partial \mathbf{w}_{a,i}^*} - \frac{3}{2s} \cdot \frac{\partial s}{\partial \mathbf{w}_{a,i}^*} - \frac{2\sqrt{2}}{2\sqrt{s}} \frac{\partial s}{\partial \mathbf{w}_{a,i}^*} \\
&= -\frac{1}{|\Sigma_{\mathbf{Y}}|} \cdot \frac{\partial |\Sigma_{\mathbf{Y}}|}{\partial \mathbf{w}_{a,i}^*} + \left[\frac{2}{\sqrt{2s} + 4s} - \frac{3}{2s} - \frac{\sqrt{2}}{\sqrt{s}} \right] \cdot \frac{\partial s}{\partial \mathbf{w}_{a,i}^*}.
\end{aligned} \quad (13.208)
$$

Using the gradient information in (13.205–13.206) and (13.207–13.208) for the Laplace and K_0 pdfs, respectively, it is straightforward to develop an algorithm for optimizing the sensor weights based, for example, on the method of conjugate gradients (Bertsekas 1995, sect. 1.6).

Note that it is also possible to develop the MinMI beamformer under the assumption of a Γ pdf. Details are provided in Kumatani *et al.* (2007).

Recall the relation between MI and negentropy described in Section 12.2.3. Given this similarity, negentropy is the preferable criterion for two reasons. Firstly, negentropy is applicable for a single as well as multiple active speakers. Secondly, it does not require the development of variates $p(Y_1, \ldots, Y_N)$ of arbitrarily high order N, as implied by (12.16), to model a total of N active speakers. Rather, only the complex univariate pdf is required regardless of the number of active speakers. This is a significant advantage, inasmuch as the high-order variates can typically only be estimated through such complicated means as the Meier G-function, as discussed in Appendix B.5.2.

13.5.5 Geometric Source Separation

Parra and Alvino (2002) proposed a *geometric source separation* algorithm with many similarities to the algorithm proposed by Kumatani *et al.* (2007). Their work was based on two beamformers with geometric constraints that made them functionally equivalent to GSC beamformers. The GSS beamformer can be likened to a MinMI beamformer under a Gaussian assumption which minimizes $|\epsilon_{12}|^2$ instead of $|\rho_{12}|^2$. If a regularization term is added as before, the GSS optimization criteria,

$$\mathcal{I}'(Y_1, Y_2; \alpha) = |\epsilon_{12}|^2 + \alpha |\mathbf{w}_{a,1}|^2 + \alpha |\mathbf{w}_{a,2}|^2, \tag{13.209}$$

is obtained. Then taking partial derivatives of (13.209) gives

$$\frac{\mathcal{I}'(Y_1, Y_2; \alpha)}{\partial \mathbf{w}_{a,1}^*} = -\mathbf{B}_1^H \boldsymbol{\Sigma}_{\mathbf{X}} (\mathbf{w}_{q,2} - \mathbf{B}_2 \mathbf{w}_{a,2}) \, \epsilon_{12}^* + \alpha \mathbf{w}_{a,1} \tag{13.210}$$

$$\frac{\mathcal{I}'(Y_1, Y_2; \alpha)}{\partial \mathbf{w}_{a,2}^*} = -\mathbf{B}_2^H \boldsymbol{\Sigma}_{\mathbf{X}} (\mathbf{w}_{q,1} - \mathbf{B}_1 \mathbf{w}_{a,1}) \, \epsilon_{12} + \alpha \mathbf{w}_{a,2}. \tag{13.211}$$

Although at first blush it may seem that a closed-form solution for $\mathbf{w}_{a,1}$ and $\mathbf{w}_{a,2}$ could be derived, the presence of ϵ_{12}^* and ϵ_{12} in (13.210) and (13.211) respectively actually makes this impossible. Hence, a numerical optimization algorithm is needed, as before.

While the difference between minimizing $|\epsilon_{12}|^2$, as in the GSS algorithm, instead of $|\rho_{12}|^2$, as in the MinMI beamformer, may seem very slight, it can in fact lead to radically different behavior. To achieve the desired optimum, both criteria will seek to place deep nulls on the unwanted source; this characteristic is associated with $|\epsilon_{12}|^2$, which also comprises the *numerator* of $|\rho_{12}|^2$. Such null steering is also observed in conventional adaptive beamformers, as discussed in Section 13.3.3. The difference between the two optimization criteria is due to the presence of the terms σ_i^2 in the denomimnator of $|\rho_{12}|^2$, which indicate that, in addition to nulling out the unwanted signal, an improvement of the objective function is also possible by *increasing* the strength of the desired signal. For acoustic beamforming in realistic environments, there are typically strong reflections from hard surfaces such as tables and walls. A conventional beamformer would attempt to null out strong reflections of an interfering signal, but strong reflections of the desired signal can lead to signal cancellation (Widrow *et al.* 1982). The GSS algorithm would attempt to null out those reflections from the unwanted signal. But in addition to nulling out reflections from the unwanted signal, the MinMI beamforming algorithm would attempt to *strengthen* those reflections from the desired source; assuming statistically independent sources, strengthening a reflection from the desired source would have little or no effect on the numerator of $|\rho_{12}|^2$, but would increase the denominator, thereby leading to an overall reduction of the optimization criterion. Of course, any reflected signal would be delayed with respect to the direct path signal. Such a delay would, however, manifest itself as a phase shift in the subband domain, and could thus be removed through a suitable choice of $\mathbf{w}_a$. Hence, the MinMI beamformer, much like the MNB and HMM–MNB beamformers considered in Sections 13.5.2 and 13.5.3, respectively, offers the possibility of steering both nulls *and* sidelobes; the former toward the undesired signal and its reflections, the latter toward reflections of the desired signal. This difference in the behavior between the MinMI and GSS beamforming algorithms was demonstrated with a simple acoustic simulation in Kumatani *et al.* (2007).

13.6 Array Shape Calibration

As a final application, we consider an algorithm proposed by McCowan *et al.* (2008) that is related to beamforming in that it provides estimates of the relative positions of several microphones. This may well be required when beamforming is to be performed using an array with an unknown geometry. Let $X_k(\omega)$ denote the subband sample reaching the kth microphone. Consider that the coherence between signals reaching microphones positioned at points $\mathbf{m}_k$ and $\mathbf{m}_l$ is defined as

$$\Gamma_{kl}(\omega) \triangleq \frac{\gamma_{kl}(\omega)}{\sqrt{\gamma_{kk}(\omega)\,\gamma_{ll}(\omega)}}, \tag{13.212}$$

where $\gamma_{kl} = \mathcal{E}\{X_k(\omega)X_l^*(\omega)\}$. The auto- and cross-spectral densities are readily obtained from a standard recursive update formula (Allen *et al.* 1977), according to

$$\hat{\gamma}_{kl}(\omega) = \alpha\hat{\gamma}_{kl}'(\omega) + (1-\alpha)X_k(\omega)X_l^*(\omega), \tag{13.213}$$

where $\hat{\gamma}_{kl}'(\omega)$ is the density estimate from the prior time step, $\alpha = \exp(-T/\tau_\alpha)$, T is the time step in seconds, and τ_α is a time constant.

As mentioned in Section 13.3.4, a diffuse noise field is a good model for many common acoustic environments, including cars and offices (Elko 2000). The diffuse noise field is characterized by the spherically isotropic coherence function (13.69). The latter implies that for a given frequency under, the coherence between the signals reaching any two microphones is a function only of the distance between them. Hence, in order to learn the distance $d_{kl} = |\mathbf{m}_k - \mathbf{m}_l|$, we can compare the coherence predicted by (13.69) to that measured by (13.213). To make such a comparison, we adopt the squared error optimization criterion

$$\epsilon_{kl}(d) = \sum_{m=0}^{M/2} \left| \mathrm{Re}\{\hat{\gamma}_{kl}(\omega_m)\} - \mathrm{sinc}\left(\frac{\omega_m d}{c}\right) \right|^2, \tag{13.214}$$

$$\hat{d}_{kl} = \arg\min_d \epsilon_{kl}(d), \tag{13.215}$$

where ω_m is the center frequency of the mth subband. The value $\hat{d}$ achieving the desired minimum can be found with a straightforward line search such as Brent's method (Press *et al.* 1992, sect. 10.2).

Once the intersensor spacing is known for all pairs of microphones, the configuration of the entire array can be inferred through the application of the *multidimensional scaling* algorithm (Birchfield and Subramanya 2005). McCowan *et al.* (2008) used additional steps, however, to achieve more robust estimates of the intersensor distances. To wit, pairs of sensor spacing and error estimates $(d_{kl}(t), \epsilon_{kl}(t))$ were calculated as in (13.214–13.215) for several time instants t for each microphone pair (k, l). The several pairs $(d_{kl}(t), \epsilon_{kl}(t))$ were then separated into two clusters using the K-means algorithm (Duda *et al.* 2001, sect. 4.4). The motivation for the latter operation is based on the observation that some frames do not fit the diffuse noise model. Hence, a more robust estimate of the intersensor spacing can be obtained by removing these frames from the estimation process. The final estimate of the intersensor spacing is taken as the d-centroid of the cluster with the smallest ϵ-centroid.

13.7   Summary and Further Reading

In this chapter, we have presented a class of techniques known collectively as beamforming by which signals from several sensors can be combined to emphasize a desired source and suppress interference from other directions. Beamforming begins with the assumption that the positions of all sensors are *known*, and that the position of the desired source is known or can be estimated. The simplest of beamforming algorithms, the DSB, uses only this geometrical knowledge to combine the signals from several sensors. More sophisticated adaptive beamformers attempt to minimize the total output power of the array under a constraint that the desired source must be unattenuated.

The total output power of the array under a conventional beamforming algorithm is minimized through the adjustment of an active weight vector, which effectively places a null on any source of interference, but can also lead to undesirable *signal cancellation* (Widrow *et al.* 1982). To avoid the latter, many algorithms based on conventional optimization criteria have been developed. Among such approaches, the following solutions have been proposed:

- *updating* the active weight vector only when noise signals are dominant (Cohen *et al.* 2003; Herbordt and Kellermann 2003; Nordholm *et al.* 1993);
- *constraining* the update formula for the active weight vector (Claesson and Nordholm 1992; Hoshuyama *et al.* 1999; Nordebo *et al.* 1994);
- *blocking* the leakage of desired signal components into the sidelobe canceler by designing the blocking matrix (Herbordt and Kellermann 2002; Herbordt *et al.* 2007; Hoshuyama *et al.* 1999; Warsitz *et al.* 2008);
- *taking* speech distortion due to the the leakage of a target signal into account using multi-channel Wiener filter which aims at minimizing a weighted sum of residual noise and speech distortion terms (Doclo *et al.* 2007); and
- *using* acoustic transfer functions from a desired source to microphones instead of just compensating for time delays (Cohen *et al.* 2003; Gannot and Cohen 2004, Sharon Gannot *et al.* 2001; Warsitz *et al.* 2008).

All of the algorithms mentioned above attempt to minimize nearly the same criterion based on the second-order statistics, the total output power, while maintaining a distortionless constraint.

Recent research has revealed that the optimization criteria used in conventional array processing are not optimal for acoustic beamforming applications. In the latter sections of this chapter we have discussed nonconventional optmization criteria for beamforming. We have discussed the negentropy optimization criterion, whereby the active weight vectors of the beamformer are adjusted in order to provide an output signal that is as non-Gaussian as possible. We also presented a beamforming algorithm with the ability to separate the voices of two or more speakers who speak simultaneously by minimizing the MI between the outputs of two GSCs based on a minimum MI criterion. Both negentropy and MI are widely-used optimization criteria in the field of ICA.

It is probably safe to say that recent interest in microphone array hardware is moving in two directions. On one side, there is extensive interest in small arrays based on standard audio equipment that can be mounted on a wide range of possibly portable devices. These include PCs, laptops, PDAs, hearing aids and may include cell phones in the near future. Recent work in designing a distant speech acquisition and recognition system for mobile devices was described by Takada et al. (2008). Indeed, this trend has been accelerated by the inclusion of basic beamforming capability in the Microsoft Windows Vista operating system for PCs (Tashev and Allred 2005). In this category also fall arrays mounted in automobiles, which often have severe limitations on where the array can be placed and its physical size; see Nordholm et al. (2001). Because of the small extent of such devices, the algorithms used to combine the signals from the several sensors become very critical, as they must bear the brunt of the task of suppressing interference and reverberation.

On the other side, there is now significant interest within the research community in building specialized hardware that offers a geometry more conducive to acoustic beamforming. Such attempts include the NIST Mark III in its several versions. A highly novel approach to acoustic beamforming is represented by the spherical microphone arrays originally proposed by Meyer and Elko (2004). Indeed this approach has enjoyed a spate of popularity, as is evident from several recent publications, including Li and Duraiswami (2005), Rafaely (2008) and Zotkin et al. (2008). Other work on the decomposition of the sound field of a regularly-shaped aperture into spatial harmonics is presented by Teutsch and Kellermann (2005) for cylindrical arrays.

Useful general references on beamforming include the massive volume by Van Trees (2002), as well as the collection edited by Brandstein and Ward (2000). Another useful reference to the most recent research is Huang and Benesty (2004). Hänsler and Schmidt (2004) is also a useful reference to the related field of acoustic echo and noise control. An interesting survey on the combination of beamforming and acoustic echo cancellation is given by Kellermann (2001).

The minimum mutal beamformer is presented in Kumatani et al. (2007). The maximum negentropy design was introduced in Kumatani et al. (2008b), and further refined through the incorporation of HMM information by Rauch et al. (2008). The use of kurtosis as a beamforming optimization criterion was introduced in Kumatani et al. (2008c).

13.8 Principal Symbols

Symbol	Description
α	step size in least mean square algorithm
$\Upsilon(\omega, \mathbf{k})$	frequency wavenumber response function
λ	wavelength
ϕ	azimuth
$\overline{\phi}$	broadside angle
τ_n	time delay of arrival to nth sensor
θ	polar angle
$\boldsymbol{\theta}(K)$	Givens' rotation
ω	angular frequency
$\Gamma_{\mathbf{N}}(\omega)$	coherence matrix
r	range to desired source
$\mathbf{a}$	direction of propagation of plane wave
$\mathbf{A}(K)$	prearray
$\mathbf{B}(K)$	postarray
$\mathbf{B}(\omega)$	generalized sidelobe canceler blocking matrix
$B(\omega : \theta, \phi)$	beam pattern
c	speed of sound
$\mathbf{C}$	constraint matrix
d	distance between microphones in a uniform linear array
D	directivity
$\mathbf{g}$	right-hand side of constraint equation
$\mathbf{h}(t)$	continuous-time impulse response of a microphone array
$\mathbf{H}(\omega)$	frequency response of a microphone array
$H(Y)$	entropy of Y
$I(Y_1, Y_2)$	mutual information between Y_1 and Y_2
$J(Y)$	negentropy of Y
$\mathbf{k}$	wavenumber
N	number of elements in a microphone array
$P(\theta, \phi)$	power pattern
$\mathbf{v_k}$	array manifold vector
$\mathbf{w}(\omega)$	beamformer weight vector
$\mathbf{w}_{\mathrm{q}}(\omega)$	generalized sidelobe canceler quiescent weight vector
$\mathbf{w}_{\mathrm{a}}(\omega)$	generalized sidelobe canceler active weight vector
u_x, u_y, u_z	direction cosines for the Cartesian coordinates

14

Hands On

In this chapter, we present empirical results demonstrating the effectiveness of many of the techniques for feature enhancement, speaker tracking, beamforming and speaker adaptation described in this volume. These empirical studies were undertaken on data captured with *real* speakers in *real* acoustic environments. As such, the results of the experiments reported here are vastly different from those reported in the great majority of the array processing, *independent component analysis* (ICA) and *blind source separation* (BSS) literature, where, if word error rates are given at all, it is common practice to report results on data that was originally captured with a *close-talking microphone* (CTM), and thereafter artificially convolved with a measured room impulse response, perhaps with the addition of noise. In the experience of the authors, the results obtained on such data almost invariably fail to carry over to data captured with real far-field sensors in real acoustic environments. Moreover, it is not difficult to understand why this is the case: If the impulse response of a room is fixed, it can be learned and compensated for provided sufficient adaptation data, typically a few seconds, is available. The great difficulty in working with real data is that the impulse response between the speaker's mouth and each individual element in a sensor array changes constantly. The opening of a door or window – or even the motion of the speaker's head by a few centimeters – is sufficient to radically alter this impulse response. This means that most, if not all, known methods of explicitly estimating the room impulse response will fail to converge on data captured in normal acoustic environments.

While the facts outlined above are well known to nearly everyone who has conducted research in the array processing, ICA, or BSS fields, they are all but unknown to the great majority of the *automatic speech recognition* (ASR) research community. This lack of knowledge is undoubtedly due in large part to the fact that the *distant speech recognition* (DSR) performance gap between experiments conducted on real and artificially convolved data is very seldom mentioned or even alluded to in publications about array processing, ICA, or BSS. In the experience of the present authors, highly experienced colleagues from the mainstream ASR community are uniformly surprised to learn that such a performance gap even exists.

For precisely the reasons outlined above, we report results on no such synthetic data here. Rather, the data we have used during our evaluations of the algorithms was captured with real microphones, in completely normal acoustic environments, and with real human speakers. Moreover, the speakers received no special instructions about how to speak or otherwise behave. The results we report in this chapter demonstrate that the algorithms described in this volume provide state-of-the-art performance. We have indeed documented this fact by either providing comparisons with other algorithms that have been proposed in the literature, or by referring to the results of open, international evaluation campaigns such as the Speech Separation Challenge, Part 2, or the CLEAR evaluations of audio-visual technologies sponsored by the *Computers in the Human Interaction Loop* (CHIL), project. That said, we must hastily add the following: Although the algorithms for which results are reported here are, as of the time of this writing, the best among those that have been proposed in the literature, we can make no guarantee that they will remain the best in several years' time, or even after the passage of a few months. Indeed, our great hope in writing this volume was and remains that the current state-of-the-art will soon be rendered obsolete, both through our own continuing efforts and those of our readers.

We now summarize the remainder of this chapter. In Section 14.1, we describe two realistic acoustic environments used to capture much of the data for the experiments reported here. The first environment was an instrumented seminar room built at the Universität Karlsruhe (TH) during the CHIL project. The second was the instrumented meeting room created at the *Center for Speech Technology Research* (CSTR) at the University of Edinburgh in connection with the *Augmented Multi-party Interaction* (AMI) and *Augmented Multi-party Interaction with Distance Access* (AMIDA) projects. Section 14.2 describes the decoding configurations of the two ASR engines used for the experiments reported here, namely, the Janus Recognition Toolkit, as well as Millennium. Our principal performance metric, word error rate, is defined in Section 14.3. Section 14.4 describes a set of feature enhancement experiments conducted with a single distant microphone. Acoustic and audio-video speaker-tracking experiments are described in Sections 14.5 and 14.6, respectively. For the experiments reported in those sections, the performance metric was tracking accuracy rather than word error rate, while the results reported in Section 14.7 demonstrate the close relation between tracking accuracy and recognition performance for DSR systems. Sections 14.8 and 14.9 report the results of a series of DSR experiments comparing the performance of several beamforming algorithms, both for a single speaker as well as for two simultaneous speakers. Section 14.10 presents a comparison of several filter bank designs for the speech separation application described in Section 14.9. Finally, Section 14.11 summarizes the results reported here a number of references for further reading.

14.1 Example Room Configurations

Here we describe the two example room configurations depicted in Figure 14.1. The first configuration is a seminar room of size $7.1 \times 5.9 \times 3$ m and a reverberation time of approximately 410 ms; the second is a meeting room with dimensions $6.5 \times 4.9 \times 3.25$ m and a reverberation time of approximately 380 ms. Most of the speaker-tracking and DSR experiments described in subsequent sections were conducted with data collected in one of these rooms.

The seminar room located at Universität Karlsruhe (TH) is equipped with a 64-channel Mark III microphone array developed at US National Institute of Standards and

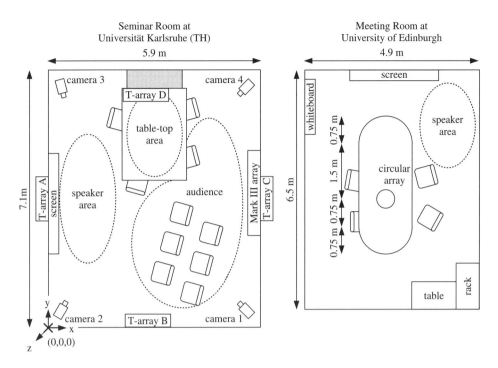

Figure 14.1 Sensor configurations in the instrumented seminar and meeting rooms at the Universität Karlsruhe (TH) and University of Edinburgh

Technologies (NIST). This large array, which was mounted on the wall opposite the lecturer, was intended primarily for beamforming. For speaker tracking, four T-shaped arrays were mounted on the walls of the seminar room. The T-shape permits three-dimensional tracking, which would be impossible with a linear configuration. The room was also equipped with commercially-available microphones, which were located on a table next to the presenter. In addition to the audio sensors, calibrated video cameras were located in each corner of the room near the ceiling. These cameras, which had been calibrated with the technique proposed by Zhang (2000), were used for audio-video speaker-tracking experiments, as well as for determining the true speaker positions. The location of the centroid of the speaker's head in the images from the four calibrated cameras was manually marked every 0.7 second. Using these hand-marked labels, the true position of the speaker's head in three-dimensional room coordinates was calculated using the technique described in Zhang (2000). These "ground truth" speaker positions were accurate to within approximately 10 cm. For details of the data collection apparatus see Macho *et al.* (2005).

The meeting room located at the CSTR was equipped with two circular, eight-channel microphone arrays with diameters of 20 cm. Video data from calibrated cameras was not captured in the CSTR meeting room. The data from the eight-channel circular arrays was, however, ideally-suited for the single-speaker beamforming and speech separation experiments described in Sections 14.8 and 14.9, respectively. Details of the data collection apparatus at the CSTR are available from Lincoln *et al.* (2005).

A precise description of the sensor and room configuration of both rooms is provided in Figure 14.1. Additionally, in both room configurations each active speaker was equipped with a CTM to capture the best possible speech signal as a reference for DSR experiments. In the CHIL seminar room, the sampling rate for the Mark III was 44.1 kHz, while that for the T-arrays, table top microphones, and CTMs was 48 kHz. After beamforming or other processing, the data was downsampled to 16 kHz prior to feature extraction and recognition. The data captured in the meeting room at the University of Edinburgh was sampled at a rate of 16 kHz, and all processing was conducted at this rate.

As both data sets were captured with real speakers in real acoustic environments, the recordings contain significant distortions from the three banes of DSR, namely, noise, reverberation and overlapping speech. Moreover, the speaking volume, head orientation and positions of the speakers changed constantly in both environments, which provided an additional challenge for effective DSR.

14.2 Automatic Speech Recognition Engines

All experiments described in the following sections were conducted with either the *Janus Recognition Toolkit*, which is developed and maintained jointly at Universität Karlsruhe (TH), in Karlsruhe, Germany and at Carnegie Mellon University in Pittsburgh, Pennsylvania, USA, or with *Millennium*, which was formerly developed at Universität Karlsruhe (TH). Millennium is structured as a set of dynamically-linkable shared-object libraries that can be accessed from the interpreter of the Python scripting language. These capabilities are contained in four modules: Automatic Speech Recognition (`asr`), the Source Localization Toolkit (`sltk`), the Beamforming Toolkit (`btk`) and Speech Feature Extraction (`sfe`).

The feature extraction used for the DSR experiments reported here was based on cepstral features estimated with a warped *minimum variance distortionless response* (MVDR) spectral envelope of model order 30, as described in Section 5.3.7. Front-end analysis involved extracting 20 cepstral coefficients per frame of speech, as explained in Section 5.4, and performing global *cepstral mean normalization* (CMN) as discussed in Section 6.6.1. The final features were obtained by concatenating 15 consecutive frames of cepstral coefficients together, as indicated in Section 5.6, then performing *linear discriminant analysis* (LDA), as described in Section 5.7.2, to obtain a feature of length 42. The LDA transformation was followed by a global semi-tied covariance transform estimated with a maximum likelihood criterion (Gales 1999).

For the experiments using the Janus recognition engine reported in subsequent sections, the SRILM-toolkit (Stolcke 2002) was used to train either tri- or four-gram language models with vocabulary sizes between 20 and 60 thousand words based on the modified Kneser and Ney discounting method proposed by Chen and Goodman (1998). The basic Kneser and Ney approach for modeling the backoff probabilities of a N-gram language model was presented in Section 7.3.1. The language models were trained on a subset of the seminar data collected by the CHIL partner sites. Additional corpora were used for *language model* (LM) training, such as conference proceedings, the *translanguage English database*, as well as data culled from the Internet through key word searches using Google. The perplexities of the language models on the development sets were between 100 and 150, and the out-of-vocabulary-rates were below 1% for all cases considered.

For the experiments with Millennium, the standard 5000-word vocabulary *Wall Street Journal* (WSJ) LM was used. The WSJ model was converted to a weighted-finite state transducer. For these experiments, acoustic training was performed on approximately 30 hours of American WSJ and 12 hours of Cambridge WSJ data. The latter dataset was necessary in order to provide coverage of the British accents for the speakers in the *Speech Separation Challenge* (SSC) development and evaluation sets (Lincoln *et al.* 2005).

The triphone or quinphone acoustic models used for the Janus system had between 3500 and 4000 context-dependent codebooks with up to 64 Gaussians with diagonal covariances each. Different *hidden Markov model* (HMM) training schemes were used to train the systems required for the several decoding passes:

- conventional *maximum likelihood* (ML) HMM training, as described in Section 8.1.2;
- conventional *maximum mutual information* (MaxMI) HMM training, as described in Section 8.2.1;
- speaker-adapted training under a ML criterion, as described in Section 8.1.3;
- speaker-adapted training under a MaxMI criterion, as described in Section 8.2.4.

The speech recognition experiments in Section 14.4 and Section 14.7 were conducted with the one pass decoder implemented in the Janus system (Soltau *et al.* 2001). The speech recognition experiments reported in Sections 14.8, 14.9 and 14.10 were conducted with a word trace recognizer in the Millennium system, as described in Section 7.1.4. For the latter, word lattices for speaker adaptation were written during recognition as described in Section 7.1.3, and then weighted finite-state transducer operations described in Section 7.2 were used to produce minimum equivalent lattices; i.e., the raw lattice from the decoder was projected onto the output side to discard all arc information save for the word identities, and then compacted through epsilon removal, determinization and minimization.

Either two or four decoding passes were performed on the waveforms obtained either by a single microphone or with a microphone array applying beamforming algorithms as indicated in the particular section. Each pass of decoding used a different speaker adaptation scheme or acoustic model, while the same language model was used for all passes.

In the Janus system, the adaptation parameters were estimated on the first best hypothesis, while in the four-pass system based on Millennium the adaptation parameters were estimated using the word lattices generated during the prior pass.

The processing steps for the two-pass decoding strategy used for experiments with the Janus system can be summarized as follows:

1. Decode with the unadapted, conventional MaxMI acoustic model.
2. Estimate *vocal tract length normalization* (VTLN) parameters, as described in Section 9.1.1, *constrained maximum likelihood linear regression* (CMLLR) parameters, as described in Section 9.1.2, and *maximum likelihood linear regression* (MLLR) parameters, as described in Section 9.2.1, for each speaker.
3. Redecode with the conventional MaxMI model.

The processing steps for the four passes of decoding used with Millennium can be summarized as follows:

1. Decode with the unadapted, conventional ML acoustic model.
2. Estimate VTLN and CMLLR parameters for each speaker.
3. Redecode with the conventional ML acoustic model.
4. Estimate VTLN, CMLLR and MLLR parameters for each speaker.
5. Redecode with the conventional model.
6. With the ML-*speaker-adapted training* (SAT) acoustic model, estimate VTLN, CMLLR, and MLLR parameters for each speaker.
7. Redecode with the ML–SAT model trained as described in Section 8.1.3.

14.3 Word Error Rate

Word error rate (WER) is the metric of first choice for determining the quality of automatically derived speech transcriptions. As mentioned in Section 1.3, transcription errors are typically grouped into three categories[1]:

- *insertion*: an extra word is added to the recognized word sequence;
- *substitution*: a correct word in the word sequence is replaced by an incorrect word;
- *deletion*: a correct word in the word sequence is omitted.

The minimum error rate can be determined by aligning the hypothesized word string with the correct reference string. This problem is known as *maximum substring matching* and can be solved by *dynamic programming* (Bellman 1957). After the alignment, the *word error rate* can readily calculated as

$$\text{WER} = \frac{\text{substitutions} + \text{deletions} + \text{insertions}}{\text{total number of word tokens in the reference}}.$$

In tabulating the results of the DSR experiments described below, the reference text was used to perform case-insensitive scoring. Case-sensitive scoring is sometimes used, however. The nonlexical tokens, such as breath or noise, were not evaluated in scoring.

Burger (2007) investigated the differences between close- and distant-talking microphone transcriptions. Burger found that in order to generate distant-talking microphone transcriptions from close-talking microphone transcriptions, transcribers had to remove an average of 4% of complete utterances, 2% of word tokens, 15% of word fragments and 12% of laughter annotations. The far-field transcriptions show an average of 60% more labels for nonidentifiable utterances and 19% more word tokens tagged as hard to identify. Burger did not investigate differences in the annotation of breath noise. Those values, of course, vary with the test set.

A very useful tool to measure the WER is provided by NIST and is freely available on the Internet (NIST).

---

[1] Note that changes in word order that do not change the meaning of a sentence are treated as errors under the WER metric. This makes WER unsuitable for evaluating automatic speech translation systems, for example.

14.4   Single-Channel Feature Enhancement Experiments

In order to emphasize the importance of speech feature enhancement in DSR under realistic conditions, this section first compares two different speech recognition front-ends and then compares different single-channel feature enhancement techniques on data recorded in the seminar room described in Section 14.1. The effect on recognition accuracy by compensation for different distortion types is emphasized by compensating for either additive or convolutive distortion. Thereafter results are presented which highlight that compensating for both kinds of distortion can lead to additional improvements, in particular when this compensation is performed jointly.

The GMM representing clean speech in the PF was trained with 64 Gaussians on clean speech. For the second pass experiments the GMM was trained on features to which vocal tract length normalization had been applied. The latter is described in Section 9.1.1. The noise GMM used to initialize the PF was trained for each individual utterance on silence regions found by voice activity detection. A dynamic autoregressive matrix (Section 6.7.4) was used to predict the evolution of noise.

We start our analysis by estimating the signal-to-additive-distortion (labeled with *additive*), signal-to-reverberation (labeled with *reverberation*) and signal-to-distortion (labeled with *overall*) ratio calculated within the joint estimation framework. Comparing the different estimates in Table 14.1 to the signal-to-noise (labeled with *SNR*) estimate based on voice activity detection we immediately observe that the distortion estimates are significantly higher within the joint estimation framework which becomes more pronounced for higher SNR values.

On the *close-talking microphone* (CTM) the energy estimates of additive distortions and the energy estimates of late reverberation are nearly alike. The distortion estimates of the lapel microphone are higher for late reverberations than for additive distortions which is also true for the table top microphone. The difference, however, between additive distortions and late reverberation energies is much smaller. On the wall-mounted microphone the energy estimates of additive distortions and late reverberation estimates again become nearly equivalent. In addition, we observe that the energy estimates of late reverberation only slightly increase between the lapel, table top and the wall-mounted microphone.

Figure 14.2 presents the average energy over all bands of the observed signal, the nonstationary additive distortion estimate, the late reverberation estimate and the enhanced signal estimate. Comparing the energies of the additive distortion and reverberation

Table 14.1 Average energy of additive nonstationary noise and reverberation vs cleaned speech estimate

Microphone	Close talk	Lapel	Table top	Wall
Distance	1 cm	20 cm	150–200 cm	300–400 cm
Estimate	Average energy vs Cleaned estimate (dB)			
SNR	24	23	17	10
Noise	15.1	13.7	12.0	11.3
Reverberation	15.5	11.6	11.5	11.1
Together	12.3	9.5	8.7	8.2

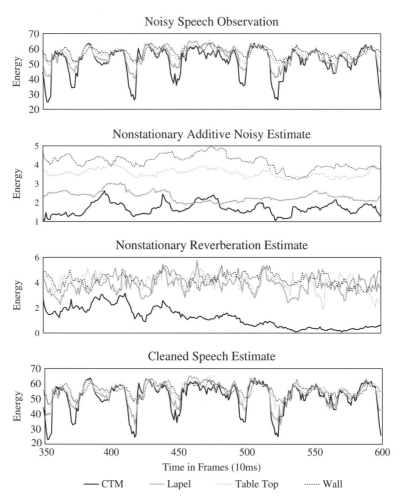

Figure 14.2   Average energy contours over all frequency bands vs time of the noisy speech frames, the noise frames, reverberant frames and cleaned speech frames

estimated over different frames we can clearly observe the time-dependent characteristics of the distortions. Furthermore, we note that the reverberation estimate has a significantly higher fluctuation than the additive distortion estimate, except for the case of the CTM.

Comparing the first two lines in Table 14.2 demonstrates that robust cepstral feature extraction, such as that based on the warped MVDR, outperforms the feature extraction based on the MFCC. Thus, warped MVDR feature extraction was used exclusively for the subsequent experiments. Comparing the second with the third lines, we observe that feature enhancement based on the *particle filter* (PF), as described in Chapter 6, improved the recognition performance in all tabulated cases. This comes as a little surprise, as it was not expected that the nearly clean close talk and lapel microphones can profit from enhancement techniques. Upon compensating for the reverberation using *multi-step linear prediction* (MSLP), as described in Section 6.6.5, a different picture emerges: With the

Table 14.2 Speech recognition experiments on single channel. After Wölfel (2008b)

Microphone			Close talk		Lapel		Table top		Wall	
Distance			5 cm		20 cm		150–200 cm		300–400 cm	
Pass			1	2	1	2	1	2	1	2
Front-end	Noise compensation		Word error rate (%)							
	Additive	Reverberant								
Power spectrum	no	no	11.3	9.5	12.3	10.3	18.0	14.2	45.9	30.0
Warped MVDR	no	no	11.2	9.1	10.9	9.2	18.6	14.0	45.4	28.6
Warped MVDR	yes	no	10.6	9.0	10.7	9.0	17.8	13.2	42.8	25.4
Warped MVDR	no	yes	14.4	9.5	15.1	9.6	17.7	13.4	39.2	23.9
Warped MVDR	yes	yes	12.1	9.3	13.4	9.5	17.7	13.3	38.3	23.3
Warped MVDR	joint		11.5	8.6	11.9	9.0	16.9	12.6	38.4	22.2

close-talk and lapel microphones, where no reverberation is expected, the word accuracy deteriorates if the acoustic models are not adapted. As is clear upon comparing the second pass results, this deterioration is less severe if unsupervised adaptation is performed. On the table top microphone, the reductions in WER are comparable to those of the PF. On the wall-mounted microphone, where more reverberation is expected, MSLP is able to significantly outperform the PF approach. Both the PF as well as the MSLP approaches are able to compensate for distortions which cannot be treated well by MLLR or CMLLR. This is apparent by comparing the second pass results. Applying both approaches, MSLP followed by PF, can either maintain or further reduce the error if the speech signal is significantly distorted. On the close talk and lapel microphones, the PF can compensate for some distortions introduced by MSLP. The last line in Table 14.2 presents results for a joint compensation approach. This approach provides equal or superior recognition performance on all channels after unsupervised model adaptation. Note that this is in contrast to a variety of feature enhancement techniques which are able to improve the accuracy on distorted signals. Such techniques, however, reduce accuracy if the signal does not contain distortions; e.g., the MSLP approach.

14.5 Acoustic Speaker-Tracking Experiments

This section reports the results of acoustic speaker-tracking experiments conducted on approximately three hours of audio and video data recorded in the seminar room described in Section 14.1. An additional hour of test data was recorded at Athens Institute of Technology in Athens, Greece, IBM at Yorktown Heights, New York, USA, Instituto Trentino di Cultura in Trento, Italy, and Universitat Politecnica de Catalunya in Barcelona, Spain. These recordings were made in connection with the European Union integrated project CHIL.

Acoustic speaker-tracking performance was evaluated on those portions of the seminars where only a single speaker was active. For these parts, it was determined whether the error between the ground truth and the estimated position was less than 50 cm. Any instance where the error exceeded this threshold was treated as a *false positive* and was not considered when calculating the *multiple object-tracking precision* (MOTP), which is defined as the average horizontal position error. If no estimate fell within 50 cm of

Table 14.3 Speaker-tracking performance for *iterated extended Kalman filter* (IEKF) and *joint probabilistic data association filter* (JPDAF) systems

Filter	Test set	MOTP (cm)	% Miss	% False positive	% MOTE
IEKF	Lecture	11.4	8.32	8.30	16.6
IEKF	Interactive	18.0	28.75	28.75	57.5
IEKF	Complete	12.1	10.37	10.35	20.7
JPDAF	Lecture	11.6	5.81	5.78	11.6
JPDAF	Interactive	17.7	19.60	19.60	39.2
JPDAF	Complete	12.3	7.19	7.16	14.3

the ground truth, it was treated as a *miss*. Letting N_{fp} and N_m, respectively, denote the total number of false positives and misses, the *multiple object-tracking error* (MOTE) is defined as $(N_{fp} + N_m)/N$, where N is the total number of ground truth positions. Performance was evaluated separately for the portion of the seminar during which only the lecturer spoke, and that during which the lecturer interacted with the audience. Shown in Table 14.3 are the experimental results reported in Gehrig *et al.* (2006); those authors used the multiple speaker-tracking algorithm described in Section 10.3, which is based on the *joint probabilistic data association filter* (JPDAF) discussed in Section 4.3.6.

These results clearly show that the JPDAF provided better tracking performance for both the lecture and interactive portions of the seminar. As one might expect, the reduction in MOTE was largest for the interactive portion, where multiple speakers were often simultaneously active.

Shown in Figure 14.3 are images from the four calibrated video cameras in the CHIL seminar room. The oval projected on the heads of the two speakers, one sitting and

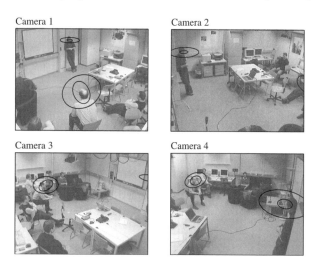

Figure 14.3 Multiple simultaneous speaker-tracking system. The ovals on the speakers' heads represent the uncertainty regions of the speakers' positions as determined by the state estimation error covariance matrices K_k and $K_{k|k-1}$ in (4.32) and (4.77), respectively. (© Photo reproduced by permission of Universität Karlsruhe (TH))

one standing next to the presentation screen, represent the regions of uncertainty of the speakers' positions estimated with the JPDAF algorithm. These regions of uncertainty are based on the predicted $\mathbf{K}_{k|k-1}$ and filtered $\mathbf{K}_k$ state estimation error covariance matrices in (4.32) and (4.77), respectively, that are propagated forward at every time step during the state estimate update of the JPDAF.

It is worth noting that the system for tracking multiple simultaneous speakers based on the JPDAF, as described in Section 10.3, provided the best performance on the acoustic speaker tracking task in both the 2006 and 2007 CLEAR Evaluations (Stiefelhagen *et al.* 2008). The other systems for which results were reported in those evaluations were based on PFs, as described in Section 10.5, or on the generalized coherence field, as proposed in Brutti *et al.* (2005). The speaker-tracking system based on the JPDAF also provided the acoustic speaker position estimates used by the system which achieved the best overall performance in the audio-video tracking portion of the CHIL evaluations.

14.6 Audio-Video Speaker-Tracking Experiments

Gehrig *et al.* (2005) used a test set of approximately 2.5 hours of audio and video data recorded during five seminars by students and faculty at the Universität Karlsruhe (TH) in order to evaluate the algorithms described in Section 10.4 for audio-video speaker tracking. As explained in Section 14.1, the video cameras in the Universität Karlsruhe (TH) seminar room enabled the true speaker positions in room coordinates to be extracted from manually-marked video images. These "ground truth" speaker positions were accurate to within 10 cm.

As the seminars took place in an open lab area, the layout of which is described in Section 14.1, used both by seminar participants as well as students and staff engaged in other activities, the recordings were optimally-suited for evaluating speaker-tracking and other technologies in a realistic, natural setting. In addition to speech from the seminar speaker, the far-field recordings contained noise from fans, computers, and doors, in addition to cross-talk from other people present in the room.

Table 14.4 shows the results reported by Gehrig *et al.* of a set of experiments that were made to compare the accuracy of a speaker-tracking system running in different modes. The columns in the table labeled X, Y and Z show the average error in speaker position for each dimension. The columns labeled 2D and 3D tabulate the root mean square error on the floor or $X-Y$ plane, and the entire $X-Y-Z$ space, respectively, as indicated on the left side of Figure 14.1. The audio-video experiment used the same

Table 14.4 Results of audio-only and audio-video speaker-tracking experiments

	Root mean square error (cm)				
Tracking mode	X	Y	Z	2D	3D
Audio only	46.7	43.5	22.8	65.1	69.4
Video only	101.5	119.3	24.4	162.6	164.6
Audio-video	41.4	36.9	12.5	56.0	58.6

parameters that were used to run the experiments on a single modality. To initialize the tracking algorithm, a common starting position was used for all seminars, so that the Kalman filter was forced to converge to the true position. The innovation sequence (4.21) of the Kalman filter was filtered using twice the standard deviation of the innovation covariance matrix (4.26) as a threshold, in order to remove outliers. The *iterated extended Kalman filter* (IEKF) described in Sections 4.3.3 and 10.2 was iterated at most fives times. Additionally, the position estimates returned by the Kalman filter were restricted to be within the physical room and the time delays to be within the bounds determined by the dimensions of the room. Moreover, a threshold of 0.18 on the maximum peak of the generalized cross-correlation, described in Section 10.1, for each microphone pair was set, and only those pairs that had correlation values that exceeded that threshold were used for *time delays of arrival* (TDOA) estimation. This was done to ensure that only the signals on the direct path from the speaker's mouth to each pair of microphones were used for TDOA estimation. As input for the audio-based speaker tracking, TDOAs estimated from all combinations of microphone pairs of the T-Arrays B and D on the left side of Figure 14.1 were used. The measurement noise (10.33) for the microphone pairs was set between 0.11 ms and 0.54 ms.

The detected positions of the speakers' faces in the image planes of Cameras 1 through 4 in Figure 14.1 were used as video features for the speaker-tracking experiments. The size of the face detector's search window was determined by the projection of a cube with an edge size of 50 cm. Additionally, the state estimation error of the Kalman filter projected onto the camera image planes was added to obtain a dynamic search window. The measurement noise of the cameras was approximately 20 pixels.

From the results reported in Table 14.4, it is clear that the audio-video speaker-tracking system provided performance superior to either modality taken individually. This was largely due to the fact that the two modalities had different and complementary failure modes. The audio-tracking system could obviously not track a speaker when he no longer spoke, and hence continually lost track during periods of silence. The video tracker, on the other hand, was unable to distinguish which face in the room was actually speaking, and hence often confused a person sitting silently in the audience for the person actually holding the seminar.

14.7 Speaker-Tracking Performance vs Word Error Rate

This section investigates the effect of tracking accuracy using a simple *delay-and-sum beamformer* (DSB) implemented in the subband domain. Once more the audio and video data were recorded in the CHIL seminar room described in Section 14.1. For these experiments, the data was captured with the 64-channel NIST Mark III. The signals from all channels were processed with the simple DSB.

As can be seen from Figure 14.4, although the video-only tracker performs considerably better than the audio-only tracker, the performance can still be significantly increased by combining both modalities. Wölfel *et al.* (2005) found that the video-only tracker had the same performance for all frames and speech-only frames, while the precision of the audio-only and the combined tracker was higher for the frames in which speech was present. This stands to reason given that, as discussed in the last section, the acoustic features could contribute nothing to the tracking accuracy whenever the desired speaker was silent.

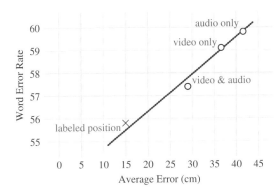

Figure 14.4 Plot comparing the average position error to the word error rate. After Wölfel *et al.* (2005)

The accuracy of a speaker-tracking system has a very significant influence on the recognition accuracy of the entire system. This can be easily observed from Figure 14.4 where the average position error of the source localization is compared to the WER. Assuming that the hand-labeled position error to the ground truth is around 15 cm (the calculated error is around 10 cm) then a nearly linear relationship between the source localization and word error becomes apparent.

14.8 Single-Speaker Beamforming Experiments

This section compares the performance in WER of several conventional and nonconventional beamforming algorithms. In order to make such comparisons on real data, Kumatani *et al.* (2008b) performed DSR experiments on the *Multi-Channel Wall Street Journal Audio Visual Corpus* which was collected by the AMI project in the meeting room described in Section 14.1. The far-field speech data used for the experiments reported in this section were recorded in the *stationary single-speaker* scenario. The speaker was asked to read sentences from six positions, four seated around the table in Seats 1–4 shown on the right side of Figure 14.1, one standing at the white board, and one standing at the presentation screen.

Prior to beamforming, Kumatani *et al.* first estimated the speaker's position relative to the microphone array with the JPDAF speaker-tracking system described in Section 10.2. Based on the average speaker position estimated for each utterance, utterance-dependent active weight vectors $\mathbf{w}_a$, as indicated in Figure 13.15, were estimated for the source. The active weight vectors for each subband were initialized to zero for estimation. Iterations of the conjugate gradients algorithm (Bertsekas 1995, sect. 1.6) were run over the entire utterance until convergence was achieved. The parameters of the *generalized Gaussian* (GG) pdf were trained as described in Section 13.5.2 with approximately 45 minutes of CTM speech data taken from the development set of the Speech Separation Challenge, Part II. Zelinski postfiltering (Simmer *et al.* 2001) was performed after all beamforming algorithms, which is a variant of the Wiener postfilter described in Sections 4.2 and 13.3.5.

Table 14.5 shows the WERs for every beamforming algorithm. The configurations of the recognition and speaker adaptation modules used for the individual decoding passes

Table 14.5 Word error rates for each beamforming algorithm after every decoding pass

Pass	1	2	3	4
Beamforming algorithm	Word error rate (%)			
DSB with postfilter	79.0	38.1	20.2	16.5
Minimum mean square error beamformer	78.6	35.4	18.8	14.8
Generalized eigenvector beamformer	78.7	35.5	18.6	14.5
Maximum negentropy beamformer with Γ pdf	75.6	34.9	19.8	15.8
Maximum negentropy beamformer with GG pdf	75.1	32.7	16.5	13.2
Single distant microphone	87.0	57.1	32.8	28.0
Close-talking microphone	52.9	21.5	9.8	6.7

reported in the figure were described in Section 14.2. As references, WERs in recognition experiments on speech data recorded with the single distant microphone and CTM are reported in Table 14.5. It is clear from Table 14.5 that the maximum negentropy beamforming algorithm provided better recognition performance than the simple DSB. It is also clear from Table 14.5 that maximum negentropy beamforming with the GG pdf assumption achieved the best recognition performance. This is because the GG pdf was able to model the subband samples of speech best, inasmuch as the shape factors were estimated for each subband independently, as described in Section 13.5.2.

Table 14.5 suggests that the Γ pdf assumption led to better noise suppression performance to some extent. The reduction in WER over the DSB was, however, limited because the Γ pdf could not model the subband components of speech as precisely as the GG pdf. DSR experiments were also performed on speech enhanced by the MVDR beamformer with Zelinski postfiltering, which, as shown in Section 13.3.5, is equivalent to the *minimum mean squared error* (MMSE) beamformer. It is clear from Table 14.5 that the MMSE beamformer provided better performance than DSB with postfilter. The MMSE beamformer could suppress correlated background noise, but could not enhance the target signal. On the other hand, as demonstrated in Section 13.5.2, the maximum negentropy beamforming algorithm can suppress both noise and interference, as well as strengthen the target signal by concentrating reflections solely based on the maximum negentropy criterion. Note that MMSE beamforming algorithms require speech activity detection in order to halt the adaptation of the weight vector when the desired speaker is active, and thereby avoid canceling the desired signal. For the adaptation of the MMSE beamformer, the first 0.1 and last 0.1 seconds in each utterance data, which contain only background noise, were used. In contrast to conventional beamforming methods, the maximum negentropy algorithm described in Section 13.5.2 did not require the start and end points of the target speech, as this method can suppress noise and reverberation without canceling the desired signal.

Table 14.5 also shows the recognition results obtained with the generalized eigenvector beamformer proposed by Warsitz *et al.* (2008). The latter algorithm achieved slightly better recognition performance than the MMSE beamformer. In this DSR task, the transfer function from the sound source to the microphone array changed in time due to movements of the speaker's head. Moreover, it was difficult to determine whether or not the signal

Table 14.6 Word error rates as a function
of the regularization parameter α

Pass	1	2	3	4
Value of	Word error rate (%)			
$\alpha = 0.0$	72.7	31.9	16.4	13.7
$\alpha = 10^{-3}$	73.9	32.2	16.6	13.6
$\alpha = 10^{-2}$	75.1	32.7	16.5	13.2
$\alpha = 10^{-1}$	76.2	32.5	17.5	13.5

observed at any given time contained both speech and noise components in each frequency bin, which is required to estimate the transfer function. Due to the difficulties inherent in this real acoustic environment, the performance improvement provided by the generalized eigenvector beamformer was limited.

Kumatani *et al.* (2008b) also examined the effect of the regularization term α in (13.173) on DSR performance. Table 14.6 shows WER as a function of the regularization parameter α. It is apparent from the table that the regularization parameter $\alpha = 10^{-2}$ provided the best result, although the overall impact on the recognition performance was slight. The requirement of a small α seems to imply that the input data are not sufficiently reliable to completely determine the active weight vector due, for example, to steering errors. As described in Section 13.3.8, a steering error occurs when the position estimated by a speaker-tracking system does not correspond to the speaker's true position.

14.9 Speech Separation Experiments

This section compares the performance in WER of the delay-and-sum and *minimum mutual information* (MinMI) beamformers in a DSR task where two speakers were speaking at the same time. The MinMI beamformer was described in Section 13.5.1. The data collection apparatus used to capture the data for the Speech Separation Challenge, Part 2 was the same as that previously described in Section 14.8. In this case, however, the dataset contained recordings of five pairs of speakers who were *simultaneously active*. Given that the room is reverberant and some recordings included significant amounts of noise in addition to the second speaker, it is obvious that this is a challenging source separation task.

Kumatani *et al.* (2007) and McDonough *et al.* (2008a) found that in addition to the speaker's position, the information when each speaker is active proved useful in segmenting the utterances of each speaker. The utterance spoken by one speaker was often much longer than that spoken simultaneously by the other. In the absence of perfect separation, running the speech recognizer over the entire waveform produced by the beamformer instead of only that portion where a given speaker was actually active would have resulted in significant insertion errors. These insertions would also have proven disastrous for the speaker adaptation algorithms described in Chapter 9, as the adaptation data from one speaker would have been contaminated with speech of the other speaker.

Table 14.7 shows WERs for every beamforming algorithm and speech recorded with the CTM after every decoding pass on the SSC development data. These results were obtained

Table 14.7 Word error rates for every beamforming algorithm after every decoding passes

Pass	1	2	3	4
Beamforming algorithm	Word error rate (%)			
Delay and Sum	85.1	77.6	72.5	70.4
MinMI: Gaussian	79.7	65.6	57.9	55.2
MinMI: Laplacian	81.1	67.9	59.3	53.8
MinMI: K_0	78.0	62.6	54.1	52.0
MinMI: Γ	80.3	63.0	56.2	53.8
CTM	37.1	24.8	23.0	21.6

with subband-domain beamforming where subband analysis and synthesis were performed with the perfect reconstruction cosine modulated filter bank described in Vaidyanathan (1993, sect. 8). After the fourth pass, the DSB had the worst recognition performance of 70.4% WER. The MinMI beamformer with a Gaussian pdf achieved a WER of 55.2%. The best performance of 52.0% WER was achieved with the MinMI beamformer by assuming that the subband samples were distributed according to the K_0 pdf. Under the Γ pdf, a WER of 53.8% was achieved. This performance was indeed better than that obtained under the Gaussian assumption, and equivalent to the WER under the Laplacian assumption pdf, but worse than that obtained with the K_0 pdf. The complex uni- and bivariates required for MinMI beamforming with super-Gaussian pdfs were calculated with the help of Meier G-function, as described in Appendix B.5.2.

14.10 Filter Bank Experiments

All of the beamforming algorithms described in Chapter 13, including the MinMI beamformer discussed in Section 13.5.4, operate in the frequency or subband domain. Hence, as previously explained in Section 11.1, the digital filter bank used for subband analysis and resynthesis is an important component of a complete DSR system. For the experiments reported in this section, four different filter bank designs were compared on the basis of the SSC, Part 2 data described earlier, including:

1. The *cosine modulated filter bank* described by Vaidyanathan (1993, sect. 8), which yields *perfect reconstruction* (PR) under optimal conditions. In such a filter bank, PR is achieved through *aliasing cancellation*, wherein the aliasing that is perforce present in one subband is canceled by the aliasing in all others. Aliasing cancellation breaks down if arbitrary complex factors are applied to the subband samples. For this reason, such a PR filter bank is not optimal for beamforming or adaptive filtering applications.
2. A *discrete Fourier transform* (DFT) filter bank based on overlap-add.
3. The uniform DFT filter bank proposed by De Haan described in Section 11.6, whereby separate analysis and synthesis prototypes are designed to minimize an error criterion consisting of a weighted combination of the total spectral response error and the aliasing distortion. This design is dependent on the use of oversampling to reduce aliasing error.

Table 14.8 DSR results on the speech separation challenge development data

Pass	1	2	3	4
Filter bank	Word error rate (%)			
Perfect reconstruction	87.7	65.2	54.0	50.7
Perfect reconstruction + postfilter + binary mask	87.1	66.6	55.7	52.5
DFT	88.5	71.1	58.8	55.5
De Haan	88.7	68.2	56.1	53.3
De Haan + postfilter + binary mask	82.7	57.7	42.7	39.6
Nyquist(M) + postfilter + binary mask	84.8	58.0	43.4	40.9

4. A uniform DFT design which differs from the De Haan filter bank in that a $Nyquist(M)$ constraint, as described in Section 11.5, is imposed on the prototype in order to ensure that the total response error vanishes. Thereafter, the remaining components of the prototype are chosen to minimize aliasing error, as with the De Haan design. As discussed in Section 11.7, the Nyquist(M) design is dependent on oversampling to reduce aliasing distortion, much like the De Haan design.

As reported by McDonough *et al.* (2008a) and Kumatani *et al.* (2008d), the WERs obtained with the four filter banks on the SSC development data are shown in Table 14.8. For these experiments, the Gaussian pdf was used exclusively. McDonough *et al.* and Kumatani *et al.* also investigated the effect of applying a Zelinski postfilter (Simmer *et al.* 2001) to the output of the beamformer in the subband domain, as well as the binary mask described in McCowan *et al.* (2005). The results indicate that the performance of a PR filter bank is actually quite competitive if no postfiltering or binary masking is applied to the output of the beamformer. For the PR design, performance *degrades* from 50.7% WER to 52.5% when such postfiltering and masking are applied, which is not surprising given that both will tend to destroy the aliasing cancellation on which this design is based. When postfiltering and masking are applied using either the De Haan or the Nyquist(M) designs, performance is greatly enhanced. With the De Haan design, the addition of postfiltering and masking reduced the WER from 53.5% to 39.6%. With postfiltering and masking, the Nyquist(M) design achieved a WER of 40.9%, which was very similar to that of the De Haan filter bank. For both the De Haan and Nyquist(M) designs, an oversampling factor of 8 was used. The simple DFT achieved significantly worse performance than all of the subband filter banks.

14.11 Summary and Further Reading

This chapter began with a brief statement of the importance of reporting results of experiments conducted on real acoustic data, captured from real speakers in real acoustic environments. Briefly, the extreme necessity of conducting and reporting the results of such experiments follows directly from the fact that only such experiments have any worth in assessing the true performance of a DSR system. Experiments conducted on data that has been artificially created through the convolution of data captured with a CTM with measured impulse responses, or through the addition of noise, have little or – more likely – no worth in predicting the performance of a DSR system under actual operating conditions. In Section 14.1, we described two realistic acoustic environments, as well as the sensor configurations used in each for far-field data capture. In Section 14.2, we described general recognizer configurations as were used in latter sections. Section 14.3 described the most important metric of the performance of a DSR system, namely, the word error rate. In Section 14.4, the results of several single-channel feature enhancement experiments based principally on particle filters were described. There, we learned that it is in fact straightforward to compensate for different types of distortions with various techniques. We also saw that compensating for several distortions jointly leads to improved DSR performance in contrast to their independent compensation. The results of a set of experiments designed to assess the performance of the speaker-tracking component in isolation from the rest of the DSR system were presented in Sections 14.5 and 14.6 for acoustic-only and audio-video tracking, respectively. Section 14.7 presented experimental results illustrating the overall effect of the speaker-tracking component on final word error rate. The results of beamforming experiments for both single speakers and two simultaneous speakers were presented in Sections 14.8 and 14.9. Finally, in Section 14.10, the effect of the filter bank design on DSR system performance was illustrated through a set of empirical studies.

The problems entailed in improving the overall performance of DSR systems for a wide variety of applications, as described in this chapter and throughout this volume, are the subjects of continuing research and study. This book by no means represents the last word toward this end, nor was it ever intended to. Rather, our intention in writing this work was to provide an accessible "snapshot" of the current state-of-the-art, as well as to assemble in one place many references into the wider literature in order to provide a starting point for further investigations and research. Those readers wishing to keep abreast of future progress in the field are advised to consult the technical journals and conference proceedings mentioned in Section 1.6. New publications by the current authors should appear on the companion website of this volume

```
http://www.distant-speech-recognition.org
```

in a timely fashion. Reference implementations and documentation of the algorithms described here can also be downloaded from the companion website.

Appendices

A

List of Abbreviations

AM	Acoustic Model
AMI	Augmented Multi-party Interaction
AMIDA	Augmented Multi-party Interaction with Distance Access
APT	All-Pass Transform
AR	AutoRegressive
ARMA	AutoRegressive Moving Average
ASR	Automatic Speech Recognition
BIBO	Bounded Input, Bounded Output
BLT	Bilinear Transformation
BSS	Blind Source Separation
CHIL	Computers in the Human Interaction Loop
CMLLR	Constrained Maximum Likelihood Linear Regression
CMN	Cepstral Mean Normalization
CTM	Close-Talking Microphone
DARPA	Defense Advanced Research Project Agency
dB	deciBel
DCT	Discrete Cosine Transform
DFS	Discrete Fourier Series
DFT	Discrete Fourier Transform
DOA	Direction Of Arrival
DSB	Delay-and-Sum Beamformer
DSR	Distant Speech Recognition
DSP	Digital Signal Processing
DTS	Discrete-Time System
EAA	European Acoustics Association
EKF	Extended Kalman Filter

Distant Speech Recognition   Matthias Wölfel and John McDonough
© 2009 John Wiley & Sons, Ltd

EM	Expectation–Maximization
FB	Filter Bank
FFT	Fast Fourier Transformation
FIR	Finite Impulse Response
FOI	Frequency Of Interest
FSA	Finite-State Automaton
FSG	Finite-State Grammar
FSM	Finite-State Machine
GCC	Generalized Cross-Correlation
GG	Generalized Gaussian
GMM	Gaussian Mixture Model
GSC	Generalized Sidelobe Canceling
GSS	Geometric Source Separation
HCI	Human–Computer Interaction
HMM	Hidden Markov Model
HMM-MNB	HMM Maximum Negentropy Beamformer
HOS	Higher Order Statistics
ICA	Independent Component Analysis
IEKF	Iterated Extended Kalman Filter
iff	if and only if
i.i.d.	independent identically distributed
IIR	Infinite Impulse Response
INR	Interference-to-Noise Ratio
IPA	International Phonetic Alphabet
JPDAF	Joint Probabilistic Data Association Filter
KF	Kalman Filter
LCBE	Logarithmic Critical Band Energies
LCMV	Linear Constraigned Minimum Variance
LDA	Linear Discriminant Analysis
LI	Linear Intersection
LM	Language Model
LMS	Least Mean - Square
LMSE	Least Mean Square Error
LP	Linear Prediction
LPC	Linear Prediction Coefficients
LTI	Linear Time Invariant
MA	Microphone Array or Moving Average
MFCC	Mel-Frequency Cepstral Coefficient

MI	Mutual Information
ML	Maximum Likelihood
MLB	Maximum Likelihood Beamformer
MLLR	Maximum Likelihood Linear Regression
MaxMI	Maximum Mutual Information
MinMI	Minimum Mutual Information
MMSE	Minimum Mean Squared Error
MNB	Maximum Negentropy Beamformer
MOTE	Multiple Object-Tracking Error
MOTP	Multiple Object-Tracking Precision
MPE	Minimum Phone Error
MSE	Mean Squared Error
MSLP	Multi-Step Linear Prediction
MVDR	Minimum Variance Distortionless Response
MWE	Minimum Word Error
NIST	National Institute of Standards and Technology
OOV	Out Of Vocabulary rate
ORC	Optimal Regression Class
PCA	Principal Component Analysis
pdf	probability density function
PDAF	Probabilistic Data Association Filter
PF	Particle Filter
PHAT	Phase Transformation
PLP	Perceptual Linear Predictive
PMF	Probability Mass Function
PR	Perfect Reconstruction
PSD	Power Spectral Density
RAM	Random Access Memory
RAPT	Rational All-Bass Transform
RASTA	RelAtive SpecTrA
RLS	Recursive Least Squares
RMS	Root Mean Square
ROC	Region of Convergence
ROVER	Recognizer Output Voting Error Reduction
RSR	Residual Systematic Resampling
RT	Rich Transcription
r.v.	random variable
SAT	Speaker-Adapted Training

SBRP	Steered Beamformer Response Power
SI	Speaker - Independent or Spherical Interpolation
SIA	Statistical Inference Approach
SIRP	Spherically Invariant Random Processes
SNR	Signal-to-Noise Ratio
SOS	Second-Order Statistics
SPL	Sound Pressure Level
SSC	Speech Separation Challenge
SR	Systematic Resampling
SRR	Signal-to-Reverberation Ratio
STC	Semi-Tied Covariance
SVD	Singular Value Decomposition
SX	Spherical Intersection
TDOA	Time Delays Of Arrival
TP	Turning Point
VAD	Voice Activity Detection
VTLN	Vocal Tract Length Normalization
VTS	Vector Taylor Series
WER	Word Error Rate
WFSA	Weighted Finite-State Acceptor
WFST	Weighted Finite-State Transducer
WSJ	*Wall Street Journal*

B

Useful Background

In this appendix we present a brief review of several topics that are useful for understanding the techniques described in the book, but which are somewhat outside the topic of speech processing or recognition. We also provide proofs of several results that would have disrupted the flow of the presentation in the main text. All of this material is standard and can be found in any one of a number of sources. We have included this material here only to save the reader the trouble of pulling out the appropriate reference.

B.1 Discrete Cosine Transform

The *discrete cosine transform* (DCT) has several applications in science and engineering. In particular, it is useful for solving certain partial differential equations under given boundary conditions. It also finds application for lossy data compression due to its good *energy compaction* properties. The latter refers to the fact that when audio signals or video images are encoded with the one- or two-dimensional DCTs, respectively, the most important information is encoded in the lowest coefficients. Hence, the low coefficients can be transmitted or stored with a relatively large number of bits, while the higher coefficients can be allocated fewer bits if not ignored entirely, without seriously compromising the fidelity of the reproduced signal or image. This use of the DCT is most similar to our intended application, whereby we seek to "encode" a log-power spectral density with as few coefficients as possible. For reasons which will shortly become apparent, this encoding will produce a cepstral sequence. Indeed, this encoding has another useful property in that truncating the cepstral sequence serves to eliminate the harmonic structure of the spectrum due to the fundamental frequency. In the automatic recognition of Western languages, the latter provides no information relevant to classification, and is thus typically eliminated.

Here we will consider the definition of two DCT matrices, the Type 2 matrix, which will prove useful in calculating cepstral sequences, and the Type 3 matrix, which will prove useful in that it can be used to obtain the inverse of the Type 2 matrix.

The components of the Type 2 DCT matrix $\mathbf{T}^{(2)}$ of dimension $N \times N$ are by definition

$$T_{m,n}^{(2)} \triangleq \cos \left[\frac{\pi}{N} \left(n + \frac{1}{2} \right) m \right] \; \forall \; m, n = 0, \ldots, N - 1. \tag{B.1}$$

Distant Speech Recognition Matthias Wölfel and John McDonough
© 2009 John Wiley & Sons, Ltd

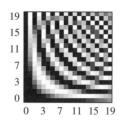

Figure B.1 Plot of the discrete cosine transformation matrix

The Type 2 DCT, which is typically referred to as simply the *discrete cosine transform*, is plotted in Figure B.1. Let us define the length-N vector $\mathbf{X}$ as the DCT of the input sequence $x[n] \,\forall\, k, n = 0, \ldots, N - 1$. The components of $\mathbf{X}$ can then be expressed as

$$X_m \triangleq \sum_{n=0}^{N-1} T_{m,n}^{(2)} \, x[n].$$

Given this definition, two important properties of the DCT come to light. Firstly, the sequence X_m will have even and odd symmetry about $m = 0$ and $m = N$, respectively. Secondly, X_m can be regarded as one-half of the DFT (3.61) of the length-$4N$ sequence given by

$$x'[n] = \begin{cases} x[(n-1)/2], & \text{for odd } n : 1 \leq n < 2N, \\ x[(4N-n)/2], & \text{for odd } n : 2N \leq n < 4N, \\ 0, & \text{for even } n. \end{cases}$$

This property enables X_m to be readily computed with the fast Fourier transform.

The Type 3 DCT matrix $\mathbf{T}^{(3)}$ has components

$$T_{m,n}^{(3)} \triangleq \begin{cases} \frac{1}{2}, & \text{for } n = 0, \\ \cos\left[\frac{\pi}{N}\left(m + \frac{1}{2}\right)n\right], & \text{otherwise.} \end{cases} \tag{B.2}$$

As mentioned previously, the Type 3 matrix is of interest because it enables the inverse of $\mathbf{T}^{(2)}$ to be calculated according to

$$\left(\mathbf{T}^{(2)}\right)^{-1} = \frac{2}{N}\,\mathbf{T}^{(3)}. \tag{B.3}$$

This property was used in deriving the hidden Markov model maximum negentropy beamformer in Section 13.5.3 and the dimension reduced logarithmic spectra as used in Section 10.6.

B.2 Matrix Inversion Lemma

We will now consider the *matrix inversion lemma*, which was applied in analyzing the performance of the minimum variance distortionless response in Section 13.3.3, and in deriving the recursive least squares beamformer in Section 13.4.3, among others. Let us make the following defintions: $\mathbf{A}$ is an $N \times N$ matrix, $\mathbf{B}$ is $N \times M$, $\mathbf{C}$ is $M \times M$, and $\mathbf{D}$ is $M \times N$. Assuming that the required inverses exist, the *matrix inversion lemma* states that

$$(\mathbf{A} + \mathbf{BCD})^{-1} = \mathbf{A}^{-1} - \mathbf{A}^{-1}\mathbf{B}(\mathbf{D}\mathbf{A}^{-1}\mathbf{B} + \mathbf{C}^{-1})^{-1}\mathbf{D}\mathbf{A}^{-1}. \qquad (\text{B.4})$$

A simplification of (B.4) is known as *Woodbury's identity*, which can be stated as

$$\left(\mathbf{A} + \mathbf{xx}^H\right)^{-1} = \mathbf{A}^{-1} - \frac{\mathbf{A}^{-1}\mathbf{xx}^H\mathbf{A}^{-1}}{1 + \mathbf{x}^H\mathbf{A}^{-1}\mathbf{x}}, \qquad (\text{B.5})$$

where $\mathbf{x}$ is a vector of length N. Several other useful relations following from (B.4) are:

$$\left(\mathbf{A}^{-1} + \mathbf{B}^H\mathbf{C}^{-1}\mathbf{B}\right)^{-1} = \mathbf{A} - \mathbf{AB}^H \left(\mathbf{BAB}^H + \mathbf{C}\right)^{-1} \mathbf{BA}, \qquad (\text{B.6})$$

$$\left(\mathbf{A}^{-1} + \mathbf{B}^H\mathbf{C}^{-1}\mathbf{B}\right)^{-1} \mathbf{B}^H\mathbf{C}^{-1} = \mathbf{AB}^H \left(\mathbf{BAB}^H + \mathbf{C}\right)^{-1}, \qquad (\text{B.7})$$

and

$$\mathbf{C}^{-1} - \left(\mathbf{BAB}^H + \mathbf{C}\right)^{-1} = \mathbf{C}^{-1}\mathbf{B} \left(\mathbf{A}^{-1} + \mathbf{B}^H\mathbf{C}^{-1}\mathbf{B}\right)^{-1} \mathbf{B}^H\mathbf{C}^{-1}. \qquad (\text{B.8})$$

B.3   Cholesky Decomposition

Given a symmetric positive-definite matrix $\mathbf{A}$, the Cholesky decomposition constructs a lower triangular matrix $\mathbf{L}$ such that $\mathbf{A} = \mathbf{LL}^T$. The matrix $\mathbf{L}$ is called the square root of $\mathbf{A}$.

The first column of $\mathbf{L}$ are determined by

$$l_{1,1} = \sqrt{a_{1,1}}, \, l_{2,1} = \frac{a_{1,2}}{l_{1,1}}, \cdots, l_{n,1} = \frac{a_{1,n}}{l_{1,1}}$$

With the first columns $i - 1$, the ith column can be determined by

$$l_{i,i} = \sqrt{a_{i,i} - \sum_{k=1}^{i-1} l_{i,k}^2}, \, l_{i+1,i} = \frac{a_{i,i+1} - \sum_{k=1}^{i-1} l_{i,k}l_{i+1,k}}{l_{i,i}}, \cdots, l_{n,i} = \frac{a_{in} - \sum_{k=1}^{i-1} l_{i,k}l_{n,k}}{l_{i,i}}.$$

B.4   Distance Measures

Distance measures are used to calculate the space between two classes. The different distance measures are defined by a combination of the class *means*

$$\mu_{(m)} = \frac{1}{N_m} \sum_{i=1}^{N_m} \mathbf{x}_{m,i} \qquad (\text{B.9})$$

and the *within-class variance* or *covariance matrix*

$$\boldsymbol{\Sigma}_{(m)} = \left(\mathbf{x}_{m,i} - \boldsymbol{\mu}_{(m)}\right)\left(\mathbf{x}_{m,i} - \boldsymbol{\mu}_{(m)}\right)^T. \tag{B.10}$$

The within-class variance represents how much the samples vary within a class. The covariance matrix itself can be classified into

- *spherical* – the covariance matrix is a scalar multiple of the identity matrix, $\boldsymbol{\Sigma}_{(m)} = \sigma_{(m)}^2 \mathbf{I}$,
- *diagonal* – the covariance matrix is diagonal, $\boldsymbol{\Sigma}_{(m)} = \mathrm{diag}(\sigma_{(m,1)}^2, \sigma_{(m,2)}^2, \ldots, \sigma_{(m,d)}^2)$,
- *full* – the covariance matrix is allowed to be any positive-definite matrix with rank $d \times d$.

With the class means and covariance matrices, the distance between the two classes $\Omega^{(p)}$ and $\Omega^{(q)}$ can be defined as

- the *Euclidean distance*

$$D_{(p,q)}^{\mathrm{Euclidean}} = \sqrt{(\boldsymbol{\mu}_{(p)} - \boldsymbol{\mu}_{(q)})^T (\boldsymbol{\mu}_{(p)} - \boldsymbol{\mu}_{(q)})}, \tag{B.11}$$

which is based upon the Pythagorean theorem
- the *Mahalanobis distance*

$$D_{(p,q)}^{\mathrm{Mahalanobis}} = \sqrt{(\boldsymbol{\mu}_{(p)} - \boldsymbol{\mu}_{(q)})^T \boldsymbol{\Sigma}_{(q)}^{-1} (\boldsymbol{\mu}_{(p)} - \boldsymbol{\mu}_{(q)})} \tag{B.12}$$

where $\boldsymbol{\Sigma}_{(p)}^{-1}$ represents the inverse of the covariance matrix of class $\Omega^{(p)}$. The Mahalanobis distance is therefore a weighted Euclidean distance where the weighting is determined by the range of variability of the sample point; expressed by the covariance matrix. The Mahalanobis distance can be extended to account for the variability of both classes $\boldsymbol{\Sigma}_{(p)}$ and $\boldsymbol{\Sigma}_{(q)}$ as

$$D_{(p,q)} = \sqrt{(\boldsymbol{\mu}_{(p)} - \boldsymbol{\mu}_{(q)})^T \left(\boldsymbol{\Sigma}_{(p)} + \boldsymbol{\Sigma}_{(q)}\right)^{-1} (\boldsymbol{\mu}_{(p)} - \boldsymbol{\mu}_{(q)})}. \tag{B.13}$$

- The *Kullback–Leibler distance* for Gaussians has already been defined in (B.46).
- the *Bhattacharya Distance*

$$D_{(p,q)}^{\mathrm{Bhattacharya}} = \frac{1}{4}(\boldsymbol{\mu}_{(p)} - \boldsymbol{\mu}_{(q)})^T \left(\boldsymbol{\Sigma}_{(p)} + \boldsymbol{\Sigma}_{(q)}\right)^{-1} (\boldsymbol{\mu}_{(p)} - \boldsymbol{\mu}_{(q)})$$

$$+ \frac{1}{2} \log \left(\frac{\left|\boldsymbol{\Sigma}_{(p)} + \boldsymbol{\Sigma}_{(q)}\right|}{2\sqrt{\left(\left|\boldsymbol{\Sigma}_{(p)}\right|\left|\boldsymbol{\Sigma}_{(q)}\right|\right)}}\right).$$

To measure the distance of more than two classes the distance measures introduced before have to be extended to a multiclass measure. One possibility is to sum over different classes under the consideration of the *a priori* probability $P(\Omega)$ of each class $\{\Omega\}_1^M$ where M is the number of classes

$$D_{\text{average}} = \sum_{m=1}^{M} \sum_{k=1}^{m-1} P(\Omega_m) P(\Omega_k) D_{(m,k)}. \tag{B.14}$$

B.5   Super-Gaussian Probability Density Functions

In this section, we present several super-Gaussian densities that have been shown to be useful in speech processing. As discussed in Section 12.2.3, *super-Gaussian* random variables are those with positive kurtosis, while those random variables with negative kurtosis are called *sub-Gaussian*.

B.5.1   Generalized Gaussian pdf

The *generalized Gaussian* (GG) pdf is well known and finds frequent application in the blind source separation and independent component analysis fields. Moreover, it subsumes the Gaussian and Laplace pdfs as special cases. This particular pdf is specified by three free parameters, a mean, a *scale factor* $\hat{\sigma}$, and a *shape factor* f. The GG pdf with zero mean for a real-valued r.v. y is by definition

$$p_{\text{GG}}(y) \triangleq \frac{1}{2\Gamma(1 + 1/f) B(f) \hat{\sigma}} \exp\left\{-\left|\frac{y}{B(f)\hat{\sigma}}\right|^f\right\}, \tag{B.15}$$

where

$$B(f) \triangleq \left[\frac{\Gamma(1/f)}{\Gamma(3/f)}\right]^{1/2}, \tag{B.16}$$

and $\Gamma(.)$ is the Gamma function (Luke 1969). As indicated in Figure 12.2, the shape factor f controls how fast the tail of the pdf decays. Note that the GG with $f = 1$ corresponds to the Laplacian pdf, and that $f = 2$ yields to the Gaussian pdf, whereas in the case of $f \to +\infty$ the GG pdf converges to a uniform distribution.

We will begin the derivation of the GG pdf of a circular complex random variable by assuming that $z = \rho\, e^{\phi}$ is such a random variable, which implies that the pdf of z is independent of ϕ and thus has the functional form

$$p(z) = p(\rho, \phi) = \frac{1}{c} \exp\left\{-\left[\frac{\rho}{\hat{\sigma}\, B_c(f)}\right]^f\right\},$$

where c is the normalization constant required to ensure that $p(z)$ is a valid pdf. Let us firstly calculate c. In polar coordinates a differential element of area ΔA can be expressed as

$$\Delta A = \rho\, d\rho\, d\phi.$$

Hence, the normalization constant must satisfy

$$
\begin{aligned}
c &= \int_{-\pi}^{\pi} \int_{0}^{\infty} \rho \exp\left\{-\left[\frac{\rho}{\hat{\sigma}\,B_{\mathrm{c}}(f)}\right]^{f}\right\} d\rho \, d\phi \\
&= 2\pi \int_{0}^{\infty} \rho \exp\left\{-\left[\frac{\rho}{\hat{\sigma}\,B_{\mathrm{c}}(f)}\right]^{f}\right\} d\rho.
\end{aligned}
\tag{B.17}
$$

Under the change of variables

$$
v = \frac{\rho}{\hat{\sigma}\,B_{\mathrm{c}}(f)},
\tag{B.18}
$$

(B.17) can be rewritten as

$$
c = 2\pi \hat{\sigma}^{2}\, B_{\mathrm{c}}^{2}(f) \int_{0}^{\infty} v \exp(-v^{f})\, dv.
\tag{B.19}
$$

Next we must calculate the variance of $z = \rho e^{j\phi}$, which is by definition

$$
\begin{aligned}
\sigma_{z}^{2} &\triangleq E\{|z|^{2}\} = \int_{-\pi}^{\pi} \int_{0}^{\infty} \rho\, e^{j\phi} \cdot \rho\, e^{-j\phi} \cdot p(\rho, \phi) \cdot \rho \, d\rho \, d\phi \\
&= \frac{2\pi}{c} \int_{0}^{\infty} \rho^{3} \exp\left\{-\left[\frac{\rho}{\hat{\sigma}\,B_{\mathrm{c}}(f)}\right]^{f}\right\} d\rho.
\end{aligned}
$$

Once more introducing the change of variables (B.18) provides

$$
\sigma_{z}^{2} = \frac{2\pi \hat{\sigma}^{4}\, B_{\mathrm{c}}^{4}(z)}{c} \int_{0}^{\infty} v^{3} \exp(-v^{f})\, dv.
\tag{B.20}
$$

The calculation of both c and σ_{z}^{2} in (B.19) and (B.20) involves integrals of the form

$$
I_{n}(f) = \int_{0}^{\infty} v^{n} \exp\left(-v^{f}\right) dv = \frac{1}{f} \cdot \Gamma\left(\frac{n+1}{f}\right) \quad \forall\, n = 1, 3.
\tag{B.21}
$$

Substituting (B.21) into (B.19) provides

$$
c = 2\pi \, \hat{\sigma}^{2}\, B_{\mathrm{c}}^{2}(z) \cdot \frac{1}{f} \cdot \Gamma\left(\frac{2}{f}\right).
\tag{B.22}
$$

Then substituting (B.21) and (B.22) into (B.20), we arrive at

$$
\sigma_{z}^{2} = \hat{\sigma}^{2}\, B_{\mathrm{c}}^{2}(f) \cdot \frac{\Gamma\left(4/f\right)}{\Gamma\left(2/f\right)}.
$$

Hence, in order to ensure that $\sigma_z^2 = \hat{\sigma}^2$, we must set

$$B_c(f) \triangleq \left[\frac{\Gamma(2/f)}{\Gamma(4/f)}\right]^{1/2}. \tag{B.23}$$

Moreover, the final GG pdf for circular, complex data can be expressed as

$$p_{gg}(z) \triangleq \frac{f}{2\pi \, \hat{\sigma}^2 \, B_c^2(f) \, \Gamma(2/f)} \exp\left\{-\left|\frac{z}{\hat{\sigma} \, B_c(f)}\right|^f\right\}. \tag{B.24}$$

Note that for the Gaussian case, $f = 2$ and the pdf reduces to

$$p_{Gauss}(z) = \frac{1}{\pi \hat{\sigma}^2} \exp\left\{-\left|\frac{z}{\hat{\sigma}}\right|^2\right\}, \tag{B.25}$$

which is the correct form for complex data (Neeser and Massey 1993). Similarly, for Laplacian random variables $f = 1$, and the pdf can be expressed as

$$p_{Laplace}(z) = \frac{3}{\pi \hat{\sigma}^2} \exp\left\{-\left|\frac{\sqrt{6}\,z}{\hat{\sigma}}\right|\right\}. \tag{B.26}$$

Among several methods for estimating the shape parameter f of the GG pdf (Kokkinakis and Nandi 2005; Varanasi and Aazhang 1989), the moment and *maximum likelihood* (ML) methods are arguably the most straightforward. Kumatani *et al.* (2008a) used the moment method to initialize the parameters of the GG pdf, and then updated them with the ML estimate (Varanasi and Aazhang 1989). The shape factors were estimated from training samples offline and were held fixed during the adaptation of the active weight vector $\mathbf{w}_a$, as described in Sections 13.5.2 and 13.5.3. The shape factor of each subband was estimated individually; hence, the optimal pdf was frequency-dependent.

Based on (B.24), the log-likelihood of the GG pdf for a complex r.v. can be expressed as

$$\log p(Y; f, \hat{\sigma}) = -\log\left\{2\pi \frac{1}{f} \Gamma(2/f) B_c^2(f) \hat{\sigma}^2\right\} - \frac{|Y|^f}{B_c^f(f) \hat{\sigma}^f}. \tag{B.27}$$

Then the derivative of $\log p(Y; f, \hat{\sigma})$ with respect to $\hat{\sigma}$ is given by

$$\frac{\partial \log p(Y; f, \hat{\sigma})}{\partial \hat{\sigma}} = \frac{f \, |Y|^f}{B_c^f(f) \hat{\sigma}^{f+1}} - \frac{2}{\hat{\sigma}}. \tag{B.28}$$

These relations were used in Section 13.5.2 in order to develop formulae for estimating optimal shape factors and active weight vectors.

B.5.2 Super-Gaussian pdfs with the Meier G-function

As explained in Brehm and Stammler (1987b), it is useful to make use of the fact that the Laplace, K_0, and Γ pdfs can be modeled with the Meijer G-functions for two principal reasons. Firstly, it implies that multivariates of all orders can be readily derived from the univariate pdf as soon as the covariance matrix is known. Secondly, such variants can be extended to the case of complex r.v.s, which is essential for our current development. In this section, we very briefly introduce the notation of Meijer G-functions, and their use in modeling *spherically invariant random processes* (SIRPs).

In this section, we very briefly introduce the notation of the *Meijer G-function*, along with the most important relations required to use G-functions to model super-Gaussian pdfs.

To denote the Meijer G-function, we will use one of the following equivalent forms

$$
G_{pq}^{mn}\left(z \left| \begin{matrix} a_p \\ b_q \end{matrix}\right.\right) = G_{pq}^{mn}\left(z \left| \begin{matrix} a_1, \ldots, a_p \\ b_1, \ldots, b_q \end{matrix}\right.\right)
$$

$$
= G_{pq}^{mn}\left(z \left| \begin{matrix} a_1, \ldots, a_n & | & a_{n+1}, \ldots, a_p \\ b_1, \ldots, b_m & | & b_{m+1}, \ldots, b_q \end{matrix}\right.\right).
$$

The G-function is defined by the contour integral

$$
G_{pq}^{mn}\left(x \left| \begin{matrix} a_1, \ldots, a_p \\ b_1, \ldots, b_q \end{matrix}\right.\right) = \frac{1}{2\pi i} \oint_{\Gamma_L} x^s ds \times \frac{\prod_{j=1}^{m} \Gamma(b_j - s) \prod_{j=1}^{n} \Gamma(1 - a_j + s)}{\prod_{j=n+1}^{p} \Gamma(a_j - s) \prod_{j=m+1}^{q} \Gamma(1 - b_j + s)},
$$

where Γ_L is a contour of integration defined as in Brehm and Stammler (1987b). The definition (B.29) implies

$$
G_{pq}^{mn}\left(z \left| \begin{matrix} a_p \\ b_q \end{matrix}\right.\right) = z^{-u} G_{pq}^{mn}\left(z \left| \begin{matrix} a_p + u \\ b_q + u \end{matrix}\right.\right) \tag{B.29}
$$

where $a_p + u$ and $b_q + u$ indicate that u is to be added to all $a_1, \ldots, a_p$ and all $b_1, \ldots, b_q$, respectively. To determine the normalizing constants of the several pdfs generated from the Meijer G-function, it will be useful to apply the *Mellin transform*

$$
M\{f(x); z\} = \int_0^\infty x^{z-1} f(x)\, dx. \tag{B.30}
$$

Under suitable conditions (Brehm and Stammler 1987b), the Mellin transform of a Meijer G-function can be expressed as

$$
M\left\{ G_{pq}^{mn}\left(z \left| \begin{matrix} a_p \\ b_q \end{matrix}\right.\right); z \right\} = \frac{\prod_{i=1}^{m} \Gamma(b_i + z) \prod_{i=1}^{n} \Gamma(1 - a_i - z)}{\prod_{i=1}^{m} \Gamma(1 - b_i - z) \prod_{i=1}^{n} \Gamma(a_i + z)}.
$$

We now show how G-functions can be used to represent SIRPs. To begin, we can express a univariate pdf of a SIRP as

$$p_1(x) = A G_{pq}^{mn} \left(\lambda x^2 \left| \begin{matrix} a_p \\ b_q \end{matrix} \right. \right) \tag{B.31}$$

for all $-\infty < x < \infty$. As can be verified by the Mellin transform relations (B.30–B.31), the normalization factor A and the constant λ, which assures unity variance, must be chosen according to

$$A = \lambda^{1/2} \frac{\prod_{i=m+1}^{q} \Gamma\left(\frac{1}{2} - b_i\right) \prod_{i=n+1}^{p} \Gamma\left(\frac{1}{2} + a_i\right)}{\prod_{i=i}^{m} \Gamma\left(\frac{1}{2} + b_i\right) \prod_{i=1}^{n} \Gamma\left(\frac{1}{2} - a_i\right)}, \tag{B.32}$$

$$\lambda = (-1)^{\epsilon} \frac{\prod_{i=1}^{q} \left(\frac{1}{2} + b_i\right)}{\prod_{i=1}^{p} \left(\frac{1}{2} + a_i\right)}, \qquad \epsilon = n - (q - m). \tag{B.33}$$

Let us now consider the G-function as $G_{p\,q}^{mn}(\lambda x^2 | b_1, b_2)$ for two real parameters b_1 and b_2. Brehm and Stammler (1987b) note that the subclass of SIRPs that are useful for modeling the statistics of speech can be expressed as

$$p_1(y) = A\, G_{0\,2}^{2\,0}(\lambda x^2 | b_1, b_2) \tag{B.34}$$

where

$$\lambda = \left(\frac{1}{2} + b_1\right)\left(\frac{1}{2} + b_2\right) \tag{B.35}$$

and

$$A = \frac{\lambda^{1/2}}{\Gamma\left(\frac{1}{2} + b_1\right) \Gamma\left(\frac{1}{2} + b_2\right)} \tag{B.36}$$

Table B.1 lists the values of these parameters for the Laplace, K_0 and Γ pdfs.

In general, the multivariate density of order v can also be expressed in terms of Meijer's G-functions according to (Brehm and Stammler 1987b)

$$p_v(\mathbf{x}) = \pi^{-v/2} f_v(s), \tag{B.37}$$

where

$$f_v = \pi^{1/2}\, A_v\, s^{(1-v)/2} \cdot G_{1\,3}^{3\,0} \left(\lambda_v s \left| \begin{matrix} 0 \\ \frac{1}{2}(v-1), b_1, b_2 \end{matrix} \right. \right) \tag{B.38}$$

Table B.1 Meijer G-function parameters for the Laplace, K_0 and Γ pdfs. After Brehm and Stammler (1987b)

pdf	$p(x)$	b_1	b_2	A	λ
Laplace	$\frac{1}{\sqrt{2}}e^{-\sqrt{2}\lvert x\rvert}$	0	$\frac{1}{2}$	$(2\pi)^{-1/2}$	$\frac{1}{2}$
K_0	$\frac{1}{\pi}K_0(\lvert x\rvert)$	0	0	$(2\pi)^{-1}$	$\frac{1}{4}$
Γ	$\frac{\sqrt{3}}{4\sqrt{\pi}}\left(\frac{\sqrt{3}\lvert x\rvert}{2}\right)^{-1/2}e^{-\sqrt{3}\lvert x\rvert/2}$	$-\frac{1}{4}$	$\frac{1}{4}$	$\frac{\sqrt{3/2}}{4\pi}$	$\frac{3}{16}$

and $s = \mathbf{x}^T\mathbf{x}$. In this case, equations (B.33–B.32) can be specialized as

$$\epsilon = 0,$$

$$\lambda_\nu = \nu \left(\tfrac{1}{2}+b_1\right)\left(\tfrac{1}{2}+b_2\right), \tag{B.39}$$

$$A_\nu = \lambda_\nu^{1/2}\frac{\Gamma\left(\tfrac{1}{2}\right)}{\Gamma\left(\tfrac{1}{2}\nu\right)\Gamma\left(\tfrac{1}{2}+b_1\right)\Gamma\left(\tfrac{1}{2}+b_2\right)}. \tag{B.40}$$

The bivariate is obtained by specializing (B.37–B.38) as,

$$p_2(\mathbf{x}) = \frac{A_2}{\sqrt{\pi s}}\cdot G_{1\,3}^{3\,0}\left(\lambda_2 s \left|\begin{array}{c} 0 \\ \tfrac{1}{2},b_1,b_2 \end{array}\right.\right). \tag{B.41}$$

For the moment, assume $\mathbf{x}$ is real-valued; this analysis will be extended to the case of complex $\mathbf{x}$ in Appendix B.5.2. If the components of $\mathbf{x}$ are correlated, we must set

$$s = \mathbf{x}^T\boldsymbol{\Sigma}_\mathbf{X}^{-1}\mathbf{x}$$

and modify (B.41) according to

$$p_2(\mathbf{x}) = \frac{A}{\sqrt{\pi s\lvert\boldsymbol{\Sigma}_\mathbf{X}\rvert}}\cdot G_{1\,3}^{3\,0}\left(\lambda s \left|\begin{array}{c} 0 \\ \tfrac{1}{2},b_1,b_2 \end{array}\right.\right). \tag{B.42}$$

where $\boldsymbol{\Sigma}_\mathbf{X} = \mathcal{E}\{\mathbf{X}\mathbf{X}^T\}$ is the covariance matrix of $\mathbf{X}$.
For the four-variate case, we have

$$p_4(\mathbf{x}) = \frac{A_4}{(\pi s)^{3/2}\lvert\boldsymbol{\Sigma}_\mathbf{X}\rvert^{1/2}}\cdot G_{1\,3}^{3\,0}\left(\lambda_4 s \left|\begin{array}{c} 0 \\ 1.5,b_1,b_2 \end{array}\right.\right). \tag{B.43}$$

Specializing the above we arrive at the bi- and four-variates shown in Table B.2.

Table B.2 Bi- and four-variates of the Laplace and K_0 pdfs

pdf	$\nu = 2$	$\nu = 4$
Laplacian	$\dfrac{1}{\sqrt{\pi\,\lvert\boldsymbol{\Sigma}_{\mathbf{X}}\rvert}}\,K_0(2\sqrt{s})$	$\dfrac{4\sqrt{2}}{\pi^{3/2}\,s^{1/2}\,\lvert\boldsymbol{\Sigma}_{\mathbf{X}}\rvert^{1/2}}\,K_1(2\sqrt{2s})$
K_0	$\dfrac{1}{\sqrt{2\pi\,s\,\lvert\boldsymbol{\Sigma}_{\mathbf{X}}\rvert}}\,e^{-\sqrt{2s}}$	$\dfrac{(1+2\sqrt{s})}{(\pi s)^{3/2}\lvert\boldsymbol{\Sigma}_{\mathbf{X}}\rvert^{1/2}}\,e^{-2\sqrt{s}}$

Complex Distributions

The multivariate pdfs derived thus far have been for real-valued random vectors. In order to extend this development for complex-valued subband samples, we make use of the following lemma.

Lemma B.5.1 *Let $\mathbf{X}_{\mathrm{c}}, \mathbf{X}_{\mathrm{s}} \in \mathbb{R}^N$ be two random vectors drawn from the same random process. Define the stacked vector*

$$\mathbf{X} = \begin{bmatrix} \mathbf{X}_{\mathrm{c}} \\ \mathbf{X}_{\mathrm{s}} \end{bmatrix}$$

and the covariance matrix of $\mathbf{X}$ as

$$\boldsymbol{\Sigma}_{\mathbf{X}} = \mathcal{E}\{\mathbf{X}\mathbf{X}^T\},$$

Now let

$$\mathbf{Y} = \mathbf{X}_{\mathrm{c}} + j\mathbf{X}_{\mathrm{s}}$$

and define the covariance of $\mathbf{Y}$ as

$$\boldsymbol{\Sigma}_{\mathbf{Y}} = \mathcal{E}\{\mathbf{Y}\mathbf{Y}^H\}.$$

Then,

$$\sqrt{\lvert\boldsymbol{\Sigma}_{\mathbf{X}}\rvert} = 2^{-N}\,\lvert\boldsymbol{\Sigma}_{\mathbf{Y}}\rvert, \tag{B.44}$$

$$\mathbf{x}^T\boldsymbol{\Sigma}_{\mathbf{X}}^{-1}\mathbf{x} = 2\,\mathbf{y}^H\boldsymbol{\Sigma}_{\mathbf{Y}}^{-1}\mathbf{y}. \tag{B.45}$$

This result is proven in the appendix of Kumatani *et al.* (2007).

Based on Lemma B.5.1, we arrive at the complex uni- and bi-variates shown in Table B.3.

Figure B.2 shows the region of the b_1-b_2 parameter space that yields valid pdfs, along with the points corresponding to the Laplace, K_0 and Γ pdfs (Brehm and Stammler 1987b).

Table B.3   Uni- and bi-variates of the complex Laplace and K_0 pdfs

pdf	$\nu = 1$	$\nu = 2$
Laplacian	$\dfrac{2}{\sqrt{\pi}\sigma_Y^2} K_0\left(\dfrac{2\sqrt{2}\,|Y|}{\sigma_Y}\right)$	$\dfrac{16}{\pi^{3/2}|\boldsymbol{\Sigma}_\mathbf{Y}|\sqrt{s}} K_1\left(4\sqrt{s}\right)$
K_0	$\dfrac{1}{\sqrt{\pi}\sigma_Y|Y|} e^{-2|Y|/\sigma_Y}$	$\dfrac{\sqrt{2}+4\sqrt{s}}{\pi^{3/2}|\boldsymbol{\Sigma}_\mathbf{Y}|\,s^{3/2}} e^{-2\sqrt{2}\,s}$

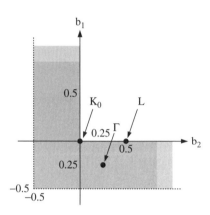

Figure B.2   The b_1–b_2 parameter space of $G_{0\,2}^{2\,0}(\lambda x^2|b_1, b_2)$. The gray region yields valid pdfs. (After Brehm and Stammler 1987b)

B.6   Entropy

The concept of *entropy* was first introduced in the now classic paper by Shannon (1948) and is defined by

$$H(X) = H(p) = -\sum_{x \in \mathcal{X}} p(x) \log_2 p(x).$$

It measures the uncertainty associated with a discrete r.v. x and thus quantifies the information contained in the data. Entropy can be measured in *bits* for the given equation. For the natural logarithms it is measured in *nits* while for the logarithm of base 10 it is measured in *hartleys*.

The *joint entropy* as defined by

$$H(X, Y) = -\sum_{x \in \mathcal{X}} \sum_{y \in \mathcal{Y}} p(x, y) \log_2 p(x, y)$$

measures how much entropy is contained in a system with two r.v.s.

The *conditional entropy*, or equivocation, as defined by

$$H(Y|X) = -\sum_{x \in \mathcal{X}} \sum_{y \in \mathcal{Y}} p(y, x) \log p(y|x) = H(Y, X) - H(X)$$

quantifies the uncertainty of a r.v. Y given that the value of a second r.v. X is known. Conditional entropy is zero iff the value of Y is completely determined by the value of X. Conversely, $H(Y|X) = H(Y)$ iff Y and X are independent r.v.s.

B.7   Relative Entropy

The *relative entropy* or *Kulback–Leibler distance* (Kullback and Leibler 1951) defines the similarity of two probability mass functions p and q as

$$D(p|q) = \sum_{x \in \mathcal{X}} p(x) \log \left(\frac{p(x)}{q(x)} \right).$$

Typically p represents data, observations or a precisely calculated probability distribution. The measure q typically represents a theory, a model, a description or an approximation of p. From the equation it is obvious that the Kulback–Leibler divergence is *not* symmetric. Kullback and Leibler themselves have defined the divergence as the symmetrical measure

$$D_S(p|q) = \frac{1}{2} \left(D(p|q) + D(q|p) \right).$$

The Kulback–Leibler distance can be solved in close form if both distributions are assumed to be Gaussian by

$$D(p|q) = \frac{1}{2} \left(\log \frac{|\Sigma_{(q)}|}{|\Sigma_{(p)}|} + \text{trace}\,\Sigma_{(q)}^{-1}\Sigma_{(p)} + \sqrt{(\mu_{(p)} - \mu_{(q)})^T \Sigma_{(q)}^{-1}(\mu_{(p)} - \mu_{(q)})} - d \right)$$

$$\text{(B.46)}$$

where d is the dimensionality.

B.8   Transformation Law of Probabilities

The pdf $p_y(u)$ of a random variable $y = f(x)$ might be a transformation with the deterministic function $f(x)$ of another random variable x with pdf $p_x(u)$. Then, provided the derivative of $f^{-1}(u)$ with respect to u is well defined, $p_y(u)$ can be calculated by the *fundamental transformation law of probabilities* as

$$p_y(u) = \left| \frac{df^{-1}(u)}{du} \right| p_x \left(f^{-1}(u) \right) \qquad \text{(B.47)}$$

if x and y are scalars. If $\mathbf{x}$ and $\mathbf{y}$ are vectors we have to use the Jacobian determinant and get

$$p_{\mathbf{y}}(u) = \left| \det \frac{df^{-1}(u)}{du} \right| p_{\mathbf{x}}\left(f^{-1}(u)\right) \tag{B.48}$$

This relationship holds since the probability density $p_y(u)$ is just the derivative of its cumulative distribution function

$$P_y(u') = \int_{-\infty}^{u'} p_y(u)du = \int_{f(-\infty)}^{f(u')} \frac{df^{-1}(u)}{du} p_x\left(f^{-1}(u)\right) du$$

Note that the absolute value of the derivative has to be taken, since probability densities are supposed to be ≥ 0.

B.9 Cascade of Warping Stages

We show that two stages of bilinear transformation with the warping parameter α_1 and α_1 are equivalent to applying one stage. To do so, let z be the variable in the original space,

$$s^{-1} = \frac{z^{-1} - \alpha_1}{1 - \alpha_1 z^{-1}} \tag{B.49}$$

the variable after one transformation and

$$u^{-1} = \frac{s^{-1} - \alpha_2}{1 - \alpha_2 s^{-1}} \tag{B.50}$$

the variable after the second bilinear transformation.
  With (B.49) and (B.50) we obtain

$$u^{-1} = \frac{\frac{z^{-1} - \alpha_1}{1 - \alpha_1 z^{-1}} - \alpha_2}{1 - \alpha_2 \frac{z^{-1} - \alpha_1}{1 - \alpha_1 z^{-1}}} = \frac{(1 + \alpha_1\alpha_2)z^{-1} - (\alpha_1 + \alpha_2)}{1 + \alpha_1\alpha_2 - (\alpha_1 + \alpha_2)z^{-1}} = \frac{z^{-1} - \alpha}{1 - \alpha z^{-1}}$$

where

$$\alpha = \frac{\alpha_1 + \alpha_2}{1 + \alpha_1\alpha_2}.$$

B.10 Taylor Series

The *Taylor series* of a function $f(x)$ that is infinitely differentiable in the neighborhood a, is defined as the power series

$$f(x) = \sum_{n=0}^{\infty} \frac{f^{(n)}(a)}{n!}(x - a)^n$$

$$= f(a) + \frac{f'(a)}{1!}(x - a) + \frac{f''(a)}{2!}(x - a)^2 + \cdots + \frac{f^{(n)}(a)}{n!}(x - a)^n + \cdots$$

The partial sums (the Taylor polynomials) of the Taylor series can be used to approximate the value of an entire function in every point for sufficiently many terms.

B.11   Correlation and Covariance

Correlation is a statistical measurement, ranging between -1 and $+1$, of the degree to which the movements of two variables are related. A negative value indicates an inverse relationship meaning that if one variable goes up, the other goes down. A positive value indicates a movement in the same direction. The higher the correlation, the closer is the value to either -1 or $+1$. If the variables are *independent* of each other the correlation is zero. Cross-correlation characterizes the causal relationship between two vectors x and y.

The *cross-correlation* matrix is defined as

$$\mathbf{R}_{xy} = \mathbf{R}_{xy} = \mathcal{E}\left\{\mathbf{xy}^T\right\}.$$

The *cross-covariance* matrix is defined with $\boldsymbol{\mu}_x = \mathcal{E}\left\{\mathbf{x}\right\}$ and $\boldsymbol{\mu}_y = \mathcal{E}\left\{\mathbf{y}\right\}$ as

$$\boldsymbol{\Sigma}_{xy} = \boldsymbol{\Sigma}_{yx} = \mathcal{E}\left\{\left(\mathbf{x} - \boldsymbol{\mu}_x\right)\left(\mathbf{y} - \boldsymbol{\mu}_y\right)^T\right\} = \mathbf{R}_{xy} - \boldsymbol{\mu}_x\boldsymbol{\mu}_y^T.$$

Similar the *auto-correlation* matrix $\mathbf{R}_{xx}$ and the *auto-covariance* matrix $\boldsymbol{\Sigma}_{xx}$ are defined as special cases of the cross-correlation and cross-covariance matrices respectively, comparing different sections of the same signal x. The auto-correlation can be interpreted as the degree of randomness in a signal.

B.12   Bessel Functions

Bessel functions[1] of the first kind $J_n(x)$, defined by Daniel Bernoulli and generalized by Friedrich Bessel, are defined as the solutions to Bessel's differential equation

$$x^2\frac{d^2y}{dx^2} + x\frac{dy}{dx} + (x^2 - \alpha^2)y = 0$$

which are nonsingular at the origin. Alternatively, they can also be defined by the contour integral

$$J_n(z) = \frac{1}{2\pi i}\oint e^{((z/2)(t-1/t))t^{(-n-1)}}\,dt$$

where the contour encloses the origin and is traversed in a counterclockwise direction.

Assuming that n is integer it is possible to define the function by its Taylor series expansion around $x = 0$ by

$$J_\alpha(x) = \sum_{m=0}^{\infty}\frac{(-1)^m}{m!\,\Gamma(m + \alpha + 1)}\left(\frac{x}{2}\right)^{2m+\alpha}$$

---

[1] Bessel functions are also known as cylinder functions or cylindrical harmonics.

where

$$\Gamma(n) = (n - 1)!$$

is the *Gamma function*.

A reasonable approximation of $J_0(x)$ is given by

$$J_0(x) = \frac{1}{\sqrt{2\pi x}} e^x$$

for large values, say $x > 5$, while it is a very bad approximation for small values.

B.13   Proof of the Nyquist–Shannon Sampling Theorem

Applying the continuous-time inverse Fourier transform (3.49) at time $t = nT_0$, we have

$$x(nT_0) = \frac{1}{2\pi} \int_{-\infty}^{\infty} X(\omega) e^{-j\omega n T_0} d\omega. \tag{B.51}$$

The infinite integral in (B.51) can be decomposed into an infinite sum of finite integrals, each of length $2\pi T_0$, such that

$$x(nT_0) = \frac{1}{2\pi} \sum_{k=-\infty}^{\infty} \int_{(2k-1)\pi/T_0}^{(2k+1)\pi/T_0} X(\omega) e^{-j\omega n T_0} d\omega.$$

Applying the variable substitution $\omega = \omega' - 2\pi/T_0$, the last equation becomes

$$x(nT_0) = \frac{1}{2\pi} \sum_{k=-\infty}^{\infty} \int_{-\pi/T_0}^{\pi/T_0} X\left(\omega' - \frac{2\pi k}{T_0}\right) \exp\left[j\left(\omega' - \frac{2\pi k}{T_0}\right) n T_0\right] d\omega'$$

$$= \frac{1}{2\pi} \sum_{k=-\infty}^{\infty} \int_{-\pi/T_0}^{\pi/T_0} X\left(\omega' - \frac{2\pi k}{T_0}\right) e^{j\omega' n T_0} d\omega'.$$

Using another variable substitution, $\omega' = \theta/T_0$ and exchanging the order of summation and integration, we find

$$x(nT_0) = \frac{1}{2\pi} \int_{-\pi}^{\pi} \left[\frac{1}{T_0} \sum_{k=-\infty}^{\infty} X\left(\frac{\theta - 2\pi k}{T_0}\right)\right] e^{j\theta n} d\theta. \tag{B.52}$$

Upon comparing (B.52) with (3.15), we realize that the term within brackets is the discrete-time Fourier transform of $x[n] = x(nT_0)$.

B.14   Proof of Equations (11.31–11.32)

Repeating (11.27) we can express the Type 1 polyphase representation of the filter function $H(z)$ as

$$H(z) = c + z^{-1} E_1(z^M) + \cdots + z^{-(M-1)} E_{M-1}(z^M), \tag{B.53}$$

where $\{E_i(z)\}$ are the Type 1 polyphase components. If we now sum over all subbands, we obtain

$$\sum_{k=0}^{M-1} H(zW_M^k)$$

$$= \sum_{k=0}^{M-1} [c + (zW_M^k)^{-1} E_1(z^M W_M^{kM}) + \cdots + (zW_M^k)^{-(M-1)} E_{M-1}(z^M W_M^{kM})]$$

$$= \sum_{k=0}^{M-1} [c + W_M^{-k} \cdot z^{-1} E_1(z^M) + \cdots + W_M^{-(M-1)k} \cdot z^{-(M-1)} E_{M-1}(z^M)]$$

$$= Mc + z^{-1} E_1(z^M) \sum_{k=0}^{M-1} W_M^{-k} + \cdots + z^{-(M-1)} E_{M-1}(z^M) \sum_{k=0}^{M-1} W_M^{-k(M-1)}. \quad \text{(B.54)}$$

Now note that

$$\sum_{k=0}^{M-1} z^k = \frac{1 - z^M}{1 - z}.$$

Hence,

$$\sum_{k=0}^{M-1} W_M^{-km} = \frac{1 - W_M^{-mM}}{1 - W_M}$$

and as $W_M^{-mM} = 1 \; \forall \; m = 0, 1, 2, \ldots, M-1$, it is clear that all terms in (B.54) vanish, save for the first. Therefore, for $H(z)$ in (B.53) with $c = 1/M$,

$$\sum_{k=0}^{M-1} H(zW_M^k) = 1.$$

Similarly, if the impulse response associated with $H(z)$ is delayed by $m_d M$ samples to obtain a causal system, then

$$H(z) = cz^{-m_d M} + z^{-1} E_1(z^M) + \cdots + z^{-(M-1)} E_{M-1}(z^M)$$

and it is readily verified that

$$\sum_{k=0}^{M-1} H(zW_M^k) = z^{-m_d M},$$

which implies the composite analysis-synthesis filter bank produces a simple delay of the input in the absence of aliasing.

B.15   Givens Rotations

Givens rotations are a convenient means for implementing a Cholesky or QR decomposition. They also find frequent application in other matrix decomposition and decomposition updating algorithms, inasmuch as they provide a convenient means of imposing a desired pattern of zeros on a given matrix. For instance, they can be used to restore the prearray described in Section 13.4.4 to lower triangular form, as is required for the square-root implementation of the *recursive least squares* (RLS) estimator.

In order to describe the use of Givens rotations, let us define the complex vector of length 2,

$$\mathbf{v} = \begin{bmatrix} v_1 \\ v_2 \end{bmatrix}.$$

For this simple case, we will define the 2×2 unitary *Givens matrix* as

$$\mathbf{G} \triangleq \begin{bmatrix} c^* & s \\ -s^* & c \end{bmatrix}. \tag{B.55}$$

It is desired that the second component of the rotated vector $\mathbf{Gv}$ is driven to zero, or *annihilated*, such that

$$\mathbf{Gv} = \begin{bmatrix} c^* v_1 + s\, v_2 \\ -s^* v_1 + c\, v_2 \end{bmatrix} \triangleq \begin{bmatrix} v_1' \\ 0 \end{bmatrix} = \mathbf{v}'.$$

Clearly, annihilating the second component of $\mathbf{v}'$ requires that

$$s^* v_1 = c\, v_2,$$

and the unitarity of the final transformation requires that

$$|c|^2 + |s|^2 = 1.$$

Both objectives are accomplished by setting

$$c = \frac{v_1}{\sqrt{|v_1|^2 + |v_2|^2}},$$

$$s = \frac{v_2^*}{\sqrt{|v_1|^2 + |v_2|^2}}.$$

In this case $\mathbf{v}'$ will have the same length as $\mathbf{v}$, but only one nonzero component.

The simple technique can readily be extended to higher dimensional vectors. Consider a $K \times 1$ complex vector

$$\mathbf{v} = \begin{bmatrix} v_0 & \cdots & v_m & \cdots & v_n & \cdots & v_{K-1}, \end{bmatrix}^T$$

and a $K \times K$ *Givens rotation matrix*

$$
\mathbf{G}(m, n) = \begin{bmatrix}
\mathbf{I} & \ddots & & & & & & 0 \\
& & c_{mm} & & & s_{mn} & & \\
& & & \ddots & \mathbf{I} & & \ddots & \\
& & s_{nm} & & & c_{nn} & & \\
& & & & & & \ddots & \\
0 & & & & & & & \mathbf{I}
\end{bmatrix}.
\tag{B.56}
$$

The cosine c_{mm} and sine s_{nm} parameters are chosen such $\mathbf{G}(m, n)$ is unitary and

$$
\mathbf{v}' = \mathbf{G}(m, n)\,\mathbf{v} = \begin{bmatrix} v_0 & \cdots & v'_m & \cdots & 0 & \cdots & v_{K-1} \end{bmatrix}^T.
$$

The effect of applying this rotation is that the nth component of $\mathbf{v}$ is *annihilated*, v_m is altered, and all other components are unchanged. Annihilating v_n requires that

$$
s^* v_m = c v_n
$$

and the unitarity of the final transformation requires that

$$
|c|^2 + |s|^2 = 1,
$$

much as in the simple case considered before. Both objectives are accomplished by setting

$$
c = c^*_{mm} = c_{nn} = \frac{v_m}{\sqrt{|v_m|^2 + |v_n|^2}},
\tag{B.57}
$$

$$
s = s_{mn} = \frac{v^*_n}{\sqrt{|v_m|^2 + |v_n|^2}},
\tag{B.58}
$$

and

$$
s_{nm} = s^*_{mn} = -s^*.
\tag{B.59}
$$

It is then readily verified that $\mathbf{G}(m, n)$ is unitary and that

$$
\mathbf{v}' = \mathbf{G}(m, n)\,\mathbf{v} = \begin{bmatrix} v_0 & \cdots & \sqrt{|v_m|^2 + |v_n|^2} & \cdots & 0 & \cdots & v_{K-1} \end{bmatrix}^T.
$$

Multiplying $\mathbf{G}(m, n)$ by any other arbitrary vector

$$
\mathbf{u} = \begin{bmatrix} u_0 & \cdots & u_m & \cdots & u_n & \cdots & u_{K-1} \end{bmatrix}^T
$$

we obtain

$$
\mathbf{u}' = \mathbf{G}(m, n)\mathbf{u} = \begin{bmatrix} u_0 & \cdots & c^* u_m + s u_n & \cdots & -s^* u_m + c u_n & \cdots & u_{K-1} \end{bmatrix}^T
$$

If it is desired to annihiliate the nth component of a row vector $\mathbf{x}$, we can simply define $\mathbf{G}(m, n)$ as in (B.56–B.59) and form the product $\mathbf{x}\,\mathbf{G}^T(m, n)$.

In the remainder of this section, we develop a procedure based on Givens rotations for performing the update inherent in the covariance form of the square-root implementation of the RLS estimator. As described in Section 13.4.4, this requires restoring the prearray

$$
\begin{bmatrix}
\mu^{1/2}\boldsymbol{\Phi}^{H/2}(K-1) & \mu^{1/2}\mathbf{X}(K) \\
\mu^{-1/2}\hat{\mathbf{w}}^H(K-1)\,\boldsymbol{\Phi}^{H/2}(K-1) & D(K),
\end{bmatrix}
$$

to lower triangular form. This in turn entails forcing a desired pattern of zeros on the first row and last column of the prearray in order to obtain the postarray.

As we have seen, a Givens rotation is completely specified by two indices, the element which is to be annihilated, and the element into which the annihilated element is to be rotated. Our convention in the following schematic illustration of a factorization update algorithm will be:

- the element annihilated by the last rotation is marked with $\cdot$
- nonzero elements that were altered by the last rotation are marked with $\otimes$
- nonzero elements that were *not* altered by the last rotation are marked with $\times$
- zero elements that were annihilated in prior rotations, or that will become nonzero, are marked with 0
- other zero elements are not shown.

The update is actually very straightforward and involves rotating the elements in the last column into the leading diagonal, as shown below.

$$
\begin{bmatrix}
\times & 0 & 0 & 0 & 0 & \times \\
0 & \times & & & & \times \\
0 & \times & \times & & & \times \\
0 & \times & \times & \times & & \times \\
0 & \times & \times & \times & \times & \times \\
0 & \times & \times & \times & \times & \times
\end{bmatrix}
\xrightarrow{\mathbf{G}_5}
\begin{bmatrix}
\otimes & 0 & 0 & 0 & 0 & \cdot \\
\otimes & \times & & & & \otimes \\
\otimes & \times & \times & & & \otimes \\
\otimes & \times & \times & \times & & \otimes \\
\otimes & \times & \times & \times & \times & \otimes \\
\otimes & \times & \times & \times & \times & \otimes
\end{bmatrix}
\xrightarrow{\mathbf{G}_6}
\begin{bmatrix}
\times & 0 & 0 & 0 & 0 & 0 \\
\times & \otimes & & & & \otimes \\
\times & \otimes & \times & & & \otimes \\
\times & \otimes & \times & \times & & \otimes \\
\times & \otimes & \times & \times & \times & \otimes \\
\times & \otimes & \times & \times & \times & \otimes
\end{bmatrix}
\xrightarrow{\mathbf{G}_7}
\begin{bmatrix}
\times & 0 & 0 & 0 & 0 & 0 \\
\times & \times & & & & 0 \\
\times & \times & \otimes & & & \cdot \\
\times & \times & \otimes & \times & & \otimes \\
\times & \times & \otimes & \times & \times & \otimes \\
\times & \times & \otimes & \times & \times & \otimes
\end{bmatrix}
\xrightarrow{\mathbf{G}_8}
\begin{bmatrix}
\times & 0 & 0 & 0 & 0 & 0 \\
\times & \times & & & & 0 \\
\times & \times & \times & & & 0 \\
\times & \times & \times & \otimes & & \otimes \\
\times & \times & \times & \otimes & \times & \otimes \\
\times & \times & \times & \otimes & \times & \otimes
\end{bmatrix}
\xrightarrow{\mathbf{G}_9}
\begin{bmatrix}
\times & 0 & 0 & 0 & 0 & 0 \\
\times & \times & & & & 0 \\
\times & \times & \times & & & 0 \\
\times & \times & \times & \times & & 0 \\
\times & \times & \times & \times & \otimes & \otimes \\
\times & \times & \times & \otimes & \times & \otimes
\end{bmatrix}
$$

Such an update can also be devised for the square-root implementation of the iterated extended Kalman filter described in Section (10.2.1).

A more detailed discussion of Givens rotations can be found in Golub and Van Loan (1996a, sect. 5.1). Another rotation that is frequently used to enforce a desired pattern of zeros on an array is the *Householder transform*, which was introduced in Householder (1958). The properties of the Householder transformation are also discussed in Golub and Van Loan (1996a, sect. 5.1).

It is worth noting that using Givens rotations to extract the Cholesky factors of a matrix can, for many important cases, serve the same purpose as extracting the inverse. Suppose, for example, we are confronted with a problem of finding that $\mathbf{x}$ satisfying $\mathbf{A}\mathbf{x} = \mathbf{b}$ for some known $\mathbf{b}$ and symmetric positive-definite $(N \times N)$ matrix $\mathbf{A}$. A naive solution would entail forming the inverse $\mathbf{A}^{-1}$, then solving as $\mathbf{x} = \mathbf{A}^{-1}\mathbf{b}$. The inverse, however, is notoriously unstable numerically (Golub and Van Loan 1996b). Thus, instead of forming the inverse, we might first extract the lower triangular Cholesky factor $\mathbf{A}^{1/2}$

either by the direct construction presented in Section B.3, or through the application of a set of Givens rotations as described above. Thereafter, we can set $\mathbf{y} = \mathbf{A}^{T/2}\mathbf{x}$ and write

$$\mathbf{A}^{1/2}\mathbf{y} = \mathbf{b}.$$

Let us represent the Cholesky factor $\mathbf{A}^{1/2}$ as

$$\mathbf{A}^{1/2} = \begin{bmatrix} a_{0,0} & 0 & \cdots & 0 \\ a_{1,0} & a_{1,1} & \cdots & 0 \\ \vdots & \vdots & \ddots & \vdots \\ a_{N-1,0} & a_{N-1,1} & \cdots & a_{N-1,N-1} \end{bmatrix}.$$

Then we can readily solve for the components of $\mathbf{y}$ through the *forward substitution*

$$y_0 = \frac{b_0}{a_{0,0}}, \tag{B.60}$$

$$y_1 = \frac{b_1 - a_{1,0}\, y_0}{a_{1,1}}, \tag{B.61}$$

$$\vdots \qquad \vdots \tag{B.62}$$

$$y_{N-1} = \frac{1}{a_{N-1,N-1}} \left(b_{N-1} - \sum_{n=0}^{N-2} a_{N-1,n}\, y_n \right). \tag{B.63}$$

The name *forward* substitution stems from the fact that the complete solution is obtained by beginning with y_0 and working forward through the rest of the components. Once $\mathbf{y}$ is known, we can write

$$\mathbf{A}^{T/2}\mathbf{x} = \mathbf{y},$$

and perform *backward substitution* to solve for the components of $\mathbf{x}$, which entails first solving for $x_{N-1} = y_{N-1}/a_{N-1,N-1}$, then working backward to solve for the rest of the components of $\mathbf{y}$.

B.16   Derivatives with Respect to Complex Vectors

As we learned in Chapter 13, adaptive beamforming algorithms typically operate in the frequency or subband domain. This implies that the weights applied to the output of each sensor are complex-valued. While subband processing offers the considerable advantage that the active weights for each subband can be designed independently of all others, it brings with it the necessity of optimizing a real-valued function with respect to a vector of complex-valued weights. Formulating an optimization algorithm typically entails calculating gradients of the optimization criterion with respect to the vector of complex-valued sensor weights. A naive approach to this problem would be to express the sensor weights $\mathbf{w}$ as

$$\mathbf{w} = \mathbf{x} + j\mathbf{y},$$

for the real-valued vectors $\mathbf{x}$ and $\mathbf{y}$, then to take partial derivatives with respect to $\mathbf{x}$ and $\mathbf{y}$. Such an approach, however, would lead to a great deal of tedium. For the optimization criteria of interest in array processing, we can more simply work directly with the vector $\mathbf{w}$. The material and presentation in this section are based on the work of Brandwood (1983) as summarized by Van Trees (2002, sect. A.7.4).

Let us denote the optimization function of interest as

$$f(z) = f(x, y)$$

where $z = x + jy$. Let $g(z, z^*)$ denote a function that is analytic with respect to z and z^* independently, and set

$$g(z, z^*) = f(x, y).$$

Brandwood shows that the partial derivative of $g(z, z^*)$ with respect to z, whereby z^* is treated as a constant in g, yields the result

$$\left. \frac{\partial g(z, z^*)}{dz} \right|_{z=x+jy} = \frac{1}{2} \left[\frac{\partial f(x, y)}{dx} - j \frac{\partial f(x, y)}{dy} \right].$$

Similarly,

$$\left. \frac{\partial g(z, z^*)}{dz^*} \right|_{z^*=x-jy} = \frac{1}{2} \left[\frac{\partial f(x, y)}{dx} + j \frac{\partial f(x, y)}{dy} \right].$$

Brandwood also demonstrates that a necessary and sufficient condition for $f(z)$ to have a stationary point is either

$$\frac{\partial g(z, z^*)}{\partial z} = 0,$$

where z^* is treated as a constant in calculating the partial derivative, or

$$\frac{\partial g(z, z^*)}{\partial z^*} = 0,$$

where z is treated as a constant in the partial derivative.

In order to manipulate vector values, it is necessary to define the complex gradient operator

$$\nabla_{\mathbf{z}} = \left[\frac{\partial}{\partial z_1} \ \frac{\partial}{\partial z_2} \ \cdots \ \frac{\partial}{\partial z_N} \right]^T$$

where

$$\frac{\partial}{\partial z_n} \triangleq \frac{\partial}{\partial x_n} - j \frac{\partial}{\partial y_n} \ \forall \ 1, \ldots, N.$$

Similarly,

$$\nabla_{\mathbf{z}^H} = \left[\frac{\partial}{\partial z_1^*} \ \frac{\partial}{\partial z_2^*} \ \cdots \ \frac{\partial}{\partial z_N^*} \right]^T$$

where

$$\frac{\partial}{\partial z_n^*} \triangleq \frac{\partial}{\partial x_n} + j\frac{\partial}{\partial y_n} \ \forall \ 1, \ldots, N.$$

Now let us define

$$f(\mathbf{z}) \triangleq f(\mathbf{x}, \mathbf{z}) = g(\mathbf{z}, \mathbf{z}^H)$$

where $g(\mathbf{z}, \mathbf{z}^H)$ is a real-value function of $\mathbf{z}$ and $\mathbf{z}^H$, which is analytic with respect to $\mathbf{z}$ and $\mathbf{z}^H$ independently. Then $f(\mathbf{z})$ will have a stationary point if either

$$\nabla_{\mathbf{z}} g(\mathbf{z}, \mathbf{z}^H) = \mathbf{0}, \tag{B.64}$$

where $\mathbf{z}^H$ is treated as a constant, or

$$\nabla_{\mathbf{z}^H} g(\mathbf{z}, \mathbf{z}^H) = \mathbf{0}, \tag{B.65}$$

where $\mathbf{z}$ is treated as a constant. Normally, we will find (B.65) more useful than (B.64) in practice.

Consider now an application taken from Section 13.3.1, namely, that of minimizing a scalar product $\mathbf{w}^H \mathbf{R} \mathbf{w}$ subject to a linear constraint $\mathbf{w}^H \mathbf{c} = a$, where $\mathbf{R}$ is a $N \times N$ Hermitian matrix, $\mathbf{w}$ and $\mathbf{c}$ are $N \times 1$ complex vectors, and a is a complex scalar. Let us define the real-valued cost function

$$\begin{aligned} g(\mathbf{w}, \mathbf{w}^H) &= \mathbf{w}^H \mathbf{R} \mathbf{w} + 2\text{Re}\left[\lambda(\mathbf{w}^H \mathbf{c} - a)\right] \\ &= \mathbf{w}^H \mathbf{R} \mathbf{w} + \lambda(\mathbf{w}^H \mathbf{c} - a) + \lambda^*(\mathbf{c}^H \mathbf{w} - a^*), \end{aligned} \tag{B.66}$$

where λ is a complex Lagrange multiplier. Taking the gradient of $g(\mathbf{w}, \mathbf{w}^H)$ with respect to $\mathbf{w}^H$ and equating it to zero yields

$$\nabla_{\mathbf{w}^H} g(\mathbf{w}, \mathbf{w}^H) = \mathbf{R} \mathbf{w}_\text{o} + \lambda \mathbf{c} = \mathbf{0}.$$

Hence, we can solve for the optimal weights $\mathbf{w}_\text{o}$ according to

$$\mathbf{w}_\text{o} = -\lambda \mathbf{R}^{-1} \mathbf{c}. \tag{B.67}$$

We can now apply the constraint $\mathbf{w}^H \mathbf{c} = a$ to calculate λ, according to

$$\mathbf{w}_\text{o}^H \mathbf{c} = -\lambda \mathbf{c}^H \mathbf{R}^{-1} \mathbf{c} = a$$

or, equivalently,

$$\lambda = -\frac{a}{\mathbf{c}^H \mathbf{R}^{-1} \mathbf{c}}. \tag{B.68}$$

Substituting (B.68) into (B.67) provides the final solution,

$$\mathbf{w}_\text{o}^H = \frac{\mathbf{c}^H \mathbf{R}^{-1} a^*}{\mathbf{c}^H \mathbf{R}^{-1} \mathbf{c}}.$$

B.17   Perpendicular Projection Operators

Consider an N-dimensional vector $\mathbf{x}$, and an $N \times M$ matrix $\mathbf{C}$ whose linearly independent columns define an M-dimensional *subspace* of the complete N-dimensional space. We wish to find the *perpendicular projection* of $\mathbf{x}$ onto the $\mathbf{C}$ subspace. The projection can be expressed as $\mathbf{Cy}$ where the M-dimensional $\mathbf{y}$ minimizes

$$|\mathbf{x} - \mathbf{Cy}|^2 = (\mathbf{x} - \mathbf{Cy})^H (\mathbf{x} - \mathbf{Cy}) \tag{B.69}$$

$$= \mathbf{x}^H \mathbf{x} - \mathbf{y}^H \mathbf{C}^H \mathbf{x} - \mathbf{x}^H \mathbf{Cy} + \mathbf{y}^H \mathbf{C}^H \mathbf{Cy}. \tag{B.70}$$

Equation (2.69) is a statement that the perpendicular projection is the point in the subspace defined by $\mathbf{C}$ with minimal Euclidean distance to $\mathbf{x}$. Taking the gradient of (B.70) with respect to $\mathbf{y}^H$ and equating it to zero, we find

$$-\mathbf{C}^H \mathbf{x} + \mathbf{C}^H \mathbf{Cy} = \mathbf{0},$$

or

$$\hat{\mathbf{y}} = (\mathbf{C}^H \mathbf{C})^{-1} \mathbf{C}^H \mathbf{x}.$$

The inverse must in the last equation exist, because the columns of $\mathbf{C}$ are linearly independent. Hence, the desired projection is

$$\mathbf{x}_\mathbf{C} = \mathbf{C}\hat{\mathbf{y}} = \mathbf{P}_\mathbf{C} \mathbf{x},$$

where the *perpendicular projection operator* is by definition

$$\mathbf{P}_\mathbf{C} \triangleq \mathbf{C}(\mathbf{C}^H \mathbf{C})^{-1} \mathbf{C}^H.$$

Moreover, it is possible to define a projection operator $\mathbf{P}_\mathbf{C}^\perp$ onto the space *orthogonal* to the columns of $\mathbf{C}$ according to

$$\mathbf{P}_\mathbf{C}^\perp \triangleq \mathbf{I} - \mathbf{P}_\mathbf{C} = \mathbf{I} - \mathbf{C}(\mathbf{C}^H \mathbf{C})^{-1} \mathbf{C}^H.$$

Bibliography

Abel, J. S. and Smith, J. O. (1987) The spherical interpolation method for closed–form passive source local-ization using range difference measurements. *Proc. of ICASSP*, pp. 471–474.

Abramowitz, M. and Stegun, I. A. (1965, 1972) *Handbook of Mathematical Functions*. Dover, New York.

Acero, A. (1990) *Acoustical and environmental robustness in automatic speech recognition*. PhD thesis, Carnegie Mellon University, Pittsburgh, PA.

Acero, A. and Stern, R. (1991) Robust speech recognition by normalization of the acoustic space. *Proc. of ICASSP*, pp. 893–896.

Ahmed, B. and Holmes, W. (2004) A voice activity detector using the chi-square test. *Proc. of ICASSP*, pp. 625–628.

Aho, A. V., Hopcroft, J. E. and Ullman, J. D. (1974) *The Design and Analysis of Computer Algorithms*. Addison Wesley, Reading, Massachusetts.

Alais, D. and Burr, D. (2004) The ventriloquist effect results from near optimal cossmodal integration. *Curr. Biol.* **14**, 257–262.

Allen, J. B. and Berkley, D. A. (1979) Image method for efficiently simulating small room acoustics. *Jour. of ASA* **65**(4), 943–950.

Allen, J. B., Berkley, D. A. and Blauert, J. (1977) Multimicrophone signal-processing technique to remove room reverberation from speech signals. *Jour. of ASA* **62**, 912–915.

Almajai, I., Milner, B., Darch, J. and Vaseghi, S. (2007) *Visually-derived Wiener filters for speech enhancement*. *Proc. of ICASSP*, vol. **4**, pp. 585–588.

Alvarado, V. M. (1990) Talker Localization and Optimal Placement of Microphones for a Linear Microphone Array Using Stochastic Region Contraction.

Amari, S., Cichocki, A. and Yang, H. H. (1996) A new learning algorithm for blind signal separation. In *Advances in Neural Information Processing D* (ed. Touretzsky D, Mozer M and Hasselmo M), vol. **8**. MIT Press, Cambridge, MA, USA.

Amari, S. I. (1998) Natural gradient works efficiently in learning. *Neural Computation* **10**, 251–276.

Amazigo, J. C. and Rubenfeld, L. A. (1980) *Advanced Calculus*. John Wiley & Sons.

AMIDA – augmented multi-party interaction with distance access. *http://www.amiproject.org*.

Anastasakos, T., McDonough, J., Schwartz, R. and Makhoul, J. (1996) *A compact model for speaker–adaptive training*. *Proc. of ICSLP*, vol. **2**, pp. 1137–1140.

Anderson, T. W. (1984) *An Introduction to Multivariate Statistical Analysis*. John Wiley & Sons.

Andreou, A., Kamm, T. and Cohen, J. (1994) Experiments in vocal tract normalization *Proc. of CAIP Workshop: Frontiers in Speech Recognition II*.

Anguera, X., Wooters, C. and Hernando, J. (2005) Speaker diarization for multi-party meetings using acoustic fusion. *Proc. of ASRU*, pp. 426–431.

Applebaum, S. P. and Chapman, D. J. (1976) Adaptive arrays with main beam constraints. *IEEE Trans. on Antennas Propagation*, **AP-24**, 650–662.

Armani, L., Matassoni, M., Omologo, M. and Svaizer, P. (2003) Use of a CSP-based voice activity detector for distant-talking ASR. *Proc. of Eurospeech*, vol. **2**, pp. 501–504.

ASA – Acoustical Society of America. *http://asa.aip.org/*.

ASJ – Acoustical Society of Japan. *http://www.jstage.jst.go.jp/browse/ast/-char/en*.

Atal, B. (1974) Effectiveness of linear prediction characteristics of the speech wave for automatic speaker identification and verification. *Jour. of ASA* **55**, 1304–1312.

Atal, B. and Schroeder, M. (1967) Predictive coding of speech signals. *Proc. of IEEE Conference on Communication and Processing*, pp. 360–361.

Atal, B. and Schroeder, M. (1984) Stochastic coding of speech signals at very low bit rates. *Proc. of International Conference on Communication*, pp. 1610–1613.

Aubert, X. (2000) A brief overview of decoding techniques for large vocabulary continuous speech recognition. *ISCA ITRW ASR 2000 Automatic Speech Recognition: Challenges for the New Millennium*, Paris France.

Axelrod, S., Goel, V., Gopinath, R., Olsen, P. and Visweswariah, K. (2007) Discriminative estimation of subspace constrained gaussian mixture models for speech recognition. *IEEE Trans. on ASLP* **15**(1), 172–189.

Baba, A., Lee, A., Saruwatari, H. and Shikano, K. (2002) Speech recognition by reverberation adapted acoustic model. *Proc. of ASJ General Meeting*, pp. 27–28.

Bar-Shalom, Y. and Fortmann, T. E. (1988) *Tracking and Data Association*. Academic Press, San Diego.

Barker, J. and Cooke, M. (1997) Modelling the recognition of spectrally reduced speech. *Proc. of Eurospeech*, pp. 2127–2130.

Bass, H., Bauer, H. J. and Evans, L. (1972) Atmospheric absorption of sound: analytical expression. *Jour. of ASA*, pp. 821–825.

Basseville, M. (1989) Distance measures for signal processing and pattern recognition. *IEEE Trans. on SP* **18**(4), 287–314.

Baum, L. E., Petrie, T. and Soules, G. (1970) A maximization technique in the statistical analysis of probabilistic function of Markov chains. *Annals of Mathematical Statistics* **41**, 164–171.

Bellman, R. (1957) *Dynamic Programming*. Princeton University Press.

Bellman, R. (1961) *Adaptive Control Processes*. Princeton University Press.

Benesty, J. (2000) Adaptive eigenvalue decomposition algorithm for passive acoustic source localization. *Jour. of ASA* **107**(1), 384–391.

Benesty, J., Makino, S. and Chen, J. (2005) *Speech Enhancement*. Springer.

Bengio, Y., Ducharme, R., Vincent, P. and Jauvin, C. (2003) A neural probabilistic language model. *Machine Learning Research* **3**(6), 1137–1155.

Beritelli, F., Casale, S. and Ruggeri, G. (2001) Performance evaluation and comparison of ITU–T/ETSI voice activity detectors. *Proc. of ICASSP*.

Bertsekas, D. P. (1995) *Nonlinear Programming*. Athena Scientific, Belmont, MA, USA.

Berzuini, C. and Gilks, W. (2001) Resample move filtering with cross-model jumps. In *Sequential Monte Carlo Methods in Practice* (ed. Doucet A, de Freitas N and Gordon N), pp. 117–138. Springer.

Birchfield, S. and Subramanya, A. (2005) Microphone array position calibration by basis-point classical multidimensional scaling. *IEEE Trans. on SAP* **13**(5), 1025–1034.

Bitzer, J. and Simmer, K. U. (2001) Superdirective microphone arrays. In *Microphone Arrays* (ed. Branstein M and Ward D). Springer, Heidelberg, pp. 19–38.

Blahut, R. E. (1987) *Principles and Practice Information Theory*. Addison-Wesley, Reading, MA, USA.

Blauert, J. (1997) *Spatial Hearing: The Psychophysics of Human Sound Localization*. MIT Press; third edition.

Bocchieri, E., Digalakis, V., Corduneanu, A. and Boulis, C. (1999) Correlation modeling of MLLR transform biases for rapid HMM adaptation to new speakers *Proc. of ICASSP*, vol. II, pp. 773–776.

Bogert, B., Healy, M. and Tukey, J. (1963) The frequency analysis of time series for echoes. *Proc. of Symposium on Time Series Analysis*.

Bolic, M., Djuric, P. M. and Hong, S. (2003) New resampling algorithms for particle filters. *Proc. of ICASSP* **2**, 589–592.

Boll, S. (1979) Suppression of acoustic noise in speech using spectral subtraction. *IEEE Trans. on ASSP* **27**, 113–120.

Braccini, C. and Oppenheim, A. V. (1974) Unequal bandwidth spectral analysis using digital frequency warping. *IEEE Trans. on ASSP* **22**, 236–244.

Bracewell, R. (1999) *The Fourier Transform and Its Applications*. 3rd edn. McGraw–Hill.

Brandstein, M. and Ward, D. (2000) *Microphone Arrays*. Springer.

Brandstein, M. S. (1995) *A framework for speech source localization using sensor arrays*. PhD thesis, Brown University, Providence, RI.

Brandstein, M. S., Adcock, J. E. and Silverman, H. F. (1997) A closed-form location estimator for use with room environment microphone arrays. *IEEE Trans. on SAP* **5**(1), 45–50.

Brandwood, D. H. (1983) A complex gradient operator and its application in adaptive array theory. *Proc. of IEEE, Special Issue on Adaptive Arrays*, pp. 11–17.

Brehm, H. and Stammler, W. (1987a) Description and generation of spherically invariant speech-model signals. *IEEE Trans. on SP* **12**, 119–141.

Brehm, H. and Stammler, W. (1987b) Description and generation of spherically invariant speech-model signals. *IEEE Trans. on SP* **12**, 119–141.

Broesch, J. D. (1997) *Digital Signal Processing Demystified*. Newnes.

Brown, P. F., Pietra, V. J. D., deSouza, P. V., Lai, J. C. and Mercer, R. L. (1992) Class-based n-gram models of natural language. *Computational Linguistics* **18**, 18–4.

Brungart, D. (2001) Information and energetic masking effects in the perception of two simultaneous talkers. *Jour. of ASA* **193**(3), 1101–1109.

Brunn, D., Sawo, F. and Hanebeck, U. D. (2006a) Efficient nonlinear Bayesian estimation based on Fourier densities. *Proc. of International Conference on Multisensor Fusion and Integration for Intelligent Systems*, pp. 317–322, Heidelberg, Germany.

Brunn, D., Sawo, F. and Hanebeck, U. D. (2006b) Nonlinear multidimensional Bayesian estimation with Fourier densities. *Proc. of IEEE Conference on Decision and Control*, San Diego, CA, USA.

Brutti, A., Omologo, M. and Svaizer, P. (2005) Oriented global coherence field for the estimation of the head orientation in smart rooms equipped with distributed microphone arrays. *Proc. of Interspeech*.

Buchner, H., Aichner, R. and Kellermann, W. (2004) Blind source seperation for convolutive mixtures: A unified treatment. *Audio Signal Processing for Next–Generation Multimedia Communication Systems*. Kluwer Academic, Boston, pp. 255–289.

Bucy, R. (1969) Bayes' theorem and digital realization of nonlinear filters. *Jour. on Astronautic Science* **87**, 493–500.

Bulyko, I., Ostendorf, M. and Stolcke, A. (2003) Getting more mileage from web text sources for conversational speech language modeling using class-dependent mixtures. *Proc. of HLT-NAACL*, pp. 7–9.

Bulyko, I., Ostendorf, M., Siu, M., Ng, T., Stolcke, A. and Cetin, O. (2007) Web resources for language modeling in conversational speech recognition. *ACM Trans. on Speech and Language Processing*.

Burg, J. (1972) The relationship between maximum entropy and maximum likelihood spectra. *Geophysics* **37**, 375–376.

Burger, S. (2007) The CHIL RT07 Evaluation data. In *Multimodal Technologies for Perception of Humans, Joint Proceedings of the Second International Evaluation workshop on Classification of Events, Activities and Relationships, CLEAR 2007 and the Spring 2007 Rich Transcription Meeting Evaluation* (ed. Stiefelhagen R, Bowers R and Fiscus J). Lecture Notes in Computer Science, No. 4625. Springer, Baltimore, USA.

Byrne, W., Gunawardana, A. and Khudanpur, S. (2000) Information geometry and em variants. *IEEE Trans. on SP*.

CALO – cognitive agent that learns and organizes. *http://caloproject.sri.com*.

Capon, J. (1969) High-resolution frequency-wavenumber spectrum analysis. *Proc. of the IEEE* **57**, 1408–1418.

Cappé, O. (1994) Elimination of the musical noise phenomenon with the Ephraim and Malah noise suppressor. *IEEE Trans. on SAP* **2**(2), 345–349.

Cardoso, (1998) Blind signal separation: Statistical principles. *Proc. of IEEE* **9**(10), 2009–2025.

Carter, G. C. (1981) Time delay estimation for passive sonar signal processing. *IEEE Trans.* on ASSP **29**, 463–469.

Caseiro, D. and Trancoso, I. (2006) A specialized on-the-fly algorithm for lexicon and language model composition. *IEEE Trans. on ASLP*.

Charniak, E. (2001) Immediate-head parsing for language models *Proc. of 39th Annual Meeting of the Association for Computational Linguistics*.

Chen, B., Chang, S. and Sivadas, S. (2003a) Learning discriminative temporal patterns in speech: Development of novel TRAPS like classifiers *Proc. of Eurospeech*.

Chen, B., Zhu, Q. and Morgan, N. (2004a) Learning long term temporal features in LVCSR using neural networks. *Proc. of ICSLP*, pp. 612–615.

Chen, J., Benesty, J. and Huang, Y. A. (2003b) Robust time delay estimation exploiting redundancy among multiple microphones. *IEEE Trans. on SAP* **11**(6), 549–57.

Chen, J., Juang, Y. A. and Benesty, J. (2004b) Time delay estimation. *Audio Signal Processing for Next-Generation Multimedia Communication Systems*. Kluwer Academic, Boston, pp. 197–228.

Chen, J. C., Hudson, R. E. and Yao, K. (2002) Maximum-likelihood source localization and unknown sensor location estimation for wideband signals in the near-field. *IEEE Trans. on SP* **50**, 1843–1854.

Chen, S. and Goodman, J. (1998) An empirical study of smoothing. *Techniques for Language Modeling*. TR–10–98, Computer Science Group, Harvard University.

Chen, S. and Goodman, J. (1999) An empirical study of smoothing techniques for language modeling. *Jour. on CSL* **13**(4), 359–394.

Chen, S. F. (2003) Compiling large-context phonetic decision trees into finite-state transducers. *Proc. of Eurospeech*, pp. 1169–1172.

Cheng, O., Dines, J. and Doss, M. M. (2007) A generalized dynamic composition algorithm of weighted finite state transducers for large vocabulary speech recognition. *Proc. of ICASSP*.

Cherry, C. (1953) Some experiments on the recognition of speech with one and two ears. *Jour. of ASA* **25**, 975–981.

CHIL – computers in the human interaction loop. *http://chil.server.de*.

Chu, W. and Warnock, A. (2002a) Detailed directivity of sound fields around human talkers. *Technical Report*, National Research Council Canada, IRC–RR–104.

Chu, W. and Warnock, A. (2002b) Voice and background noise levels measured in open offices.. *Internal Report*, National Research Council Canada, IR–837.

Churchill, R. V. and Brown, J. W. (1990) *Complex Variables and Applications*, fifth edn. McGraw-Hill, New York.

Cirillo, A., Parisi, R. and Uncini, A. (2008) Sound mapping in reverberant rooms by a robust direct method *Proc. of ICASSP*, Las Vegas, NV, USA.

Claesson, I. and Nordholm, S. (1992) A spatial filtering approach to robust adaptive beaming. *IEEE Trans. on Antennas and Propagation* **19**, 1093–1096.

Cohen, I., Gannot, S. and Berdugo, B. (2003) An integrated real-time beamforming and postfiltering system for nonstationary noise environments. *Jour. on AppSP* **2003**, 1064–1073.

Comon, P. (1994) Independent component analysis – a new concept? *IEEE Trans. on SP* **36**, 287–314.

Cooley, J. and Tukey, J. (1965) An algorithm for the machine calculation of complex fourier series. *Math. Comput.* **19**, 297–301.

Cormen, T. H., Leiserson, C. E., Rivest, R. L. and Stein, C. (2001) *Introduction to Algorithms*, 2 edn. MIT Press, Cambridge, MA, USA.

COSY – cognitive systems for cognitive assistants. *http://www.cognitivesystems.org*.

Cover, T. M. and Thomas, J. A. (1991) *Elements of Information Theory*. John Wiley & Sons.

Cox, H., Zeskind, R. M. and Owen, M. M. (1987) Robust adaptive beamforming. *IEEE Trans. on ASSP* **35**, 1365–1376.

Crawford, C. R. (1976) A stable generalized eigenvalue problem. *SIAM Jour. of Numerical Analysis* **13**(6), 854–860.

Cron, B. F. and Sherman, C. H. (1962) Spatial-correlation functions using small microphone arrays with optimized directivity. *Jour. of ASA* **34**, 1732–1736.

Davis, S. B. and Mermelstein, P. (1980) Comparison of parametric representations for monosyllabic word recognition in continuously spoken sentences. *IEEE Trans. on ASSP* **28**, 357–366.

de Figueiredo, R. and Jan, J. (1971) Spline filters *Proc. of the 2nd Symposium on Nonlinear Estimation Theory and Its Applications*, pp. 127–138.

De Haan, J. M. (2001) *Filter bank design for subband adaptive filtering*. PhD thesis, Karlskrona. Blekinge Institute of Technology.

De Haan, J. M., Grbic, N., Claesson, I. and Nordholm, S. E. (2003) Filter bank design for subband adaptive microphone arrays. *IEEE Trans. on SAP* **11**(1), 14–23.

Deller, J., Hansen, J. and Proakis, J. (1999) *Discrete-Time Processing of Speech Signals*. Wiley – IEEE Press.

Deller, Jr J., Hansen Jr, J. and Proakis, J. (2000) *Discrete-Time Processing of Speech Signals*. IEEE Press.

Deller, Jr J., Proakis, J. and Hansen, J. (1993) *Discrete-Time Processing of Speech Signals*. Macmillan.

Dempster, A. P., Laird, N. M. and Rubin, D. B. (1977) Maximum likelihoood from incomplete data via the EM algorithm. *Jour. of the Royal Statistical Society* **39 B**, 1–38.

Deng, L., Acero, A., Plumpe, M. and Huang, X. D. (2000) Large vocabulary speech recognition under adverse acoustic environments *Proc. of ICSLP*.

Deng, L. and O'Shaughnessy, D. (2003) *Speech processing: A Dynamic and Optimization-Oriented Approach*. Marcel Dekker, Inc.

Deng, L., Droppo, J. and Acero, A. (2002) A Bayesian approach to speech feature enhancement unsing the dynamic cepstral prior. *Proc. of ICASSP*.

Deng, L., Droppo, J. and Acero, A. (2004a) Enhancement of log mel power spectra of speech using a phase-sensitive model of the acoustic environment and sequential estimation of the corrupting noise. *IEEE Trans. on SAP* **12**, No. 2, 133–143.

Deng, L., Wu, J., Droppo, J. and Acero, A. (2005) Analysis and comparison of two speech feature exaction/compensation algorithms. *IEEE SPL* **12**(6), 477–480.

Deng, L., Yu, D. and Acero, A. (2004b) A quantitative model for formant dynamics and contextually assimilated reduction in fluent speech. *Proc. of ICSLP*.

Dharanipragada, S., Yapanel, U. and Rao, B. (2007) Robust feature extraction for continuous speech recognition using the MVDR spectrum estimation method. *IEEE Trans. on SAP* **15**(1), 224–234.

DiBiase, J. H., Silverman, H. F. and Brandstein, M. S. (2001) Robust localization in reverberant rooms. In *Microphone Arrays* (ed. Brandstein M and Ward D). Springer Verlag, Heidelberg, Germany, chapter 4

Digalakis, V., Berkowitz, S., Bocchieri, E., Boulis, C., Byrne, W., Collier, H., Corduneanu, A., Kannan, A., Khudanpur, S. and Sankar, A. (1996) Rapid speech recognizer adaptation to new speakers. *Proc. of ICASSP*, vol. I, pp. 339–341.

Digalakis, V., Rtischev, D. and Neumeyer, L. (1995) Fast speaker adaptation using constrained-estimation of Gaussian mixtures. *IEEE Trans. on SAP* **3**, 357–366.

Doblinger, G. (1995) Computationally efficient speech enhancement by spectral minima tracking in subbands. *Proc. of Eurospeech* **2**, 1513–1516.

Doclo, S., Spriet, A., Wouters, J. and Moonen, M. (2007) Frequency-domain criterion for the speech distortion weighted multichannel Wiener filter for robust noise reduction. *Speech Communication, special issue on Speech Enhancement* **49**, 636–656.

Doerbecker, M. (1997) Speech enhancement using small microphone arrays with optimized directivity. *Proc. of IWAENC*, London, UK.

Dolfing, H. and Hetherington, I. (2001) Incremental language models for speech recognition using finite-state transducers *Proc. of ASRU*.

Douc, R. and Cappe, O. (2005) Comparison of resampling schemes for particle filtering. *Proc. of Image and Signal Processing and Analysis*, pp. 64–69.

Doucet, A. (1998) *On Sequential Simulation-Based Methods for Bayesian Filtering,* Technical report CUED/F–INFENG/TR 310. Cambridge University, Department of Engineering.

Doucet, A., Godsill, S. and Andrieu, C. (2000) On sequential Monte Carlo sampling methods for Bayesian filtering. *Statistics and Computing.*

Douglas, S. C. (2001) Blind separation of acoustic signals. In *Microphone Arrays* (ed. Brandstein M and Ward D). Springer, chapter 10

Douglas, S. C. (2002) Blind signal separation and blind deconvolution. In *Handbook of Neural Network Signal Processing* (ed. Hu Y and Hwang J), CRC Press, chapter 7

Douglas, S. C. and Amari, S. I. (2000) Natural gradient adaptation. *Unsupervised Adaptive Filtering, vol. I: Blind Signal Separation* John Wiley & Sons, pp. 13–61.

Droppo, J., Acero, A., Deng L, L. (2002) A nonlinear observation model for removing noise from corrupted speech log mel-spectral energies. *Proc. of ICSLP*, pp. 182–185.

Duda, R. O., Hart, P. E. and Stork, D. G. (2001) *Pattern Classification*, second edn. John Wiley & Sons.

Dunteman, G. (1989) *Principal Component Analysis.* Sage Publications.

Durlach, N. (1972) Binaural signal detection: Equalization and cancellation theory *Foundations of Modern Auditory Theory.* Academic Press.

EAA – European Acoustics Association *http://www.european-acoustics.org.*

Edler, B. and Schuller, G. (2000) Audio coding using a psychoacoustic pre- and postfilter. *Proc. of ICASSP*, vol. **2**, pp. 881–884.

Eide, E. and Gish, H. (1996) A parametric approach to vocal tract length normalization. *Proc. of ICASSP*, vol. I, pp. 346–348.

Eisele, T., Häb-Umbach, R. and Langmann, D. (1996) A comparative study of linear feature transformation techniques for automatic speech recognition. *Proc. of ICSLP.*

Elko, G. W. (2000) Superdirectional microphone arrays. *Acoustic Signal Processing for Telecommunication.* Kluwer Academic, Boston, chapter 10, pp. 181–235.

Elko, G. W. (2001) Spatial coherence functions. In *Microphone Arrays* (ed. Brandstein M and Ward D) Springer, chapter 4.

Eneman, K., Duchateau, J., Moonen, M., van Compernolle, D. and van Hamme, H. (2003) Assessment of dereverberation algorithms for large vocabulary speech recognition systems, *Proc. of Eurospeech.*

Ephraim, Y. and Malah, D. (1984) Speech enhancement using a minimum mean-square error short-time spectral amplitude estimator. *IEEE Trans. on ASSP* **32**(6), 1109–1121.

Ephraim, Y. and Malah, D. (1985) Speech enhancement using a minimum mean-square error log-spectral amplitude estimator. *IEEE Trans. on ASSP* **33**, No. 2, 443–445.

Ephraim, Y. and Van Trees, H. (1995) A signal subspace approach for speech enhancement. *IEEE Trans. on SAP* **3**, 251–266.

Ephraim, Y., Malah, D. and Juang, B. H. (1989) On the application of hidden Markov models for enhancing noisy speech. *IEEE Trans. on ASSP* **37**(12), 1846–1856.

Er, M. H. and Cantoni, A. (1983) Derivative constraints for broad-band element space antenna array processors. *IEEE Trans. on ASSP* **31**, 1378–1393.

Ernst, M. and Bülthoff, H. (2004) Merging the senses into a robust perception. *Trends in Cognitive Sciences.* Elsevier, **8**(4), 162–169.

Eskénazi, M. (1993) Trends in speaking style research *Proc. of Eurospeech.*

EURASIP European Association for Signal Processing. *http://www.eurasip.org.*

Evans, N., Mason, J., Liu, W. and Fauve, B. (2006) An assessment on the fundamental limitations of spectral subtraction. *Proc. of ICASSP.*

Fancourt, C. L. and Parra, L. (2001) The coherence function in blind source separation of convolutive mixtures of non-stationary signals *Proc. of Int. Workshop on Neural Networks for Signal Processing (NNSP).*

Faubel, F. (2006) *Speech feature enhancement for speech recognition by sequential Monte Carlo methods.* Diploma thesis, Universität Karlsruhe (TH), Germany.

Faubel, F. and Wölfel, M. (2006) Coupling particle filters with automatic speech recognition for speech feature enhancement. *Proc. of Interspeech*.

Faubel, F. and Wölfel, M. (2007) Overcoming the vector tailor series approximation in speech feature enhancement – a particle filter approach. *Proc. of ICASSP*.

Fiscus, J. (1997) A post-processing system to yield reduced word error rates: Recognizer output voting error reduction (ROVER). *Proc. of ASRU*.

Fish, J. and Belytschko, T. (2007) *A First Course in Finite Elements*. John Wiley & Sons.

Fisher, R. (1936) The use of multiple measures in taxonomic problems. *Eugen* **7**, 179–188.

Fletcher, H. (1940) Auditory patterns. *Rev. Mod. Phys.* **12**, 47–65.

Fletcher, H. (1953) *Speech and Hearing in Communication*. Van Nostrand, New York.

Fletcher, H. and Munson, W. (1933) Loudness, its definition, measurement, and calculations. *Jour. of ASA* **5**, 82–108.

Friedman, J. H. and Tukey, J. W. (1974) A projection pursuit algorithm for exploratory data analysis. *IEEE Trans. on Computers* **C–23**(9), 881–890.

Frost, O. L. (1972) *An algorithm for linearly constrained adaptive array processing*. *Proc. of IEEE*, vol. **60**, pp. 926–935.

Fujimoto, M. and Nakamura, S. (2005a) Particle filter and polyak averaging-based non-stationary noise tracking for ASR in noise. *Proc. of ASRU*.

Fujimoto, M. and Nakamura, S. (2005b) Particle filter based non-stationary noise tracking for robust speech feature enhancement. *Proc. of ICASSP*.

Fujimoto, M. and Nakamura, S. (2006) Sequential non-stationary noise tracking using particle filtering with switching dynamical system. *Proc. of ICASSP*.

Fukunaga, K. (1990) *Introduction to Statistical Pattern Recognition*. Academic Press.

Furui, S. (1981) Cepstral analysis technique for automatic speaker verification. *IEEE Trans. on ASSP* **29**(2), 254–272.

Furui, S. (1986) Speaker-independent isolated word recognition using dynamic features of speech spectrum. *IEEE Trans. on ASSP* **34**, 52–59.

Gales, M. and Young, S. (1992) An improved approach to the hidden markov model decomposition of speech and noise. *Proc. of ICASSP*, pp. 233–236.

Gales, M. and Young, S. (1993) HMM recognition in noise using parallel model combination. *Proc. of Eurospeech*, vol. **2**, pp. 837–840.

Gales, M. J. F. (1995) *Model based techniques for noise robust speech recognition*. PhD thesis, Cambridge University.

Gales, M. J. F. (1996) *The generation and use of regression class trees for MLLR adaptation*. Technical Report CUED/F–INFENG/TR263, Cambridge University.

Gales, M. J. F. (1998) Maximum likelihood linear transformations for HMM-based speech recognition. *Jour. on CSL* **12**, 75–98.

Gales, M. J. F. (1999) Semi-tied covariance matrices for hidden Markov models. *IEEE Trans. on SAP* **7**, 272–281.

Gales, M. J. F. and Woodland, P. C. (1996) Mean and variance adaptation within the MLLR framework. *Jour. on CSL* **10**, 249–264.

Gallager, R. G. (1968) *Information Theory and Reliable Communication*. John Wiley & Sons.

Gannot, S. and Cohen, I. (2004) Speech enhancement based on the general transfer function GSC and postfiltering. *IEEE Trans. on SAP* **12**, 561–571.

Gehrig, T., Klee, U., McDonough, J., Ikbal, S., Wölfel, M. and Fügen, C. (2006) Tracking and beamforming for multiple simultaneous speakers with probabilistic data association filters. *Interspeech*.

Gehrig, T., Nickel, K., Ekenel, H., Klee, U. and McDonough, J. (2005) Kalman Filters for Audio-Video Source Localization. *Proc. of IEEE workshop on applications of signal processing to audio and acoustics*.

Gespert, D. and Duhamel, P. (1997) Robust blind channel identification and equalization based on multi–step predictors. *Proc. of ICASSP*.

Gilkey, R. H. and Anderson, T. R. (1997) *Binaural and Spatial Hearing in Real and Virtual Environments*. Lawrence Erlbaum Associates, Mahwah, New Jersey.

Gill, P. E., Murray, W. and Wright, M. (1981) *Practical Optimization*. Academic Press, London.

Gillespie, B. and Atlas, L. (2003) Strategies for improving audible quality and speech recognition accuracy of reverberant speech *Proc. of ICASSP*, pp. 676–679.

Glentis, G. O., Berberidis, K. and Theodoridis, S. (1999) Efficient least squares adaptive algorithsm for FIR transversal filtering. *IEEE SPM* **16**, 13–41.

Golub, G. H. and Van Loan, C. F. (1996) *Matrix Computations*, third edn. The Johns Hopkins University Press, Baltimore.

Goodman, J. (2001) A bit of progress in language modeling. *Computer Speech and Language* **15**(4), 403–434.

Goodwin, G. and Sin, K. (1984) *Adaptive Filtering, Prediction and Control*. Prentice-Hall, Englewood Cliffs, NJ, USA.

Gopalakrishnan, P., Kanevsky, D., Nádas, A. and Nahamoo, D. (1991) An inequality for rational functions with applications to some statistical estimation problems. *IEEE Trans. on Information Theory* **37**, 107–113.

Gordon, N., Salmond, D. and Smith, A. (1993) Novel approach to nonlinear/non-Gaussian Bayesian state estimation. *IEEE Proc. on Radar and Signal Processing* **140**, 107–113.

Greene, R. E. and Krantz, S. G. (1997) *Function Theory of One Complex Variable*. John Wiley & Sons, New York.

Grewal, M. S. and Andrews, A. P. (1993) *Kalman Filtering: Theory and Practice*. Prentice Hall, Upper Saddle River, NJ.

Griesinger, D. (1996) Beyond M. L. S. – Occupied hall measurement with FFT techniques. *Convention of the Audio Engineering Society, 101st Convention*.

Griffiths, L. J. (1969) *A simple adaptive algorithm for real-time processing in antenna arrays*. *Proc. of IEEE*, vol. **57**, pp. 1696–1704.

Griffiths, L. J. and Jim, C. W. (1982) An alternative approach to linearly constrained adaptive beamforming. *IEEE Trans. on Antennas Propagation* **AP–30**, 27–34.

Gunawardana, A. (2001) *Maximum mutual information estimation of acoustic HMM emission densities*. Technical Report 40, Center for Language and Speech Processing, The Johns Hopkins University, 3400N. Charles St., Baltimore, MD 21218, USA.

Gunawardana, A., Mahajan, M., Acero, A. and Platt, J. C. (2005) Hidden conditional random fields for phone classification. *Proc. of Interspeech*, Lisbon, Portugal.

Häb-Umbach, R. and Ney, H. (1992) Linear discriminant analysis for improved large vocabulary continuous speech recognition. *Proc. of ICASSP*, vol. **1**, pp. 13–16.

Häb-Umbach, R. and Schmalenströer, J. (2005) A comparison of particle filtering variants for speech feature enhancement. *Proc. of Interspeech*.

Hall, E. (1963) A system for the notation of proxemic behaviour. *American Anthropologist* **65**, 1003–1026.

Handel, S. (1989) *Listening: An Introduction to the Perception of Auditory Events*. MIT Press, New York.

Handshin, J. and Mayne, D. (1969) Monte Carlo techniques to estimate the condicional expectation in multi-stage non-linear filtering. *International Journal of Control* **9**, 547–559.

Hänsler, E. and Schmidt, G. (2004) *Acoustic Echo and Noise Control: A Practical Approach*. John Wiley & Sons.

Hänsler, E. and Schmidt, G. (2008) *Speech and Audio Processing in Adverse Environments: Signals and Communication Technologie*. Springer.

Härmä, A. and Laine, U. (2001) A comparison of warped and conventional linear predictive coding. *IEEE Trans. on SAP* **9**(5), 579–588.

Hastie, T., Tibshirani, R. and Friedman, J. (2001) *The Elemens of Statistical Learning*. Springer.

Hawley, M., Litovsky, R. and Cutting, J. (2004) The benefit of binaural hearing in a cocktail party: Effect of localization and type of interferer. *Jour. of ASA* **115**(2), 833–843.

Haykin, S. (2002) *Adaptive Filter Theory*, fourth edn. Prentice Hall, New York.

He, X., Deng, L. and Chou, W. (2008) Discriminative learning in sequential pattern recognition. *IEEE SPM*, pp. 14–36.

Herbordt, W. and Kellermann, W. (2002) Frequency-domain integration of acoustic echo cancellation and a generalized sidelobe canceller with improved robustness. *European Trans. on Telecommunications (ETT)* **13**, 123–132.

Herbordt, W. and Kellermann, W. (2003) Adaptive beamforming for audio signal acquisition In *Adaptive Signal Processing – Applications to Real-World Problems* (ed. Benesty J and Huang Y). Springer, Berlin, Germany, pp. 155–194.

Herbordt, W., Buchner, H., Nakamura, S. and Kellermann, W. (2007) Multichannel bin-wise robust frequency-domain adaptive filtering and its application to adaptive beamforming. *IEEE Trans. on ASLP* **15**, 1340–1351.

Hermansky, H. (1990) Perceptual linear predictive (PLP) analysis of speech. *Jour. of ASA* **87**(4), 1738–1752.

Hermansky, H. and Morgan, N. (1994) RASTA processing of speech. *IEEE Trans. on SAP* **2**(4), 578–589.

Hermansky, H. and Sharma, S. (1998) TRAPS – Classifiers of temporal patterns. *Proc. of ICSLP*, pp. 1003–1006.

Hermus, K., Wambacq, P. and Van Hamme, H. (2007) A review of signal subspace speech enhancement and its application to noise robust speech recognition. *Jour. on AppSP* **1**, 195–209.

Hirsch, H. and Ehrlicher, E. (1995) Noise estimation techniques for robust speech recognition. *Proc. of ICASSP*, pp. 153–156.

Hopcroft, J. E. and Ullman, J. D. (1979) *Introduction to Automata Theory, Languages and Computation*. Addison-Wesley Publishing, Reading, MA, USA.

Hori, T. and Nakamura, A. (2005) Generalized fast on-the-fly composition algorithm for WFST-based speech recognition. *Proc. of Interspeech*.

Hoshuyama, O., Sugiyama, A. and Hirano, A. (1999) A robust adaptive beamformer for microphone arrays with a blocking matrix using constrained adaptive filters. *IEEE Trans. on ASLP* **47**, 2677–2684.

Householder, A. S. (1958) Unitary triangularization of a non-symmetric matrix. *Jour. of Assoc. Computer Mathematics* **5**, 204–243.

Hu, Y. and Loizou, P. C. (2007) Subjective comparison and evaluation of speech enhancement algorithms. *Jour. on SC* **49**, 588–601.

Huang, J. and Zhao, Y. (2000) A DCT-based fast signal subspace technique for robust speech recognition. *IEEE Trans. on SAP* **8**, 747–751.

Huang, X., Acero, A. and Hon, H. W. (2001) *Spoken Language Processing*. Prentice Hall.

Huang, Y. A. and Benesty, J. (2004) *Audio Signal Processing for Next-Generation Multimedia Communication Systems*. Kluwer, Boston, MA.

Huang, Y. A., Benesty, J. and Elko, G. W. (2004) Source localization. *Audio Signal Processing for Next-Generation Multimedia Communication Systems*. Kluwer Academic, Boston, pp. 229–254.

Hugonnet, C. and Walder, P. (1998) *Stereophonic Sound Recording: Theory and Practice*. John Wiley & Sons.

Hyvärinen, A. (1998) New approximations of differential for independent component analysis and projection pursuit. *Advances in Neural Information Processing Systems*, vol. **10**. MIT Press, Cambridge, MA, USA, pp. 273–279.

Hyvärinen, A. (1999) Survey on independent component analysis. *Neural Computing Surveys* **2**, 94–128.

Hyvärinen, A. and Oja, E. (2000) Independent component analysis: Algorithms and applications. *Neural Networks* **13**, 411–430.

IEEE – Institute of Electrical and Electronics Engineers. *http://www.ieee.org*.

Indrebo, K., Povinelli, R. and Johnson, M. (2005) Third-order moments of filtered speech signals for robust speech recognition *ITRW on Non-Linear Speech Processing (NOLISP)*.

IPA (1999) *Handbook of the International Phonetic Association: A Guide to the Use of the International Phonetic Alphabet*. International Phonetic Association.

ISCA – International Speech Communication Association. *http://www.isca-speech.org*.

Jacobs, R. (2002) What determines visual cue reliability? *Trends Cogn. Sci.* **6**, 345–350.

JASA – *The Journal of the Acoustical Society of America*. http://scitation.aip.org/jasa.

Jayant, N. S. and Noll, P. (1984) *Digital Coding of Waveforms: Principles and Applications to Speech and Video*. Prentice Hall, Englewood Cliffs.

Jazwinski, A. H. (1970) *Stochastic Processes and Filtering Theory*. Academic Press, New York.

Jelinek, F. (1998) *Statistical Methods for Speech Recognition*. Bradford Books.

Johnson, D. and Dudgeon, D. (1993) *Array Signal Processing*. Prentice Hall.

Jones, M. and Viola, P. (2003) Fast multi-view face detection. *Proc. of IEEE Conference on Computer Vison and Pattern Recognition*.

Julier, S. and Uhlmann, J. (2004) Unscented filtering and nonlinear estimation. *IEEE Review* **92**(3), 401–421.

Junqua, J. (1993) The Lombard reflex and its role on human listeners and automatic speech recognizers. *Jour. of ASA* **93**(1), 510–24.

Kalman, R. (1960) A new approach to linear filtering and prediction problems. *Trans. of the ASME – Journal of Basic Engineering* **82** (Series D), 35–45.

Kaminski, P., Bryson, A. and Schmidt, S. (1971) Discrete square root filtering: A survey of current techniques. *IEEE Transactions on Automatic Control* **AC-16**, 727–736.

Kannan, A. and Khudanpur, S. (1999) Tree-structured models of parameter dependence for rapid adaptation in large vocabulary conversational speech recognition. *Proc. of ICASSP*, vol. II, pp. 769–772.

Kanthak, S., Ney, H., Riley, M. and Mohri, M. (2002) A comparison of two LVR search optimization techniques. *Proc. of ICSLP*, Denver, CO, USA.

Katz, S. (1987) Estimation of probabilities from sparse data for the language model component of a speech recognizer. *Proc. of IEEE Trans. on ASSP* **35**, 400–401.

Kay, S. (1993) *Fundamentals of Statistical Signal Processing: Estimation Theory*. Prentice-Hall, Englewood Cliffs, NJ.

Kellermann, W. (2001) Acoustic echo cancellation for beamforming microphone arrays. In *Microphone Arrays* (ed. Branstein M and Ward D). Springer, Heidelberg, pp. 281–306.

Kim, N. S. (1998) IMM-based estimation for slowly evolving environments. *IEEE SPL* **5**, No. 6, 146–149.

Kinoshita, K., Nakatani, T. and Miyoshi, M. (2005) Efficient dereverberation framework for automatic speech recognition. *Proc. of Interspeech*, pp. 3145–3148.

Kinoshita, K., Nakatani, T. and Miyoshi, M. (2006) Spectral subtraction steered by multi-step forward linear prediction for single channel speech dereverberation. *Proc. of ICASSP*, pp. 817–820.

Kitagawa, G. (1996) Monte Carlo filter and smoother for non-Gaussian nonlinear state space models. *Jour. of Computational and Graphical Statistics* **5**, 1, 1–25.

Kjellberg, A., Ljung, R. and Hallman, D. (2007) Recall of words heard in noise. *Applied Cognitive Psychology* **21**, 1–11.

Klee, U., Gehrig, G. and McDonough, J. (2005a) Kalman filters for time delay of arrival-based source localization. *Proc. of Eurospeech*.

Klee, U., Gehrig, T. and McDonough, J. (2005b) Kalman filters for time delay of arrival-based source localization. *Jour. of AppSP, Special Issue on Multi-Channel Speech Processing*.

Kneser, R. and Ney, H. (1995) Improved backing-off for *m*-gram language modeling *Proc. of ICASSP*, Detroit, MI, USA.

Koenig, W., Dunn, H. and Lacy, L. (1946) The sound spectrograph. *Jour. of ASA* **18**, 19–49.

Kokkinakis, K. and Nandi, A. K. (2005) Exponent parameter estimation for generalized Gaussian probability density functions with application to speech modeling. *Signal Processing* **85**, 1852–1858.

Kolmogorov, A. (1941a) Interpolation and extrapolationof stationary random sequences. *Bull. Math. Univ. Moskow (in Russian) Izv. Akad. Nauk USSR, Ser. Math.* **5**(5), 3–14.

Kolmogorov, A. (1941b) Stationary sequences in Hilbert space. *Bull. Math. Univ. Moskow (in Russian)*.

Kong, A., Liu, J. S. and Wong, W. H. (1994) Sequential imputations and Bayesian missing data problems. *Jour. of the American Statistical Association*.

Kuhn, R., Junqua, J. C., Nguyen, P. and Niedzielski, N. (2000) Speaker adaptation in eigenvoice space. *Trans. on SAP* **8**(6), 695–707.

Kullback, S. and Leibler, R. A. (1951) On information and sufficiency. *Annals of Mathematical Statistics* **22**, 79–86.

Kumar, N. and Andreou, A. (1996) *A generalization of linear discriminant analysis in maximum likelihood framework*. Tech. Rep. JHU–CLSP Technical Report No. 16, Johns Hopkins University.

Kumar, N. and Andreou, A. (1998) Heteroscedastic discriminant analysis and reduced rank HMMS for improved speech recognition. *Jour. on SC* **26**, 283–97.

Kumatani, K., Gehrig, T., Mayer, U., Stoimenov, E., McDonough, J. and Wölfel, M. (2007) Adaptive beamforming with a minimum mutual information criterion. *Trans. on SAP* **15**, 2527–2541.

Kumatani, K., McDonough, J., Klakow, D., Garner, P. N. and Li, W. (2008a) Adaptive beamforming with a maximum negentropy criterion. *Trans. on SAP*.

Kumatani, K., McDonough, J., Klakow, D., Garner, P. N. and Li, W. (2008b) Adaptive beamforming with a maximum negentropy criterion. *Proc. of HSCMA*, Trento, Italy.

Kumatani, K., McDonough, J., Rauch, B., Garner, P. N., Li, W. and Dines, J. (2008c) Maximum kurtosis beamforming with the generalized sidelobe canceller. *Proc. of Interspeech*, Sydney, Australia.

Kumatani, K., McDonough, J., Schact, S., Klakow, D., Garner, P. N. and Li, W. (2008d) Filter bank design based on minimization of individual aliasing terms for minimum mutual information subband adaptive beamforming. *Proc. of ICASSP*, Las Vegas, NV, USA.

Kuttruff, H. (1997) Sound in enclosures. *Encyclopedia of Acoustics*.

Kuttruff, H. (2000) *Room Acoustics*. Elsevier Applied Science.

Lacoss, R. (1971) Data adaptive spectral analysis methods. *Geophysics* **36**(4), 661–675.

Lamel, L. and Gauvain, J. (2005) Phone models for conversational speech. *Proc. of ICASSP*.

Lebart, K., Boucher, J. and Denbigh, P. (2001) A new method based on spectral subtraction for speech dereverberation. *Acta Acustica united with Acustica* **87**(3), 359–366.

Lee, C. H. and Gauvain, J. L. (1993) Speaker adaptation based on map estimation of HMM parameters. *Proc. of ICASSP*, pp. 558–561.

Lee, K., Lee, B. G. and Ann, S. (1997) Adaptiv filtering for speech enhancement in colored noise. *IEEE SPL* **4**, 277–279.

Lee, L. and Rose, R. C. (1996) Speaker normalization using efficient frequency warping procedures. *Proc. of ICASSP*, vol. I, pp. 353–356.

Leggetter, C. J. and Woodland, P. C. (1995a) Maximum likelihood linear regression for speaker adaptation of continuous density hidden markov models. *Jour. on CSL*, pp. 171–185.

Leggetter, C. J. and Woodland, P. C. (1995b) Maximum likelihood linear regression for speaker adaptation of continuous density hidden markov models. *Jour. on CSL*, pp. 171–185.

Li, Z. and Duraiswami, R. (2005) Hemispherical microphone arrays for sound capture and beamforming. *Proc. of WASPAA*, New Paltz, NY, USA.

Lienhart, R. and Maydt, J. (2002) An extended set of Haar–like features for rapid object detection. *Proc. of International Conference on Image Processing*.

Lim, J. and Oppenheim, A. (1978) All pole modeling of degraded speech. *IEEE Trans. on ASSP* **26**, 197–209.

Lim, J. and Oppenheim, A. (1979) Enhancement and bandwidth compresson of noisy speech. *Proc. of IEEE* **67**(12), 1586–1604.

Lincoln, M., McCowan, I., Vepa, I. and Maganti, H. K. (2005) The multi-channel Wall Street Journal audio visual corpus (MC–WSJ–AV): Specification and initial experiments *Proc. of ASRU*, pp. 357–362.

Liu, C., Jiang, H. and Li, X. (2005) Discriminative training of CDHMMs for maximum relative separation margin *Proc. of ICASSP*, Philadelphia, Pennsylvania, USA.

Ljolje, A., Pereira, F. and Riley, M. (1999) Efficient general lattice generation and rescoring *Proc. of Eurospeech*, Budapest, Hungary.

Llisterri, J. (1992) Speaking styles in speech research *In ELSENET/ESCA/SALT Workshop on Integrating Speech and Natural Language*, pp. 17–37.

Lockwood, P. and Boudy, J. (1992) Experiments with a nonlinear spectral subtraction (NSS), hidden Markov models and the projection, for robust speech recognition in car. *Jour. on SC*, pp. 215–228.

Loizou, P. (2007) *Speech Enhancement*. CRC Press, Taylor & Francis Group.

Lombard, E. (1911) Le signe de l'élévation de la voix. *Ann. Maladies Oreille, Larynx, Nez, Pharynx* **37**, 101–119.

Lotter, T. and Vary, P. (2005) Speech enhancement by MAP spectral amplitude estimation using a super–Gaussian speech model. *Jour. on AppSP* **7**, 1110–1126.

Luenberger, D. G. (1984) *Linear and Nonlinear Programming*, second edn. Addison-Wesley, New York.

Luke, Y. L. (1969) *The Special Functions and their Approximations*. Academic Press, New York.

Lyons, R. G. (2004) *Understanding Digital Signal Processing*, second edn. Prentice Hall, Upper Saddle River, NJ.

Maccormick, J. and Blake, A. (2000) A probabilistic exclusion principle for tracking multiple objects. *International Jour. of Computer Vision* **39**(1), 57–71.

Macho, D., Padrell, J., Abad, A., Nadeu, C., Hernando, J., McDonough, J., Wölfel, M., Klee, U., Omologo, M., Brutti, A., Svaizer, P., Potamianos, G. and Chu, S. (2005) Automatic speech activity detection, source localization, and speech recognition on the CHIL seminar corpus. *Proc. of ICME*.

Mahajan, M., Gunawardana, A. and Acero, A. (2006) Training algorithms for hidden conditional random fields. *Proc. of ICASSP*, Toulouse, France.

Makhoul, J. (1975) Linear prediction: A tutorial review. *Proc. of the IEEE* **63**(4), 561–580.

Mangu, L., Brill, E. and Stolcke, A. (2000) Finding consensus in speech recognition: Word error minimization and other applications of confusion networks. *Jour. on CSL* vol. **14**, no. 4 **14**(4), 373–400.

Manning, C. D. and Schütze, H. (1999) *Foundations of Statistical Natural Language Processing*. The MIT Press, Cambridge, MA, USA.

Manning, C. D., Raghavan, P. and Schütze, H. (2008) *Introduction to Information Retrieval*. Cambridge University Press, Cambridge, England.

Markel, J. and Gray, A. H. J. (1980) *Linear Prediction of Speech*. Springer.

Martin, R. (1994) Spectral subtraction based on minimum statistics. *European Signal Processing Conference*, pp. 1182–1185.

Martin, R. (2005) Speech enhancement based on minimum square error estimation and super-Gaussian priors. *Trans. on SAP* **13**(5), 845–856.

Masry, E., Stieglitz, K. and Liu, B. (1968) Bases in Hilbert space related to the representationof stationary operators. *SIAM Jour. on Appl. Math.* **16**, 552–562.

Matsumoto, H. and Moroto, M. (2001) Evaluation of mel-LPC cepstrum in a large vocabulary continuous speech recognition. *Proc. of ICASSP* **1**, 117–120.

Matsumoto, M., Nakatoh, Y. and Furuhata, Y. (1998) An efficient mel–LPC analysis method for speech recognition. *Proc. of ICSLP*, pp. 1051–1054.

Matsumura, M., Yamane, H. and Fujii, K. (2007) Recognition of vowels and a semivowel using formant locus extracted by cubic spline. *Electronics and Communications in Japan* **72**(11), 73–86.

McAulay, R. and Malpass, M. (1980) Speech enhancement using a soft-decision noise suppression filter. *IEEE Trans. on ASSP* **28**(2), 137–145.

McCowan, I., Hari-Krishna, M., Gatica-Perez, D., Moore, D. and Ba, S. (2005) Speech acquisition in meetings with an audio-visual sensor array. *Proceedings of the IEEE International Conference on Multimedia and Expo (ICME)*.

McCowan, I., Lincoln, M. and Himawan, I. (2008) Microphone array shape calibration in diffuse noise fields. *IEEE Trans. on Audio Speech Language Processing* **16**(3), 666–670.

McDonough, J. (2000) *Speaker compensation with all-pass transforms*. PhD thesis, Johns Hopkins University, Baltimore, Maryland, USA.

McDonough, J. and Stoimenov, E. (2007) An algorithm for fast composition with weighted finite-state transducers. *Proc. of ASRU*, Kyoto, Japan.

McDonough, J. and Waibel, A. (2004) Performance comparisons of all-pass transform adaptation with maximum likelihood linear regression. *Proc. of ICASSP*, Miami, FL, USA.

McDonough, J. and Wölfel, M. (2008) Distant speech recognition: Bridging the gaps. *Proc. of HSCMA*, Trento, Italy.

McDonough, J., Byrne, W. and Luo, X. (1998) Speaker normalization with all-pass transforms. *Proc. of ICSLP*.

McDonough, J., Kumatani, K., Gehrig, T., Stoimenov, E., Mayer, U., Schacht, S., Wölfel, M. and Klakow, D. (2008a) To separate speech! A system for recognizing simultaneous speech. *Proc. of MLMI*.

McDonough, J., Wölfel, M. and Waibel, A. (2007) On maximum mutual information speaker-adapted training. *Jour. on CSL*.

McDonough, J., Wölfel, M., Kumatani, K., Rauch, B., Faubel, F. and Klakow, D. (2008b) Distant speech recognition: No black boxes allowed. *Proc. of ITG-Fachtagung*, Aachen, Germany.

McDonough, J. W. (1998a) On the estimation of optimal regression classesfor speaker adaptation.

McDonough, J. W. (1998b) Transformation of discrete-time sequences with analytic functions.

McGurk, H. and MacDonald, J. (1976) Hearing lips and seeing voices. *Nature* **264**, 746–748.

Meyer, J. and Elko, G. W. (2004) Spherical microphone arrays for 3D sound recording *Audio Signal Processing for Next-Generation Multimedia Communication Systems*. Technical Report 131, Center for Language and Speech Processing, The Johns Hopkins University, Kluwer Academic, Boston, pp. 67–90.

Meyer, J., Simmer, K. and Kammeyer, K. (1997) Comparison of one and two-channel noise-estimation techniques. *Proc. of IWAENC*. Technical Report 145, Center for Language and Speech Processing, The Johns Hopkins University, pp. 17–20.

Meyn, S. P. and Tweedie, R. L. (1993) Markov chains and stochastic stability. *Communication and Control Engineering*, second edn. Springer.

Middlebrooks, J. and Green, D. (1991) Sound localization by human listeners. *Annual Review of Psychology* **42**, 135–159.

Miller, G. A. and Licklider, J. C. R. (1950) The intelligibility of interrupted speech. *Jour. of ASA* **22**, 167–173.

Mohri, M. (1997) Finite-state transducers in language and speech processing. *Computational Linguistics*.

Mohri, M. (2002) Generic epsilon-removal and input epsilon-normalization algorithms for weighted transducers. *International Jour. of Foundations of Computer Science* **13**(1), 129–143.

Mohri, M. and Riley, M. (1998) Network optimizations for large vocabulary speech recognition. *Jour. on SC*.

Mohri, M. and Riley, M. (2001) A weight pushing algorithm for large vocabulary speech recognition. *Proc. of ASRU*, pp. 1603–1606, Aarlborg, Denmark.

Mohri, M. and Riley, M. (2002) An efficient algorithm for the n-best-strings problem. *Proc. of ICSLP*.

Mohri, M., Pereira, F. and Riley, M. (2000) The design principles of a weighted finite-state transducer library. *Theoretical Computer Science* **231**(1), 17–32.

Mohri, M., Pereira, F. and Riley, M. (2002) Weighted finite-state transducers in speech recognition. *Jour. on CSL* **16**, 69–88.

Mohri, M., Pereira, F. C. and Riley, M. (1998) A rational design for a weighted finite-state transducer library. *Second International Workshop on Implementing Automata*. Lecture Notes in Computer Science, No. 1436, Heidelberg.

Moreno, A. and Pfretzschner, J. (1978) Human head directivity in speech emission: A new approach. *Acoustics Letters* **1**, pp. 78–84.

Moreno, P. (1996) *Speech Recognition in noisy environments*. PhD thesis, Carnegie Mellon University.

Moreno, P., Raj, B. and Stern, R. (1995) Multivariate-Gaussian-based cepstral normalization for robust speech recognition. *Proc. of ICASSP*.

Moreno, P., Raj, B. and Stern, R. (1996) A vector Taylor series approach for environment-independent speech recognition. *Proc. of ICASSP*.

Morgan, N. and Bourlard, H. (1995) Continuous speech recognition: An introduction to the hybrid HMM/connectionist approach. *IEEE SPM*, pp. 25–42.

Morii, S. (1988) *Spectral subtraction in the sphinx system*. Unpublished.

Möser, M. (2004) *Engineering Acoustics*. Springer.

Murthi, M. and Rao, B. (1997) Minimum variance distortionless response (MVDR) modelling of voiced speech. *Proc. of ICASSP*.

Murthi, M. and Rao, B. (2000) All-pole modeling of speech based on the minimum variance distortionless response spectrum. *IEEE Trans. on SAP* **8**(3), 221–239.

Musicus, B. (1985) Fast MLM power spectrum estimation from uniformly spaced correlations. *IEEE Trans. on ASSP* **33**, 1333–1335.

Nakatoh, Y., Nishizaki, M., Yoshizawa, S. and Yamada, M. (2004) An adaptive mel–LP analysis for speech recognition. *Proc. of ICSLP*.

Neely, S. and Allen, J. (1979) Invertibility of a room impulse. response. *Jour. of the Acoustical Society* **66**, 165–169.

Neeser, F. D. and Massey, J. L. (1993) Proper complex random processes with applications to information theory. *IEEE Trans. on Information Theory* **39**(4), 1293–1302.

Nemer, E., Goubran, R. and Mahmoud, S. (2002) Speech enhancement using fourth–order cumulants and optimum filters in the subband domain. *Jour. on SC* **36**(3), 219–246.

Newell, Fea, (2001) Viewpoint dependence in visual and haptic object recognition. *Psychol. Sci.* **12**, 37–42.

Ney, H., Essen, U. and Kneser, R. (1994) On structuring probabilistic dependences in stochastic language modelling. *Jour. on CSL* **8**, 1–38.

Nishiura, T., Hirano, Y., Denda, Y. and Nakayama, M. (2007) Investigations into early and late reflections on distant–talking speech recognition toward suitable reverberation criteria *Proc. of Interspeech*.

NIST Speech Tools. *http://www.nist.gov/speech/tools*.

Nocerino, N., Soong, F., Rabiner, L. and Klatt, D. (1985) Comparative study of several distortion measures for speech recognition. *Proc. of ICASSP*.

NOISEX – noisex. *http://www.spib.rice.edu/spib/select_noise.html*.

Noll, A. (1964) Short-time spectrum and 'cepstrum' technique for vocal-pitch detection. *Jour. of ASA* **36**, 296–302.

Nordebo, S., Claesson, I. and Nordholm, S. (1994) Adaptive beamforming: spatial filter designed blocking matrix. *IEEE Jour. of Oceanic Engineering* **19**, 583–590.

Nordholm, S., Claessen, I. and Grbić, N. (2001) Optimal and adaptive microphone arrays for speech input in automobiles. In *Microphone Arrays* (ed. Brandstein M and Ward D). Springer Verlag, Heidelberg, Germany, chapter 4.

Nordholm, S., Claesson, I. and Bengtsson, B. (1993) Adaptive array noise suppression of handsfree speaker input in cars. *IEEE Trans. on Vehicular Technology* **42**, 514–518.

Normandin, Y. (1991) *Hidden Markov models, maximum mutual information, and the speech recognition problem*. PhD thesis, McGill University.

Obuchi, Y. (2004) Multiple-microphone robust speech recognition using decoder-based channel selection. *Workshop on Statistical and Perceptual Audio Processing, Jeju, Korea*.

Olive, J. (1993) *Acoustics of American English Speech: A Dynamic Approach*. Springer.

Omologo, M. and Svaizer, P. (1994) Acoustic event localization using a crosspower-spectrum phase based technique. *Proc. of ICASSP*, vol. II, pp. 273–276.

Oppenheim, A. and Schafer, R. (1989) *Discrete-Time Signal Processing*. Prentice-Hall Inc.

Oppenheim, A., Johnson, D., and Steiglitz, K. (February 1971) Computation of spectra with unequal resolution using the fast Fourier transform. IEEE Proceedings Letters, Vol. **59**, No. 2, 229–301.

Oppenheim, A. V. and Johnson, D. H. (1972) Discrete-time representation of signals. *Proc. of IEEE* **60**(6), 681–691.

Owsley, N. L. (1971) *Source location with an adaptive antenna array*. Technical report, Naval Underwater Systems Center, National Technical Information Service.

Paliwal, K. K. and Basu, A. (1987) A speech enhancement method based on kalman filtering. *IEEE Trans. on ASSP* **12**, 177–180.

Pan, Y. and Waibel, A. (2000) The effects of room acoustics on MFCC speech parameter. *Proc. of ICSLP*.

Papoulis, A. (1984) *Probability, Random Variables, and Stochastic Processes*, second edn. McGraw-Hill, New York.

Parra, L. C. and Alvino, C. V. (2002) Geometric source separation: Merging convolutive source separation with geometric beamforming. *IEEE Trans. on SAP* **10**(6), 352–362.

Paul, D., Lippmann, R., Chen, Y. and Weinstein C, C.J. (1986) Robust HMM-based techniques for recognition of speech produced under stress and in noise. *Proc. of DARPA Speech Recognition Workshop*, pp. 81–92.

Pearsons, K., Bennett, R. and Fidell, S. (1977) *Speech levels in various noise environments*. Technical Report EPA–600/1–77–025, U.S. Environment Protection Agency.

Pereira, F. and Riley, M. (1997) Speech recognition by composition of weighted finite automata. In *Finite-State Language Processing* (ed. Roche E and Schabes Y). MIT Press, Cambridge, MA, pp. 431–453.

Pereira, F. C. N. and Wright, R. N. (1991) Finite-state approximation of phrase structure grammars. *Proc. of the 29th Annual Meeting on Association for Computational Linguistics*, Berkeley, CA, USA.

Petrick, R., Lohde, K., Wolff, M. and Hoffmann, R. (2007) The harming part of room acoustics in automatic speech recognition. *Proc. of Interspeech*, pp. 1094–1097.

Picone, J. (1993) Signal modeling techniques in speech recognition. *Proc. of the IEEE* **81**(9), 1215–1247.

Pitt, M. and Shephard, N. (1999) Filtering via simulation: Auxiliary particle filters. *Jour. of the American Statistical Association.*

Pollefeys, M. (2000) *Tutorial on 3D Modeling from Images*. Katholieke Universiteit, Leuven.

Potamianos, G., Neti, C., Luettin, J. and Matthews, I. (2004) In *Issues in Visual and Audio-Visual Speech Processing* (ed. Bailly G, Vatikiotis-Bateson, E and Perrier P). MIT Press, chapter Audio–Visual Automatic Speech Recognition: An Overview.

Povey, D. and Woodland, P. (2002) Minimum phone error and I-smoothing for improved discriminative training. *Proc. of ICASSP*, Orlando, Florida, USA.

Povey, D., Kingsbury, B., Mangu, L., Saon, G., Soltau, H. and Zweig, G. (2004) FMPE: Discriminatively trained features for speech recognition. *Rich Transcription 2004 Workshop.*

Povey, D., Kingsbury, B., Mangu, L., Saon, G., Soltau, H. and Zweig, G. (2005) fMPE: Discriminatively trained features for speech recognition. *Proc. of ICASSP*, Philadelphia, Pennsylvania, USA.

Press, W. H., Teukolsky, S. A., Vetterling, W. T. and Flannery, B. P. (1992) *Numerical Recipes in C*, second edn. Cambridge University Press, Cambridge, UK.

Proakis, J. G. and Manolakis, D. G. (2007) *Digital Signal Processing*, fourth edn. Prentice Hall, Upper Saddle River, NJ.

Project54. *http://www.project54.unh.edu.*

Pye, D. and Woodland, P. C. (1997) Experiments in speaker normalisation and adaptation for large vocabulary speech recognition. *Proc. of ICASSP*, vol. II, pp. 1047–1050.

Quinlan, A. and Asano, F. (2008) Tracking a varying number of speakers using particle filtering. *Proc. of ICASSP*, Las Vegas, NV, USA.

Rabiner, L. R. (1989) A tutorial on hidden markov models and selected applications in speech recognition. *Proc. of the IEEE* **77**(2), 257–286.

Rabiner, L. R. and Schafer, R. W. (1978) *Digital Processing of Speech Signals*. Prentice-Hall.

Rabinkin, D., Renomeron, R., French, J. and Flanagan, J. (1996) Estimation of wavefront arrival delay using the cross-power spectrum phase technique. *132nd Meeting of the ASA*, Honolulu.

Rabinkin, D., Renomeron, R., French, J. and Flanagan, J. (1997) Optimum microphone placement for array sound capture. *133nd Meeting of the ASA*, State College, PA.

Rafaely, B. (2008) Spherical microphone array with multiple nulls for analysis of directional room impulse responses. *Proc. of ICASSP*, Las Vegas, NV, USA.

Raj, B., Singh, R. and Stern, R. (2004) On tracking noise with linear dynamical system models. *Proc. of ICASSP.*

Rajasekaran, P. and Doddington, G. (1986) Robust speech recognition: Initial results and progress *Proc. of DARPA Speech Recognition Workshop*, pp. 73–80.

Raub, D., McDonough, J. and Wölfel, M. (2004) A cepstral domain maximum likelihood beamformer for speech recognition. *Proc. of ICSLP*, Jeju Island, Korea.

Rauch, B., Kumatani, K., Faubel, F., McDonough, J. and Klakow, D. (2008) On hidden Markov model maximum negentropy beamforming. *Proc. of IWAENC*, Seattle, WA, USA.

Ristic, B., Arulampalam, S. and Gordon, N. (2004) *Beyond the Kalman Filter: Particle Filters for Tracking Applications*. Artech House, Boston, MA.

Robert, C. and Casella, G. (2004) *Monte Carlo Statistical Methods*. *Springer Texts in Statistics*, second edn., Springer.

Rosen, S. and Howell, P. (1990) *Signals and Systems for Speech and Hearing*. Academic Press, London.

Rosenfeld, R. (2000) Two decades of statistical language modeling: Where do we go from here? *Proc. of IEEE* **88**, 1270–1278.

RT – rich transcription evaluation project. *http://www.nist.gov/speech/tests/rt*.

Ryan, J. (1998) Criterion for the minimum source distance at which plane-wave beamforming can be applied. *Jour. of. ASA* **104**(1), 595–598.

Saito, S. and Nakata, K. (1985) *Fundamentals of Speech Signal Processing*. Academic Press.

Salavedra, J., Masgrau, E., Moreno, A., Estarellas, J. and Jove, X. (1994) Robust coefficients of a higher order ar modelling in a speech enhancement system using parameterized Wiener filtering. *Electrotechnical Conference*, pp. 12–14.

Sankar, A. and Lee, C. H. (1996) A maximum-likelihood approach to stochastic matching for robust speech recognition. *IEEE Trans. on SAP* **4**(3), 190–201.

Saon, G., Povey, D. and Zweig, G. (2005) Anatomy of an extremely fast LVCSR decoder. *Proc. of Interspeech*, Lisbon, Portugal.

Sayed, A. H. and Kailath, T. (1994) A state-space approach to adaptive RLS filtering. *IEEE SPM*, pp. 18–60.

Schau, H. C. and Robinson, A. Z. (1987) Passive source localization employing intersecting spherical surfaces from time-of-arrival differences. *IEEE Trans. on ASSP*, **35**(8), 1223–5.

Schlüter, R. (2000) *Investigations on discriminative training criteria*. PhD thesis, Rheinisch–Westfälische Technische Hoschschule, Aachen.

Schouten, J. (1940) The residue, a new component in subjective sound analysis. *Proc. of Koninklijke Nederlandse Akademie Wetenschappen* **43**, 356–365.

Schrempf, O. C., Brunn, D. and Hanebeck, U. D. (2006) Density approximation based on Dirac mixtures with regard to nonlinear estimation and filtering. *Proc. of IEEE Conf. on Decision and Control*, San Diego, CA, USA.

Schröder, M. R. (1965) U.S. Patent No 3,180,936, filed Dec. 1, 1960, issued Apr. 27, 1965.

Schröder, M. R. (1968) U.S. Patent No 3,403,224, filed May 28, 1965, issued Sept. 24, 1968.

Schröder, M. R. (1981) Direct (nonrecursive) relations between cepstrum and predictor coefficients. *IEEE Trans. on ASSP* **29**(2), 297–301.

Schuler, G. (2004) Audio coding. In *Audio Signal Processing for Next–Generation Multimedia Communication Systems* (ed. Huang Y and Benesty J). Kluwer, chapter 10.

Schuster, M. and Hori, T. (2005) Efficient generation of high-order context-dependent weighted finite state transducers for speech recognition. *Proc. of ICASSP*, pp. 201–204.

Schwenk, H. and Gauvain, J. (2004) Neural network language models for conversational speech recognition *Proc. of ICSLP*, pp. 1215–1218, Jeju Island, Korea.

Sehr, A. and Kellermann, W. (2007) A new concept for feature-domain dereverberation for robust distant-talking ASR. *Proc. of ICASSP*.

Sehr, A. and Kellermann, W. (2008) Towards robust distant-talking automatic speech recognition in reverberant environments. In *Topics in Speech and Audio Processing in Adverse Environments* (ed. Hänsler E and Schmidt G). Springer.

Sellars, P. (2000) Behind the mask – perceptual coding: How MP3 compression works. *Cambridge, Sound on Sound*.

Seltzer, M. L., Raj, B. and Stern, R. M. (2004) Likelihood-maximizing beamforming for robust hands-free speech recognition. *Trans. of SAP* **12**(5), 489–498.

Seymore, K. and Rosenfeld, R. (1996) Scalable backoff language models. *Proc. of ICSLP*, Philadelphia, PA, USA.

Sha, F. and Saul, L. K. (2006) Large margin Gaussian mixture modeling for phonetic classification and recognition *Proc. of ICASSP*, Toulouse, France.

Shannon, C. (1948) A mathematical theory of communication. *Bell System Technical Journal* **27**, 379–423 and 623–656.

Sharon Gannot, D., Burshtein Weinstein, E. (2001) Signal enhancement using beamforming and nonstationarity with applications to speech. *IEEE Trans. on SP* **49**, 1614–1626.

Shikano, K. (1986) *Evaluation of LPC spectral matching measures for phonetic unit recognition*. Technical report, Computer Science Department, Carnegie Mellon University, Pittsburgh, Pennsylvania.

Shimizu, Y., Kajita, S., Takeda, K. and Itakura, F. (2000) Speech recognition based on space diversity using distributed multi-microphones. *Proc. of ICASSP*.

Shriberg, E., Stolcke, A. and Baron, D. (2001) Observations on overlap: findings and implications for automatic processing of multi-party conversation. *Proc. of Eurospeech 2001*, vol. II.

Silverman, H. (1987) Some analysis of microphone arrays for speech data acquisition. *Trans. on ASSP* **35**, 1699–1712.

Simmer, K. U., Bitzer, J. and Marro, C. (2001) Post-filtering techniques. In *Microphone Arrays* (ed. Branstein M and Ward D). Springer, Heidelberg, pp. 39–60.

Simon, D. (2006) *Optimal State Estimation: Kalman, H_∞, and Nonlinear Aproaches*. John Wiley & Sons, Inc., New York.

Singh, R. and Raj, B. (2003) Tracking noise via dynamical systems with a continuum of states. *Proc. of ICASSP*.

Siohan, O., Ramabhadran, B. and Kingsbury, B. (2005) Constructing ensembles of ASR systems using randomized decision trees. *Proc. of ICASSP*.

Smith III, J. O. and Abel, J. S. (1999) Bark and ERB bilinear transforms. *IEEE Trans. on SAP* **7**(6), 697–708.

Smith, J. O. and Abel, J. S. (1987) Closed-form least-squares source location estimation from range-difference measurements. *Trans. on ASSP* **35**(12), 1661–9.

Söderström, T. and Stoica, P. (1989) *System Identification*. Prentice-Hall, Englewood Cliffs, NJ, USA.

Soltau, H., Metze, F., Fügen, C. and Waibel, A. (2001) A one pass-decoder based on polymorphic linguistic context assignment. *Proc. of ASRU*.

Sorenson, H. and Alspach, D. (1971) Recursive bayesian estimation using gaussian sums. *Automatica* **7**, 465–479.

Sproat, R. and Riley, M. (1996) Compilation of weighted finite-state transducers from decision trees. *Proc. of ACL*, Santa Cruz, California.

Stahl, V., Fischer, A. and Bippus, R. (2000) Quantile based noise estimation for spectral subtraction and Wiener filtering. *Proc. of ICASSP*.

Stark, H. and Woods, J. W. (1994) *Probability, Random Processes, and Estimation Theory for Engineers*, second edn. Prentice–Hall, Upper Saddle River, NJ.

Stein, J. Y. (2000) *Digital Signal Processing: A Computer Science Perspective*. Wiley-Interscience.

Stevens, S., Volkman, J. and Newman, E. (1937) The mel scale equates the magnitude of perceived differences in pitch at different frequencies. *Jour. of ASA* **8**(3), 185–190.

Stiefelhagen, R., Bernardin, K., Bowers, R., Rose, R. T., Michel, M. and Garofolo, J. (2008) The CLEAR 2007 evaluation. *Multimodal Technologies for Perception of Humans: International Evaluation Workshops CLEAR 2007 and RT 2007*, vol. LNCS 4625. Springer, Heidelberg, pp. 3–34.

Stockham, T. (1966) High-speed convolution and correlation. *Proc. of Spring Joint Computer Conference (AFIPS)*.

Stoica, P. and Moses, R. (2005) *Spectral Analysis of Signals*. Prentice Hall.

Stoimenov, E. and McDonough, J. (2006) Modeling polyphone context with weighted finite-state transducers. *Proc. of ICASSP*.

Stoimenov, E. and McDonough, J. (2007) Memory efficient modeling of polyphone context with weighted finite-state transducers. *Proc. of Interspeech*.

Stolcke, A. (2002) SRILM – an extensible language modeling toolkit. *Proc. of ICSLP*.

Stolcke, A., Anguera, X., Boakye, K., Çetin, O., Grezl, F., Janin, A., Mandal, A., Peskin, B. Wooters, C. and Zheng, J. (2005) Further progress in meeting recognition: The ICSI–SRI spring 2005 speech-to-text evaluation system. *Proc. of the Rich Transcription 2005 Spring Meeting Recognition Evaluation*.

Strang, G. (1980) *Linear Algebra and Its Applications*, second edn. Academic Press, New York.

Strobach, P. (1990) *Linear Prediction Theory: A Mathematical Basis for Adaptive Systems*. Springer.

Strobel, N., Spors, S. and Rabenstein, R. (2001) Joint audio-video signal processing for object localization and tracking. In *Microphone Arrays* (ed. Brandstein M and Ward D). Springer, chapter 10.

Strube, H. (1980) Linear prediction on a warped frequency scale. *Jour. of ASA* **68**(8), 1071–1076.

Stuart, A. and Ord, J. K. (1994) *Kendall's Advanced Theory of Statistics*. Edward Arnold, London.

Stüker, S., Fügen, C., Burger, S. and Wölfel, M. (2006) Cross-system adaptation and combination for continuous speech recognition: The influence of phoneme set and acoustic front-end. *Proc. of Interspeech*.

Swokowski, E. (1979) *Calculus with Analytic Geometry*. Prindle, Weber & Schmidt, Boston.

Takada, S., Ogawa, T., Akagiri, K. and Kobayashi, T. (2008) Speech enhancement using square microphone array for mobile devices *Proc. of ICASSP*, Las Vegas, NV, USA.

Talantzis, F., Constantinides, A. G. and Polymenakos, L. C. (2005) Estimation of direction of arrival using information theory. *IEEE SPL* **12**(8), 561–564.

Tashev, I. and Allred, D. (2005) Revereberation reduction for improved speech recognition *Proc. of HSCMA*.

Terhardt, E. (1972) Zur tonhöhenwahrnehmung von klängen. *Psychoakustische Grundlagen. Acustica* **26**, 173–186.

Terhardt, E. (1974) Pitch, consonance, and harmony. *Jour. of ASA* **55**(5), 1061–1069.

Teutsch, H. and Kellermann, W. (2005) Estimation of the number of wideband sources in an acoustic wavefield using eigen-beam processing for circular apertures. *Proc. of WASPAA*, New Paltz, NY, USA.

Titterington, D. M. (1984) Recursive parameter estimation using incomplete data. *Jour. of the Royal Statistical Society Series B* **46**, 256–267.

Tokuda, K., Kobayashi, T. and Imai, S. (1995) Adaptive cepstral analysis of speech. *IEEE Trans. on SAP* **3**(6), 481–489.

Uebel, L. and Woodland, P. (2001) Improvements in linear transform based speaker adaptation. *Proc. of ICASSP*.

Vaidyanathan, P. P. (1993) *Multirate Systems and Filter Banks*. Prentice Hall, Englewood Cliffs.

Valtchev, V., Odell, J. J., Woodland, P. C. and Young, S. J. (1997) MMIE training of large vocbulary speech recognition systems. *Jour. on SC* **22**, 303–314.

Van Trees, H. L. (2002) *Optimum Array Processing*. Wiley-Interscience, New York.

Varanasi, M. K. and Aazhang, B. (1989) Parametric generalized gaussian density estimation. *Jour. of ASA* **86**, 1404–1415.

Varga, A. and Moore, R. (1990) Hidden Markov model decomposition of speech and noise. *Proc. of ICASSP* **2**, 845–848.

Vaseghi, S. (2000) *Advanced Digital Signal Processing and Noise Reduction*. John Wiley & Sons.

Vaseghi, S. and Frayling-Cork, R. (1992) Restoration of old gramophone recordings. *Jour. Audio Engineering* **40**(10), 791–801.

Vermaak, J. and Blake, A. (2001) Nonlinear filtering for speaker tracking in noisy and reverberant environments. *Proc. of ICASSP*, Salt Lake City, UT.

Wang, W. and Stolcke, A. (2007) Integrating map, marginals, and unsupervised language model adaptation. *Proc. of Interspeech*, pp. 618–621, Antwerp, Belgium.

Ward, D. B. and Williamson, R. C. (2002) Particle filter beamforming for acoustic source localization in a reverberant environment. *Proc. of ICASSP*, Orlando, FL.

Ward, D. B., Lehmann, A. and Williamson, R. C. (2003) Particle filtering algorithms for tracking an acoustic source in a reverberant environment. *IEEE Trans. on SAP* **11**, 826–36.

Warren, R. M. (1970) Perceptual restoration of missing speech sounds. *Science* **167**, 392–393.

Warren, R. M., Riener-Hainsworth, K., Brubaker, B. S., Bashford, J. A. J. and Healy, E. W. (1997) Spectral restoration of speech: Intelligibility is increased by inserting noise in spectral gaps. *Percept. Psychophys.* **59**, 275–283.

Warsitz, E., Krueger, A. and Haeb-Umbach, R. (2008) Speech enhancement with a new generalized eigenvector blocking matrix for application in a generalized sidelobe canceller. *Proc. of ICASSP*.

Wegmann, S., McAllaster, D., Orloff, J. and Peskin, B. (1996) Speaker normalization on conversational telephone speech. *Proc. of ICASSP*, vol. I, pp. 339–341.

Weintraub, M., Taussig, K. and Hunicke-Smith, K. and Snodgrass, A. (1996) Effect of speaking style on LVCSR performance. *Proc. of ICSLP*.

Weiss, M., Aschkenasy, E. and Parsons, T. (1974) *Study and the development of the INTEL technique for improving speech intelligibility*. Technical Report NSC–FR/4023, Nicolet Scientific Corporation.

Welch, G. and Bishop, G. (1997) SCAAT: Incremental tracking with incomplete information. *Proc. of Computer Graphics and Interactive Techniques*.

Welch, G. F. (1996) *SCAAT: Incremental tracking with incomplete information*. PhD thesis, University of North Carolina, Chapel Hill, NC.

Widrow, B., Duvall, K. M., Gooch, R. P. and Newman, W. C. (1982) Signal cancellation phenomena in adaptive antennas: Causes and cures. *IEEE Trans. on Antennas and Propagation* **AP-30**, 469–478.

Widrow, B., Mantey, P. E., Griffiths, L. J. and Goode, B. B. (1967) Adaptive antenna systems. *Proc. of IEEE*, pp. 2143–2159.

Wiener, N. (1949) *Extrapolation, Interpolation and Smoothing of Time Series, with Engineering Applications*. Vol. originally appeared in 1942 as a classified National Defence Research Council Report. John Wiley & Sons, UK.

Wiener, N. and Hopf, E. (1931) On a class of singular integral equations. *Proc. of Russian Acad. Math. – Phys. Ser.*

Willett, D. and Katagiri, S. (2002) Recent advances in efficient decoding combining on-line transducer composition and smoothed language model incorporation. *Proc. of ICASSP*.

Wölfel, M. (2003) Mel-Frequenzanpassung der Minimum Varianz Distortionless Response Einhüllenden. *Proc. of ESSV*.

Wölfel, M. (2004) Speaker dependent model order selection of spectral envelopes. *Proc. of ICSLP*.

Wölfel, M. (2006) Warped-twice minimum variance distortionless response spectral estimation. *Proc. of EUSIPCO*.

Wölfel, M. (2007) Channel selection by class separability measures for automatic transcriptions on distant microphones. *Proc. of Interspeech*.

Wölfel, M. (2008a) Integration of the predicted walk model estimate into the particle filter framework. *Proc. of ICASSP*.

Wölfel, M. (2008b) A joint particle filter and multi-step linear prediction framework to provide enhanced speech features prior to automatic recognition. *Proc. of HSCMA*.

Wölfel, M. (2008c) Predicted walk with correlation in particle filter speech feature enhancement for robust automatic speech recognition. *Proc. of ICASSP*.

Wölfel, M. and McDonough, J. (2005) Minimum variance distortionless response spectral estimation, review and refinements. *IEEE SPM* **22**(5), 117–126.

Wölfel, M., Fügen, C., Ikbal, S. and McDonough, J. (2006) Multi-source far-distance microphone selection and combination for automatic transcription of lectures. *Proc. of Interspeech*.

Wölfel, M., McDonough, J. and Waibel, A. (2003) Minimum variance distortionless response on a warped frequency scale. *Proc. of Eurospeech*, pp. 1021–1024.

Wölfel, M., Nickel, K. and McDonough, J. (2005) Microphone array driven speech recognition: Influence of localization on the word error rate. *Proc. of MLMI*.

Woodland, P. and Povey, D. (2000) Large scale discriminative training for speech recognition. *ISCA ITRW Automatic Speech Recognition: Challenges for the Millenium*, pp. 7–16.

Woodland, P. C. and Povey, D. (2002) Large scale discriminative training of hidden markov models for speech recognition. *Jour. on CSL* **16**, 25–47.

Wright, D. and Wareham, G. (2005) Mixing sound and vision: The interaction of auditory and visual information for earwitnesses of a crime scene. *Legal and Criminological Psychology* **10**, 103–108.

Wu, J. (2004) *Disciminative speaker adaptation and environment robustness in automatic speech recognition*. PhD thesis, Univ. of Hong Kong.

Wu, J. and Huo, Q. (2002) An environment compensation minimum classification error training approach and its evaluation on Aurora2 database. *Proc. of ICSLP*, pp. 453–456.

Wu, M. and Wang, D. (2006) A two-stage algorithm for one-microphone reverberant speech enhancement. *Trans. on ASLP* **14**(3), 774–784.

Yan, Z., Soong, F. and Wang, R. (2007) Word graph based feature enhancement for noisy speech recognition. *Proc. of ICASSP*.

Yang, C., Soong, F. and Lee, T. (2005) Static and dynamic spectral features: Their noise robustness and optimal weights for ASR. *Proc. of ICASSP*.

Yang, H., Van Vuuren, S., Sharma, S. and Hermansky, H. (2000) Relevance of time–frequency features for phonetic and speaker-channel classification. *Jour. on SC* **31**(1), 35–50.

Yao, K. and Nakamura, S. (2002) Sequential noise compensation by sequential Monte Carlo methods. *Adv. Neural Inform. Process. Syst.*

Yost, W. and Gourevitch, G. (1987) *Directional Hearing*. Springer.

Young, S. J., Odell, J. J. and Woodland, P. C. (1994) Tree-based state tying for high accuracy acoustic modelling *Proc. of HLT*, pp. 307–312, Plainsboro, NJ, USA.

Young, S. J., Russell, N. H. and Thornton, J. H. S. (1989) *Token passing: A simple conceptual model for connected speech recognition systems*. Cambridge University Engineering Department Technical Report.

Yu, H., Tam, Y., Schaaf, T., Stüker, S., Jin, Q., Noamany, M. and Schultz, T. (2004) The ISL RT04 mandarin broadcast news evaluation system. *EARS Rich Transcription Workshop*.

Zelinski, R. (1988) A microphone array with adaptive post-filtering for noise reduction in reverberant rooms *Proc. of ICASSP*, New York, NY, USA.

Zhang, B. and Matsoukas, S. (2005) Minimum phoneme error based heteroscedastic linear discriminant analysis for speech recognition *Proc. of ICASSP*.

Zhang, Z. (2000) A flexible new technique for camera calibration. *IEEE Trans. on Pattern Analysis Machine Intel.* **22**, 1330–1334.

Zheng, J., Butzberger, J., Franco, H. and Stolcke, A. (2001) Improved maximum mutual information estimation training of continuous density HMMs. *Proc. of Eurospeech*.

Zhong, X. and Hopgood, J. R. (2008) Noncurrent multiple speakers tracking based on extended kalman particle filter. *Proc. of ICASSP*, Las Vegas, NV, USA.

Zotkin, D. N., Duraiswami, R. and Gumerov, N. A. (2008) Sound field decomposition using spherical microphone arrays. *Proc. of ICASSP*, Las Vegas, NV, USA.

Zue, V. (1971) *Translation of divers' speech using digital frequency warping*. Technical Report 101, Res. Lab. Eltron., Massachusetts Institute of Technology, Cambridge, Massachusetts.

Zwicker, E. (1961) Subdivision of the audible frequency range into critical bands (Frequenzgruppen). *Jour. of ASA* **33**, 248.

Zwicker, E. and Fastl, H. (1999) *Psychoacoustics*. Springer, second edition.

Index